# Insects
## and
# Plant Defence Dynamics

# Insects
# and
# Plant Defence Dynamics

*Editor*

**T.N. Ananthakrishnan**

**SCIENCE PUBLISHERS, Inc.**
Enfield (NH), USA       Plymouth, UK

SCIENCE PUBLISHERS, Inc.
Post Office Box 699
Enfield, New Hampshire 03748
United States of America

Internet site: *http://www.scipub.net*

*sales@scipub.net* (marketing department)
*editor@scipub.net* (editorial department)
*info@scipub.net* (for all other enquiries)

**Library of Congress Cataloging-in-Publication Data**

Insects and plant defense dynamics/editors, T.N. Ananthakrishnan.
    p. cm.
   Includes bibliographical references.
   ISBN 1-57808-155-6
   1. Plants--Insects resistance, 2. Plant defenses.

SB933.2.147 2001
632'.96--dc21

                                      00-053345

ISBN 1-57808-155-6

Published by Science Publishers, Inc., Enfield, NH, USA
Printed in India

# Preface

The latter half of the 20th century witnessed the emergence of chemical ecology emphasizing the chemical interaction between animals and plants. Implications of the co-evolutionary process in herbivore-plant interactions suggest their ability to overcome each other's struggle through the development of a bewildering diversity of defense mechanisms, greenplants having manipulated their natural enemies since the earliest evolutionary times through the production of an unimaginably large number of chemical compounds. While the diversity of biochemical pathways in plants involved in the production of a host of chemicals as well as mechanisms of overcoming their effects by insects engaged the attention of entomologists in the sixties and the seventies of the last century, the last two decades provided new dimensions to insect-plant interactions through an understanding of chemical signals or chemical information channels with adequate emphasis on induced defense reactions. Volatile chemicals from infested plants may not only alert chemicals defense responses in neighbouring plants, but also insect-damaged plants which tend to emit chemical signals attracting natural enemies. In this tritrophic system are links in nature's energy chain connected by chemical informational channels, resulting in considerable communication interplays. This was already sensed by Charles Darwin as early as 1859 when he indicated that plants and animals are bound together by a web of complex relations. A whole new dimension, with regard to their ecological function, is evident when the role of secondary compounds of plants are examined in this more extended context. An important area of future research has to deal with the dynamic aspects of interspecific communication, be it plant-plant, plant-insects or plant-insect-natural enemies and there is a built-in plasticity in the relationships based upon allelochemicals. Natural defense systems in plants are therefore complex, forming an intricate ecological system.

The present volume envisages an integration of diverse aspects of plant defence system against insects, aspects which are very vital in promoting sustained productivity in agriculture, particularly in view of chemical ecology playing an essential role in Integrated Pest Management.

The willing cooperation of the authors of various chapters is much appreciated and it is hoped that this volume will kindle as well as sustain continued interest on diverse aspects of Insect-plant interaction.

T.N. Ananthakrishnan

Chennai.
15.6.2000

# Contents

# Introduction

*T.N. Ananthakrishnan*

It is well known that the secondary metabolism of plants has been essentially responsible for deterring or destroying herbivores and pathogens through the production of many biologically active chemical substances. Intrinsic immunity to insect adaptation is evident in several instances, notably pest resistance crops. Plant genotype is a major determinant of quality and quantity of secondary compounds, besides leaf age which modifies many of these interactions between insects and plants. While plants use defense mechanisms—physical and chemical—the latter involving toxins and repellents in the form of phenols, terpenoids, alkaloids, cyanogens to mention a few, the fact that insects can effect or disrupt normal patterns of plant resistance allocation cannot be overlooked. The ability of insects to metabolize some of these secondary compounds, such as alkaloids and cardenolides, is equally well known and the sequestration of toxic plant substances has been well exemplified in many highly adapted aposematic species with the sequestered material stored for future defense.

An aspect of increasing relevance in defense related studies in plants pertains to induced responses or the capacity of plants to exhibit increased resistance subsequent to feeding stress. Different degrees of resistance or biochemical responses tend to result, so that the role of phytochemical induction in plant defense has emerged as an area of current interest in insect-plant interactions. Changes in chemical properties of a plant, following insect damage including secondary metabolism, results in deterrent or antibiotic effects on insects. Genetic variation involved in such induction may also account for the difference between resistance and susceptible variation. Increased involvement in studies on proteinase inhibitors as a means of defense for the damaged plant has augmented interest in biotechnology. Plants, therefore, play an important role in affecting the population of herbivores and interest in the application of induced resistance has been very evident over the last decade or two.

The title of this volume, *Insect and plant defence dynamics*, while essentially reflecting the diverse aspects of plant defence, also examines the defense tactics of caterpillars against predators, parasitoids as well as the possible evolution of monophagy in insects with sluggish movements as in the case of aposematic insects.

The introductory chapter highlights the diversity of plant defense, which promotes plant fitness depending upon the effectiveness of the defense systems, which tend to have a deterrent or antibiotic effect on insects. Discussing various aspects of plant defenses—constitutive and induced—it has been shown that plants are considered as a dynamic component of insect-plant interactions. Increasing chemical diversity of plant resources interferes with the exploitation of host plants by insects. In turn, insects adopt diverse physiological adaptations to overcome the biochemical barriers so that plant chemicals tend to influence both the behavioural and physiological processes of insects. The adaptive value of plant chemicals is considerable, since their multifaceted roles have enabled them to become the primary force in the ecological and evolutionary dynamics of insect-plant interactions. This has been exemplified through adequate examples related to the chemical defense strategy of plants and the adaptive diversity of specialist/generalist insects. Induced defenses of plants, signal diversity as well as the role of plant galls in defensive studies are also discussed.

Chapter II, dealing with role of essential nutrients and minerals in insect resistance in crop plants, reviews some of the biotic and abiotic interactions. Supply of essential plant nutrients in adequate amounts and proportions at appropriate timings during crop growth cycle can contribute substantially in improving plant resistance to some insects. The use of resistant crop species and cultivars within species has been indicated to have special economic and ecological significance.

Chapter III discusses the establishment of antibiosis in maize silk, suggesting that the oxidative reactions of C-glycosylflavones are responsible for the browning reaction of fresh cut silk in maize. The discovery of maysin as the major antibiotic factor against the corn ear worm and the initiation of studies to account for the amount of variability in maysin concentration in maize germplasm and the influence of the environment on silk maysin concentrations are discussed. Several antibiotic compounds have also been identified in maize silk, the most important being isoorientin. The diverse genes responsible for effecting the synthesis of biologically active flavones in maize silk and results of mapping studies of maysin with analogues to identify several molecular markers are also elucidated.

Chapter IV discusses the semiochemistry of crucifers and their herbivores, pinpointing the pervasiveness and mulitifunctionality of glycosinolates produced by crucifers, which affect numerous species in various trophic levels across a wide variety of taxa. Interestingly enough, they assume diverse functions as they pass from species to species earning for them the name of molecular migrants. Chemical ecological aspects of Brassicaceae, rich in glucosinolates, are discussed. Interestingly, glucosinolate breakdown products vary in their sensitivity to the secondary compounds of plants. This chapter also examines how crucifer feeding insects use these plant substances for host location, phagostimulation and chemical defense.

Chapter V discusses the adaptive significance of constitutive plant chemistry in relation to the triterpenoid quassinoids of Rutales, in particular the remarkable species and chemical diversity of Rutales and their defensive action against a host of insects. As quassinoids show insecticidal, antifeedant and growth inhibitory properties, the bioactivity of these compounds against insects is suggestive of the primary selective advantage of the production of triterpenoids in host plant defense against insects. Such diversity towards multifactor defense has been well illustrated by the elaboration of limonoid and quassinoids.

Chapter VI examines the role of plant surface waxes and trichomes, which act as mechanical and chemical barriers against insects. While as physical barriers, surface waxes impede attachment and location of insects, they also act as chemical barriers by stimulating or deterring landing, feeding and oviposition. Chemical compounds in the epicuticular wax layer are the first chemosensory stimuli that an insect perceives when landing or initiating feeding and suggestion is also made regarding the nature of the surface wax being 'recognized' by the insect as an indicator of the internal constitutents of a plant. Trichomes, as a resistance mechanism, tend to minimize the herbivore load by producing a physical and/or chemical barrier. While trichome diversity is exploited in insect resistant cultivars of several crops, the toxic and deterrent production of the exudation of glandular trichomes are equally effective in providing resistance against insects.

Chapter VII discusses the various aspects of insect-plant interactions in relation to measurement, mechanisms and insect-plant environment interactions. While the expression of resistance is dependent on environmental factors in time and space, besides induced and associative resistance mechanisms, the selection of plants with accumulated genes conferring resistance to insects will form the backbone of pest management programmes for suitable crop production and environmental conservation. Some of these aspects are discussed in relation to the shootfly resistance in land races of sorghum and Sorghum midge resistance.

Chapter VIII discusses the defensive tactics of caterpillars against predators and parasitoids. While most caterpillars feed on plants and could greatly increase their feeding, assimilation and metabolic rates, they have managed to evolve a very impressive diversity of defense capabilities. This chapter, while not intending to cover the diversity of caterpillar defenses, attempts to pinpoint some of the varied and astounding strategies to thwart the development of predators and caterpillars. Plants, caterpillars and natural enemies appear to be linked together as co-evolutionary units, each evolving their own strategies in response to the adaptation of others. The diversity of tactics discussed relates to morphological, mutualistic, chemical, physiological and behavioural aspects.

Chapter IX deals with the sluggish movements of aposematic insects as defense mechanisms against motion-oriented predators. Citing diverse examples, relating to different sluggish movements, the authors hypothesize that sluggish movement will interact positively with diverse defenses tending to increase the protection value of chemical defenses, gregarious behaviour and aposematic colourations. They also imply that monophagy could be more likely to evolve in sluggish individuals, since they would not move to new host plants.

Chapter X examines the role of genetic engineering in the production of insect resistant genetically modified crops. Characterization and modification of genes at the molecular level and DNA recombinant technology allows engineering and transfer of genes. Besides, genetically engineered plants are capable of enhanced production of very potent chemical resistant factors, which could pose a real threat to insects.

Chapter XI briefly discusses an overall assessment of the implications of insects and plant defence dynamics for future research.

# Chapter 1

# Phytochemical Defence Profiles in Insect-Plant Interactions

*T.N. Ananthakrishnan*

## INTRODUCTION

The last three or four decades have witnessed increasing and sustained interest in insect-plant interactions and incisive studies on diverse aspects have emphasised the role of a bewildering number of secondary plant chemicals, calling for a basic knowledge of physiology and biochemistry to fully appreciate the significance of these interactions. Increasing chemical diversity of plant resources interferes with the exploitation of host plants by insects, which are faced with the problem of adaptation to plant hosts. In turn, this has resulted in equally diverse physiological adaptations to overcome the barriers of insect feeding, so that plant chemicals tend to influence both the behavioural and physiological processes of insects. Whether insect feeding is responsible for reorganization of plant communities or plant community structure tends to influence insect populations has been a point of discussion, but it has been more than realised today that the adaptive value of host plant chemicals is considerable and their multi-faceted roles have enabled them to become the primary force in the ecological and evolutionary dynamics in insect-plant interactions (Dirzo, 1984). Diverse textural effects on insect feeding are known, with the host plant quality being influenced by plant community structure. Studies on insect adaptations to plant chemistry have shown that chemicals are clearly an important predominant pressure, affecting the distribution patterns of insects (Zangerl and Berenbaum, 1993), not to mention the view of Bernays and Graham (1988) that plant chemistry is one of the many potential pressures influencing arthropods. Deterioration of leaf quality being age-dependent, insect feeding could be expressed as a function of defence level and age. As has been indicated by Schoonhoven (1998) "plants although suffering from herbivores due to their phytochemicals, are at the same time the masters of herbivores".

The co-evolutionary arms race hypothesis of Ehrlich and Raven (1964), highlighting that adaptive chemical defences of plants resulting from natural selection by herbivores, laid the foundation for a better appreciation of insect-plant interactions. Subsequent hypotheses, the optimal defence hypothesis, which assumes that herbivory is the primary selective force shaping quantitative patterns of secondary metabolism (Cates and Orians, 1975; Cates and Rhoades, 1977; Feeny, 1976; Rhoades and Cates, 1976; Rhoades, 1979) and the resource availability hypothesis predicting that in resource-rich habitats competition favours fast-growing plants with low levels of defence, have explained qualitative and quantitative patterns of plant defence. Such modifications of these hypotheses as that of Coley *et al* (1985), Bryant *et al* (1988), assumed that physiological responses of plants to resource availability govern quantitative levels of plant defence, making a distinction between qualitative or mobile and quantitative of immobile defences, the former being more effective against herbivores in smaller concentrations (Herms and Mattson, 1992). The biochemical barrier hypothesis (Jones and Lawton, 1991) predicts a negative correlation between insect species richness and chemical diversity. In this coevolutionary struggle engaged in overcoming each other's defence arsenals, plants tend to produce new deterrents through adopting diverse metabolic pathways and insects overcoming them by diverse detoxifying mechanisms. While insects have the ability to exert selection on plant secondary chemistry, phylogenetic patterns in angiosperms indicate that the trend has been towards the evolution of more toxic secondary metabolites (Gottleib, 1989; Harbourne, 1991, 1993; Herms and Mattson, 1992). Needless to emphasise that phytochemical opportunism is responsible for a certain degree of diversity in secondary metabolites in many species sequestering secondary compounds from host plants (Ananthakrishnan, 1998a, b; Berenbaum and Siegler, 1992). The ability of insects to feed on plants containing pyrolizzidine alkaloids and store them, is remarkable (Bopre, 1990). Of equal importance are cyanogenic glycosides occuring in different plant families and release of hydrogen cyanide taking place due to enzyme activity when tissues are damaged.

## CHEMICAL RESISTANCE STRATEGIES OF PLANTS

Plant resistant chemicals can often function as double agents, individual plants varying in their chemical levels over time and using resistance chemicals when needed. Plants adopt chemical resistance strategy against herbivores through storing less toxic precursors, which are transformed into toxins when needed. Natural defence systems are, therefore, complex forming part of an intricate ecological system. Being biologically active compounds, secondary metabolites aid in competitive interactions against other plants, and act as feeding stimulants and deterrents. Such secondary compounds as terpenoids, alkaloids, coumarins, cyanogenic glycosides, glucosinolates sequestered in different sites such as epidermal vacuoles or in cell walls or oil glands or on leaf surface waxes are important elements in plant defence against insects, besides morphological barriers. Common toxic elements like glycoalkaloids in potato, tomatine, rutin and chlorogeneic acid in tomato and gossypol in cotton are toxic to insects. Increased phenolic and Tannin production and decreased proteins and nutritional suitability are commonly evident as a response to insect attack. Within an interval of several hours after feeding, several plants release protease inhibitors, cucurbitacins in squash and terpenes in sweet potato (Tallamy, 1985). Allelochemicals, therefore, tend to occur at effective concentrations in a plant (Constitutive) or are synthesised *de novo* or translocate when needed (induced). Each species of plants has its own unique bouquet of secondary compounds, which generally exhibit high intraspecific variations. The high "degree of chemical freedom" of secondary metabolism, which in contrast to primary metabolism allows structural modifications with almost no restriction (Hartmann, 1996) (Fig. 1).

The antibiotic effects of resorcinol, gallic and phloroglucinol in various cotton cultivars against *Helicoverpa armigera* are well known (Ananthakrishnan *et al.* 1990) (Table 1). Another interesting example relates to host plant resistance to insects feeding on corn silks resulting in the production of c-glycosyl flavones, maysin, apimaysin, methoxymaysin and chlorogenic acid, which restrict larval growth of *Helicoverpa zea* (Widstrom *et al.* 1977; Guo *et al.* 1999). Besides these, isoorientin found in corn silk is also known to inhibit the growth of the corn earworm larvae. Adding isoorientin to other compounds increases the effectiveness of plant resistance to *Helicoverpa* (Widstrom and Snook, 1998). So feeding by phytophagous insects cause qualitative or quantitative changes, resulting in specific or non specific, direct or indirect modification, low nutritive quality, high concentrations of secondary metabolites and tough foliage interacting to provide formidable barriers. Many instances of allelochemicals providing protection for young tissues are known, which tend to become less abundant with the age of the tissues. All the same, enhanced synthesis of such

**Table 1.** Some anti-insect secondary chemicals from the cotton plant

| Phenolic acids | Terpenes |
|---|---|
| *Benzoic acids* | *Monoterpenes* |
| Gentisic | Limonene |
| Salicylic | Myrcene |
| Vanillic | Ocimene |
| *Cinnamic acids* | *Sesquiiterpenes* |
| Caffeic | Caryophyllene |
| Chlorgenic | Bisabolene |
| Ferulic | *Terpenoid aldhydes* |
| Sinapic | Gossypol |
| *Flavonoids* | Hemigossypol |
| Gossypetin | *Fatty acids* |
| Kaempferol | Malvalic acid |
| Quercetin | Sterculic acid |
| Apigenin | |
| Cyanidin | |
| Delphinidin | |
| Catechin | |

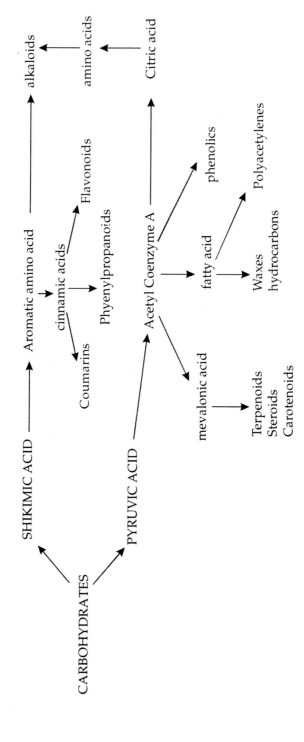

**Fig. 1.** Pathways of Secondary Metabolism in Plants.

allelochemicals is a part of the integrated defence mechanisms of plants. Increasing prospects are, therefore, available for exploiting plant secondary chemicals for crop protection. The role of glycoalkaloids such as solanine, chaconine, demissine and leptines against *Leptinotarsa decemlineata* and the potato leaf hopper *Empoasca fabae* are well known in resistance against potato pests (Tingey, 1984). Increasing concentrations of alkaloids in various cultivars of potato appear positively correlated with reduction in infestation. Because of the wide activities of alkaloids, they are considered to be multipurpose defence substances.

When a plant produces a compound that interferes with a molecular target in a herbivore, it gains selective advantage since the plant will not be eaten and can transfer its genes to the next generation (Wink, Schmeller and Latz-Bruning, 1999). Knowledge of resistance of natural host plant toxins may be applied to prolong the efficiency of genetically engineered plants (Gould, 1988). Transgenic tobacco plants containing both the cowpea trypsin inhibitor gene and lectin gene are known to express the two insecticidal genes, so that plant-derived insect resistant genes play a significant role in plant defence. Plant lectins can also interfere with the digestive system of insects through binding of lectin/glycoprotein to digestive enzymes. These lectins can form complexes with digestive enzymes and inactivate them (Peumans and Van Damm, 1995). Plant genetic manipulation as a part of the breeding program can make a significant contribution in the production of insect-resistant crops (Gatehouse *et al.* 1992). Recent advances in genetic engineering have shown that expression of different trypsin inhibitory genes in foreign plants can provide resistance against insects. For example, transformed tobacco plants with genes coding for tomato and potato inhibitor proteinases having chymotrypsin and trypsin inhibiting activity have been isolated and characterised.

While plants are in a position to tailor chemical defences for specific purposes such as defence or stress, insects are equally fascile with the capability to detoxify plant toxins. But the uniqueness of plants in diverting resources and energy from one part to another is astounding, with their ability to switch defence options because of their diverse interconnected metabolic pathways involving many regulatory mechanisms (Berenbaum and Feeny, 1981; Berenbaum *et al.* 1989), as has been adequately demonstrated by the *Papilio*-furanocoumarin synthesis in umbellifers (Berenbaum, 1981 a, b; 1983) (Fig. 2). Tannin concentrations in plants is a key feature in determining the feeding behaviour of insects (Bernays, 1981).

The leaf surface of plants, a complex world of host cuticular waxes and sugars, has alkanes, alkanols and alkanoic acids and very specific amino acids. In many families, such as Solanaceae, Labiatae and Compositae, several glandular hairs or trichomes produce a host of different mono, sesqui and diterpenes imparting a structural as well as chemical defence. The diversity of secondary compounds is enormous, some plants containing mixtures of many compounds as in the terpenoid mixtures in the leaves of essential oil plants (Capri and Tabashnik, 1992). Host selection in the carrot fly *Psila rosae* is

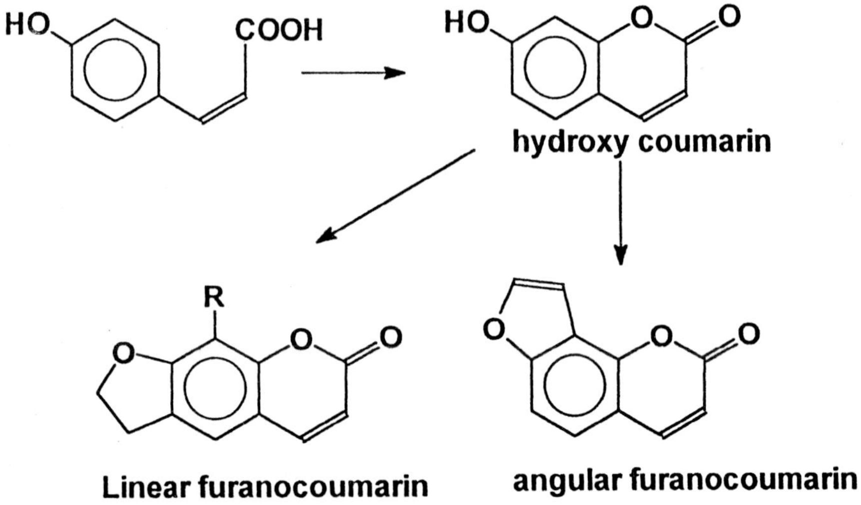

**Fig. 2** Interrelationships of coumarins in Umbellifers

influenced by physical and chemical plant cues. Semio-chemicals in the surface wax of carrot leaves stimulate oviposition, the compounds being phenyl-propenes, furanocumarins and polyacetylenes, which are in combination in umbellifers (Stadler, 1986). High trichome density, glossy leaf and internode length are morphological characters associated with insect resistance in sorghum in relation to the sorghum shootfly (*Atherigona soccata*) (Sharma *et al.* 1997). While these are external, internal leaf defences also include an array of structures of diverse forms and functions such as thickened cell walls, special cellular receptacles, intercellular pockets and ramifying channels varying in the chemistry of their contained fluids (Dussourd, 1993). Tannins have been found to be greater in some midge-resistance lines than in susceptible ones, while soluble sugars are lesser in midge-resistance lines (Sharma *et al.* 1991). Increased herbivore pressure tends to increase the benefit of defence, thus, favouring increased allocation of resources towards resistance. Production costs of secondary compounds are high, such as terpenoids, which are expensive to produce because of high enzymatic conversions involved (Mopper and Strauss, 1998). Evolution of insect abundance depends on the combination of such dimensions as host plant species, host plant parts, abundance, size and architecture and efficacy of defence mechanisms, while plant chemistry determines the composition of assemblages of phytophagous insects (Schoonhoven *et al.* 1998).

## SYNERGISTIC REACTION OF ALLELOCHEMICALS

Synergism is usually coalitive, wherein a system of two components of which one alone causes no remarkable impact, but together provide a catapulting effect. A notable instance is gossypol, a phenolic sesquiterpene toxic to a variety of insects. It causes significant decrease in the survival, growth and development of several cotton pests. Pigment glands of cotton plant produce another sesquiterpene, caryophyllene oxide, (Bernays and Chapman, 1994) serving to synergise the growth-inhibiting effect of gossypol. Increased vigour of the boll weevil is evident through the ingestion of gossypol, which suppresses the growth of gut bacteria. An enhanced phagostimulatory effect is seen when singrin and glucose interact as evident in *Plutella xylostella*.

Synergistic effects of gallic, vanillic and Salicyclic acids are evident not only in their interaction with other plant allelochemicals, but also in their interaction with the endotoxins of *Bacillus thuringiensis* (Sivamani *et al.* 1992). Gallic acid occupies an important position in the overall phenolic metabolism of higher plants exhibiting synergistic interactions. Interestingly enough, majority of hydrobenzoic acids (p-hydroxybenzoic, salicylic, vanillic, syringic, proto-catechuic) are relatively rarely encountered in plants, particularly in the herbaceous dicots (Fig. 3). Noteworthy instances of Synergism relate to glucose and sinigrin, xanthotoxin and myristicin (Berenbaum and Neal, 1985), fusaric acid and gossypol (Dowd, 1989), cis-asarone and menthol (Koul *et al.* 1990). Synergism, therefore, plays a pivotal role in increasing the bioefficacy of allelochemicals as a form of multichannel chemical defences against insects.

## HORMONAL ANALOGUES IN PLANTS

The use of hormonal analogues and the increasing promise of plant-derived antihormones for insect control has generated increasing interest in insect growth regulators. Plants have developed substances that target the insect endocrine system and the discovery of extremely active phytojuvenoids called juvocimenes from *Ocimum basilicum*, tagetone (*Tagetes erecta*), sterculic acid (*Sterculia*), Juvadecine (Piperaceae), have been very significant. Antijuvenile hormonal agents were identified from *Ageratum houstonianum* and identified as simple chromenes and called precocenes I and II. Precocene treatment of insects proved to be cytotoxic to corpora allata, causing cell death and invasion of the allata by connective tissue (Bowers, 1991).

## ADAPTIVE DIVERSITY IN SPECIALIST AND GENERALIST INSECTS

Since plants are able to synthesise a variety of chemical substances such as non-protein amino acids, alkaloids, terpenes, flavonoids and the like, their structural diversity has increased greatly during the course of evolution along with periodical changes in insect feeding pressure. This diversity, in turn, created a selection process leading to behavioural and biochemical adaptation in insects and resulting in specialists and generalists. Contrasting selection pressure exerted by them possibly explains the degree of variation in the concentration of allelochemicals in natural plant populations, very high concentrations deterring the generalists, but not the specialists. Further, specialists lack adequate variation to adapt to a greater range of host species,

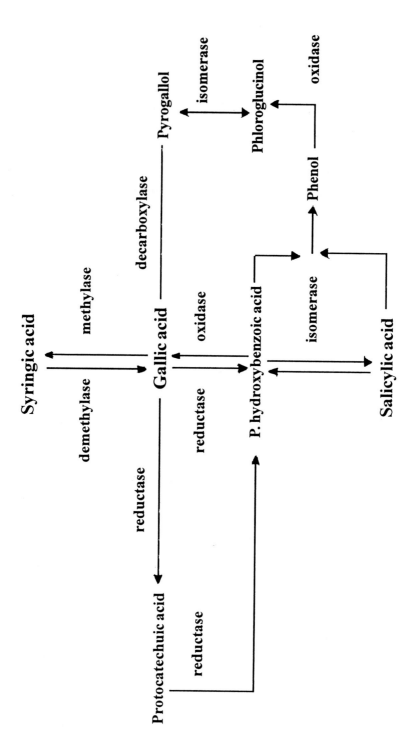

**Fig. 3**   Conversions of phenolic acids

while generalists are invariably polyphagous, a new host plant being an extension of its diet breadth, the direction of host shifts being genetically constrained. Young leaves often contain higher concentrations of defence chemicals than mature leaves and so are better protected, increasing resource availability and adding to the effectiveness of defence. In some cases, as in *Cynoglossum officinale*, the youngest leaves have 50 to 190 times higher levels of pyrolizzidine alkaloids than the oldest, so that younger leaves are avoided by generalist insects (Van Damn *et al.* 1996). Among several instances of host shifts, an interesting case is that of *Retithrips syriacus* (Thysanoptera), a polyphagous species heavily infesting castor (*Ricinus communis*), (Euphorbiaceae), and which has successfully adapted to hosts like *Eucalyptus globulus* (Myrtaceae) and *Manihot utilitissima* (Euphorbiaceae), over a period of time, overcoming the phenolic, terpenoid and cyanogenic compounds, respectively (Ananthakrishnan and Gopichandran, 1996). Papilionoid butterflies offer excellent instances of specialist and generalist species, with *Battacus philenor*, a specialist on species of *Aristolochia* (encountering aristolochic acid) and *Papilio polyxenes*, also a specialist on species of *Apiaceae* and *Rutaceae* (encountering furanocoumaric acid) and the generalist *Papilio glaucus* feeding on plants of more than 20 natural orders (encountering phenolic and cyanogenic glycosides), so that they adopt different approaches to detoxification or Sequestration of host plant compounds. Detoxification enzymes of specialists respond selectively to a narrow range of substrates than generalists (Dowd *et al.* 1983; Berenbaum *et al.* 1989; Berenbaum and Mopper, 1998).

Plant families are characterised by a particular group of allelochemicals such as glucosinolates (*Cruciferae*), cardenolides (*Asclepiadeceae*), iridoid glycosides (*Scrophulariaceae*) with variations among populations of single species. Quantitative variation in allelochemical profiles of plants tends to be an important determinant of patterns of herbivory. An interesting instance of growth inhibition in *Helicoverpa zea* larvae in corn has been shown earlier to be due to different silk maysins, a c-glycosyl flavone. Related compounds, like iridoid glycosides, may exert considerable diversity in feeding preferences in a generalist/specialist. While the generalist survived poorly on diets with the iridoids, the specialist performed best on diets with two iridoid glycosides, such as catalpol and acubin in *Plantago lanceolata*, the relative amounts of different iridoid glycosides tending to vary substantially from one population to another (Bowers and Puttick, 1986). The wild silk moth *Callosamia* (Saturniidae) is composed of three closely-related species that display various degrees of host specialisation in relation to feeding on *Magnolia* that are toxic to a number of insects mediated by host phytochemistry (Johnson, 1999).

Instances where even within a population, both generalist and specialist phenotypes are maintained, are exemplified by the butterfly *Euphydras editha*. Evidence of preference-performance correlation has been shown by offspring of specialist females of this species surviving better on plants chosen by their mother than on rejected plants. On the other hand, offspring from generalist females performed equally well on both categories of plants (Simms, 1998).

Resource texture is, therefore, an aspect to which insects respond in various ways. In a mixed crop situation, a specialist used to host specific cues might be confused or repelled by nearby non-host species, while a polyphagous generalist species may perceive diverse plant mixtures without adverse effects. Many species use non-specific chemical cues as part of their host selection mechanisms and specialists are more sensitive than generalists to deterrents from non-hosts (Bernays and Chapman, 1987, 1994).

## INDUCED DEFENCES AND SIGNAL DIVERSITY

Damage to the plant tissue due to insect feeding, leads to the translocation of a signal from the damaged site to other parts of a plant resulting in induced resistance, the expression of which tends to vary with the age, season and growth habits (Agarwal and Karban, 1999). Since the damaged plant tissues respond in diverse ways, with many of the responses occurring away from the site of damage, the terms talking or listening plants were often used. The responses may be either short term and rapid or long term and delayed, the former tending to be more effective against generalist insects and the latter against both generalist and specialist insects (Haukioja, 1980, 1990). While qualitative or mobile defences are easily translocated among plant tissues, the same is not true of quantitative or immobile defences. Some of the best examples of rapidly inducible phytochemical responses are the increase of cucurbitacins in the attacked leaves of cucurbits, the production of diverse furanocoumarins as a result of feeding by danaiid caterpillars on umbellifers (Berenbaum, 1981a). Nicotine production, wherein the amount of nicotine accumulated by the plant in response to damage tends to vary as much as 100% depending on how much damage cue is exported from the leaf before it is removed (Baldwin and Schmelz, 1994). Of equal interest is the emission by disturbed foliage of eucalyptus species of isoprene, $\alpha$-pinene, cineole, limonene and $\beta$-terpene, while undisturbed leaves released only isoprene, adequately demonstrating thereby that induced chemical responses tend to reduce plant attractiveness to herbivores (Rhoades, 1985). Plants respond to physical and chemical changes associated with insect feeding through the accumulation of phenolic compounds and in accordance with the kind and degree of damage, diverse phenols are induced, which involve enzymes such as phenylalanine ammonia lyase (PAL), tyrosine ammonia lyase (TAL), polyphenol oxidases (PPO) and peroxidases. While PAL activity is stimulated differentially in accordance with insect damage, the enzyme is also known to exist in multiple forms, which are induced differently by different cues. Most of the higher plant phenols are produced through the intermediation of phenylalanine, which involves multienzyme complexes including PAL. The number of genes coding for PAL also varies differentially and each of these tend to produce varying amounts of the enzyme. A signal passes from the site of damage to undamaged leaves inducing PAL activation (Karban and Baldwin, 1997). Other enzymes have also been observed to be produced differentially by different elicitors. It is well known today that a local stimulus triggers signalling effects that affects the whole plant. Defence responses induced by elicitors relate to biochemical and molecular mechanisms of which Proteinase inhibitors are well known (Ryan and Green, 1974). Serine proteinase inhibitors have been well investigated as markers of systemic wound responses in tomato and potato plants and are expressed in the aerial regions of the plants. Increased exposure to Proteinase inhibitors have a detrimental effect on the digestive physiology of several larval Lepidoptera (Broadway and Duffey, 1986). The polypeptide systemin produced close to the damage sites of plants acts as a plant hormone inducing Proteinase inhibitors (Pearce *et al.* 1991). Systemin, released by wounding, translocates throughout the plant and interacts with the distal tissues of plants where it stimulates synthesis of jasmonic acid, which activates proteinase inhibitor genes. Responses to Proteinase inhibitors by insects involve production of proteinases, which result in the degradation of essential aminoacids (Reinbothe *et al.* 1994). Synthesis of diverse plant proteins believed to be of importance in defense is also known. Treatment with methyl jasmonate induces accumulation of other chemicals involved in plant defense, such as ehylene, PAL, systemin and several alkaloids. Both short-distance and long-distance signal pathways are thought to activate proteinase inhibitor genes by means of a secondary signal pathway. The octodecenoid pathway, which results in the synthesis of jasmonic acid and methyl jasmonate, are known to elicit different chemical and physiological responses and provide a potential mechanism for communication in or between plants (Farmer and Ryan, 1990; Ryan, 1991; Thaler, 1999).

Feeding damage by thrips enhances production of ehylene, which in turn is known to enhance production of defense chemicals (Fig. 4). Plant phenolic levels increase in response to ehylene and initiates PAL activity. While oligosaccharide fragments of cells are known to act as systemic signals, producing rapid induction of Proteinase inhibitors, ehylene results in the production of enzymes that release these oligosaccharide fragments. Similarly, abscissic acid acts as a signal for induced accumulation of Proteinase inhibitors. Methyl salicylate found in the wounded plant tissues induces production of a variety of chemicals involved in plant defense. Thus, an intensive network of signals is mobilised through the activation of local systemic events (Bowles, 1990, 1992). The diversity of such local signals arising from tissue damage, which lead to systemic changes in gene expression, are promising areas of future research.

Increasing instances are evident in several insects-plant systems, wherein induced defences play a useful role. Damage caused by insects can induce production of tissue-based phenols, and additional compounds tend to be produced over a longer period. Systemic resistance to noctuids, aphids, leaf miners and mites result through localised damage in young tomato plants releasing atleast a dozen chemicals (Schaller *et al.* 1995). Of the chemical changes that take place during induced damage, the increase of Proteinase inhibitors and the oxidative enzyme PPO appear to be most critical to the induction of resistance (Stout *et al.* 1996) (Fig. 5). Continued attack by flea beetles on Crucifers can change the total concentration of glucosinolates in

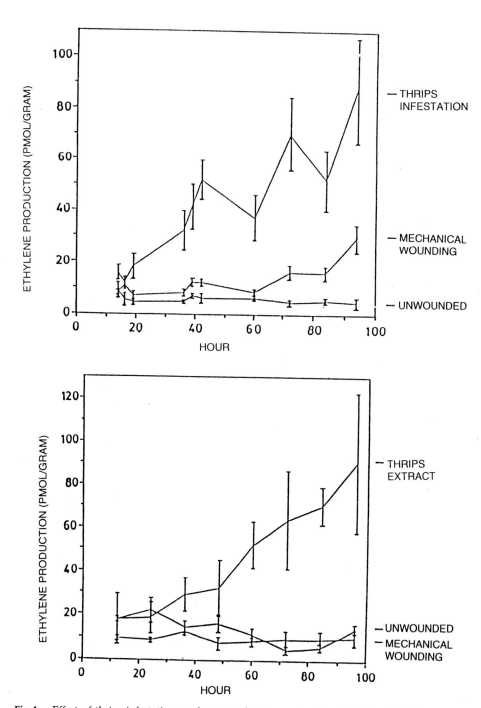

**Fig 4.**   Effect of thrips infestations and extract eliciting production of ehylene in *Allium cepa* (Courtesy: Kendall & Bjostad, 1990)

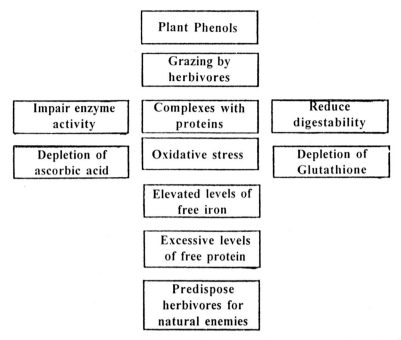

Physical and Chemical Basis of resistance to Insects in wild pigeonpea (after Annadurai)

**Fig 5.**   Mode of action of phenols in herbivores.

different plant tissues, leading to a two to four-fold increased production of glucosinolates such as glucobrassicin and neoglucobrassicin (Renwick *et al.* 1992). When the plant tissues are disrupted, the enzyme myrosinase hydrolyses the glucosinolate to produce a variety of products collectively called mustard oils (Bennet and Wallsgrove, 1994). As glucosinolate accumulation can be induced by insect attack, the mechanics of induction need more intensive studies. Similarly, allyldisulphide and propyldisulphide of onion are released from the bond formed in the plants especially after their damage. The speed and intensity of biosynthesis of these chemicals may depend on the distance from the site of injury, amount of damage done and the kind of elicitors involved (Bernays and Chapman, 1994).

As earlier indicated, resistant cultivars of cotton evincing increased PAL and TAL enzymes when attacked by *Helicoverpa armigera* result in increased production of gallic, Salicylic acids, besides resorcinol and phloroglucinol. A significant shift in the oxidative status of the host plants is typical of induced resistance, and several oxidative enzymes are induced in *Helicoverpa zea*. Increased levels of syringic acid, chlorogenic acid, ferrulic acid, and rutin resulted in foliage fed by *H. zea*. A significant decline was also seen in the host nutritional quality with most aminoacids, in the damaged tissue (Bi *et al.* 1994). Induced resistance also involves Lignification and cell wall strengthening, increased Lignification resulting from changes in ferrulic and p-coumaric acid levels in damaged squares. Inhibition of damage-induced ehylene synthesis is associated with increased ferrulic acid in foliage.

## IMPACT OF INDUCTION ON TRITROPHIC INTERACTIONS

Plants have evolved to use parasites and predators as a defence against phytophagous insects, phytochemicals of the host plants acting as long-range cues for attracting natural enemies. Variation in plant traits, whether physical, morphological or chemical, can affect the fitness of natural enemies. Effects of induced responses on natural enemies may also act additively or synergistically as is evident from the prolonged development of insect larvae-induced allelochemicals arresting or reducing growth, thereby attracting natural enemies, which have evolved the capability to learn to use induced responses for host location (Barbosa and Saunders, 1985). Thus, plants use the third trophic levels as another line of defence against herbivores. To be more effective, there should be stronger selection pressure on plants to release chemical signals that attract natural enemies. Besides, the release of volatile chemicals, herbivory may change non-volatile allelochemicals, which may further increase natural enemy activity (Ananthakrishnan, 1999). The emission of volatiles can be adaptive to plants when they attract natural enemies that reduce herbivore damage, the volatiles enhancing the foraging efficiency of natural enemies. It is well known that β-glucosidase is an effective elicitor of herbivore-induced plant odour, which attracts parasitic wasps. β-glucosidase is found in the regurgitated saliva of caterpillars (Mattiacci *et al.* 1995). Being an elicitor of defence responses to herbivory injury, it induces the emission of volatiles used by natural enemies to locate their victims. Herbivore-induced plant volatiles may provide natural enemies with the information about plant species and they are more reliable than general damage-related plant volatiles (Pickett *et al.* 1999). Release of green leaf volatiles in damaged plants includes a mixture of C6 alcohols, aldehydes and esters and these volatiles, by diffusing through the air, produce behavioural responses. One of the best examples of tritrophic systems relates to the lima bean plant and the two-spotted mite, *Tetranychus urticae*, and the predatory mite, *Phytoseiulus persimilis* (Karban and Baldwin, 1997). Similarly, herbivore-induced plant synomones are beneficial to the signaler (plant) and the receiver (natural enemy) as in the case of some anthocorids, which were attracted to pear trees infested by the pear psyllid (Vet and Dicke, 1992). The conducive niche offered by squares and flowers of cotton provide pollen and their nutritional sources to natural enemies, not to mention the kairomonal compounds such as eicosane, pentacosane and dodecenoic acid in plant sources tending to adequately increase the activity of natural enemies in a variety of crops (Annadurai *et al.* 1990). Release of plant volatiles, after feeding by phytophagous insects, increases the fitness of the entomophagous insects especially through such volatile fractions as caryophyllene, hexanoic, tetradecenoic, hexadecenoic and pentadecenoic acids, which also emanate from larval frass and larval cuticle resulting from feeding on different plant sources, appreciably attracting natural enemies (Ananthakrishnan and Senrayan, 1992). Turlings *et al.* (1990) have shown that partly damaged leaves by *Spodoptera exigua* release terpenoids such as a-transbergamotene, (E)-farnesene and (E)-nerolidol. Similarly, cotton seedlings produce volatile cues when infested with herbivorous mites and elicit attraction of predatory mites. Volatile chemicals from host plants, host insects or frass are, therefore, informational cues which augment the activity of larval parasites, such as *Microplitis croceipes*.

## PLANT GALLS AS DEFENSIVE STRUCTURES

The assumption relating to the role of plant galls as defensive structures appears sound, gall insects modifying minute areas of host plants by soliciting gene expression from adjacent cells in such a way that new developmental events result. New morphological potentialities of the host plant are evincsed, enabling differentiation of new cell types unknown in the normal organ and by organising new differentiation centres ultimately expressing a gall system with well-defined nutritional as well as a defensive system. A vast majority of gall-inducing insects are highly host specific, and although gall-inducing behaviour is confined to a small number of unrelated families, such insects spend a major part of their life on the host plant within the protected environment of the gall. The gall insect's option for a particular plant cannot be a matter of chance and invariably they encounter diverse plant species in a heterogenous natural situation (Ananthakrishnan, 1998 a). Gall formers have the ability to control and manipulate the growth of a plant, the degree of such manipulation varying widely from simple cell proliferation to the production of complex structures (Fig. 6) not normally produced by the plant (Rohfritsch and Anthony, 1992).

The gall insects use tactics that are more sophisticated than their free-living relatives, by directing their movements towards their plant source either by positive chemotaxis or by moving in a particular direction with respect to environmental cues. Host plant choice by the gall insect needs to be seen in conjunction with the phenology and quality of the plant (Abrahamson and McCrea, 1986; How *et al.* 1993). Incidence patterns of host-plant fructose and its effects are enhanced by phylloplane proline, glucose and sucrose correlate with oviposition preference. Gall formers may utilise a wide range of mechanisms to induce gall growth, including behavioural, mechanical, chemical or genetic manipulation of the host plant to produce galls (Hori, 1992).

The role of amino acids and the plant growth hormone Indole Acetic Acid (IAA) have been extensively implicated as gall-inducing factors (Miles, 1989). Amino acids, essentially lysine, histidine and tryptophan, might function as conditioners for gall induction, cell-conditioning preceding gall induction (Hori, 1992). Cell enlargement is influenced by the role of IAA in the synthesis of RNA and proteins in plants. Increased interest is currently developing to focus on relationship between the uptake of oxygen in galls and the vitalisation of auxin activity. While the vigourous uptake of oxygen is a critical stress-related response, it is normal metabolic response of the galled plant to strengthen the involved tissue site (Meyer and Maresquelle, 1983), and, therefore, becomes a

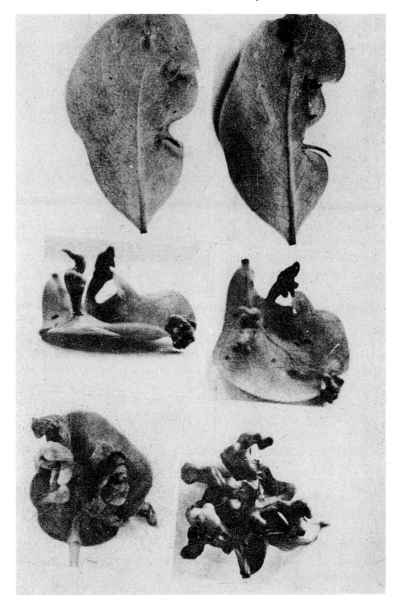

**Fig 6.**   Gall complexities from a single horn gall to a tortuose form.

defensive reaction. This is all the more critical when it is realised that all tissues do not react, while some of them are hypersensitive. As galls are formed, secondary chemicals get isolated from the nutritive tissues and are concentrated in the peripheral tissues upto precipitation ten times higher than in non galled leaves.

Host response to herbivory is evident in galls, as observed in the increasing levels of total phenols and associated activity of the phenolic enzymes. The extent of receptivity of tissues and visible changes in gall complexity also change as a result of the internal cellular dynamics. The impact of enhancing levels of total phenols severely restricts further feeding against the backdrop of diminishing nutrients with progressing gall age (Gopichandran *et al.* 1992). Host acceptability relates to the degree of resistance essentially linked to the mobilisation of phenolics and the impact depends on the biochemical characteristics of the individual phenols. While the profiles of individual phenolic acids and flavonoids may determine the suitability of hosts (Rozenthal and Janzen, 1978), with gallic acid known to be the most toxic phenolic substance contibuting to host resistance, the fact that they tend to disrupt normal life cycle patterns, can also be considered as defensive responses. The role of hydrolysable Tannins in several galls suggest the significance of these compounds not only in defense but also in tissue repair processes.

**CONCLUSION**

Phytophagy tends to alter much of the chemical profiles of a plant including plant secondary metabolites, which tend to have deterrent or antibiotic effects on insects. Phytophagy and induced-resistance to insects are prone to be important aspects promoting plant fitness, which depends on the effectiveness of the induced responses and the degree of resistance of plant to insect damage. The responses to wounding, activated by the plant signal transduction pathway, considerably influence feeding by the concerned insects. Oligosaccharide fragments from plant cell walls, systemin, Salicyclic acid, ehylene, abscissic acid and jasmonic acids are implicated as signals in systemic induction. Being of the nature of signals, a chemical must induce the observed responses. Gallic and Salicyclic acids are known to play an important role in insect-plant and plant-pathogen interactions. Salicyclic acid, besides activating as a defensive chemical, also acts as a signal which moves through the plant, having the attributes of the plant hormone. Plants are, therefore, to be considered as a dynamic component of insect-plant interactions. Secondary metabolites cannot be manipulated using transgenesis in view of the complicated biosynthetic reactions, which require several enzyme-mediated steps. While plant-insect interactions result in marked changes in plant secondary chemicals, nutrient status and physiology, which might affect subsequent resistance to herbivores, there is also a need to examine the possible impact of pathogens in this process to be able to confidently conclude regarding the specificity of induced defenses. Reduction or elimination of defense plant allelochemicals through breeding programs could have a significant but varying effect on feeding behaviour and other host plant-insect interactions in which allelochemicals play key roles. An equally important aspect relates to the need for more intensive studies on the diversity of induced responses on insects to appreciate the useful application of such responses in plants. Prospects for genetically engineered crops are exciting because resistant traits can be made to express, following attack.

## REFERENCES

Abrahamson, W.G. and McGrea, K.D. 1986. The impact of galls and gall makers on plants. *Proc. Ent. Soc. Wash.*, **88**: 364-67.

Agarwal, A.A and Karban, R. 1999. Why induced defences may be favoured over Constitutive strategies of plants. *In* R. Tollrian and C.D. Harwell (Eds) *The Ecology and Evolution of Induced defences.* University Press, New Jersey: 45-61.

Ananthakrishnan, T.N. 1998a. Insect gall systems: Patterns, processes and adaptive diversity. *Current Science*, **75**: 672-676.

Ananthakrishnan, T.N. 1998b. Insect herbivory, opportunistic interactions and strategic responses. *Pest Management in Horticultural Systems*, **4**: 1-6.

Ananthakrishnan, T.N. 1999. Behavioural dynamics in the biological control of insects: Role of infochemicals. *Current Science*, **77**: 33-37

Ananthakrishnan, T.N. and Gopichandran, R. 1996. *Chemical Ecology of Thrips.* Oxford & IBH, New Delhi.

Ananthakrishnan, T.N. and Senrayan, R. 1992. Phytochemical induced responses, a vital factor governing insect-host-parasite/predator interactions. *Phytophaga*, **4**: 87-94.

Ananthakrishnan, T.N., Senrayan, R., Annadurai, R.S. and Murugesan, S. 1990. Antibiotic effects of resorcinol, gallic and Phloroglucinol, on Heliothis armigera Hubner (Insecta: Noctuidae). *Proc. Indian Acad. Sci. (Anim. Sci.)*, **99**: 39-52.

Annadurai, R.S., Murugesan, S. and Senrayan, R.S. 1990. Age correlated tissue preferences of *Heliothis armigera* (Hubner) and *Spodoptera litura* (F) with special reference to phenolic substances. *Proc. Indian Acad. Sci.*, **99**: 317-325.

Baldwin, T. and Schmelz, E.A. 1994. Constraints on an induced defence: the role of leaf area. *Oecologia*, **97**: 424-430.

Barbosa, P. and Saunders, J.A. 1985. Plant allelochemicals: Linkage between herbivores and their natural enemies. *Recent Adv. Phytochem*, **19**: 107-137.

Bennet R.N. and Wallgrove, R.M. 1994. Secondary metabolites in plant defence mechanisms. *New Phytol.*, **127**: 39-52.

Berenbaum, M.R. 1981a. Effects of linear furanocoumarins on an adapted specialist insect (*Papilio polyxenes*). *Ecol. Entomol.*, **6**: 345.

Berenbaum, M.R. 1981b. Furanocoumarin distribution and insect herbivory in the Umbelliferae: plant chemistry and community structure. *Ecology*, **62**: 1254-1266.

Berenbaum, M.R. 1983. Coumarins and caterpillars: a case for coevolution. *Evolution*, **37**: 163-179.

Berenbaum, M.R. 1991. Comparative processing of allelochemicals in the Papilionidae (Lepidoptera). *Arch. Insec. Biochem. Physiol*, **17**: 213-221.

Berenbaum, M.R. and Feeny, P. 1981. Toxicity of angular furanocoumarins in swallowtails: escalation of the coevolutionary arms race. *Science*, **212**: 927-929.

Berenbaum, M.R. and Mopper, S. 1998. Local adaptation in specialist behaviour. *In* Mopper, S. and Strauss, S.Y. (Eds). *Genetic Structure and Local Adaptation in Natural Insect Population*, Chapman and Hall, N.Y.: 64-88.

Berenbaum, M.R. and Neal, J. 1985. Synergism between Myristian and Xanthotoxin, a naturally occuring plant toxicants. *J. Chem. Ecol.*, **11**: 1349-58.

Berenbaum, M.R. and Siegler, D. 1992. Biochemicals: engineering problems for natural selection. *In* B.T. Roitberg and M.B. Isman (Eds) *Insect Chemical Ecology*, Chapman & Hall, New York: 89-121.

Berenbaum, M.R., Zangerl, A.R. and Lee, K. 1989. Chemical barriers to adaptation by a specialist herbivore. *Oecologia*, **80**: 501-506.

Bernays, E.A. 1981. Plant Tannins and insect herbivores: an appraisal. *Ecol. Entomol.*, **6**: 353-360.

Bernays, E.A. and Chapman, R.F. 1987. Chemical deterrence of plants. *In* J.H. Law (Eds) *Molecular Entomology*, **49**: 108-116, Alan R. Liss, N.Y.

Bernays, E.A. and Chapman, R.F. 1994. *Host Plant Selection by Phytophagous Insects*, Chapman and Hall, London.

Bernays, E.A. and Graham, M. 1988. On the evolution of host specificity in phytophagous arthropods. *Ecology*, **69**: 886-892.

Bi, J.L. Felton, G.N. and Mueller, A.J. 1994. Induced resistance in soyabean to Helicoverpa zea: Role of plant protein quality. *J. Chem. Ecol.*, **20:** 183-198.

Boppré, M. 1990. Lepidoptera and pyrolizzidine alkaloids. *J. Chem. Ecol.*, **16:** 165-186.

Bowers, W.S. 1991. Insect hormones and antihormones in plants. *In* G.A. Rosenthal and M.R. Berenbaum (Eds) *Herbivores: Their Interaction with Plant Secondary Metabolites*. Academic Press, New York: 431-456.

Bowers, M.D. and Puttick, G.M. 1986. The fate of ingested iridoid glycosides in Lepidopteran herbivores. *J. Chem. Ecol.*, **12:** 169-178.

Bowers, M.D. and Puttick, G.M. 1988. Response of specialist and generalist insects to quantitative allelochemical variation. *J. Chem. Ecol.*, **14:** 319-324.

Bowles, D. 1990. Defence related proteins in higher plants. *Ann. Rev. Biochem.*, **59:** 879-907.

Bowles, D. 1992. Signals in the wounded plant. *In* P.G. Ayars (Eds) *Pest and Pathogens, Biol.* Oxford, UK: 33-38.

Broadway, R.M. and Duffey, S.S. 1986. Plant Proteinase inhibitors: Mechanism of action on the growth and digestive physiology of larval *Heliothis zea* and *Spodoptera exigua*. *J. Insect Physiol.*, **32:** 827-833.

Bryant, J.P, Toumi, J. and Niemela, P. 1988. Environmental constraint of Constitutive and long-term inducible defences in woody plants. *In* K.C. Spencer (Eds) *Chemical Mediation in Coevolution*. Academic Press, San Diego: 367-389.

Capri, M.A. and Tabashnik, B.E. 1992. Evolution of plant chemicals in insects. *In* B.D. Roitberg and M.B. Isman (Eds) *Insect Chemical Ecology*, Chapman and Hall, New York: 177-215.

Cates, R.G. and Orians, G.H. 1975. Successional status and the palatability of plants to generalized herbivores. *Ecology*, **56:** 410-418.

Cates, R.G. and Rhoades, D.F. 1977. Patterns in the production of antiherbivore defences in plant communities. *Biochem. Synt. Ecol.*, **5:** 185-193.

Coley, P.D., Bryant, J.P and Chapin, F.S. 1985. Resource availability and plant antiherbivore defense. *Science*, **230:** 895-899.

Cornell, H.V. 1983. The secondary chemistry and complex morphology of galls formed by the Cynipinae (Hymenoptera): Why and how? *Amer. Mid. Natur*, **110:** 225-234.

Dirzo, R. 1984. Herbivory: A phytocentric overview *In* Dirzo, R. and Sarukhan (Eds.) *Perspectives in Plant Population Biology*, Sinauer, Sunderland: 141-165.

Dussourd, D.E. 1993. Foraging with Finess: Caterpillar adaptation for circumventing plant defences. *In* N.E. Stamp and T.M. Casey (Eds) *Caterpillars*, Chapman & Hall, London: 92-131.

Dowd, P.F. 1989. Fusaric acid: A secondary fungal metabolite that synergises toxicity of co-occurring allelochemicals to the corn earworm, *Heliothis zea* (Lepidoptera). *J. Chem. Ecol.*, **15:** 249-254.

Dowd, P.F., Smith, C.M. and Sparks, T.C. 1983. Detoxification of plant toxin by insects. *Insect Biochem*, **13:** 453-468.

Ehrlich, P.R. and Raven, P.H. 1964. Butterflies and plants: a study in coevolution. *Evolution*, **18:** 586-608.

Farmer, E.E. and Ryan, C.A. 1990. Interplant communication: Airborne methyl jasmonate induces synthesis of Proteinase inhibitors in plant leaves. *Proc. Natl. Acad. Sci. USA.*, **92:** 7713-7716.

Feeny, P. 1976. Plant apparency and chemical defenses. *Recent Adv. Phytochem*, **10:** 1-40.

Gatehouse, A.M.R., Vaugham, A.H., Powell, K., Boulter, D. and Gatehouse, J.A. 1992. Potential of plant-derived genes in the genetic manipulation of crops for insect resistance. Proc. 8th Int. Symp. Insect-Plant Relationships, Dordrecht: Kluwer Acad. Publ: 221-233.

Gopichandran, R., Peter, A.J. and Subramaniam, V.R. 1992. Age-correlated biochemical profiles of thrips galls in relation to population density of Thrips. *J. Nat. Hist.*, **26:** 609-619.

Gottleib, O.R. 1990. Phytochemicals: Differentiation and function. *Phytochemistry*, **29:** 1715-1724.

Gould, F. 1988. Genetic engineering, Integrated pest management and evolution of pests. *TREE*. 3/*Biotech*,: S15-S19.

Guo, B.Z., Widstrom, N.W., Wiseman, B.R., Snook, M.E., Lynchr, R.E. and Plaisted, D. 1979. Comparison of silk maysin, antibiosis to corn ear worm larvae (Lepidoptera: Noctuidae) and silk browning in crosses Dent X sweet corn. *J. Econ. Ent.*, **92:** 746-753.

Harbourne, J.B. 1991. Role of secondary metabolism in chemical defense mechanisms in plants. *In* Chadwick, D.J. and Marsh, J. (Eds) *Bioactive Compounds from Plants.* Chichester: John Wiley and Sons: 126-139.

Harbourne, J.B. 1993. *Ecological Biochemistry,* Academic Press, London: 317 pp.

Hartmann, T. 1996. Diversity and variability of plant secondary metabolism: A mechanistic view. *Entomol. Exp. Applicata,* **80:** 177-188.

Haukioja, E. 1980. On the role of plant defenses in the fluctuation of herbivore population. *Oikos,* **35:** 202-213.

Haukioja, E. 1990. Inducible defenses in trees. *Ann. Rev. Entomol.,* **36:** 25-42.

Herms, D.A. and Mattson, W.J. 1992. The dilemma of plants: To grow or defend. *Quart. Rev. Biol.,* **67:** 283-335.

How, S.T., Abrahamson, W.E. and Craig, T.P. 1993. Role of host plant phenology in host use by *Eurosta solidaginis* (Diptera: Tephritidae) on Solidago (Compositae). *Env. Ent.,* **22:** 388-396.

Hori, K. 1992. Insect secretions and their effect on plant growth, with special refrence to Hemipterans. *In Biology of Insect Induced Galls.* Oxford University Press, New York: 157-170.

Johnson, K.S. 1999. Comparative detoxification of plant (*Magnolia virginiana*) allelochemicals by generalists and specialists saturniid silk moths. *J. Chem. Ecol.,* **25:** 253-269.

Jones, C.G. and Lawton, J.H. 1991. Plant chemistry and insect species richness in British Umbellifers. *J. Anim. Ecol.,* **60:** 767-778.

Karban, R. and Baldwin, I.T. 1997. *Induced Responses to Herbivory.* University of Chicago Press, Chicago. 319 pp.

Koul, O., Smirle, M.J., Isman, M.B. and Szeto, Y. 1990. Synergism of a natural insect growth inhibitor is mediated by bioactivation. *Experientia,* **46:** 1082-84.

Meyer, J and Maresquelle, H.J. 1983. *Anatomide es Galles.* Gebriider Borntrager, Stuttgart.

Mattiacci, L., Dicke, M. and Posthumus, M.A. 1995. *Proc. Natl. Acad. Sci. U.S.A.,* **92:** 2036-2040.

Miles, P.N. 1989. The response of plants to the feeding of Aphidoidea: Principles. *In* Minks, A.K. and Harrewijn, P. (Eds) *Aphids—Their Biology, Natural enemies and Control.* Elsivier, Amsterdam: 1-21.

Mopper, S. and Strauss, S.Y. 1998. *Genetic Structure and Local Adaptation in Natural Insect Populations,* Chapman and Hall, NY.

Pearce, G.D., Strydom, D., Johnson, S. and Ryan, C.A. 1991. A polypeptide from tomato leaves induces wound-inducible proteinase inhibitor proteins. *Science,* **253:** 895-898.

Peumans, N.J. and Van Damme, E.J.M. 1995. The role of lectins in plant defence. *Histochemical Journal,* **27:** 253-271.

Pickett, J.A., Smiley, D.W.M. and Woodcock, C.M. 1999. Secondary metabolites in Plant-Insect Interaction: Dynamic System of induced and adaptive response. *Adv. Bot. Res.,* **30:** 91-115.

Renwick, J, A.A., Radke, C.D., Sachdev Gupta, K. and Stadler, E. 1992. Leaf surface chemicals stimulating oviposition by *Pieris rapae* (Lepidoptera: Pieridae) on cabbage. *Chem. Ecology.,* **3:** 33-38.

Reinbothe, S., Mollenbauer, R. and Reinbothe, C. 1994. JIPs and RIPs: The regulations of plant gene expression by jasmonates in response to environmental cues ard pathogens. *Plant Cell.,* **6:** 1187-1209.

Rhoades, D.F. 1979. Evolution of plant chemical defenses against herbivores. *In* G.A. Rosenthal and D.H. Janzen (Eds) *Herbivores: Their Interaction with Secondary Plant Metabolism.* Academic Press, London: 3-54.

Rhoades, D.F. 1985. Offensive-defensive interactions between herbivores and plants: Their relevance in herbivore population dynamics and ecological theory. *Amer. Nat.* **125:** 205-238.

Rhoades, D.F. and Cates, R. G. 1976. Toward a general theme of plant antiherbivore chemistry. *Recent Adv. Phytochem,* **10:** 168-213.

Rohfritsch, O. and Anthony, M. 1992. Strategies in gall induction by two groups of Homopterans. *In* J.D. Shorthouse and O. Rohfritsch (Eds). *Biology of Insect-induced galls,* Oxford University Press, New York: 162-117.

Rosenthal, G.A. and Janzen, D.H. 1979. *Herbivores: Their Interactions with Plant Secondary Metabolites.* Academic Press, New York, 285 pp.

Ryan, C.A. and Green T.R. 1974. Proteinase inhibitors in natural plant protection. *Rec. Adv. Phytochem,* **8:** 123-140.

Ryan, C.A. 1991. Protease inhibitors in plants: Genes for improving defense against insects and pathogens. *Ann. Rev. Phytopathol,* **28:** 425-459.

Schaller, A. Bergey, D.R. and Ryan, C.A. 1995. Induction of wound response genes in tomato leaves by Bestatin, an inhibitor of aminopeptidases. *Plant cell.,* **7:** 1893-98.

Schoonhoven, L.M., Jermy, T. and Van Loon, J.J.A. 1998. *Insect-Plant Biology.* Chapman and Hall, London: pp.409.

Sharma, H.C., Vidyasagar, P. and Leuschner, K. 1991. Compensational analysis of the factors associated with resistance in Sorghum midge, *Contarinia sorghicola. Ann. Appl. Biol.,* **123:** 469-483.

Sharma, H.C., Nwanze, K.F. and Subramanian, V. 1997. Mechanism of resistance to insects and their usefulness in Sorghum improvement. ICRISAT conference paper no. CP 1164: 81-100.

Simms, E.L. 1998. *In* S. Mopper, and S.Y. Strauss, (Eds) *Genetic Structure and Local Adaptation in Natural Insect Populations,* Chapman and Hall, N.Y.

Sivamani, E., Rajendran, N., Senrayan, R., Ananthakrishnan, T.N. and Kunthala Jayaraman. 1992. Influence of some plant phenolics on the activity of δ endotoxin of *Bacillus thuringiensis* var. galleriae on *Heliothis armigera. Entomol. Exp. Appl.,* **63:** 243-248.

Stadler, E. 1986. Oviposition and feeding stimuli in leaf surface waxes. *In* B.E. Juniper and T.R.E. Southwood (Eds) *Insects and Plant Surface.* Edward Arnold, London.

Stout, M.J., Workman, K.V. and Duffey, W.W. 1996. Identity, spatial distribution and variability of induced chemical responses in tomato plants. *Ent. Exp. Appl.,* **79:** 255-271.

Tallamy, D. 1985. Squash beetle feeding lechavior: an adaptation against induced cucurbit defences. *Ecology,* **66:** 1574-1579.

Tallamy, D.W. and Raupp, M.J. 1991. *Phytochemical Induction by Herbivores.* John Wiley, New York: 155-181.

Thaler, J.S. 1999. Jasmonate-inducible defense cause increased parasitisation of herbivores. *Nature,* **399:** 686-688.

Tingey, V.M. 1984. Glycoalkaloids as plant resistance factors. *Am. Potato Journal,* **61:** 157-168.

Turlings, T.C.J., Tumlinson, J.H. and Lewis, W.J. 1990. Exploitation of herbivore induced plant odours by host seeking parasite wasps. *Science,* **250:** 1251-1253.

Van Dam, N.M. de Jong, T.J., Iwasa, Y. and Kuto, T. 1996. Optimal distribution of defenses: One plants smart investor. *Functional Ecology,* **10:** 128-136.

Van Dam, N.M., Vuiser, L.N.M., Bergshoeff, C., Deton, H. and Van der Meijden. 1995. The "raison etre" of Srolizzidine alkaloids in *Cynoglossum officinalis*: Deterrent effects against generalist herbivores. *J. Chem. Ecol.,* **21:** 507-523.

Vet, L.E.M. and Dicke, M. 1992. Ecology of infochemical use by natural enemies in a tritrophic context. *Ann. Rev. Entomol,* **37:** 141-172.

Widstrom, N.W., Wiseman, B.R. and McMilliam, N.W. 1977. Responses of corn earworm larvae to maize silks. *Agron. J.,* **69:** 815-817.

Widstrom, N.W. and Snook, M.E. 1998. A gene controlling biosynthesis of isoorientin, a compound in corn silk antibiotic to the corn silk worm. *Ent. Exp. Appl.,* **89:** 119-124.

Wink, M., Schmellar, T. and Latz-Briining, 1998. Mode of action of allelochemical alkaloids: Interaction with neuroreceptors, DNA and other molecular targets. *J. Chem. Ecol.,* **24:** 1881-1937.

Zangerl, L. and Berenbaum, M.R. 1993. Plant chemistry, insect adaptation to plant chemistry and host plant utilization pattern. *Ecology,* **74:** 47-54.

## Chapter 2

# The Role of Essential Nutrients and Minerals in Insect Resistance in Crop Plants

*N.K. Fageria[1] and J. Mark Scriber[2]*

## INTRODUCTION

Minerals and primary metabolites that are involved in basic plant processes are rarely considered as determinants of plant resistance to insects despite the major effects they have on insect behavior, physiology and ecology (Moran and Hamilton, 1980; Mattson and Scriber, 1985; Berenbaum, 1995a; Zangerl and Berenbaum, 1998, Scriber, 2001). Indeed, as with the growth of primary references on herbivore-plant interactions from only 20 publications per year in 1970 to annually more than 300 in 1998 (Fig. 1; Scriber and Ayres, 1988), chemical defense theories in plants appear to have proliferated from the plant stress hypothesis (in the early 1970s) through approximately 10 subsequent theories into the 1990s. No theory is all-encompassing enough to be able to account for defense allocations in plants (Berenbaum, 1995b; Price, 1997; see also Stamp, 1992). This highly complex set of interactions involves much more than plant chemistry and host suitability for herbivores and has generated theories and immense amounts of empirical data in thousands of publications in the last two decades (Fig. 1). The interactions of herbivores and plants must involve the dynamics of allelochemicals and nutrients, as well as abiotic (e.g. thermal) and biotic (e.g. natural enemies) factors affecting behavior and physiology (Herms and Mattson, 1992; Scriber and Lederhouse, 1992; Stamp and Yang 1996). This has resulted in a proliferation of books on this subject in the last two decades (see reviews in Scriber and Ayres, 1988; Schoonhoven, 1999).

[1] National Rice and Bean Research Center of EMBRAPA, Santo Antonio de Goias, GO, 75375-000 Brazil, email: fageria@cnpaf.embrapa.br
[2] Department of Entomology, Michigan State University, 243 Natural Science Bldg., East Lansing, MI 48824, email: scriber@pilot.msu.edu

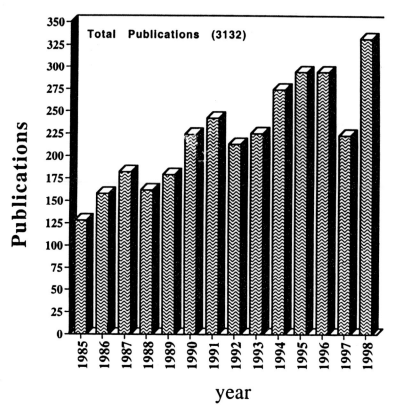

**Fig. 1.** Increase of primary references dealing with interactions between herbivorous insects and their host plants. Drawn from a computerized search of the BIOSIS database, this figure includes all indexed publications that use "herbivore" or "phytophag" or "host plant resistance" in the title, abstract, or key words and that are cross-referenced in the biosystematic index under Insecta. The total publications (N=3,132) represented here (1985-1998) is more than twice the total of the period from 1969-1985 reported by Scriber and Ayres (1988). Only 13 references turned up in this search that also included "mineral(s)".

Our understanding of the basic physiological molecular and genetic processes involved in herbivore-plant interactions during the last 10-15 years has increased rapidly. However, we have still not seen a synthesis of evolutionary ecology, pesticide engineering, phytopathology and natural products chemistry that provides predictive potential in host plant resistance and environmental impacts under any given set of conditions. Consequently, we must continue to study many cropping systems and combinations of control measures in our quest to forage and produce food more effectively in the face of millions of invertebrate competitors.

We believe that the roles of Minerals and trace elements have been understudied and warrant additional study with regard to their direct and indirect roles in herbivore-plant interactions and host plant resistance programs.

Perhaps some of the variable responses of secondary chemicals (allelochemicals) may be due to their interactions in the plant (or herbivore) with Minerals. Whether or not Minerals are a primary mechanism of resistance, their subtle and synergistic functions are woefully understudied, as this review emphasizes.

In modern agriculture, maximizing and sustaining crop yields are the main objectives. Use of adequate levels of essential plant nutrients is one of the most important factors for increasing crop yields. According to Loneragan (1997), as much as 50 per cent of the increase in crop yields worldwide during this century is due to the adoption of chemical fertilizers and the single most important factor limiting crop yields worldwide and especially among resource poor farmers may be soil infertility.

To minimize the dependence upon chemicals as the sole insect control strategy and to overcome various problems associated with the use of insecticides (Panda and Khush, 1995; Carmel and Pannell, 1996; Brummer, 1998; Casida and Quistad, 1998), research scientists have developed an ecological approach for pest control, popularly known as Integrated Pest Management or IPM (Stern *et al.* 1959). The IPM is a pest management system that, in the context of the associated environment and the population dynamics of the pest species, utilizes all suitable techniques and methods in as compatible a manner as possible and maintains the pest population at levels below those causing economic injury (Food and Agriculture Organization, 1975). It should be noted that judicious use of insecticides is likely to remain an important part of IPM systems (Wilson *et al.* 1995; Casida and Quistad, 1998). A common form of the economic-injury level has five components; two are biological variables, two are economic variables and one is a tactical variable. The general model is as follows (Pedigo, 1996; Hutchins and Pedigo, 1998): $EIL=C/VIDK$ where $EIL$=economic injury level (insects per hectare); $C$=control costs ($\$$ per hectare); $V$=market value ($\$$ per unit of yield); $I$-unit of injury per insects; $D$=yield loss per unit of injury; and $K$=proportion of total damage averted.

Despite the relatively long history of IPM, definitions and descriptions still vary widely (Cate and Hinkle, 1993). Holtzer *et al.* (1996) defined IPM as the judicious use and integration of various pest control tactics, in the context of the associated environment of the pests, in ways that complement and facilitate biological control and other ecological processes that reduce pest impact, to meet economic and environmental goals. IPM addresses the basic causes of pest problems in a holistic manner. Kogan (1998) recently defined IPM as a decision support system for the selection and use of pest control tactics, singly or harmoniously coordinated into a management strategy, based on cost/benefit analyses that take into account the interests of and impacts on producers, society and the environment. This approach uses a combination of host plant resistance, cultural, biological and chemical methods. The IPM approach is designed to suppress pest numbers to below crop-damaging levels. One salient feature of IPM is the manipulation of the environment to make it unfavorable to insects and less harmful to their natural enemies. Use of adequate quantity of essential nutrients at appropriate timing for healthy crop growth is one of the most important components of plant resistance to insects and consequently for IPM.

## Host-Plant resistance to insects

The mechanisms by which insect resistance traits are selected in agriculture or evolve in plant populations and the causes of environmentally induced variance in phenotypic resistance traits has been the subject of numerous books and reviews (e.g. Painter, 1951; Russell, 1978; Panda, 1979; Maxwell and Jennings 1980; Smith, 1989; Khush and Brar, 1991; Fritz and Simms, 1992; Smith *et al.* 1994; Panda, 1995; Clement and Quisenberry, 1999). The multiple trophic level interactions of plants with plants, pathogens, parasites (e.g. herbivores) and microbes are mediated by nutrients and allelochemicals (Whittaker and Feeny, 1971; Rosenthal and Janzen, 1979; Spencer, 1988; Roitberg, 1992; Rosenthal and Berenbaum, 1992; Carde and Bell, 1995), and are spatially (i.e. geographically variable; Denno and McClue, 1983; Hunter, 1992; Kim, 1993; Johnson and Scriber, 1994) and temporally (Feeny, 1976; Crawley, 1983; Tallamy and Raupp, 1991; Romeo *et al.* 1996; Karban and Baldwin, 1997; Tollrian and Harvell, 1999) dynamic.

Even without the defensive allelochemicals, plant tissues are not the ideal food for insects since plant parts greatly differ in nutrient composition from insects (Mattson, 1980; Scriber and Slansky, 1981; Scriber, 1984a, b; Slansky and Scriber, 1985; Mattson and Scriber, 1987; Schoonhoven *et al.* 1998). Of the nutrients or elements in plants that are normally in short supply (lower concentrations than in insects) nitrogen has most often been shown to be the most critical other than water (Southwood, 1972; Scriber, 1977; Mattson, 1980; Scriber and Slansky, 1981). However, the average concentrations of several other elements are also significantly lower in plant parts than insect bodies (including Na, S, P, Fe, Zn, and Cu) with equal or greater concentrations in plants for Mg, K, Ca, and Mn (Schoonhoven *et al.* 1998).

Seasonal variation and interspecific variation is well known in plants (Russell, 1947; Scriber, 1984a, b; Mattson and Scriber, 1987) as well as insects (Mattson and Scriber, 1987). Of course, these nutritional concentrations are also affected by other environmental factors like soil type or fertilization regimes which will alter their ratios. Other than nitrogen, there is, however, essentially no information about how mineral nutrition alters the degree or mechanisms of plant resistance to insect herbivores. In fact, in a large literature search dealing with phytophagous insect chemical ecology, most references deal with nitrogen and/or allelochemicals (Fig. 1 and Scriber and Ayres, 1988). However, only a rare mention is made of other Minerals or elements. We review this deficit and argue for additional study.

## Constitutive and Induced Chemical Defenses

Plants are faced with a dilemma in nature; to grow, reproduce, or to defend (Herms and Mattson, 1992). In addition to numerous genetically based phytochemical defenses that may segregate out in natural plant hybrid zones

(Whitham, 1989; Floate and Whitham, 1993; Floate *et al.* 1993; Fritz *et al.* 1994; Strauss, 1994; Whitham *et al.* 1994; Christensen *et al.* 1995) as well as with plant breeding programs and genetic engineering (Ives and Bedford, 1998; Rechcigl and Rechcigl, 1999; van Emden, 1999) there are also environmentally (e.g. acid rain, Riemer and Whittaker, 1989; Minerals, Heliövaara and Vaisänen, 1993; global warming, Scriber and Gage, 1995) or herbivore-induced plant chamicals (Karban and Baldwin, 1997) that may result in differential acceptability or suitability for insect herbivores and their community of natural enemies.

Plants may compensate to some degree for arthropod herbivory (McNaughton, 1979; Owen, 1980; Belsky, 1986; Crawley, 1987; Paige and Whitham, 1987; Trumble *et al.* 1993; Rosenthal and Kotanen, 1994). The degree of plant compensation or tolerance for herbivore damage is influenced by the fertilization nutrient regimes (including adequate water) that the plant is exposed to (Herms and Mattson, 1992; Rosenthal and Kotanen, 1994; Dankert *et al.* 1997; Koricheva *et al.* 1998) as well as the degree of multiple-species herbivory (Gehring and Whitham, 1994; Denno *et al.* 1995; Scriber, 1998; Scriber *et al.* 1999; including plant pathogens, Hammerschmidt and Schultz, 1996). In addition, exogenous factors such as the timing of defoliation, environmental stresses (e.g. temperature and pollutants) will alter the response of plants to herbivory.

These results and the spatial, temporal, genetic, ecological (community) and environmental (abiotic) variation that affect plant phytochemistry make it difficult to determine whether a nutritionally healthy plant is more generally resistant to herbivory, more tolerant of herbivory, or more susceptible to increased herbivory (since it may be more nutritious). Plant resistance to herbivores is a temporary and very specific phenomenon. Musick (1985) and Herzog and Funderburk (1986) concluded that each crop and pest situation must be evaluated individually and pest control decisions for each specific geographical location should be made with regard to specific insects that attack a specific crop. Such geographic variation in phytochemistry of natural plant defenses is poorly known (Johnson and Scriber, 1994).

## Mechanisms of Resistance

Among insect control measures, plant resistance is one of the most important tactics available because management is achieved with little cost to the grower and without potential health risks and environmental contamination from pesticides. Plant resistance also is generally compatible with other management tactics, including biological management (Starks *et al.* 1972; Boethel and Eikenbary, 1986; Smith, 1989). According to Panda and Khush (1995), resistance can be assessed by four characteristics: 1) plant resistance may be heridikary and controlled by one or more genes; 2) resistance is relative and can be measured only by comparison with a susceptible cultivar of the same plant species; 3) resistance is measurable (i.e. it's magnitude can be qualitatively determined by analysis of the standard scoring systems, or quantitatively by insect establishment); and 4) resistance is variable and is likely to be modified by the biotic and abiotic factors.

The mechanisms of plant resistance are antixenosis, antibiosis and tolerance (Dicke and Guthrie, 1988; Malvar *et al.* 1993; Niks *et al.* 1993; Butron *et al.* 1998; Smith, 1989). Antixenosis operates by disrupting normal arthropod behavior (non-preference), whereas antibiosis adversely affects a pest's physiological processes, thereby impairing arthropod survival, growth, development and behavior. Tolerance is a plant response to injury where acceptable plant yield is achieved inspite of injury by a pest (Funderburk *et al.* 1993). Plant breeders generally prefer the antibiosis mechanism of resistance for developing resistant cultivars due to more stability and effective control under field conditions. When this mechanism is conferred by one or a few dominant genes, selection usually can be done easily in the greenhouse or field. Antibiosis can be mediated by toxic allelochemicals, such as various types of alkaloids, arthropod growth inhibitors, reduced plant nutrient levels that inhibit arthropod growth, plant hypersensitivity and plant structural factors such as glandular trichomes that secrete allelochemicals or adhesive substances that inhibit pest movement. Twenty sources of antibiotic resistance have been identified in small grains to the Hessian fly, *Mayetiola destructor* (Say) (Patterson *et al.* 1992). Virtually all field corn hybrids grown in the Midwest United States produce elevated levels of the allelochemical DIMBOA (2, 4-dihydroxy-7-methoxy-1, 4-benzoxazin-3-one), which has antibiotic effects on European corn borer larvae during the whorl stage (Klun *et al.* 1967). However, other mechanisms of resistance in maize to European corn borers exist (Scriber *et al.* 1975) and, furthermore, it is important to note that this resistance varies with plant phenology (Scriber, 1984a), with light intensity, fertilizers and degree of inbreeding in the populations of corn borer (Manuwoto and Scriber, 1985a, 1985b, 1985c).

Antixenotic resistance consists of plant morphological or chemical factors that reduce the attractiveness of a plant as a host, resulting in the selection of an alternate host (Smith, 1989). Specific defenses include presence or absence of trichomes and leaf surface waxes, increased thickness of plant tissue and chemical repellents or feeding deterrents. Tolerance usually is the result of complex physiological changes in plant response to injury that alters the relationship between injury and damage in a plant (Funderburk *et al.* 1993). Procedures for screening tolerance are described by Morgan *et al.* (1980), Panda and Heinrichs (1983) and Smith (1989).

During the last five decades, screening techniques for evaluating germplasm for insect resistance have been developed and sources of resistance to major insects in several crop species have been identified (Fritz and Simms, 1992; Smith *et al.* 1994; Clement and Quisenberry, 1999). Resistant entries from germplasm collections have served as resistance sources in crop improvement programs. Resistant varieties of major crops such as wheat, rice, sorghum, corn and barley are now grown on millions of hectares annually. The most significant example of the success of host resistance programs are the control of the Hessian fly through breeding resistant varieties of wheat in the United States and the control of brown planthopper of rice in Asia through resistant varieties (Khush and Brar 1991). Advances in the identification and development of soybean breeding lines resistant to phytophagous insects, especially lepidopterous species and the Mexican bean beetle (*Epilachna varivestis* Mulsant) have occurred during the past 25 years (Bowers *et al.* 1999). Breeding efforts have produced four insect resistant soybean cultivars in the USA: Crocket (Bowers, 1999), Lamar (Hartwig *et al.* 1990), Lyon (Hartwig *et al.* 1994), and Shore (Smith *et al.* 1975). Other lines are being considered for release (McPherson *et al.* 1996; Sij *et al.* 1999).

Economic yield reduction is the main criterion to compare the resistant and nonresistant crop genotypes. On the basis of yield of infested and uninfested plants, the percentage of yield loss can be computed using the following formula (Butron *et al.* 1998): Yield loss (%)–(1–yield of infested plants/yield of uninfested plants) X 100.

## Transgenic Crop Plants and Insect Resistance

Transgenic plants may be defined as plant having received a foreign or modified gene by one of the various methods of transformation (Crop Science Society of America, 1992; see also Ives and Bedford, 1998). Crop yield and quality could be affected in transgenic crop plants carrying foreign genes, which confer resistance to plant pests. Transgenic lines of cotton, which carry one of two insect-control protein genes, showed high levels of resistance to several Lepidopteran insects in field trials (Wilson *et al.* 1992, 1994). Wilson *et al.* (1994) reported also that somaclonal variation, insertion of the insect-control protein genes and expression of the insecticidal proteins caused no general reduction to earliness, yield, yield components or fiber properties in the selected transgenic cotton lines. These authors also reported that several of the transgenic lines yielded significantly more lint and had superior fiber properties when compared with the parental cultivars. Yellow stem borer has been identified as a major insect pest of deepwater rice grown on about 9 million hectares of flooded lands in Asia and Africa, causing severe yield losses. Bt gene(s) from *Bacillus thuringiensis* have been proven very effective in pest resistance programs. Relative to broad-spectrum insecticides, the use of transgenic plants expressing Bt gene(s) is now considered an effective approach to control insect infestation (Alam *et al.* 1998).

The development of plant/cultivars with transgenes usually involves two distinct phases. The first phase involves selection of transformation events (transformed plants) that express the transgene at the desired level without any major negative effects on agronomic properties linked to the transgene. The second phase involves hybridization of the selected transformed plant with the elite germplasm followed by selection and evaluation of progeny to determine the expression of the transgene and the agronomic properties of the selected lines. Ideally, both laboratory and field evaluations would be conducted during each of these two phases of breeding (Jenkins *et al.* 1997). Nonetheless, any single resistance factor, whether selected and bred classically in the crop or inserted with molecular techniques, is likely to succumb to insect detoxification "resistance" of its own (van Emden, 1997, 1999; Gould, 1998; Hilder and Boulter, 1999).

More than 100 examples of plants that are transgenically resistant to insects have recently been reviewed (Pierpoint and Shewry 1996; Schuler *et al.* 1998). In only one case does the resistance involve a second gene, and at least one third of these 116 cases involve a gene for the production of *Bacillus thuringiensis* (Bt) toxins (van Emden 1999).

The environmental impacts of transgenic plants must be carefully considered, especially in regard to non-target Lepidoptera (Monarch butterflies) that are impacted by corn pollen that falls on non-target host plants (Losey *et al.* 1999). Other research done with transgenic (Btk) corn and non-Bt corn suggests that even the non-Bt pollen (i.e. not just the transformed Bt pollen) may be toxic to non-target *Papilio* on their host tree leaves in the Midwest region of the USA (Scriber 2000). In any case, since the pollen-dusted leaved are strongly avoided by one species (*P. glaucus*), it is unlikely that negative impacts of pollen (Bt and non-Bt) would be experienced except along a narrow hedgerow or adjacent forest edge.

## Macronutrients

Macronutrients essential for plant growth are nitrogen, phosphorus, potassium, calcium, magnesium and sulfur. Based on available information in literature, the role of these nutrients on insect resistance in crop plants is reviewed. Numerous observations relate the patterns of insect incidence on plants to the nutritional status of soils on which the host plants grow. The consequent change in abundance of insect herbivores is variable, depending on the impact of altered host nutrition on insect fecundity and immigration. Host nutrition can also modify the plant's reaction to insects and its susceptibility to infestation and create environmental conditions more or less favorable to insects or their natural enemies (which can also be killed by "chemically defended" plants; van Emden 1999).

### *Nitrogen*

Nitrogen plays an important role in plant-insect interactions. It is a limiting factor for many herbivores, which, as a consequence of selection pressure, have evolved behavioral, physiological, morphological and other adaptations to fully utilize the available N in their host plants (McNeil and Southwood 1978; Mattson 1980). Nitrogen deficiency reduced plant growth, dry weight and N and K content in rice plants (Yoshida 1981), while increased application of N decreased the C/N ratio and rendered rice plants soft, succulent and susceptible to pests (Regupathy and Subramanian 1972). Nitrogen concentration of plant tissue, a major determinant of insect growth (Scriber and Slansky 1981), is inversely related to the concentration of cell wall structural components in many plant species (Van Soest 1982). However, Salim and Saxena (1991) reported that resistance to whitebacked planthopper (*Sogatella furcifera*) in rice cultivars may be enhanced by applying moderate rates of N and high doses of K.

Numerous studies have shown that the growth, reproduction and survival of phytophagous insects are positively correlated with the nitrogen content in their food (McNeill and Southwood 1978, Mattson 1980, Scriber and Slansky 1981) or with nitrogen content in the soil (Tingey and Singh 1980). However, nitrogen is also a component of a variety of plant antifeedants and toxins and it may be present as nitrates, which are not readily utilizable (Jansson and Silowitz 1986). As a result, some studies have reported the absence of a correlation or a negative correlation between insect performance and host plant nitrogen content or soil nitrogen (Scriber 1984; Tingey and Singh 1980).

In 23 studies involving nine crops in India, nitrogen fertilization increased insect incidence in 17 trials, decreased incidence in one, and had no effect in five trails (Singh and Agarwal 1983). Although, N can modify the mechanisms of host plant resistance, it is usually not certain whether these effects are due to changes in the contents of allelochemicals, nutrients, or water (Scriber 1984a, 1984b). The principal role that N plays in pest management is to help plants compensate for lost or damaged parts, rapidly pass critical development stages and escape attack (Jones 1976).

Saroja *et al.* (1981) reported that a significant increase in percentage of leaffolder damage (*Cnaphalocrocis medinalis*) damage to lowland rice leaves was achieved only by a very high level of nitrogen (20 kg N ha$^{-1}$). The N was applied in three doses. One-half at sowing and the remaining half in two doses, 25 and 45 days after transplanting. Swaminathan *et al.* (1985) studied the effect of ammonium chloride and urea N at 50, 100, 150 and 200 Kg N ha$^{-1}$ on rice pest incidence. Increasing N application increased leaffolder and stem borer incidence, but N source did not affect pest populations. Gall midge intensity was greater at low N application. Urea N plots had significantly higher gall midge incidence than ammonium chloride plots. All pest incidence was low at the recommended application of 100 kg N ha$^{-1}$.

Purohit *et al.* (1986) studied influence of nitrogen fertilizer level and timing on rice stem borer. Stem borer damage was significantly lower when N was applied as ammonium sulfate. Incidence of stem borer was also reduced when N was applied in two split doses rather than total at sowing. Ukwungwu (1985) studied the effect of N fertilizer applied as ammonium sulfate at 0, 50, 100 and 150 kg N/ha as 50% basal and 25% each at active tillering and panicle initiation on gall midge (*Orseolia oryzivora*) and white stem borer (*Maliarpha septaratella*) infestation on rice plants. Gall midge and stem borer damage increased with applied N. Heinricha and Medrana (1985) studied the influence of N fertilizer on the population of brown planthopper (*Nilaparvata lugens*) on rice and concluded that brown planthopper weight, feeding rate and population growth increased with N application but varied among cultivars. Alagarsamy *et al.* (1985) studied the effect of slow release N fertilizer on rice stem borer and sheath rot incidence and concluded that incidence of both insects were significantly lower with neem cake-coated urea at 75 and 100 kg N/ha levels as compared to urea supergranules.

Hunt *et al.* (1992) found that Colorado potato beetle (*Leptinotarsa decemlineata*) on greenhouse grown tomatoes fertilized with nitrogen rates of 0, 25, 50 and 75 mg N kg$^{-1}$ soil resulted in respectively higher nitrogen concentration in the plants, greater plant yields and greater nitrogen uptake. The insects that developed on plants receiving more nitrogen showed significantly higher percentage survival from first instar to adult and a smaller number of insects remaining as larvae at the termination of the experiment. Higher nitrogen treatments also resulted in significantly more rapid insect development and greater pupal mass.

In part, substitution of organic nitrogen sources for inorganic ones is an important strategy of supplying essential plant nutrients in sustainable agriculture. In this situation, it is important to know the effect of organic manures on insect infestation in crop plants. Allee and Davis (1996) studied the effect of farmyard manure on western corn rootworm (*Diabrotica vigifera*) and concluded that larval counts did not differ by manure rate but adult emergence was reduced an average of 45% in manured plants compared with nonmanured plots. Blumberg *et al.* (1997) studied the effects of tillage practices and nitrogen source on arthropod biomass using sorghum as a test plant. The source of N was more important than tillage type in affecting arthropod biomass. Plots receiving inorganic N had significant increases in arthropod biomass, whereas, plots with organic sources of N had little or no increase in biomass of arthropods. Similarly, both the source of N and the temperature were shown to alter the resistance of tomatoes to nematodes (Melakeberhan 1998). The effect of nematodes on plant growth is worse in nutrient-poor soils, which led to the suggestion that soil nutrient manipulation may provide a tool for nematode management (Melakeberhan 1997, 1999).

Spike and Tollefson (1991) reported that N application rates, early in the growth duration of corn, helps in recovering the corn plants from corn rootworm (*Diabrotica virgifera*) damage. They also reported that the most likely time to find aboveground physiological differences between infested and control plants is from 2 to 3 weeks after silking to physiological maturity. Several authors speculated that root growth was enhanced by the N fertilizer. Spike and Tollefson (1988) also found that increased rates of N-fertilizer resulted in large root system dry weight and less lodging in rootworm damaged plants compared with plants given no N fertilizer. Reidell *et al.* (1996) reported that if tolerance to rootworm larval feeding damage is defined by the presence of a large root system and by the ability of the stem to stand erect, then banded N fertilizer placement can help improve tolerance to corn rootworm larval feeding damage.

It is unfortunate that we still do not know the mechanisms of resistance that involved N-fertilization (or naturally fertile soil) as opposed to little or no N-fertilization (or naturally infertile soil). Natural ecosystems contain plants that exist on nutrient-poor and nutrient-rich soils. It has been theorized that carbon-rich defensive compounds, such as Tannins and terpenes, should occur in greater concentrations in low-nutrient or high light environments (Bryant *et al.* 1983, 1993), whereas under high-nutrient conditions carbon/nutrient ratios are lower and pathways associated with growth and reproduction favored over defense (see also review by Herms and Mattson 1992). This carbon/nutrient balance hypothesis has a number of studies that support it, but also remains controversial (Gershenzon 1994a; Clancy *et al.* 1995).

Another hypothesis dealing with nutrient availability has been called the "plant stress" hypothesis for insect outbreaks (White 1993). If plants respond to stress with increased levels of soluble nitrogen and amino acids in their tissues, herbivores could invade susceptible plants and grow well, such as forest and range species in drought (Mattson and Haack 1987). However, other plants respond differently to stress with tougher leaves, resin production, or increased allelochemical content as DIMBOA in corn which repels insects and pathogens (Richardson and Bacon 1993). It is interesting that 2-week old corn (B49 genotype) allocated carbohydrates and nitrogen for maximal growth rather than for the production of DIMBOA even where fertilized plants had plenty of nitrogen (Manuwoto and Scriber 1985b). Irrigated oak trees attract more leaf miners than drough-stressed trees (Bultman and Faeth 1987), which is also contrary to the "plant-stress" hypothesis. Gradients in water availability of only 100 feet in prairies can result in predictable concentrations of the Cruciferae family allelochemical mythyglucosinate, which correlate with leaf damage and seed predation by insects (Louda *et al.* 1987). Different insects respond to different plant parts in variable fashion and these interactions are complicated by temporal and special variations in genetic and environmental factors. For example, while the borer and sucker guilds of insects (Slansky and Scriber 1985) may do better on stressed plants, leaf-chewers and gall-formers may do worse (Koricheva *et al.* 1998).

*Phosphorus*

Phosphorus is associated with many structural components of plants and affects metabolism of sugar phosphates, nucleic acids, nucleotides, coenzymes and phospholipids. There is very little literature on the plant-mediated effects of phosphorus on insect pests. Singh and Agarwal (1983) reported results of seven studies in India, indicate that in six, phosphorus did not have a negative effect on the shootfly *Atherigona soccata* Rond in sorghum. Mebrahtu *et al.* (1988) reported that phosphorus content in soybean lines had a significantly negative correlation with Mexican bean beetle pupal weight. Elden and Kenworth (1994) reported that the foliar concentration of P was significantly lower in Mexican bean beetle (*Epilachna varivestis*) resistant soybean genotypes as compared to susceptible genotypes. However, the laboratory study of Mebrahtu *et al.* (1988) demonstrated no association between Mexican bean beetle larval feeding damage and soybean foliar concentration.

Funderburk *et al.* (1991) studied population dynamics of soybean insect pests under 0, 30, 60 and 120 kg p ha⁻¹ as a triple superphosphate for 6 years. They concluded that the higher levels of P increased seed yields of each year and population densities of larval velvetbean caterpillars (*Anticarsia gemmatalis*) at the higher levels of P were greater than at the lower levels of P. However, a three generation study with five levels of Minerals showed that spruce budworms (*Choristoneura occidentalis*) performed best when P is at moderate concentrations and Mg is at low concentrations, suggesting that mineral ratios are important as well as single Minerals (Clancy and King 1993).

### Potassium

Potassium is needed by plants for the synthesis of sugars, proteins and starch. Deficiency of K increased soluble N and accumulation of soluble sugars, amides and amino acids, while increased K enhanced deamination of amino acids and reduced sugars in the plant sap (Chaboussou 1972). Increasing levels of potassium may generally have a negative influence on insect populations. Generally, when soil K is low, the addition of K improves the vigor of many plant species and has been shown, in some cases, to decrease the susceptibility of the plant to insect feeding and damage (Perrenoud 1977). A possible reason for the reduction of pests by increased potassium may be due to a higher proteogenesis in plants, a physiological phenomenon correlated with elimination of amino acids, and reducing sugars in the sap which are otherwise favorable for the reproduction of sucking pests (Chaboussou 1972). Further, increase in the sclerenchymatous layer and Silica content may also act as a mechanical barrier in plants that receive high potassium rates (Vaithilingam 1975). Subramanian and Balasubramanian (1976) reported that high doses (200 to 250 kg ha) of muriate of potash reduced incidence of all the insects of rice they have studies. Elden and Kenworthy (1994) showed that foliar K in soybean genotypes had no relationships with Mexican bean beetle damage. Mebrahtu *et al.* (1988) reported that potassium content in soybean lines had significant negative correlation with Mexican bean beetle pupal weight.

Funderburk *et al.* (1991) studied influence of three potassium levels, i.e. 0, 210 and 420 kg K ha$^{-1}$, and concluded that population density of soybean larval velvetbean caterpillars (*Anticarsia gemmatalis*) was increased in one year but there was no increase in the second year.

### Calcium, magnesium and sulfur

Johnson and Campbell (1982) reported that a significant ($P < 0.05$) negative correction of 0.69 was found between spider mite (*Tetranychus urticae* Koch.) damage and levels of carbohydrates and calcium in peanut (*Arachis hypogaea* L.) lines. LeRoux (1959) also reported a negative correlation between *Tetrany urticae* reproduction and the calcium level in cucumbers (*Cucumis sativus* L.) but a positive correlation with sulfur. In peanut cultivars, differences exist in the level of plant calcium. Therefore, the potential exists for induced mite resistance by selecting those cultivars with the potential for above average calcium levels (Hallock and Martens 1974). Mebrahtu *et al.* (1988) reported no association between Mexican bean beetle larval feeding damage and soybean concentration of Ca and Mg.

Elden and Kenworthy (1994) reported significant positive linear contrasts between soybean genotype foliar Ca concentration and Mexican bean beetle damage. This means that as foliar concentration of Ca increased, damage increased. Funderburk *et al.* (1991) studied the influence of Mg levels on soybean larval velvetbean caterpillars (*Anticarsia gemmatalis*) and reported that Mg levels did not affect larval populations.

Plant chemicals such as gossypol, a phenolic in cottonseed oil, and phytates in many cereals, can form insoluble complexes with Minerals (eg., Ca, Cu, Mg, Fe, and Zn; Shieh *et al.* 1968; Mattson and Scriber 1987; Slansky 1992). This complex may reduce the negative effects of the allelochemical, but would also render less available the Minerals already in short supply.

## Micronutrients

Information is limited in the literature about influence of micronutrients on insect damage in annual crop plants. However, some available data indicate positive to negative or no effect of micronutrients on insect damage in crop plants. Cannon and Terriere (1966) reported no effect of the minor elements iron, manganese, zinc, and cobalt on egg production of the two-spotted spider mite (*Tetranychus urticae* Koch.) on bean (*Phaseolus vulgaris* L.).

### Iron (Fe)

Oviposition rates of the mustard beetle decreased significantly on watercress leaves deficient in iron (Allen and Selman 1955). Elden and Kenworth (1994) reported that foliar Fe concentration of soybean genotypes increased, damage by Mexican bean beetle increased. These differences suggested that relationship between concentrations of foliar nutrients and arthropod damage differ among arthropod species as well as plant species. These results all point out that there is no single, simple solution for agricultural scientists to develop resistant cultivar of a crop species. Iron is an essential element for aerobic organisms as a catalyst (Conrad *et al.* 1980). Plants need a constant supply of iron as a chloroplast constituent (Ferredoxin; Possingham 1971) because iron is not translocated from the older to newer leaves (Oertti and Jacobson 1960; Brown 1978). It has been reported that spruce budworm growth is negatively linked to Fe and K (Mattson 1983). Furthermore, iron is the only element negatively correlated with insect body size.

Since many microbes, in response to iron-deficiency stress, excrete chelating agents (most of these compounds are hydroxamic acids), which form stable complexes with $Fe^{+++}$ (Nielands 1967; Raymond 1977), low iron fertilizer treatments were imposed upon high-DIMBOA and low-DIMBOA corn to observe effects on insect herbivores (Manuwoto and Scriber 1985a). For southern armyworms fed these corn treatments, there was little effect of the Fe-deficient treatment for either the high DIMBOA genotype or the low DIMBOA genotype, however, N-deficient treatments severely constrained growth and development of larvae on both maize genotypes (Manuwoto and Scriber 1985a).

Plants deficient in Minerals are also unlikely to be normal physiologically and may contain atypical concentrations of organic or inorganic compounds that also affect herbivores consuming them (Friend 1958). Depressed activity of aconitase in iron-deficient corn may lead to higher citric acid and lower malic acid than normal corn leaves as well as higher K: Ca ratios (Palmer *et al.* 1963). Iron-deficiency causes accumulation of asparagine in citrus, corn and rye (Bar-Akiva 1971), which could also alter insect responses.

One of the most interesting entomological detective stories involves the abnormal performance syndrome of gypsy moths *Lymantria dispar* at Federal Research facilities in Hamden, Connecticut U.S.A. (Keena 1993). Researchers checked carefully for inbreeding, container size, contamination, larvae density, rearing environment and a variety of potential pathogens as causes of larval death (inside the eggs) combined with slow larval growth and high mortality. However, it was one form of one mineral ingredient in the diet that was the primary problem. When crystalline ferric phosphate was included in the Wesson salt mixture (used in commercial meridic diets), the abnormal performance syndrome was noticed, but when the amorphous (non-crystalline) form of ferric phosphate was used, there were no problems (Keena 1993). The other Wesson mixture salts included calcium carbonate, manganous sulfate, magnesium sulfate, aluminium sulfate and potassium chloride and was originally developed for vertebrates. The iron deficiency with crystalline ferric phosphate was the source of malperformance, not the other Minerals (or the vitamins, agar, wheat germ, sorbic acid, methyl paraben or casein diet ingredients).

### Zinc

When seven levels of nitrogen were tested, both with and without mineral supplements, over three generations of spruce budworms, regression analysis estimated that the ratio of Zn to N was the best predication of fitness, not the amount of nitrogen itself in the diet (Clancy 1992). Not only is the zinc-nitrogen interaction important, but high fiber levels in plant tissues may also reduce bioavailability of zinc and iron and alter the effects of these mineral concentrations (Reinhold 1982; Kies *et al.* 1983).

Another observation that hints at the critical importance of conserving zinc for some insects is that male *Heliothis virescens* transfer zinc (but not manganese, copper, or other major metal ions) in their copulations with females (Engebretson and Mason 1980).

### Beneficial mineral elements

Mineral elements, which are not necessarily essential but which are beneficial for certain plant or insect species under certain conditions, include: silicon, Selenium, cobalt, sodium and nickel. For example, boron, cobalt and molybdenum improved growth of pea aphids in synthetic diets (Auclair and Srivastava 1972).

### Silicon

Silica (Si) is not an essential plant nutrient but its role in disease and insect control is widely reported. Silicon also benefits crop plants in several other ways, including improved photosynthetic activity, improvement in water use efficiency, increased straw strength and leaf turgor, improved P metabolism, increased grain and milling yields in rice and improvement of overall plant nutrition (Elwad and Green 1979). It is absorbed as Si $(OH)_4$ and like B, is taken up as an undissociated molecule. Silicon deficiencies normally occur on low-Si, organic soils or on highly weathered soils that have been depleted of Si (Wells *et al.* 1993). Subramanian and Gopalaswamy (1988) studied the effect of application of 1 t $SiO_2$/ha on incidence of rice thrips (*Stenchaetothrips biformis*), rice gall midge (*Orseolia oryzae*) and leaffolder (*Cnaphalocrocis medinalis*). Silicate application significantly reduced the thrips population at tillering and also significantly reduced gall midge and leaffolder incidence at panicle initiation.

Lignification and the deposition of Silica in the leaf sheath of inbred corn are associated with European corn borer (*Ostrinia nubilalis*) (Rojanaridpiched *et al.* 1984; Ostrander and Coors 1997). It has been suggested that thicker or lignified plant cell walls increase the bulk density of the European corn borer diet and may lead to greater energy expenditure to meet larval nutritional and water requirements (Rojanaridpiched *et al.* 1984; Buendgen *et al.* 1990; Bergvinson *et al.* 1994). As Silica (and cellulose) were increased in the artificial diets for *Spodoptera* (armyworms) from 0 to 20%, the digestibility decreased from 67% to 57% (and 62% to 52%) respectively (Peterson *et al.* 1988). Detergent fiber, lignin and biogenic Silica of corn leaf sheaths have been associated with second-generation European corn borer resistance (Coors 1987; Buendgen *et al.* 1990). Cell wall composition may affect insect feeding for both nutritional and physical reasons. Since the energy and nutrients contained in cell walls are largely unavailable to most foraging insects, plants high in cell wall structural components may be undesirable (Scriber and Slansky 1981). Lignified cell walls and Silica deposits may also produce tougher tissues that are more resistant to the tearing action of mandibles (Swain, 1979; Raupp, 1985). Also, as critical nutrient concentrations are diluted by Silica, cellulose, lignins, compensatory larval feeding often occurs (Scriber and Slansky, 1981; Scriber, 1984b; Simpson and Simpson, 1990; Slansky, 1993), which may result in a toxic dose of allelochemicals in the diet (Slansky and Wheeler, 1992).

## Selenium

Selenium is a trace element that has attracted a great deal of experimental attention because of its abundance in the earth's crust and widespread distribution and potential accumulation in the plant and animal kingdoms. Selenium is taken up as selenate ($SeO_4^{2-}$) or selenite ($SeO_3^{2-}$). Selenate and sulfate ($SO_4^{2-}$) compete for the same binding sites at the plasma membrane of root cells. The average Selenium content of crop plants varies between 0.01 to 1.0 mg/kg dry weight (Marschner 1986).

Hogan and Razniak (1991) studied survival of *Tenebrio molitor* under different concentrations of sodium selenite (0.125; 0.25, and 0.5%) and reported a progressive decrease in survival beginning on approximately day 7 through day 28. As an estimate from the slope of the control survival curve, ≈2.2 insects died per day during this period.

### Cadmium

Physical defenses such as plant trichomes and hairiness (Hoffman and McEvoy 1985; Kennedy *et al.* 1991; Gange 1995) may protect plants from insect herbivores or protect insect herbivores from their enemies (Van Lenteren *et al.* 1995). However, they may also help conserve water (Grammatikopoulous and Manetas 1994; Scriber and Margraf 1999). In addition, trichomes of *Brassica juncea* provide sites at which toxins from the soil such as cadmium may accumulate (Salt *et al.* 1995). The specific role of cadmium in insect resistance is not known.

### Salinity

Salinity is a problem in many regions of the world. Global distribution of salt affected soils is estimated to be about 0.9 billion hectares (Lal *et al.* 1989). In south and southeast Asia, nearly 60 million ha of land are affected by Salinity ranging from 0.4 to 1.8 S/m (Salim *et al.* 1990). Plants grown under physical stresses frequently become more susceptible to insects (Rhoades 1983). Salinity stress is known to increase or decrease the mineral content in rice plants. The concentration of P and K decreased on the tops of rice cultivars with increasing soil Salinity from 0.34 to 15 dS/m. But the concentrations of Na, Zn, Cu and Mn increased (Fageria 1985). Salinity stress increased N, decreased K and decreased the quantity of allelochemicals extracted as steam distillates from rice plants (Salim *et al.* 1990). The increase or decrease of Minerals and other soluble metabolites, such as proline, sugar, glycerol, malate and shikimiate may render the plants more susceptible to insects (Levitt 1972; Harborne 1977). Salim *et al.* (1990) reported that intake and assimilation of food, growth, adult longevity, fecundity and population of rice whitebacked planthopper (*Sogatella furcifera* Horvath) were significantly greater on plants grown at Salinity level of 12 dS/m than on unstressed control plants. These authors also reported that regardless of the level of stress, the difference between susceptibility of Taichung Native 1 and resistance of IR2035 cultivars remained distinct. This means effects of physiochemical stresses, such as Salinity, must be considered when breeding insect resistant cultivars for Salinity prone areas.

Sodium (Na) itself is generally low enough in most terrestrial plant tissues (Whittaker *et al.* 1979; Mattson and Scriber 1987) that it may actually limit animal physiology and alter herbivore behavior. For example, salt licks are provided for moose and many species of butterflies puddle, apparently to imbibe sodium, which is apparently needed (Arms *et al.* 1974; Pivnick and McNeil 1987). The amounts of sodium can affect insect growth, survival, and reproduction (Lederhouse *et al.* 1990; Stamp and Harmon, 1991).

It is known that unlike upland deciduous trees which are very low in sodium (11-18 ppm), aquatic vegetation and pasture herbage have sodium levels that may range from 100-1,000 ppm (Butler and Jones 1973). Moose, which are limited by sodium in terrestrial vegatation, often feed on aquatic plants which are high in sodium (Botkin *et al.* 1973). Saltbush, *Atriplex* spp., accumulates NaCl to levels higher than 10% of the dry weight (Bernays and Chapman 1994). Whether this is of defensive value to insects is not known. However, several studies have shown that specially formulated salt mixtures do indeed result in marked improvements in insect growth (Beck *et al.* 1968; Dadd 1985).

**Minerals: General significance for insect phytophagy**

While many mineral elements are essential for the healthy growth of animals, the minimal needs and balance of any one relative to others basically remains unknown for insects (House 1961; Rodriguez 1972; Dadd 1985; Reinecke 1985; Mattson and Scriber 1987). Nutrient differences between herbs and trees and between deciduous and evergreen trees show season trends and vary with a number of other factors (Mattson and Scriber 1987; Clancy *et al.* 1995). The metal ions Fe, K, Mg, Mn, Cu, Co and Mb are all important in catalyzing a diversity of enzyme reactions, and others such as Na, Ca, K and Mg are involved in physiological control mechanisms and maintaining the structure of cell walls. It has been proposed that the minimal mineral needs and their optimal balance for growth of insects may determine the intensity of herbivory on plants with variable levels of nutrients (Mattson *et al.* 1982). Nematode feeding interferes with absorption in plants (Ritter 1976). Manipulation of soil nutrition, to benefit plants and adversely affect nematodes, has been suggested as a tool in IPM (Melakeberhan 1997). While certain fertilizers can have a direct nematicidal effect, the particular sensitivities of the insect and host will give variable results (Ritter 1976).

   It is important to realize that numerous factors in addition to artificial fertilization will determine mineral and nutrient levels in plants including: water stress (Mattson and Haack 1987; Mattson *et al.* 1991) nutrient stress (Bryant *et al.* 1983; Dale 1988; Lorio 1988; Kozlowski *et al.* 1991; Herms and Mattson 1992), air pollution including acid rain (Riemer and Whittaker 1989; Heliövaara and Vaisanen 1993; Clancy *et al.* 1995), plant tissue age (Mattson and Scriber 1987; Clancy *et al.* 1995), stand age or density (Barbosa and Wagner 1989; Velazquez-Martinez *et al.* 1992) as well as the plant species and genotype (Clancy 1991, 1993; Clancy *et al.* 1993; Fritz and Simms 1992). The interactions of nutrients and allelochemicals in plants under different environmental conditions, complicates the search for simple, stable mechanisms of resistance or causes of susceptibility to insects (Scriber and Slansky 1981; Coley *et al.* 1985; Herms and Mattson 1992; Slansky 1992; Clancy *et al.* 1993; Hammerschmidt and Schultz 1996). Tin and bismuth are trace elements (micronutrients) that are reported to increase the quantity of Tannin produced in *Acacia catuchu*, but iron does not (Karunanithy and Kapel 1985). Lack of boron reduces the production of phenolic allelochemicals by oil palm seedlings, which may make them more susceptible to spider mite damage (Rajaratnan and Hook 1975).

The differences in corn borer oviposition in corn were in part mediated by the plant mineral balance, which involved both the absolute levels and ratios of Minerals with borers preferring higher Zn, Al, and N (Phelan *et al.* 1996). Attraction of some weevils to younger waterhyacinth leaves and stimulation to feed may be medicated by the higher concentration of nitrogen, P, K, and Mg (or total phenolics), which are higher concentrations in young leaves, but these young leaves repel generalists like wooly bear caterpillars, *Spilosoma virginica* (Center and Wright 1991) emphasizing the need to look at specific interactions. For example, Colorado potato beetles are attracted to tomato plants fertilized to increase leaf tissue nitrogen, but not with P and K concentrations (Hunt *et al.* 1994).

In oats, the content of N, P, K, Ca and Mg all decreased after flowering. However when interplanted with faba beans in a mixed cropping regime, all of these nutrients (except Ca) increased as P and K in the beans decreased, which reduced aphid damage (Helenius 1990). Since the relationships between host-plant Minerals (Ca, Mg, N, P and K) and phytophagous insect feeding are so complex, a new statistical approach has been suggested using "size and shape" analysis of the insects and the nutrients (Boecklen *et al.* 1991). However, insects are not "fixed" in their regimes and they can compensate in feeding to grow well over a wide range of mineral and nutrient levels (Scriber 1984a; Janssen 1993). These interactions are also modified by allelochemical concentrations and temperatures (Stamp 1994).

With so many co-varying natural factors affecting the nutrient, mineral and allelochemical interactions in plant tissues, artificial diets have been used by some entomologists to determine the roles of these numerous factors. This approach allows the amounts and kinds of ingredients to be experimentally varied individually (Reinecke 1985). Since plants are high in potassium, phosphate and magnesium, whereas they may be generally lower in sodium, calcium and chloride, insect diets are tailored in this direction, with total salt mixtures rarely exceeding 1% of the total dry weight (Singh 1977). Total mineral composition of deciduous tree leaves (P, K, Mg, Ca, Na, Zn, Fe, Cu, Mn and B is 5% to 8%), which is twice as high as conifers (Mattson and Scriber 1987). Herbaceous plants have higher total levels of mineral elements (>10% dry weight), probably because of less cellulose and lignin than tree leaves.

The improvement in growth and development of some insects with the right balance of copper, iron, manganese, zinc, boron, cobalt and molybdenum may actually be due to direct benefits to the symbionts rather than the insect (Reinecke 1985). However, insects have been shown to benefit from the proper balance of Minerals and nutrients, which may be the most important factor in the nutritional ecology of insect herbivores (House 1961; Clancy 1992; Slansky 1993) which can be understood as a response surface design of the "nutritional niche" (Clancy and King 1993) and the insect integration (Raubenheimer and Simpson 1999).

The role of leaf surface microbes in insect herbivory basically remains another unknown factor in resistance (Juniper and Southwood 1986). Pest management may benefit by using natural microbes as catalysts of toxicity from *Bacillus thuringiensis* (Haas and Scriber 1998; Scriber 1998). Another unusual means by which phytophagous insects may readily achieve mineral balance in their diets is by opportunistic consumption of wounded or Molting insects or animal carrion (Whitman *et al.* 1994).

Mineral limitation is not generally considered a central force in the foraging theory of insect herbivores, yet we have outlined a number of cases which suggest otherwise. Quantitative studies of dietary mineral requirements or availability are still very scarce (Mattson and Scriber 1987; Clancy *et al.* 1995). Mineral absorption in herbivorous insects also remains to be studied with regard to the impact of the pH (availability of hydrogen ions) and Eh (electrons) on biochemical reactions in the gut lumen (Appel 1994). The mineral form will vary greatly under different pH and Eh levels: Mn, Fe and S are oxides at neutral to alkaline pH's under oxidizing conditions and metals may be complexed to phenolics, lipids, or carbohydrates, which only emphasizes the variability in mineral forms that are possible under different gut conditions of any insect on any plant species.

Finally, it is important to keep in mind that comprehensive pest management programs using plant breeding for resistance must consider not only the target pest, but all possible guilds of pests (Scriber 1984a, 1984b; Hammerschmidt and Schultz 1996). For example, a genotype (CI 31A) of corn bred for its high DIMBOA (cyclic hydroxamate) and resistance to the European corn borer *Ostrinia nubilalis* is consumed at faster rates and results in faster caterpillar growth for the southern armyworm (*Spodoptera eridania*) than the low DIMBOA genotype (WF9), which is corn borer susceptible (Manuwoto and Scriber 1985a).

Various insect feeding guilds (Slansky and Scriber 1985; Slansky and Rodriguez 1987) and their interactions with other insects herbivores, pathogens and parasites (Denno *et al.* 1995) will be variable even on the same individual plant. Plant resistance is a unique phenomenon that exists under certain environmental conditions, but also has some genetic basis upon which we can select for use in IPM programs. While difficult to understand precisely, this complexity of such "resistance" may convey long-term ecological and evolutionary stability (Fritz and Simms 1992), which single mechanism molecularly-engineered (transgenic) plants may lack (van Emden 1999; Hails 1999).

## CONCLUSIONS

### The Balance of Nature in Nutrients

The fundamental problem with plant-feeding insects is that they like to eat what we as humans also like to eat or use (Southwood 1973; Slansky 1993) and they, therefore, become classified as "pests". Insects that eat what we are not interested in are not considered pests.

The foraging herbivore is faced with basic nutritional needs (Dadd 1985; Reinecke 1985; Slansky and Scriber 1985; Slansky and Rodriguez 1987). Since there are so many critical proteins, carbohydrates, lipids and Minerals, there is often a critical chemical in short supply that limits efficiencies and/or growth rates, or survival. This need to eat more of one abundant ingredient to get enough of the less abundant critical ingredient is central to the foraging theory and host plant resistance. Herbivorous insects concentrate protein, lipids, and Minerals in their bodies (compared to their host plant leaves). In fact, a recent comparison of larvae of the emperor moth (or mopane worm), *Gonimbrasia belina* (Lepidoptera: Saturnidae), used as a delicacy in southern Africa by humans, shows the nutritional value of the insect body to be higher in 17 or 18 amino acids relative to soybean seeds and similar or higher than native soybean seeds in the following Minerals (Cu, Fe, K, Mg, Mn, Na, P, and Zn; Glew *et al.* 1999).

Leaf water and Minerals are often forgotten or omitted when the chemical ecology of insect-plant interactions are discussed (e.g. Scriber 1984a, 1984b; Raubenheimer and Simpson 1999). However, there are many other intrinsic and extrinsic factors that affect host plant resistance and the effectiveness of specific traits or resistance mechanisms including the interaction of allelochemicals and nutrients (Slansky 1992) with different abiotic and biotic variables (Strong *et al.* 1984; Heliövaara and Väisänen 1993; Stamp and Casey 1993; Scriber 1996; Schoonhoven *et al.* 1998).

Plant mineral content has been repeatedly shown to change under the influence of acid rain or pollution with reductions often seen in Mg, Ca, and K and increased levels of S, F, Pb, Zn, Cu, Fe, Mn, and Al in the foliage (Riemer and Whittaker 1989). In addition, ammonium sulfate (often found in acid rain) reduces the feeding of fall armyworm larvae, *Spodoptera frugiperda* on grass. It is not surprising, therefore, that these insect-plant chemical interactions vary not only temporally (daily, seasonally: Mattson and Scriber 1987; Karban and Baldwin 1997), but also geographically (Johnson and Scriber 1994), even for the same species, genotype or cultivar of plants.

While it is important to understand the underlying genetic basis of insect resistance traits and mechanisms in plants, it is also important to evaluate the complete set of environmental conditions to which these plants will be exposed to as well as the behavioral and genetic plasticity in the insects (Howard and Berlocher 1998). Our understanding of host plant resistance (antixenosis, antibiosis and tolerance) requires such a perspective.

### Toxicity of metals/Minerals/trace elements

Any compound at high enough concentration can become a toxin. This is also true with the very Minerals that provide essential role in the biochemical processes of life, both in plants and/or insects. For example, while cadmium, arsenic, and mercury chlorides were most toxic to *Drosophila*, so too were nickel, silver, copper, cobalt, chromium, and zinc (Williams *et al.* 1982). The physiological mechanisms of these toxicities and many other insect examples are reviewed by Heliövaara and Väisänen (1993). Bees may be especially sensitive to atmospheric pollution because pollen becomes heavily contaminated with a surface deposit of metals, and honey has been suggested as an indicator of environmental metal contaminates (Jones 1987).

Iron and manganese appear to be relatively harmless to insects, probably because they are essential to their life process. While Molting is a major way of accomplishing heavy metal excretion by Collembola, iron is retained (unlike manganese which is expelled with the gut epithelium at each molt). However, iron pollution can decrease litter decomposition rates (Nottrot *et al.* 1987). Iron in insect bodies has been considered as part of a mechanism by which bees navigate or Monarch butterflies (*Danaus plexippus*) use the earth's magnetic field in long-range migration (Jones and McFadden 1982).

The important feature in this dual role of Minerals (beneficial/toxic) is that the amount relative to other diet ingredients is very important. This desirable balance is very poorly understood for healthy insect mineral nutrition (or for pest management and toxicity) and deserves additional study, especially since much more than the "target" pest is often affected, whether natural pollution or genetically engineered plant traits such as corn pollen toxic to Lepidoptera are involved.

## ACKNOWLEDGEMENTS

We are grateful for the support of our research institutions during the preparation of this review, and for continuing research support. At Michigan State University, thanks are extended to the Agricultural Experiment Station (Project 1644) and in part to the Natural Science Foundation Program (L.T.E.R. "Organisms in the agricultural landscape"; and the Ecosystems Program "Nitrogen dynamics in defoliated forests" DEB-9510044).

## REFERENCES

Agrawal, A.A., S. Tuzun, and E. Bent. 1999. *Induced Plant Defenses against Pathogens and Herbivores: Biochemistry, Ecology, and Agriculture.* APS Press, St. Paul, MN, 319 pp.

Alagarsamy, G., M. Velusamy, S. Rajagopalan and S. Palanisamy. 1985. Effect of slow-release N fertilizers on stem borer and sheath rot incidence and rice grain yield. *IRRI Newletter,* **10**: 19.

Alam, M.F., K. Datta, E. Abrigo, A. Vasquez, D. Senadhira and S.K. Datta. 1998. Production of transgenic deepwater indica rice plants expressing a synthetic *Bacillus thuringiensis* crylA(b) gene with enhanced resistance to yellow stem borer. *Plant Science,* **135**: 25-30.

Allee, L.L. and P.M. Davis. 1996. Effect of manure and corn hybrid on survival of western corn rootworm (Coleoptera: Chrysomelidae). *Environ. Entomol.,* **25**: 801-809.

Allen, M.D. and I.W. Selman. 1955. Egg-production in the mustard beetle, *Phaedon cochleariae* (F.) in relation to diets of mineral-deficient leaves. *Bull. Entomol. Res.,* **46**: 393-397.

Appel, H. 1994. The chewing herbivore gut lumen: Physiochemical conditions and their impact on plant nutrients, allelochemicals, and insect pathogens. *In* E.A. Bernays (Ed.) *Insect-Plant Interactions* Vol. V. CRC Press, Boca Raton, FL. pp. 209-223.

Auclair, J.L. and P.N. Srivastava. 1972. Some mineral requirements of the pea aphid, *Acyrthosiphum pisum* (Homoptera: Aphididae) *Canad. Entom.,* **104**: 920-936.

Arms, K., P. Feeny, and R. Lederhouse. 1974. Sodium: Stimulus for puddling behavior by swallowtail butterflies, *Papilio glaucus. Science,* **185**: 372-374.

Bar-Akiva, A. 1971. Fundamental aspects of mineral nutrients in use for the elevation of plant nutrient requirement. *Rec. Adv. Plant Nutr.,* **1**: 1125-142.

Barbosa, P. and M.R. Wagner. 1989. *Introduction to Forest and Shade Tree Insects.* Academic Press, San Diego, CA.

Beck, S.D., G.M. Chippendale, and D.E. Swinton. 1968. Nutrition of the European corn borer VI. A larval-rearing medium without crude plant functions. *Annals Entom. Soc. Amer.,* **61:** 459-462.

Belsky, A.J. 1986. Does herbivory benefit plants? A review of the evidence. *American Naturalist* **127:** 870-892.

Berenbaum, M.R. 1995a. Turnabout is fair play: Secondary roles for primary compounds. *J. Chemical Ecol.,* **21:** 925-940.

Berenbaum, M.R. 1995b. The chemistry of defense: Theory and practice. *Proc. National Acad. Sci.,* (US) **92:** 2-8.

Bergvinson, D.J., J.T. Arnason and L.N. Pietrgak. 1994. Localization and quantification of cell wall phenolics in european corn borer resistant and susceptible maize inbreds. *Can. J. Bot.,* **72:** 1243-1249.

Bernays, E.A. and R.F. Chapman. 1994. *Host-plant Selection by Phytophagous Insects.* Chapman and Hall, NY, 312 pp.

Blumberg, A.J. Y, P.F. Hendrix and D.A. Crossley, Jr. 1997. Effects of nitrogen source on arthropod biomass in no-tillage and conventional tillage grain sorghum agroecosystems. *Environ. Entomol,* **26:** 31-37.

Boecklen, W.J., S. Mopper and P. Price. 1991. Size and Shape analysis of mineral resrouces in arroyo willow and their relation to sawfly densities. *Ecological Research,* **6:** 317-331.

Boethel, D.J. and R.D. Eikenbary. 1986. *Interactions of Plant Resistance and Parasitoids and Predators of Insects.* Ellis Horwood, Chichester, England, 224 pp.

Botkin, D.B., P.A. Jordan, A.S. Dominski, H.S. Lowendorf, and G.E. Hutchinson. 1973. Sodium Dynamics in a Northern Ecosystem. *Proc. Natl. Acad. Science* (USA), **70:** 2745-2748.

Bowers, G.R. Jr. 1990. Registration of "Crockett" Soybean. *Crop Sci.,* **30:** 427.

Bowers, G.R., M.M. Kenty, M.O. Way, J.E. Funderburk and J.R. Strayer. 1999. Comparison of Three Methods for Estimating Defoliation in Soybean Breeding Programs. *Agron. J.,* **91:** 242-247.

Brown, J.C. 1978. Mechanism of iron uptake by plants. *Plant Cell. Environ.* **1:** 249-258.

Brummer, E.C. 1998. Diversity Stability, and Sustainable American Agriculture. *Agron. J.,* **90:** 1-2.

Bryant, J.P., F.S. Chapin, and D.R. Klein. 1983. Carbon/Nutrient Balance of Boreal Plants in Relation to Vertebrate Herbivory. *Oilos,* **40:** 357-368.

Bryant, J.P., P.B. Reichardt. T.P. Clausen, and R.A. Werner, 1993. Effects of mineral nutrition on delayed inducible resistance in Alaskan paper birch. *Ecology,* **74:** 2072-2084.

Buendgen, M.R., J.G. Coors, A.W. Grombacher and W.A. Russell. 1990. European corn borer resistance and cell wall composition of three maize populations. *Crop Sci.,* **30:** 505-510.

Bultman, T.L. and S.H. Faeth. 1987. Impact of Irrigation and Experimental Drought Stress on Leaf-mining Insects of Emory Oak. *Oikos,* **48:** 5-10.

Butler, G.W. and D.I.H. Jones. 1973. Mineral biochemistry of herbage. *In* G.W. Butler and R.W. Bailey (Eds) *Chemistry and Biochemistry of Herbage.* **Vol. II.** Academic, NY. pp. 127-162.

Butron, A., R.A. Malvar, P. Velasco, P. Revilla and A. Ordas. 1998. Defence mechanisms of maize against pink stem borer. *Crop Sci.,* **38:** 1159-1163.

Cannon, W.H. Jr. and L.C. Terriere. 1966. Egg production of the two spotted spider mite on bean plants supplied nutrient solutions containing various concentrations of iron, manganese, zinc, and cobalt. *J. Econ. Entomol,* **59:** 89-93.

Carde, R.T. and Bell. 1995. *Chemical Ecology of Insects* 2. Chapman, NY, 384 pp.

Carmel, P. and D.J. Pannell. 1996. Economic issues in management of herbicide-resistant weeds. *Ver. Market. Agric. Econ.,* **64:** 301-308.

Casida, J.E. and G.B. Quistad. 1998. Golden age of insecticide research: Past, present, or future? *Annu. Rev. Entomol,* **43:** 1-16.

Cate, J.R. and M.K. Hinkle. 1993. Integrated pest management: The path of a paradigm. *Natl. Audubon Soc. Spec. Rep.*

Center, T.D. and A.D. Wright. 1991. Age and phytochemical composition of waterhyacinth (Pontederiaceae) leaves determine their acceptability to *Neochetina eichhorniae* (Coleoptera: Curculionidae). *Environmental Entomology,* **20:** 323-334.

Chaboussou, R. 1972. The role of potassium and cation equilibrium in the resistance of plants towards diseases. *Potash Rev.,* **23:** 18

Christensen, K.M., T.G. Whitham, and P. Klein. 1995. Herbivory and tree mortality across a pinyon pine hybrid zone. *Oecologia,* **101:** 29-36.

Clancy, K.M. 1991. Douglas-fir nutrients and terpense as potential factors influecing western spruce budworm defoliation. *In* Y.N. Baranchikov, W.J. Mattson, F. Hain, and T.L. Payne (Eds) *Forest Insect Guild: Patterns of Interaction with Host Trees.,* USDA Forest Service Gen. Tech. Rep., NE-153. p. 124-135.

Clancy, K.M. 1992. Response of western spruce budworm (Lepidoptera: Tortricidae) to increased nitrogen in artificial diets. *Environmental Entomology,* **21:** 331-344.

Clancy, K.M. and M. King. 1993. Defining the wester spruce budworm's nutritional niche with response surface methodology. *Ecology,* **74:** 442-454.

Clancy, K.M., J.K. Itami, and D.P. Huebner, 1993. Douglas-fir nutrients and terpense: Potential resistance factors to western spruce budworm defoliation. *Forest Science,* **39:** 78-94.

Clancy, K.M., M.R. Wagner, and P.B. Reich. 1995. Ecophysiology and insect herbivory. *In* W.K. Smith and T.M. Hinkley (Eds) *Ecophysiology of Coniferous Forest.* Academic Press, San Diego, CA. **180:** 125-180.

Clancy, K.M., M.R. Wagner, and R.W. Tinus. 1988. Variation in host foliage nutrient concentrations in relation to western spruce budworm herbivory. *Can. J. For. Res.,* **18:** 530-539.

Clement, S.L. and S.S. Quisenberry. 1999. *Global Plant Genetic resources for Insect-resistant Crops.* CRC Press. Boca Raton, FL.

Coley, P.D., J.P. Bryant, and F.S. Chapin III. 1985. Resource availability and plant antiherbivore defense. *Science,* **230:** 895-899.

Conrad, H.R., D.R. Zimmerman, and G.F. Combs. 1980. *Literature Review on Iron in Animals and Poultry Nutrition.* National Feed Ingredient Assoc. Des Moines, IL.

Coors, J.G. 1987. Resistance to the European corn borer, *Ostrinia nubilalis* (Hubner), in maize, *Zea mays L.,* as affected by soil Silica, plant Silica, structural carbohydrates, and lignin. *In* W.H. Gabelman and B.C. Loughman (Ed) *Genetic Aspects of Plant Mineral Nutrition. Nijhoff Publ.* The Hague, The Netherlands. pp. 445-456.

Crawley, M.J. 1983. Herbivory. *The Dynamics of Animals-plant Interactions.* Blackwell, Oxford, England.

Crawley, M.J. 1987. Benevolent herbivores? Trends in Ecology and Evolution 2: 167-168.

Crop Science Society of America. 1992. *Glossary of Crop Science Terms.* CSSA, Madison, WI.

Dale, D. 1988. Plant-mediated effects of soil mineral stresses on insects. *In* E.A. Heinrichs (Ed.) *Plant-stress-insect Interactions.* John Wiley and Sons, NY. pp. 35-110.

Dankert, B.A., D.A. Herms, D. Parry, J.M. Scriber, and L.A. Haas. 1997. Mediation of competition between folivores through defoliated-induced changes in host quality. *In* A. Raman (Ed.) *Ecology and Evolution of Plant-feeding Insects in Natural and Man-made Environments.* Backhuys Publ. Leiden, Netherlands. pp. 71-88.

Denno, R.F. and M.S. McClure. 1983. *Variable Plants and Herbivores in Natural and Managed Ecosystems.* Academic Press, NY.

Denno, R.F., M.S. McClure, and J.R. Ott. 1995. Interspecific interactions in phytophagy, insects. *Annu. Rev. Entomol,* **40:** 297-332.

Dicke, F.F. and W.D. Guthrie. 1988. *The most important corn insects. In* G.F. Sprague and J.W. Dudley (Ed). *Corn and Corn Improvement.* Agron. Monogr. 18. ASA, CSSA, and SSSA, Madison, WI. pp. 767-867.

Elden, T.C. and W.J. Kenworthy. 1994. Foliar nutrient concentrations of insect susceptible and resistant soybean germplasm. *Crop Sci.,* **34:** 695-699.

Elwad, S.H. and E.V. Green, Jr. 1979. Silicon and The Rice Plant Environment: A Review of Recent Work., RISO **28:** 235-253.

Engebretson, J.A. and W.H. Mason. 1980. Transfer of $65_{zn}$ in *Heliothis virescens. Environmental Entomology,* **9:** 119-121.

Fageria, N.K. 1985. Salt tolerance of rice cultivars. *Plant Soil,* **88:** 237-243.

Feeny, P.P. 1976. Plant apparency and chemical defense. *Rec. Adv. Phytochem,* **10:** 1-40.

Floate, K.D. and T.G. Whitham. 1993. The "hybrid bridge" hypothesis: Host shifting via plant hybrid swarms. *American Naturalist,* **141:** 651-662.

Floate, K.D. and T.G. Whitham. 1995. Insects as traits in plant systematics: their use in discriminating between hybrid cottonwoods. *Canadian J. Botany,* **73:** 1-13.

Food and Agriculture Orgnisation. 1975. Rep. FAO Panel of Experts on Integrated Pest Control, 5th, Oct. 15-25, 1974. Rome, Italy: FAO-UN, Meeting Rep. 1975/M/2. 41 pp.

Friend, W.G. 1958. Nutritional requirements of phytophagous insects. *Ann. Rev. Entomol,* **3:** 57-74.

Fritz, R.S. 1999. Resistance of hybrid plants to herbivores: genes, environmental or both? *Ecology,* **80:** 382-391.

Fritz, R.S. and E. Simms. 1992. *Plant Resistance to Herbivores and Pathogens: Ecology, Evolution, and Genetics,* University of Chicago Press, Chicago, IL, 590 pp.

Fritz, R.S., C.M. Nichols-Orians and S.J. Brunsfeld. 1994. Interspecific hybridization of plants and resistance to herbivores: Hypothesis, genetics and variable responses in a diverse herbivore community. *Oecologia,* **97:** 106-117.

Funderburk, J.E., I.D. Teare and F.M. Rhoads. 1991. Population dynamics of soybean insect pests vs. soil nutrient levels. *Crop Sci.,* **31:** 1629-1633.

Funderburk, J.E., J.L. Higley, and D. Buntin. 1993. Arthropod pest management. *Adv. Agron.,* **51:** 126-172.

Gange, A.C. 1995. Aphid performance in an alder (*Alnus*) hybrid zone. *Ecology,* **76:** 2074-2083.

Gershenzon, J. 1994a. Metabolic costs of terpenoid accumulation in higher plants. *Jounral of Chemical Ecology,* **20:** 1281-1328.

Gershenzon, J. 1994b. The cost of chemical defense against herbivory: A biochemical perspective *In* E.A. Bernays (Ed.) *Insect-plant Interactions.* Vol. V, CRC Press, Boca Raton, FL. pp. 105-173.

Glew, R.H., D. Jackson, L. Sena, D.J. vander Jagt, A. Pastuszyn and M. Wilson. 1999. *Gonimbrasia belina* (Lepidoptera: Saturniidae): A nutritional food source rich in protein, fatty acids, and Minerals. *American Entomologist,* **45:** 250-253.

Gould, F. 1998. Sustainability of transgenic insecticidal cultivars-integrating pest genetics and ecology. *Annual Review of Entomology,* **43:** 701-726.

Grammatikopoulous, G. and Manetas., Y. 1994. Direct absorption of water by hairy leaves of *Phlomis fruticosa* and its contribution to drought avoidance. *Canadian Journal of Botany,* **72:** 1805-1811.

Hass, L. and J.M. Scriber. 1998. Phyllophane sterilization with bleach does not reduce Btk toxicity for *Papilio glaucus* larvae. *Great Lakes Entomologist,* **31:** 49-57.

Hails, R.S. 1999. Genetically modified plants—the debate continues. *Trends in Ecology and Evolution,* **15:** 14-18.

Hallock, D.L. and D.C. Martens. 1974. Contents of eight nutrients in central stem leaf segment of tempeanut cultivars and lines. *Peanut Sci.,* **1:** 53-56.

Hammerschmidt. R. and J.C. Schultz. 1996. Multiple defenses and signals in plant defense against pathogens and herbivores. *In* J.T. Romeo, J.A. Saunders, and P. Barbosa (Eds.) *Phytochemical Diversity and Redundancy in Ecological Interactions.* Plenum Press, NY. (Res. Adv. Phytochem. 30: 121-154).

Harborne, J.B. 1977. *Introduction to Ecological Biochemistry.* Academic Press, New York.

Hartwig, E.E., L. Lambert, and T.C. Kilen 1990. Registration of "Lamar" soybean. *Crop Sci.,* **30:** 231.

Hartwig, E.E., T.C. Kilen L.D. Young. 1994. Registration of "Lyon" soybean. *Crop Sci.,* **34:** 1412.

Heinricha, E.A. and F.G. Medrano. 1985. Influence of N fertilizer on the population development of brown planthopper. *IRRI Newsletter,* **10:** 20-21.

Helenius, J. 1990. Plant size, nutrient composition and biomass productivity of oats and faba bean in intercropping and the effect of controlling *Rhopalosiphum padi* (Homoptera: Aphidae) on these properties. *J. Agricultural Science in Finland,* **62:** 21-32.

Heliövaara, K. and R. Väisänen. 1993. *Insects and Pollution.* CRC Press, Boca Raton, FL, 393 pp.

Herms, D.A. and W.J. Mattson. 1992. The dilemma of plants: To grow or to defend. *Quarterly Review of Biology,* **67:** 283-335.

Herzog, D.C. and J.E. Funderburk. 1986. Ecological bases for habitat management and cultural control. *In* M. Kogan (Ed.) *Ecological Theory and Integrated Pest Management Practices.* Wiley Interscience, NY. p. 217-250.

Hilder, V.A. and D. Boulter. 1999. Genetic engineering of crop plants for insect resistance—a critical review. *Crop Protection,* **18:** 177-191.

Hoffman, G.D. and P.B. McEvoy. 1985. Mechanical limitations of feeding by meadow spittlebugs, *Philaenus spumarius* (Homoptera: Cercopidae) on wild and cultivated host plants. *Ecological Entomology,* **10:** 415-426.

Hogan, R. and H.G. Razniak. 1991. Selenium-induced mortality and tissue distribution studies in *Tenebrio molitor* (Coleoptera: Tenebrionidae). *Environmental Entomology,* **20:** 790-794.

Holtzer, T.O, R.L. Anderson, M.P. McMullen and F.B. Peairs. 1996. Integraäted pest management of insects, plant pathogens, and weeds in dryland cropping systems of the Great Plains. *J. Prod. Agric.,* **9:** 200-208.

House, H.L. 1961. Insect nutrition. *Annual Review of Entomology,* **6:** 13-26.

Howard, D.J. and S.H. Berlocher. 1998. *Endless Forms: Species and Speciation.* Oxford Univ. Press, NY. 470 pp.

Hunt, D.W.A., C.F. Drury and H.E.L. Maw. 1992. Influence of nitrogen on the performance of Colorado potato beetle (Coleoptera: Chrysomelidae) on tomato. *Environ. Entomol,* **21:** 817-821.

Hunt, D.W.A., A. Liptay, and C.F. Drury. 1994. Nitrogen supply during production of tomato transplants affects preference by Colorado potato beetle. *Hortscience,* **29:** 1326-1328.

Hunter, M.D. 1992. Interactions within herbivore communities mediated by the hostplant: The keystone herbivore concept. *In* M.D. Hunter, T. Ohgushi, P.W. Price (Eds) *Effects of Resource Distribution on Animal-plant Interactions.* Academic Press, NY. pp. 287-325.

Hutchins, S.H. and L.P. Pedigo. 1998. Feed-value approach for establishing economic injury levels. *J. Econ. Entomol,* **91:** 347-351.

Ives, C.L. and B.M. Bedford. 1998. *Agricultural Biotechnology in International Development.* CABI Publishing, NY, 354 pp.

Jansson, J.A.M. 1993. Effects of the mineral composition and water content of intact plants on the fitness of the African armyworm. *Oecologia,* **95:** 401-409.

Jansson, R.K. and Z. Smilowitz. 1986. Influence of potato persistence, foliar biomass, and foliar nitrogen on abundance of abundance of *Leptinotrasa decemlineata* (Coleoptera: Chysomelidade). on potato. *Environ. Entomol,* **15:** 726-732.

Jenkins, J.N., J.C. McCarty, Jr., R.E. Buehler, J. Kiser, C. Williams, and T. Wofford. 1997. Resistance of cotton with δ-endotoxin genes from *Bacillus thuringiensis* var. kurstaki on selected Lepidopteran insects. *Agron. J.,* **89:** 768-780.

Johnson, D.R. and W.V. Campbell. 1982. Variation in the foliage nutrients of several peanut lines and their association with damage received by the twospotted spider mite. *Tetranychus urticae. J. Georgia Entomol. Soc.,* **41:** 69-72.

Johnson, K and J.M. Scriber. 1994. Geographic variation in plant allelochemicals of significance in insect herbivores. *In* T.N. Ananthakrishnan (Ed) *Functional Dynamics of Phytophagous Insects.* Oxford and IBH Publishing. New Delhi, India. pp. 7-31.

Jones, F.G.W. 1976. Pests resistance and fertilizers. *In* International Potash Institute (Ed) *Fertilizer Use and Plant Health.* Bern, Switzerland, pp. 233-258.

Jones, R. 1987. Honey as an indicator or heavy metal contamination. *Water Air Soil Pollut,* **33:** 179.

Jones, D.S. and B.J. McFadden. 1982. Induced magnetism in the Monarch butterfly. *Danaus plexippus* (Insecta: Lepidoptera) *J. Exp. Biol.,* **96:** 1.

Juniper, B.E. and T.R.E. Southwood. (eds.) 1986. *Insects and the Plant Surface.* Edward Arnold, London. 360 pp.

Karban, R. and I.T. Baldwin. 1997. *Induced Responses to Herbivory.* Univ. Chicago Press, Chicago, IL, 319 pp.

Karunanithy, R. and M. Kapel. 1985. Effect of hard Lewis acids on Tannin synthesis in plants-relationship of tin, busmuth and iron in *Acacia catechu. J. Pharm. Pharmac.* **37:** 44.

Keena, M. 1993. The abnormal performance syndrome case. *In* S.L.C. Fosbroke and K.W. Gottshalk (Eds) 1993 *USDA Interagency Gypsy Moth Research Forum Proceedings. General Technical Report NE-179.* USDA Forest Service, Washington, DC. pp. 13-21.

Kennedy, G.G., R.R. Farrar, and R.K. Kashap. 1991. Tridecanone glandular trichome-mediated insect resistance in tomato-effect on parasitoids and predators of Heliothis zea. *American Chemical Society Symposium Series,* **449:** 150-165.

Khush, G.S. and D.S. Brar. 1991. Genetics of resistance to insects in crop plants. *Adv. Agron.*, **45**: 223-274.

Kies, C., E. Young, and L. McEndree. 1983. Zinc bioavailability from vegetarian diets: Influence of dietary fiber, ascorbic acid, and post-dietary practices. *In* G.E. Inglett (Ed) *Nutritional bioavailability of Zinc*. Amer. Chem. Soc. Symp. 210. Washington, DC. p. 115-126.

Kim, K. 1993. *Evolution of Insect Pests: Patterns of Variation*. Wiley, NY.

Klun, J.A., C.L. Tipton, and T.A. Brindley. 1967. 2, 4-Dihydroxy-7-methyoxy-1, 4-benzoxazin-3-one (DIMBOA), an active agent in the resistance of maize to the European corn borer. *J. Econ. Entomol.*, **60**: 1529-1533.

Kogan, M. 1998. Integrated pest management: Historical perspective and contemporary developments. *Annu. Rev. Entomol.*, **43**: 243-270.

Koricheva, J., S. Larsson, and E. Haukioja. 1998. Insect performance on experimentally stressed woody plants. *Annu. Rev. Entomol.*, **43**: 195-216.

Kozlowski, T.T., P.J. Kramer, and S.G. Pallardy. 1991. *Physiology of Woody Plants*. Academic Press, NY.

Lal, R., G.F. Hall, and F.P. Miller. 1989. Soil degradation. I. Basic processes. *Land degradation and Rehabilitation*, **1**: 51-69.

Lederhouse, R.C., M.P. Ayres, and J.M. Scriber. 1990. Adult nutrition affects male virility in *Papilio glaucus* L. *Functional Ecology*, **4**: 743-751.

Levitt, J. 1972. *Responses of Plants to Environmental Stresses*. Academic Press, NY.

Lilly, J.H. and H. Gunderson. 1952. Fighting the corn rootworm. *Iowa Farm Sci.*, **6**: 18-19.

Loneragan, J.F. 1997. Plant nutrition in the 20th and perspectives for the 21st centuary. *In* T. Ando et al. (Ed) *Plant Nutrition for Sustainable Food Production and Environment*. Kluwer Academic Publishers, Dordrecht, The Netherlands. p. 3-14.

Lorio, P.L. 1988. Growth differentiation-balance relationships in pines affect their resistance to bark beetles (Coleoptera: Scolytidae). *In* W.J. Mattson, J. Levieux, and C. Bernard-Dagan (Eds) *Mechanisms of Woody Plant Defense Againts Insects: Search For Pattern.*, Springer-Verlag, NY. p. 73-92.

Losey, J.E., L.S. Raynor, and M.E. Carter. 1999. Transgenic pollen harms Monarch larvae. *Nature*, **388**: 214.

Louda, S.M., M.A. Farris, and M.J. Blua. 1987. Variation in methylglucosinolate and insect damage to *Cleome serrulata* (Capparaceae) along a natural soil moisture gradient. *Journal of Chemical Ecology*, **13**: 569-581.

Malvar, R.A., M.E. Cartea, P. Revilla, A. Ordas, A. Alvarez and J.P. Mansilla. 1993. Sources of resistance to pink stem borer and european corn borer in maize. *Maydica*, **38**: 313-319.

Manuwoto, S. and J.M. Scriber. 1985a. Effects of iron deficient and nitrogen deficient nutrient solutions on consumption and utilization of two U.S. inbred corn genotypes by the southern armyworm. *J. Chemical Ecology*, **11**: 1469-1483.

Manuwoto, S. and J.M. Scriber. 1985b. Differential effects of nitrogen fertilization on biomass and nitrogen utilization of 3 corn genotypes by the southern armyworm, *Spodoptera eridania*. *Agriculture, Ecosystems and Environment*, **14**: 25-40.

Manuwoto, S. and J.M. Scriber. 1985c. Neonate larval survival of European corn borers, *Ostrinia nubilalus*, on maize: Effects of light intensity and degree of insect inbreeding. *Agriculture, Ecosystems, and Environment*, **14**: 2211-2236.

Marschner, H. 1986. *Mineral Nutrition of Higher Plants*. Academic Press, NY.

Mattson, W.J. 1980. Herbivory in relation to plant nitrogen content. *Annu. Rev. Ecol.*, **11**: 119-161.

Mattson, W.J. 1983. Spruce budworm (*Choristoneura fumiferana*) performance in relation to foliar chemistry of its host plants. *In* R.L. Talerica and M. Montgomery (Eds) *Proc. Forest Defoliator-Host Interactions: A comparison between Gypsy Moth and Spruce budworms*. USDA Forest Service, General Technical Report NE-85. Washington, DC. p 55-65.

Mattson, W.J. and R.A. Haack. 1987. The role of drought stress in provoking outbreaks of phytophagous insects. *In* Barbosa and Schultz (Eds). *Insect Outbreaks*. Academic Press, NY. p. 365-407.

Mattson, W.J. and J.M. Scriber. 1987. Nutritional ecology of insect folivores of woody plants: Water nitrogen, fiber, and mineral considerations. *In* F. Slansky Jr. and J.G. Rodriguiez (Eds) *Nutritional Ecology of Insects, Mites and Spiders and Related Invertebrates*. Wiley, NY. p. 105-146.

Mattson, W.J., N. Lorimer, and R.A. Leary. 1982. Role of plant variability (trait vector dynamics and diversity) in plant/herbivore interactions. *In* H. Heybroek, B.R. Stephan, and K. von Weissenberg (Eds). *Resistance to Diseases and Pests in Forest Tress*. Pudoc, Wageningen, The Netherlands. p. 295-303.

Mattson, W.J., R.A. Haack, R.K. Lawrence, and S.S. Slocum. 1991. Considering the nutritional ecology of the spruce budworm in its management. *For. Ecol. Manage*, **39**: 182-210.

Maxwell, F.G. and P.R. Jennings. 1980. *Breeding Plant Resistant to Insects*. Wiley, New York, 683 pp.

McNaughton, L.J. 1979. Grazing as an optimisation process: Grass-ungulate relationships in the Serengeti. *American Naturalist*, **113**: 81691-1703.

McNeill, S. and T.R.E. Southwood. 1978. The role of nitrogen in the development of insect plant relationships. *In* J.B. Harborne (Ed.) *Biochemical Aspects of Plant and Animal Coevolution*. Academic Press, London. p. 77-98.

McPherson, R.M., T.P. Mack, J.E. Funderburk, D.J. Boethel, C.G. Helm, and M. Kogan. 1996. *A Multisite Field Evaluation of Arthropod Pest Resistance in Soybean and Impact on Natural enemies*. Spec. Publ. 88. Ga. Agric. Exp., Stat., Tifton, GA.

Mebrahtu, T., W.J. Kenworthy and T.C. Elden. 1988. Inorganic nutrient analysis of leaf tissue from soybean lines screened for Mexican bean beetles resistance. *J. Entomol. Sci.* **23**: 44-51.

Melakeberhan, H. 1997b. Role of plant nutrition on alleviating nematode parasitism. *In:* T. Ando, K. Fujita, T. Mae, H. Matsumoto, S. Mori and J. Sekiya (Eds.) *Plant Nutrition for Sustainable Food Production and Environment*. Kluwer Academic Publishers, London, pp. 759-760.

Melakeberhan, H. 1998a. Effects of temperature and nitrogen source on tomato genotypes response to *Meloidogyne incognita*. *Fundamental and Applied Nematology*, **21**: 25-32.

Melakeberhan, H. 1999. Effect of nutrient source on the physiological mechanisms of *Heterodera glycines* and soybean genotype interactions. *Nematology*, **1**: 113-120.

Moran, N. and W.D. Hamilton. 1980. Low nutritive quality as defense against herbivores. *J. Theoret. Biol.*, **86**: 247-254.

Morgan, J.G. Wilde and D. Johnson. 1980. Greenbug resistance in commercial sorghum hybrids in the seedling stage. *J. Econ. Entomol.*, **73**: 510-514.

Musick, G.J. 1985. Management of arthropods pests in conservation-tillage systems in the southeastern U. S. *In* W.L. Hargrove *et al.* (Ed.) *Proc. 1985 South Region no-Till Conf.*, 16-17 July 1985. Griffin, GA. p. 191-204.

Neilands, J.B. 1967. Hydroxamic acids in nature. *Science*, **156**: 1443-1447.

Niks, R.E., P.R. Ellis and J.E. Parlevliet. 1993. Resistance to parasites. *In* M.D. Hayward *et al.* (Ed) *Plant Breeding: Principles and Prospects*. Chapman and Hall, London. p. 423-447.

Nottrot, F. E.N.G. Joose, and N.M. van Straalen. 1987. Subtle effects of iron and manganese soil pollution on *Orchesella cincta* (Collembola). *Pedobiologia*, **30**: 45.

Oertti, J.C. and L. Jacobson. 1960. Some quantitative consideration in iron nutrition of higher plants. *Plant Physiology*, **35**: 683-688.

Ostrander, B.M. and J.G. Coors. 1997. Relationship between plant composition and European corn borer resistance in three maize populaitons. *Crop Sci.*, **37**: 1741-1745.

Owen, D.F. 1980. How many plants may benefit from animals that eat them? *Oikos*, **35**: 230-235.

Paige, K.N. and T.G. Whitham. 1987. Overcompensation in response to mammalian herbivory: The advantage of being eaten. *American Naturalist*, **129**: 407-416.

Painter, R.H. 1951. *Insect Resistance in Crop Plants*. Macmillan, New York, 570 pp.

Palmer M.J, P.C. DeKock, and J.S.D. Bacon. 1963. Changes in the concentration of malic acid, citric acid, calcium and potassium in the leaves during the growth or normal and iron-deficient mustard plants (*Sinapsis alba*). *Biochem. J.*, **86**: 484-494.

Panda, N. 1979. *Principles of Host-plant Resistance in Insect Pests*. Allanheld, Osmun, and Universe Books, NY.

Panda, N. 1995. *Host Plant resistance to insects*. Oxford Univ. Press. Oxford.

Panda, N. and E.A. Heinrichs. 1983. Levels of tolerance and antibiosis in rice varieties having moderate resistance to the brown planthopper. *Environ. Entomol.* **12**: 1204-1214.

Panda, N. and G.S. Khush. 1995. *Host Plant resistance to insects.* CAB International/International Rice Research Institute, Wallingford/Los Bānos, U.K.-Philippines.

Patterson, F.L., J.E. Foster, H.W. Ohm, J.H. Hatchett and P.L. Taylor. 1992. Proposed system of nomenclature for biotypes of Hessian fly in North America. *J. Econ. Entomol.* **85**: 307-311.

Pedigo, L.P. 1996. *Entomology and Pest Management.* Prentice Hall, Upper saddle River, NJ.

Perrenoud, S. 1977. *Potassium and Plant Health. International Potash Institute of Research.* Topics No. 3. Int. Potash Inst., Bern-Worblaufen, Switzerland.

Peterson, S.S., J.M. Scriber, and J.G. Coors. 1988. Silica, Cellulose and their Interactive Effects on the Feeding Performance of the Southern armywormm Spodoptera eridania (Cramer) (*Lepidoptera: Noctuidae*). *J. Kansas Entomol. Soc.* **61**: 169-177.

Phelan, P.L., K.K. Norris and J.F. Mason. 1996. Soil management history and host preference by Ostrinia nubilalis: Evidence for plant mineral balance mediating insect-plant interactions. *Environmental Entomology,* **25**: 1329-1336.

Pierpoint, W.S. and P.R. Shewry. 1996. *Genetic engineering of Crop Plants for Resistance to Pests and Diseases.* British Crop Protection Council. Farnham, Surrey, United Kingdom.

Pivnick, K.A. and J.H. McNeil. 1987. Puddling in butterflies: sodium affects reproductive success in Thymelicus lineola. *Physiological Entomology,* **12**: 461-472.

Possingham, J.V. 1971. Some effects of mineral nutrient deficiencies on the chloroplasts of higher plants. *Recent Adv. Plant Nutr.,* **1**: 155-165.

Price, P.W. 1997. *Insect Ecology.* 3rd edition. Wiley, NY, 874 pp.

Purohit, M.S., P.M. Bhatt, A.H. Shah and S. Raman. 1986. Influence of nitrogen fertilizer level and timing on stem borer incidence in rice. *IRRI Newletter* **11**: 3

Rajaratnam, J.A. and L.I. Hook. 1975. Effect of boron nutrition on intensity of red spider mite attack on oil palm seedlings. *Exp. Agric.,* **11**: 59.

Raubenheimer, D. and S.J. Simpson. 1999. Integrating nutrition: A geometrical approach. *Entomologia Experimentalist et Applicato,* **91**: 67-82.

Raupp, M.J. 1985. Effects of leaf toughness on mandibular wear of the leaf beetle, *Plagiodera versicolora. Ecol. Entomol.,* **10**: 73-79.

Raymond, K.N. 1977. Kinetically inert complexes of the siderophores in studies of microbial iron transport. *In* K.N. Raymond (Ed) *Bioinorganic Chemistry.* II *Advances in Chemistry Series, 162.* American Chem. Soc., Washington, DC. pp. 33-55.

Rechcigl, J.E. and N.A. Rechcigl. 1999. *Biological and Biotechnological Control of Insect Pests.* Lewis Publ. Boca Raton, FL.

Regupathy, A. and A. Subramanian. 1972. Effect of different doses of fertilizers on the mineral metabolism of IR8 rice in relation to its susceptibility to gall fly, *Pachydiplosis oryzae* Wood-Mason, and leafroller, *Cnaphalocrocis medinalis* Guenee. *Oryza,* **9**: 81-85.

Reinecke, J.P. 1985. Nutrition: Artificial diets. *In:* G.A. Kerkut and L.I. Gilbert (Eds.) *Comprehensive Insect Physiology, Biochemistry and Pharmacology,* Vol. 4, Pergamon, Oxford, pp. 391-419.

Reinhold, J.G. 1982. Dietary fiber and the bioavailability of iron. *In* C. Keis (Ed) *Nutritional Bioavailability of Iron.* Am. Chem. Soc. Symp. 203. Washington, DC. pp. 143-162.

Reroux, E.J. 1959. Effects of various levels of calcium, magnesium, and sulfur in nutrient solutions on fecundity of the twospotted spider mite. *Tetranychus telarius L.* reared on cucumbers. *Can. J. Sci.,* **39**: 92-97.

Rhoades, D.F. 1983. Herbivore population dynamics and plant chemistry. *In* R.F. Denno and M.S. McClure (Ed) *Variable Plants and Herbivores in Natural and Managed Systems.* Academic Press, NY. p. 155-220.

Richardson, M.D. and C.W. Bacon. 1993. Cyclic hydroxamic acid accumulation in corn seedlings exposed to reduced water potentials before, during, and after germination. *Journal of Chemical Ecology,* **19**: 1613-1624.

Riedell, W.E., T.E. Schumacher, and P.D. Evenson. 1996. Nitrogen fertilizer management to improve crop tolerance to corn rootworm larval feeding damage. *Agron. J.,* **88**: 27-32.

Riemer, J. and J.B. Whittaker. 1989. Air pollution and insect herbivores: observed interactions and possible mechanisms. *In* E.A. Bernays (Ed.) *Insect-plant Interactions.* Vol. I. CRC Press. Boca Raton, FL. pp. 73-106.

Ritter, M.P. 1976. The interaction between nutrients and host resistance to nematodes with reference to Mediterranean crops. Proc. 12th Colloq. International Potash Institute. pp. 291-299.

Rodriguez, J.G. 1972. *Insect and Mite Nutrition.* Elselivier, Amsterdam, The Netherlands.

Roitberg, B. 1992. *Insect Chemical Ecology: An Evolutionary Approach.* Chapman. NY.

Rojanaridpiched. C., V.E. Gracen, H.L. Everett, J.G. Coors, B.F. Pugh and P. Bouthyette. 1984. Multiple factor resistance in maize to European corn borer. *Maydica,* **29:** 305-315.

Romeo, J.T., J.A. Saunders, and P. Barbosa. 1996. Phytochemical diversity and redundancy in ecological interactions. Plenum, NY. *Recent Advances in Phytochemistry,* **30:** 1-319.

Rosenthal, G.A. and M.R. Berenbaum. 1992. *Herbivores: Their Interactions with Secondary Plant Metabolites.* Vol. 1 and Vol. 2. Academic Press, NY, 493 pp.

Rosenthal, G.A. and D.H. Janzen. 1979. *Herbivores: Their Interactions with Secondary Plant Metabolites.* Academic Press, NY, 718 pp.

Rosenthal, J.P. and P.M. Kotanen. 1994. Terrestrial plant tolerance to herbivory. *Trends in Ecology and Evolution,* **9:** 145-148.

Russell, G.E. 1947. The chemical composition and digestibility of fodder shrubs and tress. Jt. Publ. Commonw. *Agric. Bur. Pastures Fields Crops For. Anim. Nutr.,* **10:** 185-231.

Russell, G.E. 1978. *Plant Breeding for Pest and Disease Resistance.* Butterworth, Boston, M.A.

Salim, M. and R.C. Saxena. 1991. Nutritional stresses and varietal resistance in rice: Effects on whitebacker planthopper. *Crop Sci.,* **31:** 797-805.

Salim, M., R.C. Saxena and M. Akbar. 1990. Salinity stress and varietal resistance in rice: Effects on whitebacked planthopper. *Crop Sci.,* **30:** 654-659.

Salt, D.E., R.C. Prince, I.J. Pickering, and I. Raskin. 1995. Mechanisms of cadmium mobility and accumulation in Indian mustard. Plant Physiology **109:** 1427-1433.

Saroja, R., S.N. Peeran and N. Raju. 1981. Effects of method of nitrogen application on the incidence of rice leaffolder. *Int. Rice Res. Newsletter,* **6:** 15-16.

Schoonhoven, L.M. 1999. Insects and plants: Two worlds come together. *Entomologia experimentalist et applicata,* **91:** 1-6.

Schoonhoven, L.M., T. Jermy, and J.J.A. van Loon. 1998. *Insect-plant Biology: From Physiology to Evolution.* Chapman and Hall, NY, 409 pp.

Schuler, T.H., G.M. Poppy, B.R. Kerry, and I. Denholm. 1998. Insect-resistant transgenic plants. *Biotechnology,* **18:** 168-175.

Scriber, J.M. 1977. Limiting effects of low leaf-water content on the nitrogen utilization, energy budget and larval growth of *Hyalophora cecropia* (Lepidoptera: Saturniidae). *Oecologia,* **28:** 269-287.

Scriber, J.M. 1984. Nitrogen nutrition of plants and insect invasion. *In:* R. Hauck (Ed.) *Nitrogen in Crop Production.* ASA, CSSA, SSSA. Madison, WI, 460 pp.

Scriber, J.M. 1984a. Nitrogen nutrition of plants and insect invasion. *In* R. Hauck (Ed) *Nitrogen in Crop Production.* ASA, CSSA, SSSA. Madison, WI. p. 441-460.

Scriber, J.M. 1984b. Insect/plant interactions: Host plant suitability. *In* W. Bell and R. Carde (Eds) *The Chemical Ecology of Insects.* Chapman and Hall, London. pp. 159-202.

Scriber, J.M. 1996. A new cold pocket hypothesis to explain local host preference shifts in *Papilio canadensis.* 9th Intern. Symp. Insects and Host Plants. Entomologia expt. & appl. 80: pp. 315-319.

Scriber, J.M. 1998. Non-target Lepidoptera impacted by Btk pesticide sprays, induced phytochemical defenses, and generalized parasitodis and predators of *Lymantria dispar. In* M.P. Zaluki, R.A.I. Drew, and G.G. White (Eds) Vo. 2: *Pest Management-Future Challenges.* Univ. Queensland, Brisbane, Australia. pp. 236-246.

Scriber, J.M. 2000. Lepidoptera as research tools for conservation and biomonitoring: The potential role for universities. *In:* R. Morgan and S. Prachl, (Eds.) *Proc. Invertebrates in Captivity Conference.* SASI, Tucson, AZ, pp. 29-37.

Scriber, J.M. 2001. Insect-Plant Interactions/Herbivory. *In:* V. Resh and R. Carde (eds.) *Encyclopedia of Insects.* Academic Press (submitted).

Scriber, J.M. and M. Ayres. 1988. Leaf chemistry as a defense against insect herbivores. Institute of Scientific Information. *In Animal and Plant Sciences,* Volume 1. pp. 117-123.

Scriber, J.M. and S. Gage. 1995. Pollution and global climate change: Plant ecotones, butterfly hybrid zones, and biodiversity. *In* J.M. Scriber, Y. Tsubaki, and R.C. Lederhouse (Eds) *The Swallowtail Butterflies: Their Ecology and Evolutionary Biology.* Scientific Publishers, Inc. Gainesville, FL. pp. 319-344.

Scriber, J.M. and R.C. Lederhouse. 1992. The thermal environment as a resource dictating geographic patterns of feeding specialization of insect herbivores. *In:* M.R. Hunter, T. Ohgushi and P.W. Price (Eds.) *Effects of Resource Distribution on Animal-Plant Interactions,* Academic Press, pp. 429-466.

Scriber, J.M. and N. Margraf. 1999. Suitability of Florida red bay (*Persea borbonia*) and silk bay (*Persea humilis*) for the *Papilio palamedes* butterfly larvae. *Holarctic Lepidoptera,* (in press).

Scriber, J.M. and F. Slansky, Jr. 1981. The nutritional ecology of immature insects. *Annu. Rev. Entomol.,* **26**: 183-211.

Scriber, J.M., W.M. Tingey, V.R. Gracen, and S.L. Sullivan. 1975. Leaf-feeding resistance to the European corn borer, *Ostrinia nubilalis,* in genotypes of tropical (low-DIMBOA) and U.S. inbred (high-DIMBOA) maize. *J. Econ. Entomol.,* **68**: 823-826.

Scriber, J.M., K. Weir, D. Parry, and J. Deering. 1999. Using hybrid and backcross larvae of *Papilio canadensis* and *P. glaucus* to detect induced chemical resistance in hybrid poplars experimentally defoliated by gypsy moths. *Entomologia experimentalis et applicata,* **91**: 233-236.

Shieh, T.R., E. Mathews, R.J. Wodzinski, and J.H. Ware. 1968. The effect of clacium and phosphate ions on the formulation of soluable iron-gossypol complex. *J. Agr. Food Chem.,* **16**: 208-211.

Simpson, S.J. and C.L. Simpson. 1990. The mechanism of nutritional compensation by phytophagous insects. *In* E.A. Bernays (Ed.) *Insect-plant Interactions.* Vol. 2, CRC Press, Boca Raton, FL. pp. 111-160.

Singh, P. 1977. *Artificial Diets for Insects.* IFI. Plenum, NY, 594 pp.

Singh, R. and R.A. Agarwal. 1983. Fertilizers and pest incidence in India. *Potash Rev.* **23**: 1-4.

Slansky, F. Jr. 1992. Allelochemical-nutrient interactions in herbivore nutritional ecology. *In* G.A. Rosenthaland and M.R. Berenbaum (Eds) *Herbivores: Their Interactions with Secondary Plant Metabolites,* Academic Press. NY. pp. 135-174.

Slansky, F. 1993. Nutritional ecology: The fundamental quest for nutrients. *In* N.E. Stamp and T.M. Casey (Eds) *Caterpillars: Ecological and Evolutionary Constraints on Foraging.* Chapman and Hall, NY. pp. 29-91.

Slansky, F. and J.G. Rodriguez. 1987. *Nutritional Ecology of Insects, Mites, Spiders, and Related Invertebrates.* Wiley, NY.

Slansky, F. and J.M. Scriber. 1985. Food consumption and utilization. *In* Volume 4 G.A. Kerkut and L.I. Gilbert (Eds) *Comprehensive Insect Physiology, Biochemistry, and Pharmacology.* Permagon Press, Oxford. pp. 87-163.

Slansky, F. and G.W. Wheeler. 1992. Caterpillar's compensatory feeding response to diluted nutrients leads to toxic allelochemical dose. *Entomologia exp. et appl.,* **39**: 3-9.

Smith, C.M. 1989. *Plant resistance to insects: A Fundamental Approach.* John Wiley, NY.

Smith, C.M., Z.R. Khan, and M.D. Pathak. 1994. *Techniques for Evaluating Insect Resistance in Crop Plants.* CRC Press, Boca Raton, FL.

Smith, T.J., H.M. Camper, Jr. and J.A. Schillinger. 1975. Registration of "Shore" soybean. *Crop. Sci.,* **15**: 100.

Southwood, T.R.E. 1973. The insect/plant relationship–an evolutionary perspective. In: H.F. van Emden (Ed) *Insect/Plant Relationships.* R. Entom. Soc. London Symposium No. 6, Blackwell Scientific, Oxford. pp. 3-30.

Spencer, K.V. 1988. *Chemical Mediation of Coevolution.* Academic Press, NY, 609 pp.

Spike, B.P. and J.J. Tollefson. 1988. Western corn rootworm (Coleoptera: Chrysomelidae) larval survival and damage potential to corn subjected to nitrogen and plant density treatments. *J. Econ. Entomol.,* **81**: 1450-1455.

Spike, B.P. and J.J. Tollefson. 1991. Response of wester corn rootworm-infested corn to nitrogen fertilization and plant density. *Crop. Sci.,* **31**: 776-785.

Stamp, N.E. 1992. Theory of plant-insect herbivores interactions on the inevitable brink of resynthesis. *Bull. Ecological Soc. Amer.,* **73**: 28-34.

Stamp, N.E. 1994. Simultaneous effects of potassium, rutin, and temperature on performance of *Manduca sexta* caterpillars. *Entomologia exp. et appl.* **72**: 135-143.

Stamp, N.E. and M. Casey. 1993. *Caterpillars: Ecological and Evolutionary Constraints on Foraging.* Chapman and Hall, NY.

Stamp, N.E. and G.D. Harmon. 1991. *Effect of Potassium and Sodium on Fecundity and Survivorship of Japanese Beetles.* Oikos, **62**: 299-305.

Stamp, N.E. and Y. Yang. 1996. Response of insect herbivores to multiple allelochemicals under different thermal regimes. *Ecology,* **77**: 1088-1098.

Starks, K.J., J.R. Muniappa and R.D. Eikenbary. 1972. Interactions between plant resistance and parasitism against greenbug and barley and sorghum. *Ann. Entomol. Soc. Am.,* **65**: 650-655.

Stern, V.M., R.F. Smith, van den Bosch and K.S. Hagen. 1959. The integrated control concept. *Hilgardia,* **29**: 81-101.

Straus, S.Y. 1994. Levels of herbivory and parasitism in host hybrid zones. *Trends in Ecology and Evolution,* **9**: 209-214.

Strong, D.R., J.H. Lawton, and T.R.F. Southwood. 1984. *Insects on Plants: Community Patterns and Mechanisms.* Blackwell Scientific, Oxford, England.

Subramanian, R. and M. Balasubramanian. 1976. Effect of potash nutrition on the incidence of certain insect pests of rice. *Madaras Agric. J.,* **63**: 561-564.

Subramanian, S. and Gopalaswamy., A. 1988. Effect of Silicate materials on rice crop pests. IRRI Newsletter **13**: 32.

Swain, T. 1979. Tannins and lignins. *In* G.A. Rosenthal and D.H. Janzen (Ed) *Herbivores: Their Interaction with Secondary Plant Metabolism.* Academic Press, NY. p. 657-682.

Swaminathan, K., R. Saroja and N. Raju. 1985. Influence of source and level of nitrogen application on pest incidence. *Int. Rice Res. Newsletter,* **10**: 24.

Tallamy, D.W. and Raupp., M.J. 1991. *Phytochemical Induction by Herbivores.* John Wiley, NY.

Tester, C.F. 1977. Constituents of soybean cultivars differing in insect resistance. *Phytochemistry,* **16**:1899-1901.

Tingey, W.M. and S.R. Singh. 1980. Environmental factors influencing the magnitude and expression of resistance. *In* F.G. Maxwell and P.R. Jennings (Ed) *Breeding Plants Resistance to Insects.* Wiley, NY. p. 87-113.

Tollrian, R. and C.D. Harvell, 1999. *The Ecology and Evolution of Inducible Defenses.* Princeton Univ. Press, NJ, 383 pp.

Trumble, J.T., D.M. Kolodny-Hirsh and I.P. Ting. 1993. Plant compensation for arthropod herbivory. *Annu. Rev. Entomol.,* **38**: 93-119.

Ukwungwu, M.N. 1985. Effect of nitrogen and carbofuran on gall midge and white stemborer infestation in Nigeria. *IRRI Newsletter,* **10**: 19-20.

Vaithilingam. C. 1975. Effect of potash nutrient on the incidence and severity of different insect pests of rice. M.S. (Ag.) Thesis. Annamalai University, Annamalai, India.

van Emden, H.F. 1997. Host plant resistance to insect pests. *In* D. Pimentel (Ed) *Techniques for Reducing Pesticide Use.* Wiley, Chichester, United Kingdom. pp. 129-152.

van Emden, H.F. 1999. Transgenic host plant resistance-some reservations. *Annals Entomol. Soc. Amer.,* **92**: 788-797.

Van Lenteren, J.C., L.Z. Hua, J.W. Kamerman, and X. Rumei. 1995. The parasite-host relationship between *Encarsia formose* (Hymenoptera: Aphelinidae) and Trialeurodes vaporariorum (Homoptera: Aleyrodidae): XXVI. Leaf hairs reduce the capacity of *Encarsia* to control greenhouse whitefly on cucumber. *J. Applied Biology,* **119**: 553-559.

Van Soest, P.J. 1982. *Nutritional Ecology of the Ruminant.* O and B Books, Corvallis, OR.

Velazquez-Martinex, A., D.A. Perry, and T.E. Bell. 1992. Response of aboveground biomass increment, growth efficiency, and foliar nutrients to thinning, fertilization and pruning in young Douglas-fir plantations in the central Oregon Cascades. *Can. J. For. Res.,* **22**: 1278-1289.

Wells, B.R., B.A. Huey, R.J. Morman and R.S. Helms. 1993. Rice. *In* W.F. Bennett (Ed) *Nutrient Deficiencies and Toxicities in Crop Plants.* The American Phytopathological Society, St. Paul, MN. p. 15-19.

White, T.C.R. 1993. *The Inadequate Environment: Nitrogen and the Abundance of Animals.* Springer, Berlin.

Whitham, T.G. 1989. Plant hybrid zones as sinks for pests. *Science,* **244:** 1490-1493.

Whitham, T.G., P.A. Morrow, and B.M. Potts. 1994. Plant hybrid zones as centers of diversity: The herbivore community of two endemic Tasmanian eucalypts. *Oecologia,* **97:** 481-490.

Whithman, D.W., M.S. Blum, and F. Slansky, Jr. 1994. Carnivory in phytophagous insects. *In* T.N. Ananthakrishnan (Ed) *Functional Dynamics of Phytophagous Insects.* Oxford and IBH Publ. New Delhi, India. pp. 161-205.

Whittaker, R.H. and P.P. Feeny. 1971. Allelochemicals: Chemical interactions between species. *Science,* **171:** 757-770:

Whittaker, R.H. and G.E. Likens, F.H. Borman, J.S. Eaton, and T.G. Siccama. 1979. The Hubbard Brook ecosystem study: Forest nutrient cycling and element behavior. *Ecology,* **60:** 203-220.

Williams, M.W., J.D. Hoeschele, J.E. Turner, K.B. Jacobson, N.T. Christie, C.L. Paton, L.H. Smith, H.R. Witschi, and E.H. Lee. 1982. Chemical softness and acute metal toxicity in mice and *Drosophila. Toxicol. Appl. Pharmacol.,* **63:** 461.

Wilson, F.D., H.M. Flint, W.R. Deaton and R.E. Buehler. 1994. Yield, yield components, and fibre properties of insect-resistant cotton lines containing *Bacillus thuringiensis* toxin gene. *Crop Sci.,* **34:** 38-41.

Wilson, F.D., H.M. Flint, W.R. Deaton, D.A. Fischhoff, F.J. Perlak, T.A. Armstrong, R.L. Fuchs, S.A. Berberich, N.J. Parks, and B.R. Stapp. 1992. Resistance of cotton lines containing a *Bacillus thuringiensis* toxin to pink bollworm (Lepidoptera: Gelechiidae) and other insects. *J. Econ. Entomol.* **85:** 1516-1521.

Yoshida, S. 1981. Fundamental of rice science. IRRI, Los banos, Laguna, Philippines.

Zangerland, A.R. and M.R. Berenbaum. 1998. Damage-inducibility of primary and secondary metabolites in the wild parsnip (*Pastinaca sativa*). *Chemoecology,* **8:** 187-193.

# Chapter 3

# Synthesis of Chemical Defenses in Maize Silks against Corn earworm and their Inheritance in the Flavonoid pathway

*N.W. Widstrom, M.E. Snook and B.Z. Guo*

## INTRODUCTION

The concept that secondary compounds produced by plants could have evolved as defense mechanisms for plants against attack by insects was postulated by Fraenkel (1959). Walter (1957) had reported a "lethal factor" against corn earworm (*Helicoverpa zea*, Boddie) in the silks of sweet corns (*Zea mays* L.), but the probable identity of this factor was not discovered until 1979 (Waiss *et al.* 1979). Several feeding studies were conducted during the decade of the 1960's (Bennett *et al.* 1967; Knapp *et al.* 1965; Starks *et al.* 1965) but none provided convincing evidence of antibiosis in maize. Studies conducted in the latter part of the decade began to produce the results necessary to establish the presence of antibiosis in maize silks (Wann and Hills, 1966; Knapp *et al.* 1967; Straub and Fairchild, 1970; Chambliss and Wann, 1971).

Synthesis of leucoanthocyanins and their colored relatives, the anthocyanins, have been a long-standing subject of interest in maize (Coe, 1955) because of their association with plant and seed pigments. The flavonoid pathway, involving anthocyanins, anthocyanidins, anthocyans, flavonols, flavanones, flavones and several other related compounds is common to numerous plant species. The compounds produced in the pathway provide color to the flowers of many ornamentals (Wagner and Mitchell, 1964) and to plant and seed color for other species like maize (Styles and Ceska, 1975, 1977, 1981, 1989). Several of the pathway compounds, specifically the anthocyanidins, produce seed and plant color when they are glycosylated at the C-3 and/or C-5 carbon position

U.S. Department of Agriculture, Agricultural Research Service, Georgia Coastal Plain Experiment Station, Insect Biology and Population Management Research Laboratory, Tifton GA 31793 of the flavonoid ring structure (Wagner and Mitchell, 1964; Stafford, 1990). However, when the flavanones are converted to flavones and glycosylation occurs, the result is often a C-glycosyl flavone with the sugar attached at the C-6 position of the flavonoid structure (Styles and Ceska, 1977, 1981). Levings and Stuber (1971) reported that the oxidation of C-glycosylflavones was responsible for the browning reaction of fresh cut silks in maize.

Measurement of antibiotic effects on the corn earworm by the silks of several maize genotypes was vigorously pursued in the 1970's, especially those for a Mexican landrace called *"Zapalote Chico"* (Wiseman *et al.* 1977, 1978). The effects were usually evaluated as differences in larval weight and maturation, but an inheritance study by Widstrom *et al.* (1977) utilized larval survival as a measure of antibiosis and heritability was found to be less than 25%. The retardation of larval growth by silks of Zapalote Chico was found to be very convincing (Wiseman *et al.* 1981).

Waiss *et al.* (1979) identified a major C-6-glycosylflavone (maysin, Fig. 1) as responsible for antibiotic activity against corn earworm in maize silks, the "lethal factor". The antibiotic activity was confirmed by Elliger *et al.* (1980b). Elliger and his co-workers (1980a) also concluded that hydroxylation of both the 3' and 4' positions of the B ring were necessary for the compound to have maximum antibiotic effect. Waiss *et al.* (1981) stated that a concentration of 0.15% (wt./wt.) maysin was required to effectively reduce larval growth by 50%. This threshold was later increased to 0.20% (wt./wt.) with a recommendation that 0.30% was most desirable, since yearly variation in maysin content often occurred for maize lines (Snook *et al.* 1993, 1994). Furthermore, maysin analogues having only a single hyroxydation at the 4' position, e.g. apimaysin and 3'-methoxymaysin (Fig. 1), have only half the activity of the 3', 4' hydroxy compound (Snook *et al.* 1997).

## CORN SILK FLAVONES: ANALYSIS, ISOLATION, AND IDENTIFICATION

Subsequent to the discovery of maysin as the major antibiosis factor in Z. Chico silks, a spectrophotometric assay for maysin in silk-methanolic extracts (Widstrom *et al.* 1982, 1983) was developed. Low correlation between maysin levels determined by this method and corn earworm larval antibiosis resulted in a high performance liquid chromatography (HPLC) method being developed that proved more accurate and reliable for flavone quantitation in silks (Snook *et al.* 1989). The mehod is simple, fast and reproducible, consisting of reversed-phase chromatography of silk methanol extracts using a linear water/methanol gradient. Virtually any octadecylsilyl (ODS or C-18) HPLC column will adequately separate the major flavones of corn silk. Silks are placed in methanol and allowed to soak for 1-2 weeks at 0°. Maysin is quantitatively extracted without the need for grinding the silks. The internal standard, chrysin, is added, mixed and aliquots removed for analysis. Column eluant is monitored at 340 nm. In the absence of an authentic sample of maysin, rutin would be an acceptable standard for response factor calibration. Gueldner *et al.* (1992) also developed a rapid thin-layer screen for silk maysin.

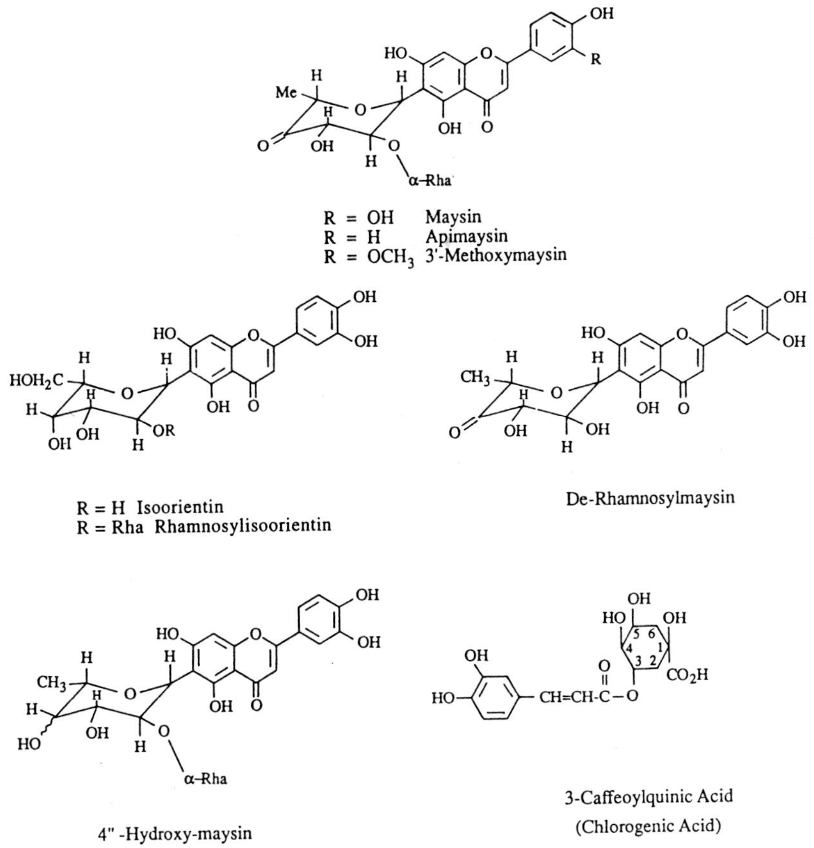

**Fig. 1.** Structures of major flavones and chlorogenic acid found in corn silks.

Silk flavones isolation entails methanol extraction, solvent partitioning, and column chromatography with Silica and preparative reverse-phase substrates (Elliger *et al.* 1980b, Snook *et al.* 1989, 1993). Identification is usually by fast atom bombardment mass spectrometry (FAB-MS), UV, hydrolysis for detection of possible glycosides, and $^{13}$C- and $^1$H-NMR.

Not only has the HPLC method resulted in precise maysin quantitation for genetic and antibiosis studies, it has allowed, for the first time, a survey of the flavone composition of silks more specific than previous thin-layer studies (Levings and Stuber, 1971; Styles and Ceska, 1989). Besides the flavone maysin (2"-*O*-aL-rhamnosyl-6-*C*-(6-deoxy-xylo-hexos-4-ulosyl) luteolin), Elliger *et al.* (1980b) reported the presence of two analogues (Fig. 2); (1) 3'-methoxymaysin and (2) 2"-0-a-L-rhamnosyl-6-*C*-(6-deoxy-xylohexos-4-ulosyl) apigenin (designated apimaysin). These three flavones contain the unique 4"-ketofucosyl sugar (Fig. 2) reported nowhere else in the plant kingdom. All high maysin lines contain minor amounts of these two analogues with 3'-methoxymaysin usually predominating. The separation of these two analogues by HPLC is particularly difficult. Only Hypersil non-endcapped phenyl columns have been found to accomplish adequate separation (Snook *et al.* 1997). Another silk flavone containing the 4"-ketofucosyl sugar structural feature is derhamnosylamysin, prepared by Elliger *et al.* (1980b) from maysin by in-vitro enzymatic degradation. It was found to occur naturally in silks of T218 by Snook *et al.* (1997). T218 was found to contain elevated levels of isoorientin (Fig. 2), whose structure represents a basic class of corn flavone glycosides containing the 4"-hydroxyl sugar moiety rather than the 4"-keto sugar of maysin. Other flavones with the 4"-hydroxyl sugar identified in silks are: rhamnosy lisoorientin (Snook *et al.* 1997), 4"-hydroxy-maysin (two isomers, equatorial and axial 4"-hydroxyls) and axial-4"-hydroxy-3'-methoxymaysin (Snook *et al.* 1995).

Snook and co-workers (1997) surveyed over 1,000 corn silks representing a diverse collection of inbreds, populations and plant introductions. From this work, it was determined that all silks could be classified into only 8 different flavone profiles (Fig. 2). Profile #1 silk types are those with moderate to high levels of maysin (>0.1% fresh wgt.) represented by GT114 and Z. Chico. Approximately, 20% of the inbreds and populations and 12% of the PI's surveyed had relatively high maysin levels. Profile #2 silk types contain low to trace quantities of maysin such as GT119. Low maysin lines represented the bulk of the silks survey, approaching 82% of all lines. Profiles #3 and #4 have high levels of apimaysin (represented by NC7, SC353 and Mp416) and 3'-methoxymaysin (represented by TX501, 9-201, SC144 and Oloton No. 1#), respectively. Profile #5 is represented by T218 containing high levels of isoorientin and derhamnosylmaysin. All lines with isoorientin have this profile. Silk profiles #6 and #7 are lines with rhamnosylisoorientin and no maysin as in PI340853 and rhamnosylisoorientin and high maysin found in line Ames 10363. The last profile silk type (profile #8) is represented by lines with high levels of 4"-hydroxy-maysin such as ESDJ1, A103 and CML131.

## CORN EARWORM AND SILK MAYSIN

The corn earworm (*Helicoverpa zea* Boddie), considered the most destructive pest on corn (*Zea mays* L.) and other crops in the southern U.S., is a silk-and-kernel-feeding insect that is among the major pests in the U.S. (McMillian *et al.* 1977; Ortega *et al.* 1980; Dicke and Guthrie 1988). In southeastern U.S., susceptible corn hybrids produce large corn earworm populations, which not only cause considerable damage to corn, but adult moths emerging from corn also migrate to other crops for oviposition, which subsequently cause substantial economic losses in cotton, peanut, soybean and many vegetable crops. Adult females lay eggs on fresh silks. After hatching, neonate larvae normally move from the exposed silks to a more protected position in the silk channel formed by the husk extension (Wiseman 1989). If silk quantity is sufficient, larval development may be completed in the silk channel. But if the amount of silk is limited, or if the husks are sufficiently loose, larvae will move to developing kernels where their feeding has the greatest economic impact.

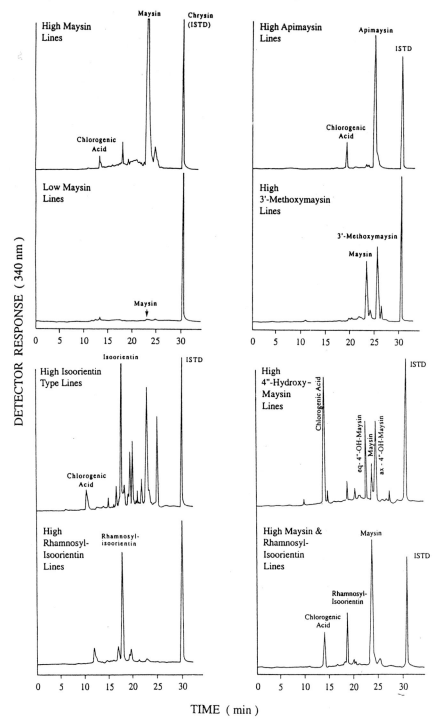

**Fig. 2.**  Characteristic HPLC flavone profiles of major corn silk types.

Maysin synthesis occurs along a branch of the flavonoid metabolic pathway (Fig. 3; Heller and Forkman 1994). In corn and other cereals, two flavonoid biosynthetic pathways have been characterized that are regulated independently by *p1* or *r1/c1* (Styles and Ceska 1989; Grotewold *et al.* 1994, 1998; Koes *et al.* 1994). One pathway results in the 3-deoxy flavonoids or C-glycosyl flavones such as phlobaphenes and maysin, whereas the other pathway produces 3-hydroxy flavonoids such as anthocyanins (Styles and Ceska, 1975; Waiss *et al.* 1979; Snyder and Nicholson 1990; Reid *et al.* 1992; Grotewold *et al.* 1994, 1998; Byrne *et al.* 1996b). C-glycosyl flavones (e.g., maysin) act as insecticidal agents through the subsequent conversion to quinones by polyphenol oxidases. In the larval gut, quinones reduce the availability of free amino acids and proteins, by binding to $-SH$ and $-NH_2$ groups, to the insect and, thus, inhibit larval growth and development (Felton *et al.* 1989; Wiseman and Carpenter 1995). This oxidization conversion also is responsible for the silk-browning reaction (Levings and Stuber 1971; Byrne *et al.* 1996a; Guo *et al.* 1999). Previous laboratory studies showed that the phenotypic correlation coefficient between silk maysin concentration and larval weight had a highly significant negative relationship (Wiseman *et al.* 1992; Byrne *et al.* 1997; Guo *et al.* 1999). Other studies also showed that silk browning was closely related to silk maysin levels (Byrne *et al.* 1996a; Guo *et al.* 1999).

## GENETIC VARIABILITY AND QUANTITATIVE GENETICS

The identification of maysin (Waiss *et al.* 1979) resulted in the initiation of studies to determine the amount of variability in maysin concentration that was available in maize germplasm (Widstrom *et al.* 1983). The studies also provided genetic information to assist in formulating a breeding program and helped to determine the influence of environment on silk maysin concentration (Widstrom *et al.* 1982). Prior to 1989, maysin concentration in silks was determined using spectrophotometry, and unusually high readings obtained for some genotypes (Widstrom *et al.* 1991, 1995) gave strong evidence that non-antibiotic compounds in addition to maysin were interfering with the optical density (O.D.) readings. Snook *et al.* (1989) had developed a reversed-phase HPLC procedure that is compound specific, eliminating the interference of other corn compounds that were inflating O.D. values obtained by spectrophotometry. The new quantitative procedure made possible the evaluation of hundreds of maize genotypes for silk maysin concentration (Snook *et al.* 1993, 1997; Wiseman *et al.* 1992), as well as the identification of several new derivatives and analogues of maysin (Snook *et al.* 1994, 1995).

Interpretation of genetic data was highly inaccurate prior to development of the HPCL procedure for accurate quantitation of maysin and related antibiotic compounds (Widstrom *et al.* 1991). The differences in measurement between O.D. and HPLC methods are not always a serious problem because interpretations are not greatly affected as long as genotypes that produce only maysin are evaluated (Henson *et al.* 1984). Difficulties in interpretation arose when attempts were made to evaluate single plant data, and when tropical or

phenylalanine

↓ *pal1, pal2, pal3*

**Cinnamaic acid**

↓

**4-coumarate**

↓

**4-coumaroyl-CoA + 3 malonyl-CoA**

↓ *c2, whp1*

chalcone

↓ *ch1, ch2, ch3*

**flavanone**

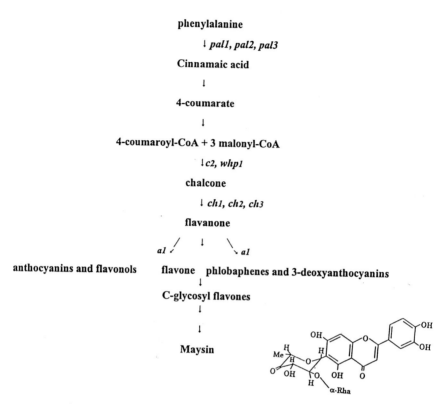

**anthocyanins and flavonols**          **flavone**   **phlobaphenes and 3-deoxyanthocyanins**

↓

**C-glycosyl flavones**

↓

↓

**Maysin**

**Fig. 3.**    Flavonoid branch of the phenylpropanoid pathway leading to maysin synthesis in corn. Genes are shown in italics. The pathway before flavanone (*c2, chi* steps) is regulated by *p1* and *rl/cl*. The steps from flavanone to phlobaphenes, 3-deoxyanthocyanins, and C-glycosyl flavones are under control of *p1*. The steps from flavanone to anthocyanins and flavonols are regulated by the *r1/c1* gene family. Chemical structure of maysin, 2"-*O-a*-l-rhamnosyl-6-c-(6 deoxy-*xylo*-hexos-4-ulosyl) luteolin, is illustrated.

flavonoid producing germplasm other than Zapalote Chico was involved (Widstrom *et al.* 1991). The new HPLC procedure resolved many of the conflicts in data interpretation that had been encountered earlier (Widstrom and Snook, 1994) and provided conclusions about specific crosses that were consistent with later studies utilizing QTL to identify gene action (Byrne *et al.* 1996b). A specific instance demonstrating the consistency is found for studies involving the cross GT114×GT119. For this cross, the conclusion drawn by Widstrom and Snook (1994) using conventional techniques was that GT114 has "one factor with dominance for low maysin expressed only when a modifier gene is also present" while Byrne *et al.* (1996b) using molecular techniques, located QTL on chromosome 9 accounting for 10.8% of the total variance and that prompted them to draw a similar conclusion, "Gene action of this region was dominant for low maysin, but was only expressed in the presence of a functional *p1* allele." Conclusions drawn from conventional quantitative studies, based on

spectrophotometric or HPLC analyses, have also been similar as long as the genotypes tested by O.D. do not produce confounding flavonoids. For example, Henson *et al.* (1984) obtained heritability estimates of 75% to 80% and concluded that maysin content can be increased through cyclic selection, nearly identical to the conclusions of Widstrom and Snook (1996, 1998a) evaluating widely different genotypes using HPLC.

Byrne *et al.* (1997) measured weights of silk-fed larvae and maysin concentrations among $F_2$ and $F_3$ families of the GT114 × GT119 cross, and concluded that chromosomes 1 and 9 together accounted for greater than 90% of the variance for each trait. Loci in the same regions of the chromosomes were responsible for the effects of each trait. QTL analyses of maysin concentrations involving GE37, another inbred line with about three times as much silk-maysin as GT114, revealed significant contributions from chromosomes 2, 6 and 8 (Byrne *et al.* 1998).

The relationship between maysin concentration and the concentration of other antibiotic compounds in silks is not clearly understood. The screening of more than 1000 germplasm sources for maysin and its analogues revealed no obvious relationship among them (Snook *et al.* 1993, 1997). A two year study of 12 inbreds and 132 crosses among them, gave a significant correlation (r=0.81) between maysin and the sum of its analogues apimaysin and 3'-methoxymaysin. Not surprisingly, the mode of inheritance for apimaysin + 3'-methoxymaysin appeared to be essentially the same as that for maysin (Widstrom and Snook, 1998a). Recent reports by McMullen *et al.* (1998) and Lee *et al.* (1998) indicate that one-third to two-thirds of the phenotypic variance for apimaysin concentrations for the cross GT114 × NC7A are attributable to a QTL found on the long arm of chromosome 5. It should be noted that few lines have high apimaysin or 3'-methoxymaysin (Snook *et al.* 1993) and that these two analogues are monohdroxylated on their B ring and have approximately one-half the biological activity of maysin against corn earworm.

Several antibiotic compounds, either found in very few maize lines or found in insignificant amounts to be biologically important, have been identified in maize silks (Snook *et al.* 1994, 1995). Among these is one of interest called isoorientin (A glycosylated isoorientin was first identified in by HPLC co-elution in corn silks by Reid *et al.* 1992). This compound is hydroxylated at both the 3' and 4' positions of the B ring and is as biologically active as maysin (Snook *et al.* 1997). The inbred T218 produces significant amounts of this compound and was investigated in two crosses by Widstrom and Snook (1998b). The isoorientin content of T218 silks was found to be controlled by a single recessive gene for high isoorientin concentration (Widstrom and Snook, 1998b).

## QUANTITATIVE TRAIT LOCI AND METABOLIC PATHWAY

Efforts have been made to study the molecular mechanisms and genetically characterize the resistance of Zapalote Chico and other corn lines reported to

have "lethal silks" (Widstrom *et al.* 1977; Wiseman *et al.* 1976). Numerous lines have been selected with resistance to corn earworm using a dried silk bioassay incorporated into an artificial diet (Wiseman, 1989) and identified with high maysin concentration in silks using HPLC procedures (Snook *et al.* 1989; Wiseman *et al.* 1992a, 1992b). The silk bioassay separates the effects of antibiosis from those of tolerance (morphological factors, tight husk), and allows genotypes of different maturities to be evaluated in the same trial. Widstrom and Snook (1994) reported that distributions of maysin concentration in several progeny generations of the cross GT114 × GT119 fit a model with a dominant allele for low maysin in GT119 and a modifier allele at another locus in GT114. Studies by Wiseman and Bondari (1992 and 1995) analyzed 8-day corn earworm larval weights from dried-silk bioassays of 6 generations each of 6 crosses. Based on generation mean analysis and other statistical genetic analyses, they concluded that gene action influencing antibiosis differed among crosses (with evidence for additive, dominant, and epistatic effects in one or more crosses), and that the number of genes involved in the inheritance of the trait was large.

Byrne *et al.* (1996b) successfully used quantitative trait locus methodology to identify corn chromosome regions associated with silk maysin concentration in the population (GT114 × GT119) F2. Because maysin is synthesized in the flavonoid metabolic pathway (Fig. 3) this study sought to explain maysin inheritance by associating phenotypic values of individual plants with genotypic variation at flavonoid pathway loci. Using RFLP techniques, they found that the *p1* region of chromosome 1 accounted for 58% of the phenotypic variance for the trait and that its effect was largely additive (Byrne *et al.* 1996b). The *p1* locus regulates the portion of the flavonoid pathway leading to maysin. Thus, *p1* also determines the silk-browning phenotype, which is associated with concentrations of maysin and its analogues (Byrne *et al.* 1996a; Guo *et al.* 1999). Synthesis of phlobaphene pigments, which are responsible for red cob and pericarp color, is also regulated by *p1*, and most information on *p1* expression concerns those two tissues (Grotewold *et al.* 1994; Styles and Ceska, 1989). In addition to *p1*, Byrne *et al.* (1996b) detected a second QTL on the short arm of chromosome 9 that showed significant epistasis with *p1*. Based on previously reported map location and a similar interaction with *p1*, they hypothesized that this locus was *brown pericarp1 (bp1)*. The major effect of *p1* on maysin concentration has been demonstrated by the importance of regulatory loci in controlling trait expression (Byrne *et al.* 1996b). The possible involvement of *bp1* suggested that genes of a competing pathway, the one leading to phlobaphenes in this case, can affect the flux of biochemical intermediates channeled toward a compound of interest.

Byrne *et al.* (1997) evaluated silks from the F3 families of the cross GT114 × GT119, corresponding to the F2 high and low maysin traits, for corn earworm antibiosis in a dried-silk assay. The same chromosome regions (*p1* on chromosome 1 and QTL on short arm of chromosome 9), significant for maysin concentration (Byrne *et al.* 1996b), were also detected in the analysis of variance for corn earworm larval weight (Byrne *et al.* 1997). A more recent study was conduced by Byrne *et al.* (1998) on the F3 families of the cross GE37×FF8, which both have phenotypically identical, functional alleles at the *p1* locus. GE37 has very high maysin levels and FF8 has low to intermediate levels of maysin in silks. Composite interval mapping revealed large effects on chromosome 1, 2, 6, and 9 for the three traits (maysin concentrations in two locations, Georgia and Missouri, and corn earworm larval weight grown on artificial diet containing oven-dried silks). Again, the large effect detected on the sort arm of chromosome 9, was revealed for maysin concentration at both locations, but was not detected in the corn earworm larval weight assay. Two of the major QTLs correspond to the locations of known flavonoid loci, *whp1* on chromosome 2L and the *sm1* on chromosome 6L. Because chalcone synthase, encoded by *whpl*, is considered a key enzyme of flavonoid synthesis, an effect on maysin concentration at this locus is feasible.

More recently, Guo *et al.* (1999) and Guo *et al.* (2001) conducted a genetic analysis in an $F_2$ segregating population derived from the high maysin dent corn inbred SC102 and a non-maysin *sh2*-sweet corn inbred B31857. Two major QTL have been identified using single-factor ANOVA, *p1* (*npi286* and *csu3* as flanking markers) on the short arm of chromosome 1 and *a1* on the long arm of chromosome 3. Two minor QTLs also were detected by this analysis, *umc66a (near c2)* on the long arm of chromosome 4 and *umc105a (brown pericarp1, bp1)* on the short arm of chromosome 9 (Meyers 1927; Emerson *et al.* 1935; Byrne *et al.* 1996b), also designated as *rem1* by Lee *et al.* (1998). Two other QTLs also were detected by MAPMAKER/QTL in the centromere region (interval of *umc67* and *asg62*) and on the long arm (*umc128* Bin 1.07) of chromosome 1, respectively. The multiple-locus model analysis retained marker *umc245* on the long arm of chromosome 7, *agrr21* on the long arm of chromosome 8, and the epistatic interaction *p1×a1, p1 × umc245, a1 × agrr21,* and *umc105a × agrr21*.

Again, we demonstrated a fundamental role of the major regulatory locus *p1* in controlling the flavone pathway branch and synthesis of maysin (Byrne *et al.* 1996b, 1998). Recently, Grotewold *et al.* (1998) demonstrated the induction of specific flavonoids and phenylpropanoids by ectopic expression of transgenes encoding regulatory transcription factors (C1/R or P) in cultured Black Mexican Sweet corn cells. The predominant compounds induced by expression of *p1* (encoding P protein) are C-glycosyl flavones closely related to maysin in silks. Additional support for the *p1* regulatory role is its additive QTL gene action and its interaction with other loci (Byrne *et al.* 1996b; Grotewold *et al.* 1998). The second maysin QTL is *a1*, which shows dominant gene action for low maysin and significant epistatic gene action with *p1*. The dominant functional allele *A1* causes anthocyanin pigments to form in the aleurone, plant and pericarp tissues; the recessive nonfunctional *a1* allele causes absence of pigment (colorless) in these tissues (Neuffer *et al.* 1997). The *A1* encodes for a dihydroflavonol 4-reductase (Reddy *et al.* 1987; Schwarz-Sommer *et al.* 1987; Bernhardt *et al.* 1998). This enzyme is involved in two branches of the flavonoid pathway (Fig. 3) and converts dihydroflavonols as well as flavanones (Styles and Ceska 1975, 1989). Dihydroflavonols are the precursors of

3-hydroxyanthocyanins, whereas flavanones are precursors of the 3-deoxyanthocyanins and phlobaphenes (Fig. 2). Styles and Ceska (1977) reported that the *A1* allele is required in the pathways of both the *c1/b1* controlled anthocyanins and the *p1* regulated 3-deoxyanthocyanins, but not in the pathway of *p1* regulated flavones (Fig. 3). Homozygous recessive *a1* alleles would block the pathways and inhibit the synthesis of the 3-hydroxyanthocyanins and 3-deoxyanthocyanins, resulting in the release of precursors or intermediates for the *p1* regulated flavone pathway (Styles and Ceska 1975, 1989). Therefore, a block at the *a1* locus could enhance C-glycosyl flavone biosynthesis and increase or enhance the end products (e.g. maysin) as demonstrated by a recent study (Pelletier *et al.* 1999), using *Arabidopsis* as a model for studying the flavonoid biosynthetic pathway regulation. Pelletier *et al.* (1999) revealed that mutant lines blocked at intermediate steps of the pathway actually accumulated higher levels of specific flavonoid enzymes and other end products. The *p1* × *a1* epistatic interaction in this study indicates the shifting of precursors or intermediates to the flavone pathway and results in high silk maysin, which agrees with a previous study (Styles and Ceska 1989). Styles and Ceska (1989) reported that, in the presence of functional *p1*, the level of flavones in developing pericarp tissue is enhanced by the homozygous recessive *a1*. This may be due to a shifting of intermediates to the flavone pathway (Pelletier *et al.* 1999).

## CONCLUSION

Styles and Cesko (1977) stated that 25 different genes were known which effect flavonoid synthesis in maize. Approximately, one-half of these, located on 7 of the 10 chromosomes, have been identified as having significant impact on the synthesis of biologically active flavones in maize silks (Widstrom and Snook, In press). The most recent gene found to be associated with maysin concentration is *sh2*, a gene closely linked to *a1* on chromosome 3. $F_2$ plants derived from *sh2* kernels of some high-maysin dent x no-maysin *shrunken-2* hybrids produce silks with two to three times as much maysin as those derived from dent kernels (Guo *et al.* 1999). The location of this gene on the long arm of chromosome 3 and its mechanism for maysin enhancement is being investigated (Guo *et al.* in press). Additionally, the consolidated results from several mapping studies of maysin and its analogues have been used to identify a total of 29 significant molecular markers, suggesting the existence of QTL with some importance for genetic control of antibiotic compounds exist on 13 of 20 chromosome arms (Widstrom and Snook, In press). Further investigation of antibiotic flavone related compounds will help to obtain an understanding of their relationship in the flavonoid pathway and their synthesis processes.

## REFERENCES

Bennett, S.E., Josephson, L.M. and Burgess., E.E. 1967. Field and laboratory studies on resistance of corn to the corn earworm. *J. Econ. Entomol.*, **60**:171-173.

Bernhardt, J., Stich, K., Schwarz-Sommer, Z., Saedler, H. and Wienand., U., 1998. Molecular analysis of a second functional A1 gene (dihydroflavonol 4-reductase) in Zea mays. *Plant J.*, **14**:483-488.

Byrne, P.F., Darrah, L.L., Snook, M.E., Wiseman, B.R., Widstrom, N.W., Moellenbeck, D.J. and Barry., B.D., 1996a. Maize silk-browning, maysin content, and antibiosis to the corn earworm, Helicoverpa zea (Boddie). *Maydica,* **41**:13-18.

Byrne, P.F., McMullen, M.D., Snook, M.E., Musket, T.A., Theuri, J.M., Widstrom, N.W., Wiseman, B.R. and Coe., E.H., 1996b. Quantitative trait loci and metabolic pathways: Genetic control of the concentration of maysin, a corn earworm resistance factor, in maize silks. Proc. Natl. *Acad. Sci.,* **93**:8820-8825.

Byrne, P.F., McMullen, M.D., Wiseman, B.R., Snook, M.E., Musket, T.A., Theuri, J.M., Widstrom, N.W. and Coe., E.H., 1997. Identification of maize chromosome regions associated with antibiosis to corn earworm (Lepidoptera: Noctuidae) Iarvae. *J. Econ. Entomol.,* **90**:1039-1045.

Byrne, P.F, McMullen, M.D., Wiseman, B.R., Snook, M.E., Musket, T.A., Theuri, J.M., Widstrom, N.W. and Coe. E.H., 1998. Maize silk maysin concentration and corn earworm antibiosis: Quantitative trait loci and genetic mechanisms. *Crop. Sci.,* **38**:461-471.

Chambliss, O.L. and Wann, E.V., 1971. Antibiosis in earworm resistant sweet corn. *J. Amer. Soc. Hort. Sci.,* **96**:273-277.

Coe, E.H. 1955. Anthocyanin synthesis in maize, the interaction of A2 and Pr in leucoanthocyanin accumulation. *Genetics,* **24**:568.

Dicke, F.F. and Guthrie, W.D., 1998. The most important corn insects. *In* Sprague GF, Dudley JW (Eds) *Corn and Corn Improvement.* American Society of Agronomy, Madison, WI pp. 767-867.

Elliger, C.A., Chan, B.G. and Waiss, Jr., A.C., 1980a. Flavonoids as larval growth inhibitors. *Naturwissenschaften,* **67**:353-360.

Elliger, C.A., Chan, B.G., Waiss, Jr., A.C., Lundin, R.E. and Haddon., W.F., 1980b. *C-glycosylflavones from Zea mays that inhibit insect development. Phytochemistry,* **19**:293-297.

Emerson, R.A., Beadle, G.W. and Fraser, A.C., 1935. *A summary of linkage studies in maize Cornell Univ. Agric. Exp. Stn. Memoir.,* **180**:1-83.

Felton, G.W., Donato, K., Del Vecchio, R.J. and Duffey, S.S., 1989. Activation of plant foliar oxidases by insect feeding reduces nutritive quality of foliage for noctuid herbivores. *J. Chem. Ecol.,* **15**:2667-2694.

Fraenkel, G.S. 1959. The raison d'être of secondary plant substances. *Science,* **129**:1466-1470.

Grotewold, E., Chamberlin, M., Snook, M., Siame, B., Butler, L., Swenson, J., Maddock, S., St. Clair, G. and Bowen, B., 1998. Engineering secondary metabolism in maize cells by ectopic expression of transcription factors. *Plant Cell.,* **10**:721-740.

Grotewold, E., Drummond, B., Bowen, B. and Peterson, T., 1994. The Myb-homologous P gene controls phlobaphene pigmentation in maize floral organs by directly activating a flavonoid biosynthetic gene subset. *Cell.,* **76**:543-553.

Gueldner, R.C., Snook, M.E., Widstrom, N.W. and Wiseman, B.R., 1992. TLC screen for maysin, chlorogenic acid, and other possible resistance factors to the fall armyworm and the corn earworm in Zea mays. *J. Agric. Food Chem.,* **40**:1211-1213.

Guo, B.Z., Widstrom, N.W., Wiseman, B.R., Snook, M.E., Lynch, R.E. and Plaisted, D., 1999. Comparison of silk maysin, antibiosis to corn earworm larvae (Lepidoptera: Noctuidae), and silk browning in crosses of dent × sweet corn. *J. Econ. Entomol.,* **92**:746-753.

Guo, B.Z., Zhang, Z.J., Widstrom, N.W., Snook, M.E., Lynch, R.E. and Plaisted, D., 2001. Quantitative trait loci analysis to identify RFLP markers associated with biosynthesis of silk maysin, an insecticidal compound, in a dent × sweet corn cross. *J. Econ. Entomol.* (in press)

Heller, W. and Forkman, G., 1994. Biosynthesis of flavonoids. *In* Harbome, J.B. (Ed) *The Flavonoids, Advances in Research since 1986.* Chapman & Hall, London pp. 499-536.

Henson, A.R., Zuber, M.S., Darrah, L.L., Barry, D., Rabin, L.B. and Waiss, A.C., 1984. Evaluation of an antibiotic factor in maize silks as a means of corn earworm (Lepidoptera: Noctuidae) suppression. *J. Econ. Entomol.,* **77**:487-490.

Koes, R.E. and Quattrocchio, F., J.N.M. Mol. 1994. The flavonoid biosynthetic pathway in plants: Function and evolution. *Bioessays,* **16**:123-132.

Knapp, J.L., Hedin, P.A. and Douglas, W.A., 1965. A chemical analysis of corn silk from single crosses of dent corn rated as resistance, intermediate, and susceptible to corn earworm. *J. Econ. Entomol.,* **59**:1062-1064.

Knapp J.L., Maxwell, F.G. and Douglas, W.A., 1967. Possible mechanisms of resistance of dent corn earworm. *J. Econ. Entomol.*, **60**:33-36.

Lee, E.A., Byrne, P.F., McMullen, M.D., Snook, M.E., Wiseman, B.R., Widstrom, N.W. and Coe, E.H., 1998. Genetic mechanisms underlying apimaysin and maysin synthesis and corn earworm antibiosis in maize (Zea mays L.). *Genetics*, **149**:1997-2006.

Levings, C.S., III, and Stuber, C.W., 1971. A maize gene controlling silk browning in response to wounding. *Genetics*, **69**:491-498.

McMillan, W.W., Wiseman, B.R., Widstrom, N.W., 1977. An evaluation of commercial sweet corn hybrids for damage by Heliothis zea. *J. Georgia Entomol. Soc.*, **12**:75-79.

McMullen, M.D., Byrne, P.E., Snook, M.E., Wiseman, B.R., Lee, E.A., Widstrom, N.W. and Coe, E.H., 1998. Quantitative trait loci and metabolic pathways. *Proc. Natl. Acad. Sci.*, **95**:1996-2000.

Meyers, M.T. 1927. A second recessive factor for brown pericarp in maize. *Ohio J. Sci.*, **27**:295-300.

Neuffer, M.G., Coe, E.H. and Wessler, S.R., 1997. *Mutants of Maize.* Cold Spring Harbor Laboratory Press, New York. 74 pp.

Ortega, A. and Vasal, S.K., Mihm, J. and Hershey, C., 1980. Breeding for insect resistance in maize. *In* Maxwell, F.G. and Jennings, P.R. (Eds) *Breeding Plant Resistant to insects.* John Wiley and Sons, Inc., New York, pp. 370-419.

Pelletier, M.K., Burbulis, J.E. and Winkel-Shirley, B., 1999. Disruption of specific flavonoid genes enhances the accumulation of flavonoid enzymes and end-products in Arabidopsis seedlings. *Plant Mol. Biol.*, **40**:45-54.

Reddy, A.R., Britsch, L., Salamini, F., Saedler, H. and Rohde, W., 1987. The A1 (anthocyanin-1) locus in Zea mays encodes dihydroquercetin reductase. *Plant Sci.*, **52**:7-14.

Reid, L.M., Mather, D.E., Arnason, J.T., Hamilton, R.I. and Bolton, A.T., 1992. Changes in phenolic constituents of maize silk infected with Fusarium graminearum. *Can. J. Bot.*, **70**:1697-1702.

Schwarz-Sommer, Z., Shepherd, N.S., Tacke, E., Gierl, A., Rohde, W., Leclerc, L., Mattes, M., Bendtgen, R., Peterson, P.A. and Saedler, H., 1987. Influence of transposable elements on the structure and function of the A1 gene of Zea mays. *EMBO J.*, **6**:287-294.

Snook, M.E., Widstrom, N.W. and Gueldner, R.C., 1989. Reversed-phase high performance liquid chromatographic procedures for the determination of maysin in corn silks. *J. Chrom.*, **477**:439-441.

Snook, M.E., Gueldner, R.C., Widstrom, N.W., Wiseman, B.R., Himmelsbach, D.S., Harwood, J.S. and Costello, C.E., 1993. Levels of maysin and maysin analogues in silks of maize germplasm. *J. Agric. Food Chem.*, **41**:1481-1485.

Snook, M.E., Widstrom, N.W., Wiseman, B.R., Byrne, P.F., Harwood, J.S. and Costello, C.E., 1995. New C-4"-Hydroxy derivatives of maysin and 3'-methoxymaysin isolated from corn silks (Zea mays) *J. Agric Food Chem.* **43**:2740-2745.

Snook, M.E., Widstrom, N.W., Wiseman, B.R., Gueldner, R.C., Wilson, R.L., Himmelsbach, D.S., Harwood, J.S. and Bostello, C.E., 1994. New flavone C-glycosides from corn (*Zea mays* L.) for the control of the corn earworm (Helicoverpa zea). P. 122-135 *In* P.A. Hedin (Ed.). ACS symp. Series #557. Am. Chem. Soc. Washington, DC.

Snook, M.E., Wiseman, B.R., Widstrom, N.W. and Wilson, R.L. 1997. Chemicals associated with maize resistance to corn earworm and fall armyworm. *In* J.A. Mihm (Ed.) *Insect Resistant Maize: Recent Advances and Utilization.* CIMMYT Mexico D.F. 27 Nov.-3 Dec. 1994. CIMMYT 37-45.

Snyder, B.A. and Nicholson, R.L., 1990. Synthesis of phytoalexins in sorghum as a site-specific response to fungal ingress. *Science*, **248**:1637-1639.

Stafford, H.A. 1990. *Flavonoid Metabolism.* CRC Press, Inc. Boca Raton, FL. pp. 101-132, 189-224, 239-259.

Starks, K.J., McMillan, W.W., Sekul, A.A. and Cox, H.C., 1965. Corn earworm larval feeding response to corn silk and kernel extracts. *Ann. Entomol. Soc. Am.*, **58**:74-76.

Straub, R.W. and Fairchild, M.L., 1970. Laboratory studies of resistance in corn to the corn earworm. *J. Econ. Entomol.*, **63**:1901-1903.

Styles, E.D. and Ceska, O., 1975. Genetic control of 3-hydroxy and 3-deoxyflavonoids in Zea mays. *Phytochemistry*, **14**:413-415.

Styles, E.D. and Ceska, O., 1977. The genetic control of flavonoid synthesis in maize. *Can. J. Genet. Cytol.*, **19**:289-302.

Styles, E.D. and Ceska, O., 1981. P and R control of flavonoids in bronze coleoptiles of maize. *Can J. Genet. Cytol.*, **23**:691-704.

Styles, E.D. and Ceska, O., 1989. Pericarp flavonoids in genetic strains of Zea mays. *Maydica*, **34**:227-237.

Wagner, R.P. and Mitchell, H.K., 1964. *Genetics and Metabolism.* 2nd Ed. John Wiley and Sons, New York, NY. pp. 561-583.

Waiss, A.C., Jr., Chan, B.G., Elliger, C.A., Dreyer, D.L., Binder, R.G. and Gueldner, R.C., 1981. Insect growth inhibitors in crop. plants. *Bull. Entomol. Soc. Am.*, **27**:217-221.

Waiss, A.C., Jr., Chan, B.G., Elliger, C.A., Wiseman, B.R., McMillan, W.W., Widstrom, N.W., Zuber, M.S. and Keaster, A.J., 1979. Maysin, a flavone glycoside from corn silks with antibiotic activity toward corn earworm. *J. Econ. Entomol.*, **72**:256-258.

Walter E.V. 1957. Corn earworm lethal factor in silks of sweet corn. *J. Econ. Entomol.*, **50**: 103-106.

Wann, E.V. and Hills, W.A., 1966. Earworm resistance in sweet corn at two stages of development. *Proc. Am. Soc. Hort. Sci.*, **89**:491-496.

Widstrom, N.W. and Snook, M.E., 1994. Inheritance of maysin content in silks of maize inbreds resistant to corn earworm. *Plant Breed.*, **112**:120-126.

Widstrom, N.W. and Snook, M.E., 1996. Inheritance of maysin and related compounds in maize adapted to the Southeastern United States. P. 2. *In* Proc. of the Plant Breeding Symposium, Potchefstroom. S. Africa. 19-21 March 1996. ARC-Grain Crops Inst. Potchefstroom, South Africa.

Widstrom, N.W. and Snook, M.E., 1998a. Genetic variation for maysin and two of its analogues in crosses among twelve corn ibreds. *Crop Science*, **38**:461-471.

Widstrom, N.W. and Snook, M.E., 1998b. A gene controlling biosynthesis of isooriention, a compound in corn silks antibiotic to the corn earworm. *Enomol. Exp. et Applic.*, **89**:119-124.

Widstrom, N.W. and Snook, M.E., 2001. Congruence of conventional and molecular studies to locate genes that contol Flavone Synthesis in maize silks. *Plant Breed.*, (In press).

Widstrom, N.W., Snook, M.E., McMillan, W.W., Waiss, Jr., A.C. and Elliger, C.A., 1991. Maize-silk maysin data: Comparison of interpretations of quantifications by spectrophotometry and HPLC. *J. Agric. Food Chem.*, **39**:182-184.

Widstrom, N.W., Snook, M.E., Wilson, D.M., Cleveland, T.E. and McMillian, W.W. 1995. Silk-maysin content and resistance of commercial corn (maize) hybrids to kernel contamination by aflatoxin. *J. Sci. Food Agric.*, **67**:317-321.

Widstrom, N.W., Waiss, Jr., A.C., McMillian, W.W., Wiseman, B.R., Elliger, C.A., Zuber, M.S., Straub, R.W., Brewbaker, J.L., Darrah, L.L., Henson, A.R., Arnold, J.M. and Overman, J.L., 1982. Maysin content of silks of nine maize genotypes grown in diverse environments. *Crop. Sci.*, **22**:953-955.

Widstrom, N.W., Wiseman, B.R. and McMillian, W.W., 1977. Responses of corn earworm larvae to maize silks. *Agron. J.*, **69**:815-817.

Widstrom, N.W., Wiseman, B.R., McMillian, W.W., Elliger, C.A. and Waiss, Jr., A.C., 1983. Genetic variability in maize for maysin content. *Crop. Sci.*, **23**:120-122.

Wiseman, B.R. 1989. Technological advances for determining resistance in maize to Heliothis zea *In* N. Russel (Ed.), *Toward Insect Resistant Maize for the Third World: Proceedings of an International Symptoms on Methodologies for Developing Host Plant Resistance to Maize Insects.* CIMMYT, Mexic, D.F. pp. 94-100.

Wiseman, B.R. and Bondari, K., 1992. Genetics of antibiotic resistance in corn silks to the earworm (Lepidopetra: Noctuidae). *J. Econ. Entomol.*, **85**:293-298.

Wiseman, B.R. and Bondari, K., 1995. Inheritance of resistance in maize silks to the corn earworm. *Entomol. Exp. Appl.*, **75**:315-321.

Wiseman, B.R. and Carpenter, J.E., 1995. Growth inhibition of corn earworm (Lepidopetra: Noctuidae) larvae reared on resistant corn silk diets. *J. Econ. Entomol.*, **88**:1037-1043.

Wiseman, B.R., McMillan, W.W. and Widstrom, N.W., 1976. Feeding of corn earworm in the laboratory on excised silks of selected corn entries with notes on Orius insidiosus. *Fla. Entomol.*, **59**:305-308.

Wiseman, B.R., Snook, M.E., Isenhour, D.J., Mihm, J.A. and Widstrom, N.W., 1992a. Relationship between growth of corn earworm and fall armyworm larvae (Lepidoptera: Noctuidae) and maysin concentration in corn silks. *J. Econ. Entomol.*, **85**:2473-2477.

Wiseman, B.R., Snook, M.E., Wilson, R.L. and Isenhour, D.J., 1992b. Allelochemical content of selected popcorn silks: Effect on growth of corn earworm larvae (Lepidoptera: Noctuidae). *J. Econ. Entomol.*, **85**:2500-2504.

Wiseman, B.R., Widstrom, N.W. and McMillian, W.W., 1977. Ear characteristics and mechanism of resistance among selected corns to corn earworm. *Fla. Entomol.*, **60**:97-103.

Wiseman, B.R., Widstrom, N.W. and McMillian, W.W., 1978. Movement of corn earworm larvae on ears of resistant and susceptible corns. *Environ. Entomol.*, **7**:777-779.

Wiseman, B.R., Widstrom, N.W. and McMillian, W.W., 1981. Influence of corn silks on corn earworm feeding response. *Fla Entomol.*, **64**:395-399.

# Chapter 4

# Semiochemistry of Crucifers and their Herbivores

*Alireza Aliabadi[1] and Douglas W. Whitman[2]*

## INTRODUCTION

Through evolutionary time, herbivores, acting as selective agents, have driven plants to evolve ever more exotic and potent defensive compounds (Ehrlich and Raven, 1964). Consequently, most terrestrial plant groups produce their own idiosyncratic blend of toxic, deterrent, or antagonistic chemicals. Such chemicals include alkyl sulfides in onions and garlic (Liliaceae), cardenolides in milkweeds (Asclepiadaceae), cyanoglucosides in *Prunus* (Rosaceae), alkaloids in Loganaceae, Solanaceae, and Apiaceae, toxic proteins in Fabaceae, sponins in *Phytolacca* (Phytolaccaceae), terpenoids in Pinaceae, and Tannins in *Quercus* (Fagaceae) (Brodnitz *et al.* 1969; Boelens *et al.* 1971; Stump and Conn, 1981; Young and Seigler, 1981; Harborne, 1988; Bennet and Wallsgrove, 1994). The end result of this long plant-herbivore interaction is that today the 350,000 species of the world's terrestrial plants represent a vast natural botanical pharmacy of unique bioactive compounds, many of which have found use in human cultures as medicines, drugs, antibiotics, spices, insecticides, poisons and scents (Simpson and Ogorzaly, 1995).

Paralleling the evolution of plant anti-herbivore chemicals is the evolution of herbivore counter-measures. Virtually every chemically defended plant has certain insects associated with it that have adapted to, or co-evolved with, that specific plant group. Those insects are often not only unharmed by the plant toxins, but, in many cases, have evolved the ability to use the chemicals to their own advantage. These herbivores use these plant chemicals as kairomones for Host plant location, identification and phagostimulation, and even employ them as allomones for defense against entomophagous predators and parasites (Bowers, 1990). Carnivores (the third trophic level) have also evolved the

[1] 4160 Chemistry, Illinois State University, Normal, Illinois 61790, USA
[2] 4120 Biological Sciences, Illinois State University, Normal, IL 61790, USA
  dwwhitm@ilstu.edu

ability to utilize these substances as synomones, to the detriment of the herbivore and the benefit of the plant. Hence, these compounds, which originally evolved in plants as defensive agents, come to serve very different ecological roles as they weave their way through the food web (Barbosa and Letourneau, 1988).

In this chapter, we explore the multi-trophic chemical ecology of the plant family Brassicaceae, the insects that feed on them, and the carnivorous natural enemies that attack those insect herbivores. We have chosen the Brassicaceae because they are rich in glucosinolates; compounds with an interesting chemistry and widespread bio-activity (Vaughan *et al.* 1976).

## THE BRASSICACEAE

The Brassicaceae (mustard or crucifer family) is a diverse group of plants found mostly in temperate regions, containing 380 genera and 3,000 species. The family has its greatest diversity in the Irano-Turanian region with 147 genera, followed by the Mediterranean region with 113 genera (Heywood, 1993). The Crucifers are mostly annual and perennial herbs. Crucifer leaves are usually alternate and without stipules, and the inflorescence is usually a raceme or corymb. Their floral structure is almost ubiquitous: four sepals, four petals in the shape of a cross (hence, the crucifer appellation), six stamens (four long and two short), and an ovary (Heywood, 1993). Crucifer fruits are linear-oblong, ovate or spherical and provide important taxonomic characters at tribal, generic, and specific levels (Heywood, 1993). The closest-related family to the Brassicaceae is the Capparaceae (caper family), which shares chemical similarities.

The Brassicaceae are of great economic importance, being used throughout the world as vegetables, animal fodder, oils, and condiments (Vaughan *et al.* 1976; Simpson and Ogorzaly, 1995). Crucifers are also moderately used as ornamentals, and in the past were used in folk medicines as emitics, poultices, pain relievers, and digestive stimulants (Lewis and Elvin-Lewis, 1977).

Among the best known Brassicaceae are the members of the genera *Brassica* (cabbage, bok-choi, turnip, kale, collard greens, cauliflower, broccoli, brussel sprouts, and kohlrabi), *Sinapis* (mustards), *Raphanus* (radishes), and *Crambe*. Other well-known Crucifers belong to the tribes Sisymbrieae (with *Alliaria*, which includes garlic mustard) and Arabideae (with *Armoracia*, which includes horseradish, and *Barbarea*, which includes winter cress). Rape cress, Crambe and other Crucifers are occasionally used as forage for livestock, and *Cakile* (sea rocket), *Cheiranthus* (wallflower), *Cardamine* (Toothwort), *Brassica oleracea* (flowering kale), *Eyrsimum*, *Alyssum*, *Aubrieta* and others are used as ornamentals. Two species, *Arabidopsis thaliana* and *Brassica rapida*, are used worldwide for genetic research.

## CRUCIFER CHEMICAL DEFENSES

Members of the Brassicaceae usually contain high titers of mustard oil glycosides, or glucosinolates, which are believed to serve as plant defensive agents (Kjaer, 1976; Feeny, 1977; Larsen, 1981). These compounds can be present in all the plant's tissues, particularly the parenchymous layer (Larsen, 1981), and their concentrations can increase following herbivory (Koritsas *et al.* 1989; Bartlet *et al.* 1999). In some cases, the concentrations can be quite high. For example, seeds of white mustard (*Sinapis alba*) can contain 128 mg/g (13%) glucosinalbin (Joseffson, 1968). A great variety of different glucosinolates have been identified, and different crucifer species produce different blends (Rodman, 1981). Although one or a few specific glucosinolates usually predominate, numerous compounds may be present in a single species. For example, at least 16 aliphatic, aromatic, and indolyglucosinolates occur in oilseed rape (Fieldsend and Milford, 1994). Although glucosinolates are most often identified with the Brassicaceae, these substances also occur sporadically throughout the class Capparales, which includes the families Capparaceae, Moringaceae, Resedaceae, Tovariaceae, and Brassicaceae (Kjaer, 1974; Larsen, 1981). Interestingly, glucosinolates also occur sporadically in the Tropaeolaceae, Limnanthaceae, Euphorbiaceae, Caricaceaea, Phytolaccaceae and Salvadoraceae; families not closely related to the Capparales (Rodman, 1981; Heywood, 1993). In addition to glucosinolates, Crucifers also contain Proteinase inhibitors with insecticidal and antifungal activities (Broadway and Colvin, 1992; Williams *et al.* 1997).

Glucosinolates are unique organic compounds because they include both a β, dextrorotatory thioglucose group and a sulfonated oxime moiety attached to a highly variable side chain, "R–" (Larsen, 1981; Robinson, 1983; Fig. 1 and Table 1). Glucosinolates are relatively stable within the intact plant. However,

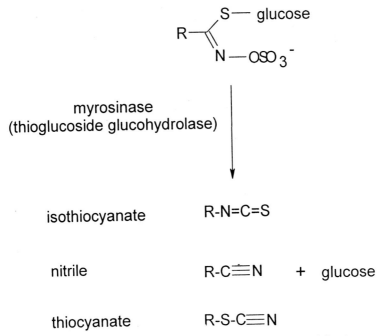

**Fig. 1.** General structure of a mustard oil glycoside and resulting products following enzymatic hydrolysis.

once the plant tissue is damaged by, for example, a chewing herbivore, myrosinase (thioglucoside glucohydrolase) enzymes are released (Rodman, 1981). These enzymes quickly hydrolize the relatively non-toxic glucosinolates into glucose and aglucone breakdown products such as thiocyanates, isothiocyanates, nitriles, or oxazolidinenthiones, with varied toxicity (Ju *et al.* 1982; Fenwick *et al.* 1983; Fig. 1 and Table 1).

**Table 1.** Some glucosinolates and their breakdown products.

| Glucosinolate | Breakdown Product |
|---|---|
| sinigrin (allyl glucosinolate) | allyl isothiocyanate |
| glucotropaeolin (benzyl glucosinolate) | benzyl isothiocyanate |
| gluconasturtiin (2-phenylethyl glucosinolate) | 2-phenylethyl isothiocyanate |
| glucosinalbin (*p*-hydroxybenzyl glucosinolate) | p-hydroxybenzyl isothiocyanate |
| progoitrin (2-hydroxy-3-butenyl glucosinolate) | goitrin:(−)–5–vinyl–2–oxazolidinethione |

Glucosinolate hydrolysis products are often highly volatile and are responsible for the pungent odor that results when crucifer leaves are lacerated (van Etten and Tookey, 1979). The polarity, toxicity, volatility and biological reactivity of the aglucone varies depending on its variable *R*-group. Furthermore, glucosinolate breakdown pathways are influenced by pH, the structure of the functional group, the crucifer species, and/or the presence of compounds that modify the enzymatic action of myrosinase (Fenwick *et al.* 1983).

Glucosinolate breakdown products vary in their toxicity, and different taxa vary in their sensitivity to these plant secondary compounds. For example, humans seem to be unharmed by allyl isothiocyanate in yellow mustard condiments and sauerkraut. However, sinigrin is considered a mutagen (Lewis and Elvin-Lewis, 1977), and crucifer-rich diets may aggravate existing hypothyroidism (Bruenton, 1999). Furthermore, as with many sulfur compounds (e.g., mercaptans from skunks and alkyl sulfides from onions), human chemosensilla are extremely sensitive to these substances. Walker and Gray (1970) found that only 50 g of mustard greens taints cow milk. This is quite extraordinary, considering that a typical milk cow weighs 650,000 g, and that only a small portion of the sulfur compounds pass through the cow's gut and into the milk glands. Clearly, these compounds are stimulating at very low doses. Indeed, humans can detect benzyl isothiocyanate (which is present in horseradish, *Armoracia rusticana*) at 5-10 ppb (Walker and Gray, 1970). Both horseradish, *Armoracia rusticana*, and the Japanese condiment wasabi, *Wasabi japonica*, trigger a strong reaction of the trigeminal nerves in human sinuses, causing pain, sweating, tearing, and rhinorrhea (runny nose), and occasionally vasomotor depression, fainting, and bloody vomiting (Lewis and Elvin-Lewis, 1977; Spitzer, 1988; Rubin and Wu, 1988). Mustard oils can also act as dermal irritants (Lewis and Elvin-Lewis, 1977).

In some non-human mammals, glucosinolates induce various pharmacological effects, including goiter (Greer, 1962; Fenwick, 1983). Goitrin, the breakdown product of the glucosinolate progoitrin (found in Crambe,

*Crambe abyssinica,* and turnip, *Brassica campestris*), blocks iodine uptake, thereby causing enlargement of the thyroid gland (Greer, 1962). In addition, erucic acids, present in the triacylglycerols of seeds of many Crucifers, act as cardiovascular poisons in rats (Lewis and Elvin-Lewis, 1977). Over-consumption of Crucifers can have severe toxic effects. For example, livestock mortality has resulted from the ingestion of rape, yellow rocket, tansy mustard, wormseed mustard, wild radish, horseradish and white mustard seeds (Holmes, 1965; Lewis and Elvin-Lewis, 1977; Cooper and Johnson, 1984).

Glucosinolate breakdown products also have antifungal and antibacterial properties (Dannenberg *et al.* 1956; Virtanen, 1962; Lewis and Elvin-Lewis, 1977). Virtanen (1962) showed the efficacy of isothiocyanates against *Styphylococcus aureus* and *Penicillium glaucum.* In addition, aromatic isothiocyanates, such as benzyl isothiocyanate, were shown by Klesse and Leukoschek (1955) and Zsolnai (1966) to strongly inhibit the growth of gram-positive bacteria. As for anti-fungal properties, allyl isothiocyanate is active against the sporangia of *Peronospora* (Greenhalgh and Mitchell, 1976), and may be the reason why certain *Brassica* are immune to cabbage downy mildew, *Peronospora parasitica* (Greenhalgh and Mitchell, 1976).

Glucosinolates deter nematodes. For example, the root lesion nematode, *Pratylenchus neglectus,* is suppressed by 2-phenylethyl glucosinolate (Potter *et al.* 1998; 2000). Likewise, the root-knot nematode, *Meloidogyne chitwoodi,* was suppressed by plowing *Brassica napus* and *B. campestris* plants into soil, a practice referred to as "green-manuring" (Mojtahedi *et al.* 1991).

Crucifer compounds also show allelopathic toxicity against plants. 1-Cyano-2-hydroxy-3-butene, a non-isothiocyanate of *Crambe abyssinica,* acts as a phytotoxin, suppressing seedling emergence and biomass accumulation of wheat, *Triticum aestivum,* and hemp sesbania, *Sesbania exaltata,* as well as inhibiting radicale elongation of wheat and velvet leaf, *Abutilon theophrasti* (Vaughn and Berhow, 1998). Vaughn and Boydston (1997) suggested that green-manuring mustard plants with high allyl isothiocyanate levels could be used for allelopathic purposes. Al-Khatib *et al.* (1997) noted that weeds in fields of green pea were suppressed by *Brassica,* while Brown and Morra (1995) found that green-manuring Rape cress, *Brassica napus,* released isothiocyanates, nitrile and other water-soluble breakdown products into the soil, inhibiting seed germination.

Insects can be strongly affected by glucosinolates and their derivatives. The general antifeedant and toxic characteristics of glucosinolates against this group are well-recognized. Perhaps, the best-known example is a study by Erickson and Feeny (1974) on black swallowtail caterpillars, *Papilio polyxenes,* which are oligophagous on members of the carrot family, Apiaceae. The larvae were raised on a diet consisting of celery leaves cultured in various sinigrin (allyl glucosinolate) solutions. The results showed a dramatic reduction in adult size, clutch size and pupal weight as sinigrin concentration was increased (Table 2). Doses above 0.01% sinigrin produced complete mortality.

Blau *et al.* (1978) found that larvae of another Lepidopteran, the generalist southern armyworm, *Spodopera eridania,* were inhibited by high (but not low)

**Table 2.** Effects of sinigrin on *P. polyxenes*, reared as larvae on celery leaves cultured in various aqueous solutions of sinigrin (16 larvae/diet) (Erickson and Feeny, 1974).

|  | Sinigrin concentration (%) | | |
| --- | --- | --- | --- |
|  | 0.0 | 0.01 | 0.1 |
| Number of pupae | 14 | 8 | _[a] |
| Mean time to pupation (days) | 16 | 20 | – |
| Men pupal weight | 227 | 164 | – |
| Survival to adult (%) | 88 | 50 | 0 |
| Mean viable eggs/female | 142 | 73 | 0 |

[a]Larvae subjected to the 0.1% sinigrin treatment failed to reach pupation.

concentrations of sinigrin. The larvae survived well on doses up to 0.87% (fresh weight) sinigrin content of leaves, but as the sinigrin dosage was increased to 1.84% and ultimately to 3.44% of fresh leaf weight, growth was significantly retarded.

Glucosinolates are also active against wireworms. These widespread and important root-feeding crop pests are larvae of Elateridae beetles (Metcalf and Flint, 1962). Brown *et al.* (1991) found that the wireworm *Limonius infuscatus* (Mots.) was repelled by the presence of isothiocyanates in soil, and possibly by ionic thiocyanates. In another study (Williams *et al.* 1993), low levels of allyl isothiocyanate reduced feeding in *Limonius californicus*, whereas higher doses caused partial to total mortality. Many other insects are known to be harmed by glucosinolate breakdown products, including vinegar flies, house flies, and confused flour beetles (Lichtenstein *et al.* 1962; Lichtenstein *et al.* 1964).

Crucifer compounds can be toxic when ingested, when encountered topically, or even as odors. For example, allyl isothiocyanate volatiles kill *Diaeretiella rapae* wasps in 3 min. (Read *et al.* 1970). Low volatile doses produced "confused" reactions and excessive antennal grooming, whereas extremely low doses elicited attraction from these parasitoids of cabbage aphids.

Even insect eggs are not immune to the activity of glucosinolates and their aglucones. Benzyl isothiocyanate was found to be toxic to eggs, as well as first instars, of three fruit fly pests (Tephritidae: *Dacus dorsalis, D. cucurbitae,* and *Ceratitis capitata)* (Seo and Tang, 1982). Furthermore, Ahman (1986) tested a number of isothiocyanates and glucosinolates against the eggs of the Brassica pod midge, *Dasineura brassicae*. Allyl isothiocyanate was found to be the most toxic, being lethal at 10 ppm. This example is interesting because *D. brassicae* is oligophagous on Crucifers, yet is harmed by glucosinolate derivatives. This suggests that neither herbivore feeding adaptations nor plant defenses are perfect or static; neither the plant nor the herbivore are ever 100% effective. Both plant and herbivore are under continual selective pressure to improve. In this sense, plants, herbivores and carnivores appear to be locked in a perpetual co-evolutionary arms race, with each enhancing its defenses in response to continued selective pressure from antagonists.

## CHEMICAL ECOLOGY OF CRUCIFERS AND CRUCIFER-FEEDING INSECTS

We have documented that the Brassicaceae produces a wide variety of glucosinolates and their derivatives, and that these compounds often have negative effects on a broad diversity of taxa, including vertebrates, insects, nematodes, plants, bacteria and fungi. However, despite the general pernicious effects of crucifer chemicals on living organisms, numerous insect groups feed on this plant family (Table 3). Table 3 represents only a few of the more common (mainy North American) agricultural pests of Brassicaceae; numerous additional species feed on wild or cultivated Crucifers (e.g., Weires and Chiang, 1973; Kirk and Gray, 1992). Each of these insects has, somehow, evolved the ability to thwart, circumvent or otherwise biochemically disarm the potent chemical weapons of the Brassicaceae. In many cases, these herbivores have evolved adaptations that enable them to exploit these plant defensive chemicals for their own use. Indeed, it is among crucifer-feeding insects that we find some of the early, classical studies in insect chemical ecology (e.g., Verschaeffelt, 1911; Thorsteinson, 1953; Kennedy, 1956; Sugiyama and Matsumoto, 1959; Gupta and Thorsteinson, 1960; David and Gardiner, 1962; Wensler, 1962; Hovanitz *et al.* 1963; Traynier, 1965, 1967; Moon, 1967; Schoonhoven, 1967; Coaker, 1969; Read *et al.* 1970). Additionally, more recent work has been done on the chemical interactions between Crucifers and their herbivores (Städler, 1978; Renwick and Radke, 1983; 1990; Chew, 1988a,b; Renwick *et al.* 1992; Roessingh *et al.* 1992; Huang *et al.* 1995). In this section, we examine how crucifer-feeding insects use these plant substances for host location, oviposition, phagostimulation, chemical defense, and even as pheromones.

### Host plant location in crucifer insects

Glucosinolate breakdown products of Crucifers are often volatile, and, thus, it is not surprising that many crucifer-feeding insects have evolved the ability to orient to these plant substances. Early research demonstrated that these insects are attracted to entire plants or crude extracts (Verschaeffelt, 1911). However, substantial progress has been made in determining the specific chemical compounds that elicit orientation behavior.

Among the first to identify specific active compounds were Gupta and Thorsteinson (1960), who noted that *Plutella maculipennis*, the congener of the diamondback moth *P. xylostella*, uses allyl isothiocyanate as a host locator. Since then, many other insects have been found to orient to various crucifer compounds (Bartlet, 1996). For example, mated female cabbage moths, *Mamestra brassicae* (Noctuidae), are stimulated by allyl, 3-butenyl, 2-phenylethyl, and benzyl isothiocyanate to fly upwind in a wind tunnel (Rojas, 1999). Surprisingly, electroantennogram evaluation of isothiocyanates elicited relatively weak responses in this species (Rojas, 1999).

Numerous beetles feed on Crucifers, and many have been shown to orient to specific crucifer compounds. Curculionids, such as the rape stem weevil (*Ceuthorhynchus napi*), cabbage seed weevil (*C. assimilis*), cabbage weevil (*C. rapae*), cabbage seedstalk weevil (*C. quadridens*), cabbage stem weevil (*C. pallidactylus*), turnip gall weevil (*C. pleurostigma*), and *C. constrictus* attack various Brassicaceae. Smart and co-workers (1997) examined chemo-orientation of *C. assimilis* in the field with the goal of developing a monitoring system for

**Table 3.** Examples of insects that feed on Crucifers.

| Insect Order | Insect Family | Insect Name |
|---|---|---|
| Heteroptera | | |
|     Homoptera | | |
| | Aphididae | Cabbage aphid (*Brevicoryne brassicae* L.)[a] |
| | | Turnip aphid (*Lipaphis erysimi* Kaltenbach)[b] |
| | | Cabbage root aphid (*Pemphigus populitransversus*) |
| | | Green peach aphid (*Myzus persicae*)[u] |
|     Hemiptera | | |
| | Lygaeidae | Northern false chinch bug (*Nysius niger* Baker)[c] |
| | Pentatomidae | Harlequin bug (*Murgantia histrionica* Hahn)[a] |
| | | *Eurydema ventrale* L.[d] |
| | | *E. oleraceum* L.[d] |
| Coleoptera | | |
| | Nitidulidae | Rape beetle (*Meligethes aeneus* F.)[e] |
| | | Blossom beetle (*M. viridescens* F.)[e] |
| | Chrysomelidae | |
| | | Red turnip beetle (*Entomoscelis americana* Brown)[f] |
| | | Daikon leaf beetle (*Phaedon brassicae* Baly)[g] |
| | | *P. cochleriae*h |
| | | Striped flea beetle (*Phyllotreta striolata* Fab.)[a] |
| | | Horseradish flea beetle (*P. Armoraciae*)[a] |
| | | Cabbage flea beetle (*P. cruciferae*)[a] |
| | | Striped turnip flea beetle (*P. nemorum*)[i] |
| | | Turnip flea beetle (*P. atra*)[a] |
| | | *P. undulata*[a] |
| | | *P. tetrastigma*[a] |
| | | Cabbage stem flea beetle (*Psylliodes chrysocephala*)[a] |
| | | *Altica oleracea* (Linné)[g] |
| | Curculionidae | |
| | | Rape stem weevil (*Ceuthorhynchus napi*)[a] |
| | | Cabbage seed weevil (*C. assimilis* Paykull)[a] |
| | | Turnip gall weevil (*C. pleurostigma* Marsham)[a] |
| | | Cabbage weevil (*C. rapae* Gyllenhal)[a] |
| | | Cabbage seedstalk weevil (*C. quadridens* Panzer)[a] |
| | | Cabbage stem weevil (*C. pallidactylus* Marsham)[a] |
| | | *C. constrictus*i |
| | | Vegetable weevil (*Listroderes costriostris*)[t] |
| Lepidoptera | Pieridae | Large white (*Pieris brassicae*)[a] |
| | | Potherb (*P. napi oleracea*)[a] |
| | | Small white (*P. rapae*)[a] |
| | | Diffuse-veined white (*P. virginiensis*)[k] |
| | | Great Basin white (*P. chloridice*)[k] |
| | | Peak white (*P. callidice*)[k] |
| | | Southern small white (*P. mannii*)[l] |
| | | Southern cabbageworm (*Pontia protodice*) |
| | | Spring white (*P. sisymbrii*)[k] |
| | | Bath white (*P. daplidice*)[l] |
| | | Western white (*P. occidentalis*)[s] |
| | | Western orange tip (*Anthocharis sara*)[k] |
| | | Desert orange tip (*A. cethura*)[k] |

contd.

**Table 3.** contd.

| Insect Order | Insect Family | Insect Name |
|---|---|---|
| | | California white tip (*A. lanceolata*)[k] |
| | | Falcate orange tip (*A. midea*)[k] |
| | | Orange tip (*A. cardamines*)[l] |
| | | Western marble (*Euchloe hyantis*)[k] |
| | | Dappled marble (*E. ausonia*)[k] |
| | | Northern marble (*E. creusa*)[k] |
| | | Rosy marble (*E. olympia*)[k] |
| | | *E. simplonia* (Freyer)[n] |
| | | Gulf white (*Ascia monuste*)[a] |
| | Noctuidae | Cabbage looper (*Trichoplusia ni* Hübner)[a] |
| | | Cabbage moth (*Mamestra brassicae* L.)[a] |
| | | Bright-line brown-eyes moth (*M. oleracea*)[a] |
| | | Silvery moth (*Autographa gamma* L.)[a] |
| | | *A. precationis* Guenée[p] |
| | | Alfalfa looper (*A. californica* Sepeyer)[p] |
| | | Turnip moth (*Agrotis segetum*)[o] |
| | | *Megalographa biloba* (Stephens)[p] |
| | | *Noctua pronuba* (L.)[q] |
| | | *Spaelotis clandestina* (Harris)[q] |
| | | *Agnorisma badinodis* (Grote)[q] |
| | | *Abagrotis alternata* (Grote)[q] |
| | Pyralidae | Cross-striped cabbageworm (*Evergestis rimosalis*)[a] |
| | | Purple-backed cabbageworm (*E. pallidata*)[a] |
| | | *E. extimalis* (Scopoli)[r] |
| | | Cabbage webworm (*Hellula rogatalis* Hulst)[a] |
| | | *Hellula phidilealis* (Walker)[r] |
| | | *Eustixia pupula* (Hübner)[r] |
| | Plutellidae | Diamondback moth (*Plutella xylostella*)[a] |
| | | *P. maculipennis*[i] |
| | | *P. porrectella* (L.)[m] |
| | | *Rhigognostis senilella* (Zetterstedt)[m] |
| | | *R. incarnatella* (Steudel)[m] |
| | | *Eidophasia messingiella* (Fischer von Röslerstamm)[m] |
| Diptera | Tortricidae | *Selania leplastriana* |
| | Cecidomyiidae | Swede midge (*Contarinia nasturtii* Kieffer)[a] |
| | | Brassica pod midge (*Dasineura brassicae*)[a] |
| | Agromyzidae | Serpentine leaf miner (*Liriomyza brassicae*)[a] |
| | Anthomyiidae | Cabbage root fly (*Delia radicum*)[a] |
| | | Turnip root fly (*D. floralis*)[a] |
| | | Adult cabbage maggot (*D. brassicae*)[a] |
| | Scaptomyzella | *Scaptomyza flava* |

References: [a]Metcalf and Flint, 1962; [b]Davidson and Lyon, 1979; [c]Pivnick *et al.* 1991; [d]Aldrich *et al.* 1996; [e]Blight and Smart, 1999; [f]Stoetzel, 1989; [g]Matsuda, 1978; [h]Tanton, 1965; [i]Nielsen, 1978; [j]Gupta and Thorsteinson, 1960; [k]Scott, 1986; [l]Higgins and Riley, 1970; [m]Agassiz, 1996; [n]Bibby *et al.* 1989; [o]Bretherton *et al.* 1979; [p]Lafontaine and Poole, 1991; [q]Lafontaine, 1998; [r]Munroe, 1972; [s]Opler, 1998; [t]Sugiyama and Matsumoto, 1959; [u]Costello and Altieri, 1995; Moyes *et al.* 2000.

this agricultural pest. Weevils were attracted to a mixture of allyl, 3-butenyl, 4-pentenyl, and 2-phenylethyl isothiocyanates, in combination with the color yellow, during their spring migratory phase. However, during the later period of host-plant colonization, the above aglucones elicited no response or were repellent, whereas a different compound, phenylacetonitrile, was attractive (Smart and Blight, 1997).

Many flea beetles (Subfam. Alticinae) of the leaf beetle family, Chrysomelidae, feed on Crucifers. Pivnick *et al.* (1992) field-tested the attractiveness of nine isothiocyanates and three nitriles to two species of flea beetles, *Phyllotreta cruciferae* and *P. striolata*, in Canada. Allyl isothiocyanate was the most attractive, although a relatively high release rate (4 mg/day) was required to elicit orientation. Because such large doses were needed, Pivnick and co-workers concluded that these beetles would normally be attracted only to *Brassica* field crops, and not to the relatively low amounts of volatiles emanating from a few individual plants. Sap beetles (Nitidulidae) also orient to crucifer compounds. Blight and Smart (1999) found that the European rape beetle, *Meligethes aeneus*, was maximally attracted to a mixture of isothiocyanates (allyl, 3-butenyl, 4-pentenyl, and 2-phenylethyl), with release rates of 25 mg/day for allyl isothiocyanate and 5 mg/day for the other three isothiocyanates. Again, these are relatively high release rates in relation to undamaged plants; however, damaged plants release volatiles at much greater rates than undamaged plants. Perhaps, these herbivores have evolved to orient preferentially to plants that are already damaged by conspecifics. This would not only bring them to plants whose defenses may already be weakened by prior herbivores, but also to potential mates. Integrated pest managers may take advantage of this attraction to relatively high odor concentrations by intercropping (thus, diluting the plant odor), or by adding repellent odors from other plants (Uvah and Coaker, 1984; Nottingham, 1987). Likewise, pest reduction has been attempted by planting a "trap crop" around the borders of crucifer fields (Luther *et al.* 1996; Bigger and Chaney, 1998).

The northern false chinch bug, *Nysius niger* (Lygaeidae: Heteropetra), is a pest of mustard crops in Canadian prairies. Pivnick *et al.* (1991) caught this seed bug in boll weevil traps baited wth mustard oils. The authors found that females predominated and that the most attractive mustard oil was ethyl 4-isothiocyanobutyrate, followed by allyl, and *n*-propyl isothiocyanates. 2-phenylethyl isothiocyanate was not attractive.

Females of the brassica pod midge, *Dasineura brassicae* (Diptera: Cecidomyiidae), apparently have a difficult time ovipositing through the tough wall of their host's seed pod, and may need to find areas of the pod that have already been damaged, for example, by the cabbage seed weevil, *Ceurohynchus assimilis* (Murchie *et al.* 1997). In field tests, both sexes of *D. brassicae* were more attracted to traps baited with allyl isothiocyanate than traps baited with 2-phenylethyl isothiocyanate. Additionally, attraction was greater at lower release rates (12-13 mg/day) of allyl isothiocyanate than at high release rates (40-53 mg/day). However, attraction to low vs. high doses may have been an artifact caused by inconsistent volatile release (Murchie *et al.* 1997). Interestingly, 100 times more males were caught than females, which was attributed to the greater flight activity of males in search of females (Williams *et al.* 1987; Murchie *et al.* 1997). Female cabbage root flies, *Delia radicum* and *D. brassicae* also orient to allyl isothiocyanate (Hawkes *et al.* 1987; Wallbank and Wheatley, 1979; Nottingham and Coaker, 1985).

**Oviposition stimulants**

Glucosinolates can be potent elicitors of oviposition for females of insect species that feed on Crucifers. For example, both the large white *(Pieris brassicae)* and potherb *(Pieris napi oleracea)* butterflies are stimulated by sinigrin to oviposit (David and Gardiner, 1962; Renwick and Radke, 1983; Huang *et al.* 1995). In addition, both species have been fooled into ovipositing on filter paper soaked in a solution of sinigrin (David and Gardiner, 1962; Huang *et al.* 1995), and on non-host plants cultured in sinigrin solutions (Ma and Schoonhoven, 1973). The small white butterfly, *Pieris rapae*, showed greatest ovipositional activity towards glucobrassicin, while showing only limited response towards sinigrin (Traynier and Truscott, 1991; Renwick *et al.* 1992). Likewise, glucosinolates stimulate oviposition in both the diamondback moth, *Plutella xylostella* (Plutellidae), and its close kin, *P. maculipennis* (Gupta and Thorsteinson, 1960; Reed *et al.* 1989; Renwick and Radke, 1990).

Gravid crucifer-feeding flies also react to crucifer kairomones. The adult cabbage maggot, *Delia (Hylemya) brassicae* (Anthomyiidae), is stimulated to oviposit in the presence of sinigrin and four other glucosinolates. Allyl isothiocyanate was most stimulatory at a dose of $2 \times 10^{-3}$ M (Nair and McEwen, 1976; Nair *et al.* 1976; Städler, 1978). The cabbage root fly, *Delia radicum*, also oviposits in response to glucosinolates (Ellis *et al.* 1980; Finch, 1980; Nottingham and Coaker, 1985; Roessingh *et al.* 1992). Roessingh and co-workers found that the "D" sensilla on segments 3 and 4 of the tarsus of *D. radicum* females contains a receptor cell sensitive to glucosinolates, especially glucobrassicin, gluconasturtiin, and glucobrassicanapin (Roessingh *et al.* 1992). Recently, a novel compound has been added to the cabbage root fly's list of oviposition stimulants, namely 1,2-dihydro-3-thia-4, 10, 10b-triaza-cyclopenta[.a.]fluorene-1-carboxylic acid, which is produced by cauliflower, *Brassica oleracea* (*cv. botrytis*) (Hurter *et al.* 1999). Additionally, de Jong and Städler (1999) found that host acceptance for oviposition by this fly seemed to result from a synergistic response to concomitantly perceived olfactory and contact chemical stimulation. Interestingly, the cabbage root fly was deterred from ovipositing when macerates from garlic mustard (*Alliaria petiolata*) were sprayed on host plants, but was stimulated to oviposit when macerates from *Brassica* leaves were sprayed on plants (Jones and Finch, 1987). Jones *et al.* (1988) also identified a repellent, sinapic (3,5-dimethoxy-4-hydroxycinnamic) acid, in the frass of the garden pebble moth caterpillar (Pyralidae) that deterred the cabbage root fly from ovipositing. This compound was relatively stable and was active at low concentrations; when formulated at 10 mM in a buffered, aqueous solution and sprayed onto crucifer leaves, it deterred oviposition for five days (Jones *et al.* 1988). In contrast, feeding by other caterpillar species stimulates oviposition by *D. radicum* (Finch and Jones, 1987).

### Feeding stimulants

Many crucifer-feeding insects use glucosinolates or their derivatives as feeding stimulants, and for some, the presence of glucosinolates is an absolute requirement (Bartlet, 1996). For example, not only are both sinigrin and methyl glucosinolate feeding stimulants for large white (*Pieris brassicae*) caterpillars (David and Gardiner, 1966), but glucosinolate presence is necessary for this species to survive: they refuse to feed, and consequently die, if glucosinolates are not present (Verschaeffelt, 1911; Feltwell, 1982). Feeding in larvae of the plutellid moth, *Plutella maculipennis,* is also stimulated by various glucosinolates, especially when accompanied by glucose and other nutrients (Thorsteinson, 1953; Nayar and Thorsteinson, 1963).

Likewise, both the cabbage aphid, *Brevicoryne brassicae,* and the turnip aphid, *Lipaphis erysimi,* use sinigrin as a feeding stimulant, and will consume a number of non-hosts if the hosts are first treated with sinigrin (Wensler, 1962; Moon, 1967; van Emden and Bashford, 1971; Nault and Styer, 1972). Interestingly, aphids often survived on these new plants, suggesting that these non-hosts were nutritionally adequate, but were not recognized as "food" by the aphids' chemosensilla (Nault and Styer, 1972). Such observations gave rise to the idea of "token stimuli." In regards to feeding, a token stimulus is a non-nutritive chemical (usually a plant defensive compound), which stimulates feeding. The advantage to the insect is simplicity and parsimony; the complex process of food identification and acceptance can be distilled down to one simple decision: is the chemical present or not? Furthermore, the insect needs to develop only one type of chemosensilla—one which responds only to the token stimulus. Accordingly, chemosensilla on the antennae of the turnip aphid respond to a variety of isothiocyanates (Dawson *et al.* 1987).

Crucifer-feeding weevils also use glucosinolates as feeding stimulants. *Ceutorhynchus constrictus* uses sinigrin, as well as other unidentified water soluble compounds, as feeding stimulants (Nielsen *et al.* 1989). The cabbage seed weevil, *Ceutorhynchus assimilis,* prefers longer chain alkenyl glucosinolates and benzyl and *p*-hydroxybenzyl glucosinolate over short-chain glucosinolates or allyl glucosinolate for feeding (Chew, 1988a, b).

Sinigrin is a feeding stimulant for the daikon flea beetle, *Phaedon brassicae,* and the cabbage flea beetle, *Phyllotreta cruciferae* (Fenny *et al.* 1970; Hicks, 1974; Matsuda, 1978). For the latter, sinigrin was active at surprisingly low concentrations of 0.01% (153 ppm) when administered topically on bean leaves (Hicks, 1974). Glucosinolates are also phagostimulatory for the chrysomelid, *P. cochleriae* (Tanton, 1965; Matsuda, 1978). In the horseradish flea beetle, *Phyllotreta Armoraciae* (which is monophagous on horseradish, *Armoracia rusticana*) sinigrin and two flavonol glycosides (kaempferol and quercetin) acted as feeding stimulants, but a mixture of all three was more stimulating than any one alone (Nielsen *et al.* 1979).

The crucifer-feeding flea beetles, *Phyllotreta undulata* and *Phyllotreta tetrastigma*, refuse to feed on most species in the crucifer genera *Chieranthus* and *Erysimum*, due to the presence of cardenolides (toxins normally found in the milkweed family, Asclepiadaceae) (Nielsen, 1978). In contrast, the oligophagous crucifer feeder, *Phyllotreta nemorum*, readily feeds on *Chieranthus* (Nielsen *et al.* 1977), but rejects the crucifer *Iberis amara* due to the presence of cucurbitacins (terpenoids common in cucurbits, family Cucurbitaceae) (Durkee and Harborne, 1973). Hence, host plant choice in these beetles, as in many insects, appears to be guided not just by token stimuli, but by both phagostimulation and phagodeterrance acting together. These examples are sometimes used to support the theory of insect-plant chemical co-evolution, in which insects co-evolve to feed on a specific group of plants that possess certain chemicals that are poisonous to most other insects. These insects become restricted to that particular plant group, in part because they are not stimulated to feed on, or are deterred or poisoned by, plants containing different secondary compounds. At the same time, selective pressure by these specialist herbivores is thought to drive the plants to produce more potent defenses.

## Use of isothiocyanates as pheromones

In at least one case, crucifer compounds appear to be used as a pheromone. Crushed turnip aphids release volatiles that act as an alarm pheromone, causing dispersal of nearby aphids. The pheromone consists of the sesquiterpene (E)-$\beta$-farnesene (a well-known aphid alarm pheromone) and a number of isothiocyanates sequestered or enzymatically altered from host plant glucosinolates (Dawson *et al.* 1987). The aphids responded poorly to (E)-$\beta$-farnesene alone, but responded strongly when these isothiocyanates were added. In electroantennogram studies, aphid sensilla responded strongly to both the sesquiterpene and the isothiocyanates (Dawson, 1987).

## Sequestration of glucosinolate products for chemical defense

Some of the insects that feed on Crucifers are diurnal, sluggish, clumped, and conspicuously colored - traits often associated with chemically defended insects (Pasteels *et al.* 1983; Evans and Schmidt, 1990). Furthermore, crucifer-feeding insects are often oligophagous and sometimes even monophagous. Insects that specialize on chemically defended plants frequently sequester plant toxins for chemical defenses (Bowers, 1990), and crucifer-feeding insects fit this model. This suggests that some crucifer insects may be chemically defended, and that their bright, conspicuous appearance serves as an aposematic warning to predators. Despite this, Sequestration of glucosinolates for defense has been demonstrated in only one group of crucifer-feeding insects, the Pieridae (Marsh and Rothschild, 1974; Aplin *et al.* 1975).

Larvae of the large white butterfly, *Pieris brassicae* (Pieridae), specialize on Crucifers and sequester glucosinolate products. The plant-derived compounds remain through the pupal stage and into the adult stage, and are even present in the brightly-colored yellow eggs (Aplin *et al.* 1975). The caterpillars are gregarious and conspicuously colored. Most predators shun the adults, which are considered the most toxic of all British Pieridae (Rothschild, 1985). Defense in the pupae depends upon diet: caterpillars reared on an artificial diet lacking glucosinolates were readily consumed by a tame magpie, whereas pupae from crucifer-fed larvae were rejected (Marsh and Rothschild, 1974). Interestingly, the adults of the related small white butterfly, *P. rapae*, lack glucosinolate products, but the pupae contain measurable amounts (Aplin *et al.* 1975).

In addition to *Pieris* spp., we believe we have found another candidate for defensive Sequestration of glucosinoaltes from the Brassicaceae (Aliabadi *et al.* unpublished data). This candidate is the harlequin bug, *Murgantia histrionica* Hahn, a common pentatomid pest of Crucifers in the New World (McPherson, 1982). Native to Central America, the harlequin bug was introduced into Texas in 1864 and has since spread through much of North America below 40° latitude (Ludwig and Kok, 1998). Harlequin bugs will feed on a variety of Crucifers and glucosinolate-containing plants outside the Brassicaceae, but prefer *Brassica*, such as cabbage, broccoli, kale, turnip, mustard, and rape (McPherson, 1982).

*Murgantia histrionica* is an ideal candidate for defensive Sequestration of host plant compounds because it exhibits a suite of characteristics often associated with chemically defended insects (Pasteels *et al.* 1983; Whitman *et al.* 1985). Harlequin bugs are relatively large, diurnal, slow-moving, resistant to flight, and conspicuously colored with black and orange, yellow, or red markings (Fig. 2). Furthermore, they often feed in high densities on the exposed upper surfaces of leaves, making them highly visible targets for predators. Despite this, they are rarely preyed upon.

We tested the hypothesis that harlequin bugs sequester glucosinolates from their host crucifer plants by rearing them on six different monophagous plant diets, which included the non-crucifer, butterfly bush (*Buddleia davidii*, Loganaceae), as a control (Table 4). Each of the six plant species showed a different blend of compounds when analyzed. The four cruciferous plant species, cabbage (*Brassica oleracea*), white mustard (*Sinapis alba*), Crambe (*Crambe abyssinica*), and horseradish (*Armoracia rusticana*), contained idiosyncratic glucosinolates, characteristic of that particualr species (Fenwick *et al.* 1983). Garden nasturtium (*Tropaeolum majus*, Tropaeolacae), a non-crucifer, contained glucotropaeolin. No glucosinolates were detected in the butterfly bush (Table 4).

We analyzed the bugs for presence of glucosinolates after they had fed on the monophagous diets for 4-9 weeks. Bugs that were raised on glucosinolate-containing plants contained glucosinolates specific to their particular host (Table 4), suggesting that the bugs sequestered the plant compounds. Interestingly, the bugs reared on butterfly bush, which lacks glucosinolates, also contained small amounts of glucosinolate. This suggests that the bugs either synthesized the glucosinolates themselves or managed to retain measurable amounts of glucosinolates during the 6 wks they fed on the butterfly

**Fig. 2.** Harlequin bugs, *Murgantia histrionica* Hahn (Heteroptera: fam. Pentatomidae). Both nymphs and adults specialize on Crucifers and are sluggish, chemically defended, and conspicuously colored black and yellow-orange.

bush (prior to being paced on the butterfly bush, the bugs had been reared on mustard plants).

In addition to apparently sequestering various noxious crucifer-derived compounds, these conspicuous, slow-moving insects were avoided by predators. During outdoor feeding trials, European starlings (*Sturnus vulgaris*) and house sparrows (*Passer domesticus*) refused to consume harlequin bugs. Summing five separate feeding experiments, the wild birds ate 91 of 91 control insects (*mealworms, Tenebrio molitor,* and crickets, *Acheta domesticus*), but only 3 of 62 offered harlequin bugs. On three occasions, we observed sparrows pick up, then drop a harlequin bug, and on three other occasions we found harlequin bugs on the ground away from the feeder, suggesting that they had been attacked, tasted, and flung away by birds. These results suggest harlequin bugs are unpalatable to these two avian predators. Given the fact that harlequin bugs appear to sequester glucosinolates, and that glucosinolates can have noxious or irritating effects, it is likely that rejection of the bugs was due to sequestered compounds.

Table 4. Selective Sequestration of crucifer compounds by harlequin bugs raised on six monophagous diets.

| Plant species | Compounds detected in: | |
| | Food plant | Harlequin bug |
| --- | --- | --- |
| Cabbage (*Brassica oleracea*) | glucobrassicin, sinigrin | sinigrin |
| White mustard (*Sinapis alba*) | glucosinalbin | glucosinalbin |
| Crambe (*Crambe abyssinica*) | sinigrin, gluconapin, progoitrin, hydroxyglucobrassicin, 2-phenylethyl glucosinolate | sinigrin, gluconapin progoitrin |
| Horseradish (*Armoracia rusticana*) | sinigrin, unknown | sinigrin |
| Garden, nasturtium (*Tropaeolum majus*) | glucotropaeolin | glucotropaeolin |
| Butterfly Bush (*Buddleia davidii*) | none detected | sinigrin |

### Glucosinolates and parasitoids

Members of the third trophic level, entomophagous carnivores, have also evolved to use glucosinolates and their derivatives. Wasp parasitoids of various crucifer herbivores orient to these compounds, to the detriment of the herbivore and the benefit of the plant. In this interaction, the plant compounds act as synomones (communication chemicals that benefit both the sender and the receiver). For example, the braconid parasitoid, *Diaeretiella rapae* (Hymenoptera), locates plants containing their cabbage aphid host by orienting to allyl isothiocyanate (Read *et al.* 1970; Akinlosotu, 1977; Sheehan and Shelton, 1989; Ayal and Green, 1993). Another braconid, *Meteorus leviventris,* which parasitizes cutworms (Lepidoptera: Noctuidae), is attracted to traps bated with allyl isothiocyanate (Pivnick, 1993). The fact that natural enemies of pest herbivores orient to the odors of their hosts' food plant poses a dilemma for pest control. Integrated pest managers have touted intercropping as a means of improving biological control; the new plants aid carnivores by providing additional nectar, refuge, and alternative hosts. However, intercropping can have negative effects because it sometimes reduces parasitization rates, possibly by diluting or confusing the plant chemical signals used by parasitoids (Costello and Altieri, 1995). Likewise, trap crops may harm biological control by trapping more carnivores than herbivores (Bigger and Chaney, 1998).

Both *Platygaster subuliformis* (Platygastridae) and *Omphale clypealis* (Eulophidae) parasitize the brassica pod midge, *Dasineura brassicae.* Traps baited with allyl isothiocyanate caught more female *O. clypealis* than traps baited with 2-phenylethyl isothiocyanates, which in turn caught more male and female *P. subuliformis* than traps baited with allyl isothiocyanate or unbaited traps (Murchie *et al.* 1997). Murchie *et al.* (1997) found that each wasp species responded preferentially to different compounds. This is important because it suggests a mechanism underlying habitat partitioning and competition reduction. The authors noted that this is the first time that any parasitoid has been shown to respond to isothiocyanates other than allyl isothiocyanate.

The above studies demonstrate that parasitoids of crucifer-feeding insects orient to glucosinolates and their derivatives, but how these compounds influence the physiology and survival of parasitoids is unknown. Glucosinolates may act as herbivore feeding stimulants, producing large, fast-growing herbivores that provide optimal nutrition for the parasitoid. Alternatively, large, healthy herbivores may be better able to behaviorally or physiologically fend off parasitoid attack. Similarly, high glucosinolate levels may harm less-adapted herbivores, reducing their ability to resist parasitoids. Conversely, the presence of sequestered glucosinolates in herbivores could harm the parasitoid.

Although we know little about the overall effects of glucosinolates on the physiology of parasitoids of crucifer-feeding insects, some interesting data have appeared. Karowe and Schoonhoven (1992) compared the suitability of different food plants to caterpillars of the large white butterlfy, *Pieris brassicae,* parasitized or unparasitized by the braconid wasp, *Cotesia glomerata.* The growth rate and pupal weight of unparasitized large whites were highest on brussel sprouts (*Brassica oleracea* var. gemmifera) and Swedish turnip (*B. napus* var. blauwkop), followed by rape (*B. napus* var. jet neuf), and lastly, garden nasturtium (*Tropaeolum majus*). In contrast, the parasitoid faired best where its host faired worst—on garden nasturtium. *Pieris brassicae* sequesters crucifer compounds for its defense (Aplin *et al.* 1975), and garden nasturtium contains relatively low titers of glucosinolates and different types of glucosinolates in comparison to the other plant species tested (Fenwick *et al.* 1983). Perhaps the wasps are more sensitive than their hosts to the crucifer-derived toxins, and benefited by low levels of glucosinolates in the tissues of the garden nasturtium-feeding caterpillars. This work suggests that host plants (via host-plant toxins) may influence the biology of the third trophic level, with subsequent fitness consequences for the plant (Duffey, 1980; Barbosa and Letourneau, 1988). Hence, the evolution of plant defensive compounds may be influenced by both herbivores *and* parasitoids.

## CONCLUSION

This chapter has demonstrated the pervasiveness and multifunctionality of one class of bio-active compounds, the glucosinolates. These substances, produced by crucifer plants, ultimately affect numerous species, in various trophic levels, across a wide diversity of taxa. In this sense, these compounds are molecular migrants, assuming different functions, as they pass from species to species. This is a common theme in the chemical ecology of plant-insect interactions (Whitman, 1988), and is illustrative of the subtle interaction and interdependence of living organisms.

## ACKNOWLEDGEMENTS

We would like to thank B. Aliabadi, B. Salazar, J.A.A. Renwick, J. Millar, B. Waters, and the Advanced Entomology Group at Illinois State University for technical assistance and comments on this manuscript.

## REFERENCES

Agassiz, D.J.L. 1996. Yponomeutidae (including Roeslerstammidae). *In* Emmet M.A. (Ed), *The Moths and Butterflies of Great Britain and Ireland.* Vol. 3 Exxes: Harley Books, pp:39-114.

Ahman, I. 1986. Toxicities of host secondary compounds to eggs of the Brassica specialist *Dasineura brassicae. J. Chem. Ecol.,* **12:** 1481-1488.

Akinlostu, T.A. 1977. Some aspects of the host finding behaviour of the female *Diaeretiella rapae* McIntosh (Hymenoptera: Braconidae). *Nigerian J. Entomol.,* **1:** 11-18.

Aldrich, J.R., Avery, J.W., Lee, C-J, Graf, J.C., Harrison, D.J. and Bin, F. 1996. Semiochemistry of cabbage bugs (Heteroptera: Pentatomidae: *Eurydema* and *Murgantia*). *J. Entomol. Sci.,* **31:** 172-182.

Al-Khatib, K., Libbey, C. and Boydston, R. 1997. Weed suppression with Brassica green manure crops in green pea. *Weed Sci.,* **45:** 439-445.

Aplin, R.T., D'Arcy Ward, T. and Rothschild, M. 1975. Examination of the large and small white butterflies (*Pieris spp.*) for the presence of mustard oil glycosides. *J. Entomol. (A).,* **50:** 73-78.

Ayal, Y. and Green, R.F. 1993. Optimal egg distribution among host patches for parasitoids subject to attack by hyperparasitoids. *Am. Nat.,* **141:** 120-128.

Barbosa, P. and Letourneau, D.K. (Eds) 1988. *Novel Aspects of Insect-Plant Interactions.* New York: John Wiley & Sons, Inc., pp. 362.

Bartlet, E. 1996. Chemical cues to host-plant selection by insect pests of oilseed rape. *Agric. Zoo. Rev.,* **7:** 89-116.

Bartlet, E., Kiddle, G., Williams, I. and Wallsgrove, R. 1999. Wound-induced increases in the glucosinolate content of oilseed rape and their effect on subsequent herbivory by a crucifer specialist. *Entomol. Exp. Appl.,* **91:** 164-167.

Bennet, R.N. and Wallsgrove, R.M. 1994. Secondary Metabolites in plant defense mechanisms. *New Phytol.,* **127:** 617-633.

Bibby, T.J., Bretherton, R.F., Dempster, J.P., Emmet, A.M., Feltweel, J.S.E., Pratt, C.R. and Warren, M.S. 1989. Pieridae. *In* M.A. Emmet and J. Health (Eds) *The Moths and Butterflies of Great Britain and Ireland,* Vol. 7, part I. Essex: Harley Books.

Bigger, D.S. and Chaney, W.E. 1998. Effects of *Iberis umbellata* (Brassicaceae) on insect pests of cabbage and on potential biological control agents. *Environm. Entomol.,* **27:** 161-167.

Blau, P.A., Feeny, P., Contrado, L. and Robson, D.S. 1978. Allylglucosinolate and herbivorous caterpillars: A contrast in toxicity and tolerance. *Science,* **200:** 1296-1298.

Blight, M.M. and Smart, L.E. 1999. Influence of visual cues and isothiocyanate lures on capture of the pollen beetle *Meligethes aeneus* in field traps. *J. Chem. Ecol.,* **25:** 1501-1516.

Boelens, M., de Valois, P.J., Wobben, H.. and van der Gen, A. 1971. Volatile flavor compounds from onion. *J. Agric. Food Chem.,* **19:** 984-991.

Bowers, M.D. 1990. Recycling plant natural products for insect defense. *In* D.L. Evans and J.O. Schmidt (Eds) *Insect Defenses: Adaptive Mechanism and Strategies of Prey and Predators.* Albany: State University of New York Press, pp: 352-386.

Bretherton, R.F., Goater, B. and Lorimer, R.I. 1979. Noctuidae: Noctuinae and Hedeninae. *In* J. Heath and M.A. Emmet (Eds) *The Moths and Butterflies of Great Britain,* **Vol. 9.** London: Curwen Books.

Broadway, R.M. and Colvin, A.A. 1992. Influence of cabbage Proteinase inhibitors in situ on the growth of larval *Trichoplusia ni* and *Pieris rapae. J. Chem. Ecol.,* **18:** 1009-1024.

Brodnitz, M.H., Pollock, C.L. and Vallon, P.P. 1969. Flavor components of onion oil. *J. Agric. Food Chem.,* **17:** 760-763.

Brown, P.D. and Morra, M.J. 1995. Glucosinolate-containing plant tissues as bioherbicides. *J. Agric. Food Chem.,* **43:** 3070-3074.

Brown, P.D., Morra, M.J., McCaffrey, J.P., Auld, D.L. and Williams, L. III. 1991. Allelochemicals produced during glucosinolate degradation in soil. *J. Chem. Ecol.,* **17:** 2021-2034.

Bruneton, J. 1999. *Toxic Plants Dangerous to Humans and Animals.* Paris: Lavoisier Publishing, pp 545.

Chew, F.S. 1988a. Biological effects of glucosinolates. *In* H.G. Cutler (Ed) *Biologically Active Natural products: Potential Uses in Agriculture,* Wash, D.C.: American Chemical Society, pp: 155-181.

Chew, F.S. 1988b. Searching for defensive chemistry in the Cruciferae, or, do glucosinolates always control interactions of Cruciferae with their potential herbivores and symbionts? No! *In* K.A. Spencer (Ed) *Chemical Mediation of Coevolution.* San Diego: Academic Press, pp: 81-112.

Coaker, T.H. 1969. New approaches to cabbage root fly control. *Proc. 5th Br. Insect Fungic. Conf.,* 3: 704-710.

Cooper, M.R. and Johnson, A.W. 1984. *Poisonous Plants in Britain and their Effects on Animals and Man,* Reference Book 161. Londres: Ministry of Agriculture, Fisheries and Food.

Costello, M.J. and Altieri, M.A. 1995. Abundance, growth rate and parasitism of *Brevicoryne brassicae* and *Myzus persicae* (Homoptera: Aphididae) on broccoli grown in living mulches. *Agric. Eco. Env.,* 52: 187-196.

Dannenberg, H., Stickl, H. and Wenzel, F. 1956. *Über den antimikrobischen wirkenden Stoff der Kapuzinerkresse (Tropaeolum majus).* Z. *Physiol. Chem.,* 303: 248.

David, W.A.L. and Gardiner, B.O.C. 1962. Oviposition and the hatching of the eggs of *Pieris brassicae* (L.) in a laboratorhy culture. *Bull. Entomol. Res.,* 53: 91-109.

David, W.A.L. and Gardiner, B.O.C. 1966. Mustard oil glucosides as feeding stimulants for *Pieris brassicae* larvae in a semi-synthetic diet. *Entomol. Exp. Appl.,* 9: 247-255.

Davidson, R.H. and Lyon, W.F. 1979. *Insect Pests of Farm, Garden and Orchard.* New York: John Wiley & Sons, 596 pp.

Dawson, G.W., Griffiths, D.C., Pickett, J.A., Wadhams, L.J. and Woodcock, C.M. 1987. Plant-derived synergists of alarm pheromone from turnip aphid, *Lipaphis (Hyadaphis) erysimi* (Homoptera: Aphididae). *J. Chem. Ecol.,* 13: 1663-1671.

de Jong, R. and Städler, E. 1999. The influence of odour on the oviposition behaviour of the cabbage root fly. *Chemoecology,* 9: 151-154.

Duffey, S.S. 1980. Sequestration of plant natural products by insects. *Ann. Rev. Entomol.,* 25: 447-477.

Durkee, A.B. and Harborne, J.B. 1973. Flavonol glycosides in *Brassica* and *Sinapis. Phytochemistry,* 12: 1085-1089.

Ehrlich, P.R. and Raven, P.H. 1964. Butterflies and plants: A study in coevolution. *Evolution,* 18: 576-608.

Ellis, P.R., Cole, R.A., Crisp, P. and Hardman, J.A. 1980. The relationship between cabbage root fly egg laying and volatile hydrolysis products of radish. *Ann. Appl. Biol.,* 95: 283-289.

Erickson, J.M. and Feeny, P.P. 1974. Sinigrin: A chemical barrier to the black swallowtail butterfly, *Papilio polyxenes. Ecology,* 55: 103-111.

Evans, D.L. and Schmidt, J.O. (Eds.) 1990. *Insect Defenses.* Albany: State University of New York Press, pp. 482.

Feeny, P. 1977. Defensive ecology of the Cruciferae. *Ann. Missouri Bot. Gard.,* 64: 221-234.

Feeny, P.P., Paauwe, K.L. and Demong, N.J. 1970. Flea beetles and mustard oils: host plant specificity of *Phyllotreta cruciferae* and *P. striolata* adults. *Ann. Ent. Soc. Am.,* 63: 832-841.

Feltwell, J. 1982. *Large White Butterfly: The Biology, Biochemistry, and Physiology of Pieris brassicae* (Linnaeus). London: Dr. W. Junk Publishers.

Fenwick, G.R., Heaney, R.K. and Mullin, W.J. 1983. Glucosinolates and their breakdown products in food and food plants. *CRC Critical Reviews in Food Science and Nutrition,* 18: 123-201.

Fieldsend, J. and Milford, G.F. 1994. Changes in glucosinolates during crop development in single-and double-low genotypes of winter oilseed rape (*Brassica napus*): II. Profiles and tissue-water concentrations in vegetative tissues and developing pods. *Ann. Appl. Biol.,* 124: 543-555.

Finch, S. 1980. Chemical attraction of plant-feeding insects to plants. *Appl. Biol.,* 5: 67-143.

Finch, S. and Jones, T.H. 1987. Interspecific competition during host-plant selection by insect pests of cruciferous crops. *In* V. Labeyrie, G. Fabres and D. Lachaise (Eds.). *Insects-Plants. Processing of the 6th International Symposium on Insect-Plant Relationship.* London: Dr. W. Junk Publishers, pp. 85-90.

Greenhalgh, J.R. and Mitchell, N.D. 1976. The involvement of flavour volatiles in the resistance to downy midew of wild and cultivated forms of *Brassica oleracea. New Phytol.,* 77: 391-398.

Greer, M.A. 1962. The natural occurrence of goitrogenic agents. *Recent Prog. Horm. Res.*, **18**: 187-219.

Gupta, P.D. and Thorsteinson, A.J. 1960. Food plant relationships of the diamond-back moth *Plutella maculipennis* (Curt.). *Entomol. Expl. Appl.*, **3**: 305-314.

Harborne, J.B. 1988. *Introduction to Ecological Biochemistry.* Cambridge: Academic Press, pp: 119-122.

Hawkes, C. Patton, S. and Coaker, T.H. 1987. Mechanisms of host-plant finding in adult cabbage root fly, *Delia brassicae. Entomol. Exp. Appl.*, **24**: 219-227.

Heywood, V.H. (Ed.). 1993. *Flowering Plants of the World.* New York: Oxford Univ. Press, pp 336.

Hicks, K.L. 1974. Mustard oil glycosides: Feeding stimulants for adult cabbage flea beetles *Phyllotreta cruciferae* (Coleoptera: Chrysomelidae). *Ann. Entomol. Soc. Am.*, **67**: 261-264.

Higgins, L.G. and Riley, N.D. 1970. *A Field Guide to the Butterflies of Britain and Europe.* London: Collins Clear-Type Press.

Holmes, R.G. 1965. A case of suspected poisoning of dairy cows by white mustard seeds (*Sinapis alba*). *Vet. Rec.*, **77**: 480.

Hovanitz, W., Chang, V.C.S. and Honch, G. 1963. The effectiveness of different isothiocyanates on attracting larvae of *Pieris rapae. J. Res. Lepidop.*, **1**: 249-259.

Haung, X.P., Renwick, J.A.A. and Chew, F.S. 1995. Oviposition stimulants and deterrents control acceptance of *Alliaria petiolata* by *Pieris rapae* and *P. napi oleracea. Chemoecology*, **2**: 79-87.

Hurter, J., Ramp, T., Patrian, B., Städler, E., Roessingh, P., Baur, R., de Jong, R., Nielsen, J.K., Winkler, T., Richter, W.J., Muller, D. and Ernst, B. 1999. Oviposition stimulants for the cabbage root fly: Isolation from cabbage leaves. *Phytochemistry*, **51**:377-382.

Jones, T.H. and Finch, S. 1987. The effect of a chemical deterrent, released from the frass of caterpillars of the garden pebble moth, on cabbage root fly oviposition. *Entomol. Exp. Appl.*, **45**: 283-288.

Jones, T.H., Cole, R.A. and Finch, S. 1988. A cabbage root fly oviposition deterrent in the frass of garden pebble moth caterpillars. *Entomol. Exp. Appl.*, **49**: 277-282.

Josefsson, E. 1968. Method for quantitative determination of p-hydroxybenzyl isothiocyanate in digest of seed meal of *Sinapis alba* L. *J. Sci. Food Agric.* **19**: 192-194.

Ju, H.-Y., Chong, C., Mullin, W.J. and Bible, B.B. 1982. Volatile isothiocyanates and nitriles from glucosinolates in rutabaga and turnip. *J. Am. Soc. Hortic. Sci.*, **107**: 1050-1054.

Karowe, D.N. and Schoonhoven, L.M. 1992. Interactions among three trophic levels: The influence of host plant on performance of *Pieris brassicae* and its parasitoid, *Cotesia glomerata. Entomol. Exp. Appl.* **62**: 241-251.

Kennedy, J.S. 1956. The experimental analysis of aphid behaviour and its bearing on current theories of instinct. *Proc. 10th Int. Congr. Entomol.*, **2**: 397-404.

Kirk, W.D. and Gray, M. 1992. *Insects on Cabbages and Oilseed Rape.* Richmond Publishing Co., Slough, U.K.

Kjaer, A. 1974. The natural distribution of glucosinolates: A uniform class of sulphur-containing glucosides. *In* G. Bendz and J. Santesson (Eds.). *Chemistry in Botanical Classification.* London: Academic Press, pp: 229-234.

Kjaer, A. 1976. Glucosinolates in the Cruciferae. *In* J.G. Vaughan, A.J. MacLeod and B.M.G. Jones (Eds.). *The Biology of the Cruciferae.* London: Academic Press, pp: 207-219.

Klesse, P. and Leukoschek, P. 1955. *Untersuchungen über die baktriostatische Wirksamkeit einiger senföle. Arzneim. Forsch.*, **5**: 505.

Koritsas, V.M., Lewis, J.A., and Fenwick, G.R. 1989. Accumulation of indole glucosinolates in *Psylliodes chrysocephala* L.—infested, or—damaged tissues of oilseed rape (*Brassica napus* L.) *Experientia.*, **45**: 493-495.

Lafontaine, J.D. 1998. Noctuoidea, Noctuidae (part). *In* R.B. Dominick *et al.* (Eds) *The Moths of America North of Mexico, fasc. 27.3.* Lawrence, Kansas: Allen Press, Inc., pp: 1-348.

Lafontaine, J.D. and Poole, R.W. 1991. Noctuoidea, Noctuidae (part). *In* R.B. Dominick *et al.* (Eds) *The Moths of America North of Mexico, fasc. 25.1.* Lawrence, Kansas: Allen Press, Inc., pp: 1-182.

Larsen, P.O. 1981. Glucosinolates. *In* P.K. Stump and E.E. Conn (Eds) *The Biochemistry of Plants.* London: Academic press, pp: 501-525.

Lewis, W.H. and Elvin-Lewis, M.P.F. 1977. *Medical Botany.* New York: John Wiley & Sons, Inc.

Lichtenstein, E.P., Strong, F.M. and Morgan, D.G. 1962. Identification of 2-phenylethylisothiocyanate as an insecticide occurring naturally in the edible part of turnips. *J. Agric. Food Chem.,* **10:** 30-33.

Lichtenstein, E.P., Morgan, D.G. and Mueller, C.H. 1964. Naturally occurring insecticides in cruciferous crops. *J. Agric. Food Chem.,* **12:** 158-161.

Ludwig, S.W. and Kok, L.T. 1998. Phenology and parasitism of harlequin bugs, *Murgantia histrionica* (Hahn) (Hemiptera: Pentatomidae), in southwest Virginia. *J. Entomol. Sci.,* **33:** 33-39.

Luther, G.C., Valenzuela, H.R. and Defrank, J. 1996. Impact of cruciferous trap crops on Lepidopteran pests of cabbage in Hawaii. *Environ. Entomol.,* **25:** 39-47.

Ma, W.C. and Schoonhoven, L.M. 1973. Tarsal contact chemosensory hairs of the large white butterfly *Pieris brassicae* and their possible role in oviposition behavior. *Entomol. Exp. Appl.,* **16:** 343-357.

Marsh, N. and Rothschild, M. 1974. Aposematic and cryptic Lepidoptera tested on the mouse. *J. Zool., Lond.,* **174:** 89-122.

Matsuda, K. 1978. Feeding stimulation of flavonoids for various leaf beetles (Coleoptera: Chrysomelidae). *Appl. Entomol. Zool.,* **13:** 228-230.

Metcalf, C.L. and Flint, W.P. (Eds.). 1962. Destructive and Useful Insects: Their Habits and Control. New York: McGraw-Hill Book Co.

McPherson, J.E. 1982. *The Pentatomoidea (Hemiptera) of Northeastern North America with Emphasis on the Fauna of Illinois.* Carbondale: Southern Illinois University Press.

Mojtahedi, H., Santo, G.S., Hang, A.N. and Wilson, J.H. 1991. Suppression of root-knot nematode population with selected rapeseed cultivars as green manure. *J. Nematol.,* **23:** 170-174.

Moon, M.S. 1967. Phagostimulation of a monophagous aphid. *Oikos,* **18:** 96-101.

Moyes, C.L., Collin, H.A., Britton, G. and Raybould, A.F. 2000. Glucosinolates and differential herbivory in wild populations of *Brassica oleracea. J. Chem. Ecol.,* **26:** 2625-2641.

Munroe, E. 1972. Pyraloidea, Pyralidae (part). *In* R.B. Dominick *et al.* (Eds) *The Moths of America North of Mexico, fasc. 13.1A.* London: Curwen Press, pp: 1-370.

Murchie, A.K., Smart, L.E. and Williams, I.H. 1997. Responses of *Dasineura brassicae* and its parasitoids *Platygaster subuliformis* and *Omphale clypealis* to field traps baited with organic isothiocyanates. *J. Chem. Ecol.,* **23:** 917-926.

Nair, K.S.S. and McEwen, F.L. 1976. Host selection by the adult cabbage maggot, *Hylemya brassicae* (Diptera: Anthomyiidae) effect of glucosinolates and common nutrients on oviposition. *Can. Entomol.,* **108:** 1021-1030.

Nair, K.S.S., McEwen, F.L. and Snieckus, V. 1976. The relationship between glucosinolate content of cruciferous plants and oviposition preferences of *Hylemya brassicae (Diptera: Anthomyiidae). Can. Entomol.,* **108:** 1031-1036.

Nault, L.R. and Styer, W.E. 1972. Effects of sinigrin on host selection by aphids. *Entomol. Exp. Appl.,* **15:** 423-437.

Nayar, J.K. and Thorsteinson, A.J. 1963. Further investigations into the chemical basis of insect-host plant relationships in an oligophagous insect, *Plutella maculipennis* (Curtis) (Lepidoptera: Plutellidae). *Can. J. Zool.,* **41:** 923-929.

Nielsen, J.K.1978. Host plant selection of monophagous and oligophagous flea beetles feeding on Crucifers. *Entomol. Expl., Appl.,* **24:** 562-569.

Nielsen, J.K., Larsen, L.M. and Sørensen, H. 1977. Cucurbitacin E and I in *Iberis amara:* Feeding inhibitors for *Phyllotreta nemorum. Phytochemistry,* **16:** 1519-1522.

Nielsen, J.K., Larsen, L.M. and Sørensen, H. 1979. Host plant selection of the horseradish flea beetle *Phyllotreta Armoraciae* (Coleoptera: Chrysomelidae): Identification of two flavonol glycosides stimulating feeding in combination with glucosinolates. *Entomol. Expl. Appl.,* **26:** 40-48.

Nielsen, J.K., Kirkeby-Thomsen, A.H. and Petersen, M.K. 1989. Host plant recognition in monophagous weevils: specificity in feeding responses of *Ceuthorhynchus constrictus* and the variable effect of sinigrin. *Entomol. Exp. Appl.,* **53:** 157-166.

Nottingham, S.F. 1987. Effects of nonhost-plant odors on anemotactic response to host-plant odor in female cabbage root fly, *Delia radicum*, and carrot rust fly *Psila rosae*. *J. Chem. Ecol.*, **5**: 1313-1318.

Nottingham, S.F. and Coaker, T.H. 1985. The olfactory response of cabbage root fly *Delia radicum* to the host plant volatile allylisothiocyanate. *Entomol. Exp. Appl.*, **39**: 307-316.

Opler, P. 1998. *A Field Guide to Eastern Butterflies*. New York: Houghton Mifflin Company.

Pasteels, J.M., Gregoire, J.-C. and Rowell-Rahier, M. 1983. The chemical ecology of defense in arthropods. *Ann. Rev. Entomol.*, **28**: 263-289.

Pivnick, K.A. 1993. Response of *Meterus leviventris* (Hymenoptera: Braconidae) to mustard oils in field trapping experiments. *J. Chem Ecol.*, **19**: 2075-2079.

Pivnick, K.A., Reed, D.W., Millar, J.G. and Underhill, E.W. 1991. Attraction of northern false chinch bug *Nysius niger* (Heteroptera: Lygaeidae) to mustard oils. *J. Chem. Ecol.*, **17**: 931-941.

Pivnick, K.A., Lamb, R.J. and Reed, D. 1992. Response of flea beetles, *Phyllotreta spp.*, to mustard oils and nitriles in field trapping experiments. *J. Chem. Ecol.*, **18**: 863-873.

Potter, M.J., Davies, K. and Rathjen, A.J. 1998. Suppressive impact of glucosinolates in *Brassica* vegetative tissues on root lesion nematode *Pratylenchus neglectus*. *J. Chem. Ecol.*, **24**: 67-80.

Potter, M.J., Vanstone, V.A., Davies, K.A. and Rathjen, A.J. 2000. Breeding to increase the concentration of 2-phenyl glucosinolate in the roots of *Brassica napus*. *J. Chem. Ecol.*, **26**: 1811-1820.

Read, D.P., Feeny, P.P. and Root, R.B. 1970. Habitat selection by the aphid parasite *Diaeretiella rapae* (Hymenoptera: Braconidae) and hyperparasite *Charips brassicae* (Hymenoptera: Cynipidae). *Can. Entomol.*, **102**: 1567-1578.

Reed, D.W., Pivnick, K.A. and Underhill, E.W. 1989. Identification of chemical oviposition stimulants for the diamond-back moth. *Plutella xylostella*, present in three species of Brassicaceae. *Entomol. Exp. Appl.*, **53**: 277-286.

Renwick, J.A.A. and Radke, C.D. 1983. Chemical recognition of host plants for oviposition by the cabbage butterfly, *Pieris rapae* (Lepidoptera: Pieridae). *Environ. Entomol.*, **12**: 446-450.

Renwick, J.A.A. and Radke, C.D. 1990. Plant constituents mediating oviposition by the diamondback moth, *Plutella xylostella* (L) (Lepidoptera: Plutellidae). *Phytophaga*, **3**: 37-46.

Renwick, J.A.A., Radke, C.D. and Sachdev-Gupta, K. 1992. Leaf surface chemicals stimulating oviposition by *Pieris rapae* (Lepidoptera: Pieridae) on cabbage. *Chemoecology*, **3**: 33-38.

Robinson, T. 1983. The Organic Constituents of Higher Plants: Their Chemistry and Relationships. North Amherst, Massachusetts: Cordus Press, pp: 318-323.

Rodman, J.E. 1981. Divergence, convergence, and parallelism in phytochemical characters: The glucosinolate-myrosinase system. *In* D.A. Young and D.A. Seigler (Eds) *Phytochemistry and Angiosperm Phyogeny*. New York: Praeger, pp: 43-79.

Roessingh, P., Städler, E., Fenwick, G.R., Lewis, J.A., Kvist, N.J., Hurter, J. and Ramp, T. 1992. Oviposition and tarsal chemoreceptors of the cabbage root fly are stimulated by glucosinolates and host plant extracts. *Entomol. Exp. Appl.*, **65**: 267-282.

Rojas, J.C. 1999. Electrophysiological and behavioral responses of the cabbage moth to plant volatiles. *J. Chem. Ecol.*, **25**: 1867-1883.

Rothschild, M. 1985. British aposematic Lepidoptera. *In* J. Health and A.M. Emmt (Eds) *The Moths and Butterflies of Great Britain and Ireland*. Essex, England: Harley Books, pp: 9-62.

Rubin, H.R. and Wu, A.W. 1988. The bitter herbs of seder: More on horseradish horrors. *JAMA*, **259**: 1943.

Schoonhoven, L.M. 1967. Chemoreception of mustard oil glucosides in larvae of *Pieris brassicae*. *Proc. K. Ned. Akad. Wet. (C).*, **70**: 556-568.

Scott, J.A. 1986. *The Butterflies of North America: A Natural History and Field Guide*. Stanford, California: Stanford University Press, 583. pp.

Seo, S.T. and Tang, C.-S. 1982. Hawaiian fruit flies (Diptera: Tephritidae): Toxicity of benzyl isothiocyanate against eggs or 1st instars of three species. *J. Econ. Entomol.*, **75**: 1132-1135.

Sheehan, W. and Shelton, A.M. 1989. Parasitoid response to concentration of herbivore food plants: finding and leaving plants. *Ecology*, **70**: 993-998.

Simpson, B.B. and Ogorzaly, M.C. 1995. *Economic Botany*. New York: McGraw Hill, 742 pp.

Smart, L.E. and Blight, M.M. 1997. Field discrimination of oilseed rape, *Brassica napus*, volatiles by cabbage seed weevil, *Ceutorhynchus assimilis. J. Chem. Ecol.*, **23**: 2555-2567.

Smart, L.E., Blight, M.M. and Hick, A.J. 1997. Effect of visual cues and a mixture of isothiocyanates on trap capture of cabbage seed weevil, *Ceutorhynchus assimilis* (Paykull) (Coleoptera: Curculionidae). *J. Chem. Ecol.*, **23**: 889-902.

Spitzer, D.E. 1988. Horseradish horrors: Sushi syncope. *JAMA*. **259**: 218-219.

Städler, E. 1978. Chemoreception of host plant chemicals by ovipositing female of *Delia (Hylemya) brassicae. Entomol. Exp. Appl.*, **24**: 711-720.

Stoetzel, M.B. 1989. *Common Names of Insects and Related Organisms*. Wash., D.C.: Entomological Society of America, 88 pp.

Stump, P.K. and Conn, E.E. (Eds) 1981. *The Biochemistry of Plants*. London: Academic Press.

Sugiyama, S. and Matsumoto, Y. 1959. Olfactory responses of the vegetable weevil larvae to various mustard oils. Studies on the host plant determination of the leaf-feeding insects II. *Nogaku Kenkyu*, **46**: 150-157.

Tanton, M.T. 1965. Agar and chemostimulant concentrations and their effect on intake of synthetic food by larvae of the mustard beetle *Phaedon cochleriae* Fab. *Entomol. Exp. Appl.*, **8**: 74-82.

Thorsteinson, A.J. 1953. The chemotactic responses that determine host specificity in an oligophagous insect (*Plutella maculipennis* Curt.; Lepidoptera). *Can. J. Zool.*, **31**: 52-72.

Traynier, R.M.M. 1965. Chemostimulation of oviposition by the cabbage rootfly *Erioischia brassicae* (Bouché). *Nature, Lond.*, 218-219.

Traynier, R.M.M. 1967. Stimulation of oviposition by the cabbage root fly *Erioischia brassicae. Entomol. Exp. Appl.*, **10**: 401-412.

Traynier, R.M.M. and Truscott, R.J.W. 1991. Potent natural egg-laying stimulant for cabbage butterfly *Pieris rapae. J. Chem. Ecol.*, **17**: 1371-1380.

Uvah, I.I.I. and Coaker, T.H. 1984. Effect of mixed cropping on some insect pests of carrots and onions. *Entomol. Exp. Appl.*, **36**: 159-167.

van Emden, H.F. and Bashford, M.A. 1971. The performance of *Brevicoryne brassicae* and *Myzus persicae* in relation to plant age and leaf amino acids. *Entomol. Exp. Appl.*, **14**: 349-360.

van Etten, C.H. and Tookey, H.L. 1979. Chemistry and Biological Effects of Glucosinolates. *In* G.A. Rosenthal and D.H. Janzen. (Eds) *Herbivores: Their Interaction with Secondary Plant Metabolites*. New York: Academic Press, pp: 471-500.

Vaughan, J.G., MacLeod, A.J. and Jones, B.M.G. (Eds). 1976. *The Biology and Chemistry of the Cruciferae*. London: Academic Press. pp. 355.

Vaughn, S.F. and Berhow, M.A. 1998. 1-cyano-2-hydroxy-3-butene, a phytotoxin from Crambe (*Crambe abyssinica*) seedmeal. *J. Chem. Ecol.*, **24**: 1117-1126.

Vaughn, S.F. and Boydston, R.A. 1997. Volatile allelochemicals released by crucifer green manures. *J. Chem. Ecol.*, **23**: 2107-2116.

Verschaeffelt, E. 1911. The cause determining the selection of food in some herbivorous insects. *Proc. Acad. Sci., Amsterdam*, **13**: 536-542.

Virtanen, A.I. 1962. Some organic sulfur compounds in vegetable and fodder plants and their significance in human nutrition. *Angew. Chem.* (Int. Ed.), **1**: 299-306.

Walker, N.J. and Gray, I.K. 1970. The glucosinolates of land cress (*Coronopus didymus*) and its enzymatic degradation products as precursors of off-flavor in milk–a review. *J. Agric. Food Chem.*, **18**: 346-352.

Wallbank, B.E. and Wheatley, G.A. 1979. Some responses of cabbage root fly (*Delia brassicae*) to allyl isothiocyanate and other volatile constituents of Crucifers. *Ann. Appl. Biol.*, **91**: 1-12.

Weires, R.W. and Chiang, H.C. 1973. Integrated control prospects of major cabbage insect pests in Minnesota-based on the faunistic, host varietal, and trophic relationships. *Univ. Minn. Agr. Exp. Sta. Tech. Bull.*, **291**: 44.

Wensler, R.J.D. 1962. Chemical factors and host selection in *Bervicoryne. Nature*, **195**: 830-831.

Whitman, D.W. 1988. Allelochemical interactions among plants, herbivores, and their predators. *In* P. Barbosa and D. Letourneau (Eds) *Novel Aspects of Insect-Plant Interactions*. New York: John Wiley & Sons, Inc., pp: 11-64.

Whitman, D.W., Blum, M.S. and Jones, C.G. 1985. Chemical defense in *Taeniopoda eques* (Orthoptera: Acrididae): Role of the metathoracic secretion. *Ann. Entomol. Soc. Am.*, **78**: 451-455.

Williams, D.L., Kain, W.C. and Broadway, R.M. 1977. Isolation and characterization of a serine proteinase inhibitor cDNA (Accession No. U18995) from cabbage. *Plant Physiol.*, **114**: 747-749.

Williams, I.H., Martin, A.P. and Kelm, M. 1987. The phenology of the emergence of the brassica pod midge (*Dasineura brassicae* Winn.) and its infestation of winter oil-seed rape (*Brassica napus* L.). *J. Agric. Sci. Cambridge*, **108**: 579-589.

Williams, L. III, Morra, M.J., Brown, P.D. and McCaffrey, J.P. 1993. Toxicity of allyl isothiocyanate–amended soil to *Limonius californicus* (Mann.) (Coleoptera: Elateridae) wireworms. *J. Chem. Ecol.*, **19**: 1033-1046.

Young, D.A. and Seigler, D.A. (Eds) 1981. *Phytochemistry and Angiosperm Phylogeny*. New York: Praeger.

Zsolnai, T. 1966. Die Antimikrobielle Wirkung von Thiocyanaten und Isothiocyanaten. *Arzneim. Forschung.*, **16**: 870.

## Chapter 5

# Tetranortriterpenoids and Quassinoids: Constitutive Chemical Defenses of the Rutales

*T.R. Govindachari, G. Suresh and Geetha Gopalakrishnan*

"A significant characteristic of plants is their ability to produce endless variations on a single chemical theme, and this is no better exemplified than in the hundreds of trepenoids derived from one simple $C_5$ biological isoprenoid unit, isopentenyl pyrophosphate. If one accepts the biosynthetic prodigality of plants as an example of evolution in progress, then gradually one can see emerging atleast one function for each main group of terpenoids" (Goodwin, 1967). It is, hence, relevant to note that a number of examples are available for the phytocentric role of terpenoid mixtures in host plant defense (Langenheim, 1994), which illustrates the need for such prodigality of terpenoid diversity. Evidence for ecological effects of terpenoids for multitrophic interactions (Harborne, 1991; Gershenzon and Croteau, 1991) and their defensive role as *raison d'etre* in host plant defense (Feeny, 1992) are well documented. Such diversity towards multifaceted defense is well-illustrated by the elaboration of limonoids and quassinoids, the two biosynthetically related terpenoids that are major constituents of the families belonging to the angiospermic Order Rutales. There are fairly consistent differences between the tetranortriterpenoids of the Rutaceae and Meliaceae with the quassinoids restricted to the Simaroubaceae (Das *et al.* 1987). The members of Rutales have been shown to have remarkable species and chemical diversity. Species belonging to this Order also showed diversity in their floral and vegetative parts, which are well-connected by intermediate ones. Most of the genera currently recognised as rutalean have been regarded as such since the time of their discovery. It is remarkable, then, that these observations are applicable to the chemistry of this Order, which is also equally diverse, with triterpenoids of extreme oxidation levels that are well connected with intermediate structures.

Tetranortriterpenoids are C26 tetracyclic triterpenoids derived from tirucallol

Centre for Natural products, SPIC Science Foundation, 111, Mount Road, Madras 600 032, India.

(20αH) [1] or euphol (20βH) precursor, which undergoes rearrangement to the corresponding apotirucallol [2] having an oxygen introduced at C-7 (Arigoni *et al.* 1960). **Loss of four carbon atoms** and modification of side chain to form the furan ring results in the intact tetranortriterpenoid [3].

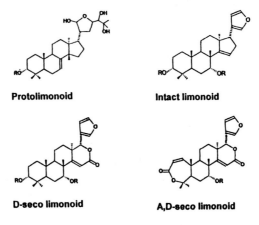

Oxidations through Baeyer-Villiger ring cleavages, rearrangements and degradation results in a wide range of structural types, such as limonoids and quassinoids that are found in genera belonging to the angiospermic Order Rutales.

The diversity and abundance of the limonoids are more pronounced in Meliaceae and Rutaceae and over 300 limonoids have been isolated till date. The limonoids are structurally quite diverse, but the extreme types could be mechanistically connected by justifiable biogenetic pathways from a common precursor (Das *et al.* 1984). On the basic limonoid skeleton, a variety of oxidations and rearrangements result in diverse structural types classified on the basis of modifications of the four rings in triterpenoid molecule as given below:

**Protolimonoid**          **Intact limonoid**

**D-seco limonoid**        **A,D-seco limonoid**

**A-seco limonoid**            **B-seco limonoid**

**B,D-seco limonoid**          **A,B-seco limonoid**

**Azadirachtin type**          **Volkensin type**

**(C-seco limonoid)**         **C-seco limonoid**

The quassinoids can also be classified into distinct groups according to their basic skeleton (see figure). By far, the majority of the numerous quassinoids known have the $C_{20}$ basic skeleton (picrasane). Quassinoids are heavily oxygenated lactones and have varying numbers of different oxygen containing groups. Except the carbons C-5, C-9 and methyl groups at C-4 and C-10, all the other carbon atoms have oxygenated functions (Polonsky, 1973, 1985).

## BASIC SKELETONS OF QUASSINOIDS

Limonoids and quassinoids are distributed in all plant tissues and the types produced by different tissues of the same plant may also differ. Not much is known on the cellular location or the biosynthetic pathways leading to the diverse structural types. The pathways suggested for biosyntheses are characterised by an evolutionary pattern of increasing oxidation and rearrangements (Das *et al.* 1984, 1987). The bioactivity of these compounds against insects, especially affecting their behaviour and physiology, is suggestive of the primary selective advantage of the production of triterpenoids is host plant defense against insect herbivory (Champagne *et al.* 1992). If such a hypothesis is true, then one would find the evolutionary trends of increasing structural complexities and oxidations correlating well with increasing activity against insect herbivores (Champagne *et al.* 1992).

The simplest of structural types among the tetranortriterpenoids (limonoids) are the protolimonoids and meliantriol and melianone from neem seed oil are

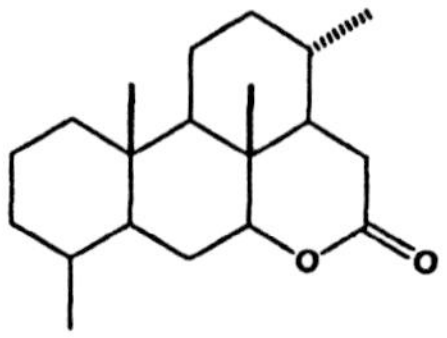

**C$_{20}$**

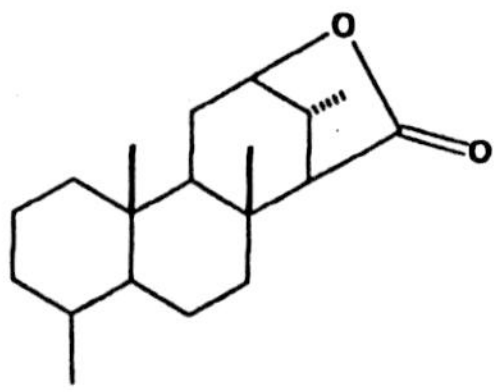

**C$_{19}$**

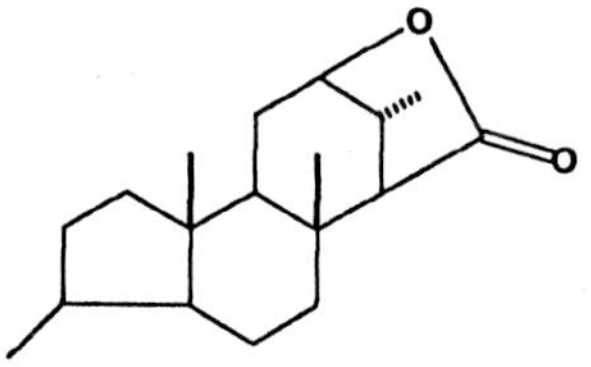

**C$_{18}$**

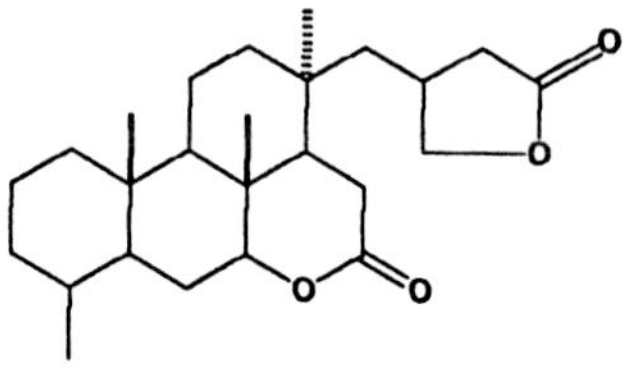

**C$_{25}$**

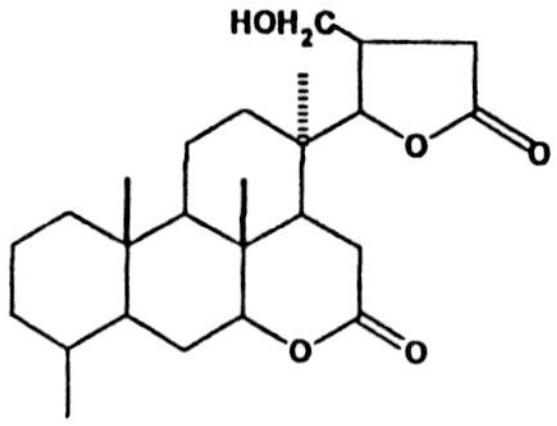

HOH$_2$C

**C$_{25}$**

well-known examples. These compounds possess excellent antifeedant activity against *Schistocerca gregaria* and *Epilachna varivestis* (Lavie *et al.* 1967). Ocotillone, dymalol and shoreic acid (the latter two being A-ring modified protolomonoids) inhibited feeding by *Spodoptera litura*, but did not affect the growth of the larval instars (Govindachari *et al.* 1995). A number of intact apoeuphol limonoids have been studied for their antifeedant and growth regulating activity against a variety of insect pests, which includes cedrelone, anthothecol and sendanin. Cedrelone has been reported to have moult inhibiting activities against *Oncopeltus fasciatus* (Champagne *et al.* 1992), *Pectinophora gossypiella* (Kubo and Klocke, 1986) although against *Peridroma saucia* (Champagne *et al.* 1992) and *Ostrinia nubilalis* (Arnason *et al.* 1987) cedrelone was ineffective. Cedrelone significantly inhibited growth of different larval instars of *Peridroma saucia* and *Mamestra configurata* and nutritional analyses revealed that growth inhibition and reduced food consumption are a consequence of post-ingestive malaise rather than pheripherally-mediated antifeedant effect. It was also suggested that the toxicity of cedrelone of these larval instars is not likely to involve endocrine system (Koul and Isman, 1992). Cedrelone showed appreciable antifeedant activity (Fl$_{50}$ = 10 ug/cm$^2$ leaf area) against *Spodoptera litura* and

**Ocotillone**                    **Sendanin**

comparison of the cedrelone analogues (modified by synthetic means) indicated the importance of the decalin ring substitution for antifeedant activity (Govindachari *et al.* 1995). Cedrelone increased larval durations of *S. litura* and decreased the pupal weights thereby decreasing growth indices (Govindachari *et al.* 1995). Against *S. littorallis*, amoorastatins, azedarachins, trichilins and their keto compounds exhibited excellent antifeedant activity (Nakatani, 1999), and these results were in conformity with the earlier reports on intact apoeuphol limonoids against a number of insect species such as *Epilachna varivestis*, *Heliothis zea*, *Heliothis virescens*, *S. frugiperda*, *Pectinophora gossypiella* and *Aedes aegypti* (Champagne *et al.* 1992). It was proposed that 14, 15-epoxide and either a 19/28 lactol bridge or a cyclohexenone A ring are features that correlated well with activity against insects. Amoorastatins, azedarachins, trichilins and their 15 keto compounds from *Melia toosendon* and meliatoxins A and B from *M. azedarach* var. *australasia*, with C-19/C-28 acetal brdiged system showed excellent antifeedant activity (Nakatani, 1999; MacLeod *et al.* 1990). In our experiments on *S. litura* treated with azadiradione and epoxy azadiradione, we did not find any appreciable difference in antifeedant and growth regulating activities between these compounds, clearly indicating that an acetoxy at C-7, instead of enone and the carbonyl at C-16, may increase the antifeedant and growth regulatory activity considerably indicating active sites (Govindachari *et al.* 1995).

Gedunin, a D-seco limonoid, occurs in a number of taxa of meliaceae (Taylor, 1983) and its 7-desacetyl and 7-keto analogues have been isolated and tested for antifeedant and growth inhibitory activities against a number of insects. Gedunin was found effective against selected insect species, such as *P. gossypiella*, *S. frugiperda* and *Heliothis zea* (Kubo and Klocke, 1986). A,D-seco limonoids, found in all the four families of the Rutales, are mainly elaborated by the Rutaceae. Compounds such as limonin, nomilin and abacunone occur frequently in the Rutaceae and appear to be less active than the D-seco limonoids such as gedunin (Champagne *et al.* 1992). Interestingly, in our experiments, limonin was equally effective as an antifeedant compared to selected protolimonoids and intact apoeuphol limonoids when tested against *S. litura;* limonin also drastically increased larval durations, decreased pupal weight and showed mortality as well (Govindachari *et al.* 1995).

**Limonin**

Only a few B-seco and A,B-seco limonoids have been examined in detail for bioactivity against insects and the assays were restricted to feeding deterrence (Champagne *et al.* 1992); much less is known on their IGR/moult inhibitory activities. Prieurianin, rohitukin and 21 (R,S) -hydroxy toonacilid are found to be less active when compared to azadirachtin A as antifeedants against *H.zea*, *S. frugiperda* and *E. varivestis* (Lidert *et al.* 1985). Nymania-3 an A,B-seco compound was isolated in quantities from the bark of *Dysoxylum malabaricum* by prep HPLC, compared well with azadirachtin-A in its antifeedancy and effects on larval durations and mortality of *S. litura* (Suresh, 1996) and *Pericallia ricini* (Govindachari *et al.* 1999). (Table 1.). Growth inhibitory effects of three A-seco limonoids (evodulone, tecleanine and 7-deacetylproceranone) against *H.zea*, *S.frugiperda* and *P.gossypiella* were evaluated and found to be less effective than some of the intact apo-euphol limonoids (Kubo and Klocke, 1986).

**Table 1.** Percentage Feeding Index (PFI) of *P.ricini* and *S. litura* when fed on Nymania-3 treated castor leaves

| Compound | Percent Feeding Index at | | | |
|---|---|---|---|---|
|  | *1µg/cm$^2$ | 5µg/cm$^2$ | 10µg/cm$^2$ | 50µg/cm$^2$ |
| Nymania-3[pr] | 33.5 ± 5.1 | 30.0 ± 4.6 | 23.4 ± 4.5 | 19.5 ± 2.6 |
| Azadirachtin A[pr] | 20.8 ± 3.5 | 15.6 ± 3.1 | 12.5 ± 3.5 | 10.0 ± 3.1 |
| Nymania - 3[sl] | 27.5 ± 4.0 | 17.9 ± 6.6 | 16.5 ± 7.8 | 11.5 ± 4.8 |
| Azadirachtin A[sl] | 38.5 ± 4.5 | 34.5 ± 7.5 | 22.5 ± 11.4 | 21.0 ± 4.5 |

*pr - Pericallia ricini, sl - Spodoptera litura.*
*concentration of compounds per cm$^2$ castor leaf area; values are mean ± S.D. ; n = 5 replicates*

Among the limonoids known in the Rutales, the most effective are the C-seco limonoids, which are characterised by a perhydro linear naphthofuran ring. The examples of this type include salannin, nimbin, ochinolide and volkensin. Interestingly, the highly oxygenated and modified C-seco limonoids, such as azadirachtins, lack this feature and are unique in having free rotation of the hydroxy tricyclic hydrofuran acetal fragment attached to C8. C-seco limonoids are found distributed only in the subfamily Melioideae represented by two genera, i.e., *Melia* and *Azadirachta*. While studying the problems connected with the antifeedant bioassays, Schwinger *et al.* (1984) and Kraus *et al.* (1987) screened several other neem triterpenoids, especially the C-seco limonoids, and showed that salannin was equally effective as azadirachtin as an insect antifeedant against *Epilachna varivestis*. Against *Pieris brassicae,* larvae salannin was found to be more active than azadirachtin-A (Luo Lin-Er *et al.* 1995). Salannin occurs in species of *Melia* and *Azadirachta* and is an effective antifeedant against *Musca domestica, Acalymma vitatum, Diabrotica undecimpunctatat, Earias insulana, Aonideilla aurantil* and *Locusta* sp. Salannin and fourteen derivatives of salannin were assayed for antifeedant activity against *Leptinotarsa decemlineata* and it was found that any modification of the furan ring, replacement of acetoxyl group, modification of tigloyl group and saponification of the methyl ester affects antifeedant activity (Yamasaki and Klocke, 1989). Govindachari *et al.* (1996) showed that salannin not only deterred feeding but also delayed molt through increased larval durations, caused larval and pupal mortalities and decreased pupal weights in two lepidopterous species and caused molt delays and nymphal mortalities in *Oxya fuscovitata*. Nimbin and desacetylnimbin were reported to be feeding deterrents, though to a lesser extent than azadirachtin - A (Schwinger *et al.* 1984; Kraus *et al.* 1987; Govindachari *et al.* 1995, 1996). Results on similar lines were also reported by Aerts and Mordue (Luntz) (1997). Among the C-seco limonoids, compounds such as ochinolide B and ochinol have little or no activity (Rouseff and Nagy 1982; Liu *et al.* 1990). Comparison of antifeedant activity of the naturally occurring azadirachtins against *S. litura* showed that substitutions in the decalin ring system at C12 and C29 are critical while substitution at C1 and C3 produce only small variations (Govindachari *et al.* 1994). For insect growth regulating activity, it is believed that the free hydroxyls at C1 and C3 are essential and it is presumed that this increases the structural resemblance to ecdysone. It is indicated that absence of the C22, 23 double bond of the dihydrofuran ring, the presence of an ethoxy side chain at C23 and the hydroxyls at C1 and C3 generally improve IGR activity, with 13, 14 epoxide group being critical (Rembold, 1989).

Quassinoids are bitter principles found distributed in the family Simaroubaceae and biogenetically regarded as degraded triterpenoids. The precursor to quassinoids is proposed to be $\Delta^7$ euphol or more probably its 20$\alpha$ isomer, $\Delta^7$ tirucallol (Polonsky, 1973). Interest in the quassinoids increased rapidly with the finding that quassinoids display marked antileukemic activity. Quassinoids also showed insecticidal, antifeedant and growth inhibitory activity effects against *H. virescens* and *S. frugiperda* (Klocke *et al.* 1985). Thirteen quassinoids of different structural types were tested against *Epilachna varivestis* and *S. eridania* and all of them showed significant activity against the Mexican bean beetle. Bruceantin, glaucarubinone, isobrucein A and simalikalactone D were effective antifeedants against the Southern armyworm (Leskinen *et al.* 1984). Eight quassinoids were tested for their antifeedant and insecticidal

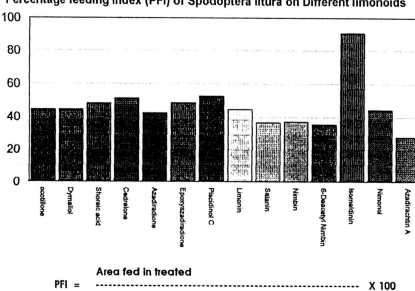

**Percentage feeding Index (PFI) of Spodoptera litura on Different limonoids**

$$PFI = \frac{\text{Area fed in treated}}{\text{Area fed in treated} + \text{Area fed in control}} \times 100$$

activity against *Locusta migratoria* and glaucarubinone and brucein B showed significant insecticidal activity (Odjo *et al.* 1981). A series of quassinoids were tested for antifeedant activity against *Myzus persicae*. Isobrucein B, brucein B and C, glaucarubinone and quassin were found effective antifeedants and isobrucein A was the most effective even at 0.01% (Polonsky *et al.* 1989).

**Brucein B**

**R = COC(Me)OHCH2Me**

**Glaucorubinone**

## CONCLUSION

Adaptive radiation of Constitutive plant secondary chemistry lay in their ecological significance in keeping at bay the insects and other herbivores and such secondary chemistry must necessarily be integrated into the total metabolic scheme and multiple functions are then to be expected (Rhoades, 1979). The twin line of evidence for such antiherbivore activity of secondary chemistry emerges from their deterrency and antibiosis. Secondary chemicals have been shown either to have a negative effect on herbivore fitness or to have a deterrent effect. In a majority of the studies, such effect of limonoids and quassinoids on ecologically relevant herbivores are not generally examined, but the circumstantial evidence is impressive. Most of these substances are end products of highly energy-demanding synthesis and it is reasonable to assume that there is a positive selection for their production. Variations in expressed chemistry within taxa is visualised as a selective deterrence to specialised herbivory spatially, seasonally, temporally and thro' evolutionary times (Spencer, 1988).

If the diversity in Constitutive secondary chemistry can be accepted as "reciprocal evolutionary change in interacting species" (Thompson, 1982), then it would be appropriate to accept the triterpenoid diversity in the Order Rutales as part of Constitutive defense strategy of these taxa. From the illustrative examples of the behavioural modifying effects, growth inhibitory effects and antifeedancy, it is possible to presume that these compounds are, after all, chemical Constitutive defenses against herbivory.

## ACKNOWLEDGEMENTS

Thanks are due to Dr. S. Narasimhan, Deputy Director and Head, Centre for Natural products, SPIC Science Foundation for encouragement. The authors are grateful to Dr. G.N. Krishna Kumari and Dr. Daniel Wesley for their help.

## REFERENCES

Aerts, R.J. and Mordue (Luntz), A.J. 1997. Feeding deterrence and toxicity of neem triterpenoids. *Journal of Chemical Ecology*, **23**: 2117-2132.

Arnason, J.T., Philogene, B.J.R., Donskov, N. and Kubo, I. 1987. Limonids from the meliaceae and Rutaceae reduce the feeding, growth and development of *Ostrinia nubilalis*. *Entomol. Exp. Appl.*, **43**: 221-226.

Champagne, D.E., Koul, O., Isman, M.B., Scudder, G.G.E. and Towers, G.H.N. 1992. Biological activity of limonoids from the Rutales. *Phytochemistry*, **31**: 377-394.

Das, M.F., Da Silva, G.F., Gottlieb, O.R. and Dreyer, D.L. 1984. Evolution of limonoids in the Meliaceae. *Biochemical Systematics and Ecology*, **12**: 299-310.

Das, M.F., Da Silva, G.F. and Gottlieb, O.R. 1987. Evolution of Quassinoids and Limonoids in the Rutales. *Biochemical Systematics and Ecology*, **15**: 85-103.

Feeny, P.P. 1992. The evolution of chemical ecology: Contributions from the study of herbivorous insects. *In* Rosenthal, G.A. and Berenbaum, M.R. (Eds) *Herbivores: Their Interactions with Secondary Metabolites*. Vol. II, Ecological and Evolutionary Process. Academic Press, New York, p. 1-44.

Gershenzon, J. and Croteau, R. 1991. Terpenoids. *In* Rosenthal, G.A. and Berenbaum, M.R. (Eds), *Herbivores, Their Interaction with Secondary Metabolites*. Vol. 1. The chemical participants. Academic Press, New York, p. 165-219.

Goodwin, T.W. 1967. The Biological significance of terpenes in Plants. *In* Pridhan, J.B. (Ed), *Terpenoids in Plants*. Academic Press, London, p. 1-23.

Govindachari, T.R., Suresh, G. and Ganeshwar Prasad, K. 1994. Structure-related antifeedant activity of azadirachtins against the tobacco cutworm *Spodoptera litura* F. *Pesticide Research Journal,* **6:** 20-25.

Govindachari, T.R., Narasimhan, N.S., Suresh, G., Partho, P.D., Geetha Gopalakrishnan and Krishna Kumari, G.N. 1995. Structure-related insect antifeedant and growth regulating activities of some limonoids. *Journal of Chemical Ecology,* **21:** 1585-1600.

Govindachari, T.R., Suresh, G., Krishna Kumari, G..N., Rajamannar, T. and Partho, P.D. 1999. Nymania-3: *a bioactive triterpenoid from Dysoxylum malabaricum. Fitoterapia,* **70:** 83-86

Harborne, J.B. 1991. Recent advances in the ecological chemistry of plant terpenoids. Harborne, J.B and F.A. Tomes - Barberam (Eds) *In Ecological Chemistry and Biochemistry of Plant Terpenoids.* Clarendon Press, Oxford, p. 399-426.

Klocke, J.A., Arisawa, M., Hunda, S.S., Kinghorn, A.D., Cordell, G.A. and Farnsworth, N.R. 1985. Growth inhibitory, insecticidal and antifeedant effects of some antileukemic and cytotoxic quassinoids on two species of agricultural pess. *Experientia,* **41:** 379-382.

Koul, O. and Isman, M.B. 1992. Toxicity of the limonoid allelochemical cedrelone to noctuid larvae. *Entomol. exp. appl.,* **64:** 281-287.

Kraus, W., Baumann, S., Bokel, M., Keller, U., Klenk, A., Klingele, M., Pohnl, H. and Schwinger, M. 1987. Control of insect feeding and development by constituents of *Melia azedarach* and *Azadirachta indica. In* Schmutterer, H. and Ascher, K.R.S. (Eds), *Natural Pesticides from the Neem Tree and other Tropical Plants.* GTZ, Eschborn. P. 111-125.

Kubo, I. and Klocke, J.A. 1986. Insect ecdysis inhibitors. *In* Green. M.B. and Hedin, P.A. (Eds.). *Natural Resistance of Plants to Pests.* ACS Symp. Ser., **296:** 206-219. American Chemical Society, Washington, D.C.

Langenheim, J.H. 1994. Higher plant terpenoids: A phytocentric overview of their ecological roles. *Journal of Chemical Ecology,* **20:** 1223-1280.

Lavie, D., Jain, M.K. and Shpan-Gabrielith. 1967. A locust phagorepellent from two *Melia* species. *Chem. Comm.,* 910-911.

Leskinen, V., Polonsky, J. and Bhatnagar, S. 1984. Antifeedant activity of quassinoids. *Journal of Chemical Ecology,* **10:** 1497-1507.

Lidert, Z., Taylor, D.A.H. and Thirugnanam, M. 1985. Insect antifeedant activity of four prieurianin type limonoids. *Journal of Natural products,* **48:** 843-845.

Liu, Y., Alford, A.R., Rajab, M.S. and Bentley, M.D. 1990. Effects and mode of action of *Citrus* limonoids against *Leptinotarsa decelineata. Physiological Entomology,* **15:** 37-45.

Luo Lin-er, Van Loon, J.J.A. and Schoonhoven L.M. 1995. Behavioural and sensory responses to some neem compounds by *Pieris brassicae* larvae. *Physiological Entomology,* **20:** 134-140.

Macleod, J.K., Moeller, P.D.R., Molinski, T.F. and Koul, O. 1990. Antifeedant activity against *Spodoptera litura* and [$^{13}$C]NMR spectral assignments of the meliatoxins. *Journal of Chemical Ecology,* **16:** 2511-2518.

Nakatani, M. 1999. Limonoids from *Melia toosendan* (Meliaceae) and their antifeedant activity. *Heterocycles,* **50:** 595-609.

Odjo, A., Piart, J., Polonsky, J. and Roth, M. 1981. Etude de l'effet insecticide de deux quassinoids sur des larves de *Locusta migratoria migratoriotdes* Ret. F. (Orthoptera: Acrididae). *C.R. Acad. Sci. Paris.,* **293:** Ser. II, 241.

Polonsky, J. 1973. Quassinoid bitter principles. *Fortschr. Chem. Org. Natrust,* **30:** 101-150.

Polonsky, J. 1985. Quassinoid bitter principles. *Fortschr. Chem. Org. Natrust,* **47:** 222-264.

Polonsky, J., Bhatnagar, S.C., Griffiths, D.C., Pickett, J.C. and Woodcock, C.M. 1989. Acitivity of quassinoids as antifeedants against aphids. *Journal of Chemical Ecology,* **15:** 993-998.

Rembold, H. 1989. Azadirachtins: their structure and mode of action. *In* Arnason, J.T., Philogene, B.J.R. and Morand, P. (Eds) *Insecticides of Plant Origin.* ACS Symp. Ser., 387, p. 150-163.

Rhoades, D.F. 1979. Evolution of plant chemical defense against herbivores. *In* Rosenthal, G.A. and Jansen, D.H. (Eds) *Herbivores: their interaction with secondary plant metabolites.* Academic Press, New York, p. 3-54.

Rouseff, R.L. and Nagy, S. 1982. Distribution of limonoids in Citrus seeds. *Phytochemistry* **21:**85.

Schwinger, M., Ehhammer, B. and Kraus, W. 1984. Methodology of the *Epilachna varivestis* bioassay of antifeedants demonstrated with some compounds from *Azadirachta indica* and *Melia azedarach*. *In* Schmutterer, H. and Ascher, K.R.S. (Eds) *Natural Pesticides from the Neem Tree and other Tropical plants.*, GTZ, Eschborn, p. 181-198.

Spencer, K.C. 1988. The chemistry of coevolution. *In* Spencer, K.C. (Ed) *Chemical mediation of Coevolution.* Academic Press, New York, p. 581-587.

Suresh, G. 1996. Structure-activity relationship analysis of bioactive limonoids from meliaceae. *In* Ananthakrishnan, T.N. (Ed) *Biotechnical Perspective in Chemical Ecology of Insects.* Oxford-IBH Publishing Co. Pvt. Ltd., New Delhi, p. 63-75.

Taylor, D.A.H. 1983. Biogenesis distribution and systematic significance of limonoids in the Meliaceae, Cneoraceae and allied Taxa. *In* Waterman, P.G. and Grundon, M.F. (Eds) *Chemistry and Chemical Taxonomy of the Rutales.* Academic Press, New York, p. 353-375.

Thompson, J.N. 1982. *Interaction and Coevollution.* Wiley, New York.

Yamasaki, R.B. and Klocke, J.A. 1989. Structure-Bioactivity relationship of salannin as an antifeedant against Colorado potato beetle (*Leptinotarsa decemlineata*). *Journal of Agricultural and Food Chemistry*, **37**: 1118-1124.

# Chapter 6

# Role of Plant Surface in Resistance to Insect Herbivores

*A. John Peter\* and Thomas G. Shanower\*\**

## INTRODUCTION

Plants and herbivorous insects have co-evolved through dynamic interactions over millions of years. These interactions have often been likened to a co-evolutionary arms race (Ehrlich and Raven, 1964; Fox, 1981). The ammunition for this arms-race includes chemical, structural and physiological compounds and processes. The production of secondary chemicals in plants, not directly related to basic metabolic activities, provided more effective defense against many herbivores (Feeny, 1975; Spencer, 1988). Co-evolved insect herbivores developed novel, functional responses to deal with these plant chemicals. For example, insect biotypes developed in response to the evolution of specific plant chemotypes (van Emden, 1991; Ananthakrishnan, 1999). Structural and morphological features were similarly influenced by co-evolutionary selection pressure. The development of chemosensory receptors and special structural modifications for attachment by insects are two examples (Chapman, 1982; Visser, 1986). Despite the considerable adaptation by insects, plants continue to withstand the pressure exerted by insects and other herbivores.

Host plant resistance, one of the most effective pest management tools, reduces insect damage by utilizing both chemical and morphological plant defenses (Painter, 1951; Kogan and Ortman, 1978; Mauricio and Rausher, 1997). Plants defend herbivorous insect invasion by producing an array of potent secondary chemicals like alkaloids, phenolics, flavonoids, terpenoids to name a few (Lukefahr and Houghtaling, 1969; Farmer and Ryan, 1990; Ananthakrishnan, 1996). Production of these non-nutritive chemicals is

---
\* Scientist, Division of Crop Protection, Nagarjuna Agricultural Research and Development Institute, 61, Nagarjuna Hills, Panjagutta, Hyderabad, Andhra Pradesh 500 082, India
\*\* Research Entomologist, Pest Management Research Unit, USDA-Agricultural Research Service Northern Plains Agricultural Research Lab., 1500 N. Central Ave., Sidney, MT 59270 USA

regulated by the enzymes Phenylalanine ammonia lyase (PAL), Tyrosine Ammonia Lyase (TAL), Peroxidases (PO) and Polyphenol Oxidases (PPO) against environmental or insect feeding injury stress (Liang *et al.* 1989; Ananthakrishnan, 1994). Other mechanisms provide physical resistance to insect herbivores using surface waxes, trichomes, stomata and cell wall thickness (David and Easwaramoorthy, 1988). Plant surface structures are the first organs to be contacted by insect herbivores when they land on a plant. Hence, the importance and functionality of these structures as resistance mechanisms.

Plant surface wax and trichomes act as mechanical and chemical barriers against insects landing on plants for food and shelter. They attract and/or repel and interfere with insect locomotion, attachment, feeding, digestion and oviposition (Eigenbrode and Espelie, 1995; Peter *et al.* 1995). Physical characters of plants have been used in a number of genetic enhancement programs. The advantage of using insect-resistant crops is lower production costs relative to other control strategies. This chapter discusses the role of these two important plant surface characters in conferring resistance in plants against insect herbivory. The disadvantages of these characters, particularly for natural enemies, are also reviewed.

## SURFACE WAXES: AN OVERVIEW

Plant surfaces are usually covered by waxes, esters formed by the linkage of long-chain fatty acids with an aliphatic alcohol group (Smith, 1989). Surface waxes protect the plant from desiccation and diseases (Jeffree, 1986), but have also been recognized as an important insect resistant factor for more than 30 years (Thompson, 1963). Only recently, however, the role and impact of surface waxes on herbivorous insects have been extensively studied (Eigenbrode and Espelie, 1995; Powell *et al.* 1999). The thin wax layer coating leaf and stem epidermis can inhibit insect attachment, egg laying, feeding and locomotion (Eigenbrode and Espelie, 1995). The mode of action may be physical, chemical or both. Surface waxes have been identified as physical and/or chemical resistance factors in a number of plant species including Crucifers (*Brassica* spp.), corn (*Zea mays*), broad beans (*Vicia faba*), wheat (*Triticum aestivum*), onion (*Allium cepa*), barley (*Hordeum vulgare*), and sorghum (*Sorghum bicolor*) (Eigenbrode and Espelie, 1995; Ni *et al.* 1998; Powell *et al.* 1999).

## WAX AS A PHYSICAL BARRIER

As a physical barrier to insect herbivores, surface waxes may impede attachment and locomotion, depending on the type and amount of wax the plant possesses. Amounts of wax in both natural and agricultural plants vary from absent to highly abundant (White, 1998). Eigenbrode and Espelie (1995) classified plant surface wax as glossy, bloomless and waxbloom (waxy), based on the availability and crystal morphology of the wax. The microcrystals may be in amorphous, plates, needles, granules, ribbons, spiky plates, or tubes forms (Jeffree, 1986). Plants with microscopic, dense and easily abraded crystals are termed as waxbloom or waxy (Stork, 1980). Glossy plants have reduced complexity of epicuticular lipid microstructure relative to waxy phenotypes with altered chemical composition (Eigenbrode and Espelie, 1995). In this chapter, we use the terms glossy and waxy in a relative and general sense to discuss the role of plant surface wax on insect herbivore behavior.

Waxy genotypes may harbor fewer eggs than the glossy genotypes. This has been demonstrated with waxy *Brassica oleracea* leaves, where the wax crystals were removed either mechanically or by detergents, and offered to the diamondback moth (DBM), *Plutella xylostella* and *Delia radicum* females. More eggs were deposited on leaves where the wax was removed as compared to leaves with natural wax crystals (Prokopy *et al.* 1983). However, survival of *P. xylostella* larvae on glossy *B. oleracea* leaves in the field was approximately 90% lower than waxy genotypes (Eigenbrode *et al.* 1990). This was mainly because of the larval feeding behavior. After hatching, larvae disperse over the entire plant in glossy genotypes and delay the initiation of feeding, as compared to larvae on waxy genotypes. Larvae on waxy genotypes begin feeding on the same leaf on which the egg was laid i.e. without dispersing (Eigenbrode and Shelton, 1990). The larval walking speed was also significantly higher on glossy cabbage than on waxy leaves. The highly mobile neonate larvae of *P. xylostella* on glossy genotypes suffer greater mortality due to desiccation and predation during dispersal, relative to larvae on waxy genotypes. Thus, infestation by DBM in glossy *B. oleracea* is one-tenth the size of those on waxy genotypes (Stoner, 1990). Secondary plant chemicals are also concentrated at higher levels in glossy leaves as compared to waxy leaves, contributing to the lower pest levels (Cole and Riggal, 1992) in addition to leaf wax chemistry and morphology (Eigenbrode *et al.* 1991).

Waxy genotypes may similarly provide resistance to insects by influencing adult (egg laying) and larval (feeding) behavior. For example, flea beetles, *Phylotreta cruciferae* on *B. oleracea* feed on the entire glossy leaf as compared to waxy leaves where only the leaf edges are consumed. This is because of their inability to adhere properly on the waxy surface (Bodnaryk, 1992). Hence, the damage caused by *Phyllotreta* spp. is two to five times lower in waxy *B. oleracea* as compared to glossy genotypes (Stoner, 1990).

## WAX AS A CHEMICAL BARRIER

The chemical composition of surface waxes may also influence insect herbivore behavior by stimulating or deterring landing, feeding and/or oviposition. Wax chemistry affects host acceptance or rejection, through stimuli perceived by plant feeding insects using highly adapted sensory structures (Stadler, 1986). The importance of the chemical composition, as opposed to the physical structure of the surface was demonstrated when *P. xylostella* larvae were stimulated to feed on glass coated with an organic extract of glossy *B. oleracea* (Eigenbrode and Shelton, 1990). The crystalline microstructure of the wax layer was destroyed when removed from the leaf surface by the solvent. The larval preference for glossy genotypes is therefore, due to differences in epicuticular lipid chemical composition. Jeffree (1986) and Kolattukudy and Espelie (1985) listed some of the major classes of plant epicuticular lipids: aldehydes, n-alkanes, β-diketones, fatty acids, fatty alcohols, ketones, secondary alcohols, triterpenoids and wax esters. These structures are also related to the lipid crystal morphology. For example, plant surfaces with >80% of n-alkanes will have an amorphous appearance (Eigenbrode and Espelie, 1995). High levels of docosanol in tobaccos, triacontanol in alfalfas, α and β-amyrin in azalea varieties, triterpenols in sorghums confer resistance to tobacco budworm, *Heliothis virescens* (Johnson and Severson, 1984), spotted alfalfa aphids, *Therioaphis maculata* (Bergman *et al.* 1991), *Stephanitis pyrioides* (Balsdon *et al.* 1994) and two species of aphids (Heupel, 1985), respectively.

Chemical compounds in the epicuticular wax layer are the first chemosensory stimuli that an insect herbivore perceives, when landing or initiating feeding on a plant. The role of these compounds in the host acceptance/rejection behavior in several insect orders including Orthoptera, Diptera, Lepidoptera, and Homoptera have been well studied (Woodhead, 1983; Foster and Harris, 1992; Eigenbrode *et al.* 1991; Powell *et al.* 1999). For example, settling by pea aphids, *Acyrthosiphon pisum* was enhanced on non-host plants treated with host-plant wax (Klingauf *et al.* 1978). Video recording demonstrated the rapid insertion of stylets by black bean aphid, *Aphis fabae* on the host plant (beans, *Vicia faba*) as compared to reluctant probing and insertion on the host plant (oats, *Avena sativa*). However, a significant difference in stylet insertion was recorded when these insects were exposed to oats that were stripped of epicuticular waxes using cellulose acetate. Gas chromatography-mass spectrometry studies and related behavioral bioassays indicate that 1-hexacosanol is a major behavior-modifying chemical that prevents aphid settling on oats (Powell *et al.* 1999).

The extracts of surface waxes from host plants may also have either feeding deterrent or stimulant properties in insects. The nymphs of migratory grasshopper, *Locusta migratoria* discriminate between paper impregnated with surface lipid extracts of the host plant, *Poa annua* from that of paper impregnated only with the organic solvent (Bernays *et al.* 1976). In this case, surface wax extract attracted the nymphs for palpation and feeding. In contrast, studies with solvent and water extracts of cultivated and wild pigeonpea pods (*Cajanus cajan, C. platycarpus* and *C. scarabaeoides*) using larvae of pod borer, *Helicoverpa armigera* demonstrated antifeedant properties from surface chemicals extracted with water (Shanower *et al.* 1997). Larval feeding was significantly reduced when presented filter paper impregnated with water extract of *C. scarabaeoides* pods. Acetone extracts from *C. cajan* and *C. platycarpus*, in contrast, significantly stimulated feeding (Table 1).

Plant surface chemicals thus, undoubtedly affect insect behavior, feeding and establishment. Chapman and Bernays (1989) suggested that the nature of the surface wax may be "recognized" by insect as indicative of internal constituents of a plant. Further studies are needed to understand the effect of plant-wax chemistry on insect behavior.

**Table 1.** Antifeedant index of pod surface extracts from three *Cajanus* spp.

| Solvent | *Cajanus* spp. | Antifeedant index, mean±SEM | T[a] |
|---|---|---|---|
| n-Hexane | *C. platycarpus* | 8.84±10.83 | 23 |
| | *C. scarabaeoides* | −15.81±8.57 | 12 |
| | *C. cajan* | −10.37±9.68 | 23 |
| Acetone | *C. platycarpus* | −41.55±14.54 | 7* |
| | *C. scarabaeoides* | 0.09±18.11 | 23 |
| | *C. cajan* | −40.49±16.29 | 7* |
| Water | *C. platycarpus* | −3.89±7.28 | 24 |
| | *C. scarabaeoides* | 27.76±12.39 | 8** |
| | *C. cajan* | −14.24±9.54 | 15 |

* Significant feeding stimulation and **significant feeding inhibition (Wilconxon signed rank test; P≤0.05).
[a] Rank total of one sign, + (antifeedant) or − (stimulant), whichever is less.
(Source: Shanower *et al.* 1997)

## DISADVANTAGES

Surface waxes may positively influence the populations of phytophagous insects if the wax impairs the mobility and effectiveness of predators and parasites. Waxy genotypes can impair the locomotion and effectiveness of the predatory insect, *Chrysoperla plorabunda* (Eigenbrode *et al.* 1999). Of nine cabbage accessions tested, all glossy types provided an advantage to *C. plorabunda* over waxy genotypes by increasing the force of attachment to the leaf surface by the larvae, which in turn increased time allocated to walking, leading to greater predation of *P. xylostella* larvae. Adhesive force produced by *C. plorabunda* on glossy cabbage is 20 to 200 folds greater than on waxy genotypes. The efficiency of predators increased several-fold on glossy genotypes indicating the possibility of reducing wax levels or using glossy genotypes for the efficient management of cabbage pests (Eigenbrode *et al.* 1999).

Cabbage supports more coccinellid adults and eggs on glossy foliage than on waxy foliage. The tarsal setae of the beetles are able to adhere better on glossy leaves than on waxy leaves, thus, increasing their predatory efficiency. In addition, the proportion of time spent walking by predators is higher on glossy leaf *B. oleracea* than on waxy varieties (Eigenbrode and Espelie, 1995; Eigenbrode *et al.* 1996). This would lead to obvious increases of the herbivorous pests *P. xylostella* and aphids on waxy cabbage.

Incorporation of plant surface lipids into the integument of phytophagous insects may also indirectly influence the performance of natural enemies. Several phytophagous insects are chemically cryptic to their predators and parasites when they incorporate the plant surface lipids into their body surface. This phenomenon was demonstrated in the migratory grasshopper, *Melanoplus sanguinipes* when fed leaves with radioactively labeled n-alkanes (Blomquist and Jackson, 1973). These lipids serve as ovipositional cues and host seeking stimulant/deterrent for selective predators and parasites (Espelie *et al.* 1991).

## PLANT TRICHOMES: AN OVERVIEW

Trichomes, or plant hair, are a common anatomical feature on vegetative and reproductive structures of many plant species. They play an important role in plant defense against herbivore damage (Levin, 1973; Jeffree, 1986; Peter *et al.* 1995). Trichomes have other important plant physiology functions including: 1) altering optical properties of the leaf surface, 2) deflecting intense solar radiation, thus reducing leaf temperature, 3) maintaining water balance, 4) acquiring nutrients and water from the atmosphere, and 5) excreting excess salts (Baldocchi *et al.* 1983; Raven *et al.* 1992; Bosabalidis and Skoula, 1998; Rensburg *et al.* 1999). Trichomes are epidermal, uni- or multi-cellular structures. They are the first plant structure that an insect physically contacts during the initial stages of host acceptance. The structure, form and function of trichomes are highly variable within and among species. This chapter discusses the role of plant trichomes as an insect resistant character and their usefulness in genetic enhancement programs.

## TYPES OF TRICHOMES

There are two general types of trichomes, glandular and non-glandular, though different classifications have been used for different plant species based on their structure, function and size (Jeffre, 1986). For example, pigeonpea *(Cajanus cajan)* has both glandular and non-glandular trichomes, which have been typically classified as Types A-E based on their size and shape (Romeis *et al.* 1999a) Type A has a long tubular neck, from which a clear viscous fluid is secreted. The base is enlarged and appears to consist of 6-10 cells. Type B is a yellowish, unsegmented globular sac. Its contents are released only after the cell wall is ruptured (Plate 1). Type C and D are nonglandular trichomes that do not secrete or contain chemicals. Type D is longer than Type C, and the longest of all pigeonpea trichomes. Type E is the smallest pigeonpea trichome, and is a multicellular, stalked glandular trichome (Romeis *et al.* 1999a). There is much variation and overlap in description of trichome types in plants (Jeffree, 1986). Developing a generalized classification system based on trichome morphology has not been possible because of their highly polymorphic nature (van Dam *et al.* 1999).

## TRICHOMES AS AN INSECT DEFENSE MECHANISM

Trichomes have been reported to confer resistance to 35 plant species (Peter *et al.* 1995). As a resistance mechanism, they minimize herbivore load by providing a physical and/or chemical barrier: 1) limiting contact with the plant surface, predominantly for small, soft-bodied insects, 2) producing toxic compounds, which poison the insect through contact, ingestion and/or inhalation, and 3) producing gummy, sticky or polymerizing chemical exudates, which impede

**Plate 1.** Micrographs of *Cajanus* spp. pod surface
a) *C. cajan* pod surface showing Type-A, C & D trichome

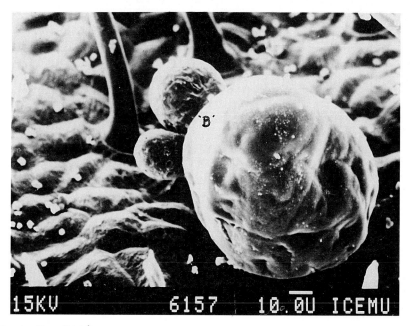

b) *C. cajan* Type-B trichome

c) *Cajanus scarabaeoides* pod—horizontal view showing the high trichome density

d) *C.cajan* pod showing glandular exudate from Type-A trichome

insect locomotion. The length, density and orientation of the trichomes influences their effectiveness as a physical barrier; the size, mode of locomotion and type of mouthparts of the specific insect are also important (Southwood, 1986). Trichomes can also interfere with insect ingestion (Shanower *et al.* 1996) and digestion (Wellso, 1973).

## TRICHOMES AS A PHYSICAL BARRIER

The impact of trichome density, length and orientation on insect behavior and performance has been well documented (Levin, 1973; Southwood, 1986; Peter et al. 1995). Trichome density is, perhaps, more important than length or orientation but all three characteristics have been considered in host plant resistance breeding programs. Wild relatives are often good sources of pubescence and have been used in several breeding programs. The use of trichomes as a physical barrier has been exploited in insect resistant cultivars for a number of crops including pigeonpea, soybean, cotton, sorghum and wheat as discussed below. Other crops in which trichomes have been utilized as an insect resistant factor are listed in Table 2.

**Table 2.** List of plants having trichomes as arthropod herbivore defense mechanisms

| Plant species | Arthropods affected | Order | Reference |
|---|---|---|---|
| Abelmoschus esculentus | Empoasca (Amrasca) devastans | Homoptera | Uthamasamy, 1985 |
| Aeschynomene brasiliana | Boophilus microplus | Acarina | Sutherst and Wilson, 1986 |
| Anaphalis margaritacea | Philaenus spumarius | Homoptera | Hoffman and McEvoy, 1985 |
| Arachis hypogea | Empoasca kerri | Homoptera | Bilderback and Mattson, 1977 |
| Avena spp. | Oscinella frit Oulema melanoplus | Diptera Coleoptera | Peregrine and Catling, 1967 Smith and Webster, 1974 |
| Brassicae oleracea | Brevicoryne brassicae Pieris rapae | Homoptera Lepidoptera | Stoner, 1992 Stoner, 1992 |
| Castanea spp. | Curculio elephas | Coleoptera | Popova, 1960 |
| Cucumis sativus | Diaphania sp. Bemisia tabaci | Lepidoptera Homoptera | Elsey and Wann, 1982 El Khidir, 1965 |
| Datura wrightii | Manduca sexta | Lepidoptera | van Dam and Hare, 1998 |
| Euphorbia (Poinsettia) | Bemisia spp. | Homoptera | Bilderback and Mattson, 1977 |
| Fragaria chiloensis | Otiorhynchus sulcatus | Coleoptera | Doss and Shanks, 1988 |
| Helianthus spp. | Homoeosoma electellum | Lepidoptera | Rogers et al. 1987 |
| Lolium spp. | Listronotus bonariensis | Coleoptera | Barker, 1989 |
| Lupinus spp. | Acyrthosiphon pisum | Homoptera | Wogerek and Dunajska, 1964 |
| Nicotiana tobaccum | Mysus persicae Manduca sexta Rhopalosiphum padi Acyrthosiphon pisum | Homoptera Lepidoptera Homoptera Homoptera | Abernathy and Thurston, 1969 Thurston, 1970 Thurston and Webster, 1962 Thurston and Webster, 1962 |
| Oryza sativa | Chilo suppressalis C. agamemnon | Lepidoptera Lepidoptera | Patanakamjorn and Pathak, 1967 Abd-El-Rahman and Salsh, 1993 |

contd.

**Table 2.** contd.

| Plant species | Arthropods affected | Order | Reference |
|---|---|---|---|
| | *Nephotettix impicticeps* | Homoptera | Petinez, 1986 |
| | *N. apicalis* | Homoptera | Petinez, 1986 |
| | *Nilaparvata lugens* | Homoptera | Petinez, 1986 |
| | *Cnaphalocrocis medinalis* | Lepidoptera | Dakshayani *et al.* 1993 |
| *Passiflora adenopoda* | *Heliconius erata* | Lepidoptera | Gilbert, 1971 |
| | *H. melpomene* | Lepidoptera | Gilbert, 1971 |
| *Pelargonium x hortorum* | *Acyrthosiphon solani* | Homoptera | Walters *et al.* 1991 |
| *P. hortorum* | *Tetranychus spp.* | Acarina | Gerhold *et al.* 1984; Grazzini *et al.* 1999 |
| *Phaseolus vulgaris* | *Empoasca fabae* | Homoptera | Wolfenbarger and Sleeman, 1963 |
| | *Liriomyza trifolli* | Diptera | Quiring *et al.* 1992 |
| | *Acyrthosiphon pisum* | Homoptera | Lampe, 1982 |
| *Phaseolus limensis* | *Empoasca kraemeri* | Homoptera | Lyman and Cardona, 1982 |
| *Saccharum officinarum* | *Aleurolobus barodensis* | Homoptera | Agarwal, 1969 |
| | *Diatraea saccharalis* | Lepidoptera | Sosa, 1990 |
| | *Melanaspis glomerata* | Heteroptera | Agarwal, 1969 |
| | *Scirpophaga nivella* | Lepidoptera | Verma and Mathur, 1950 |
| *Salix borealis* | *Melasoma lapponica* | Coleoptera | Zvereva *et al.* 1998 |
| *Solanum melongena* | *Amrasca biguttula* | Homoptera | Schreiner, 1990 |
| | *Trialeurodes vaporariorum* | Homoptera | Malausa *et al.* 1988 |
| *S. mammosum* | *Aphis gossypi* | Homoptera | Sambandam *et al.* 1969 |
| | *Epilachna vigintioctopunctata* | Coleoptera | Sambandam *et al.* 1969 |
| *Stylosanthes* spp. | *Boophilus microplus* | Acarina | Sutherst *et al.* 1982 |
| *Vigna radiata* | *Porthesia taiwana* | Diptera | Talekar *et al.* 1988 |
| | *Helicoverpa armigera* | Lepidoptera | Talekar *et al.* 1988 |
| *V. unguiculata* | *Ophiomya phaseoli* | Diptera | Uengprasertporn, 1985 |
| | *Maruca testulalis* | Lepidoptera | Oghiakhe *et al.* 1992 |
| *V. vexillata* | *Clavigralla tomentosicollis* | Homoptera | Chiang and Singh, 1988 |
| | *Maruca testulalis* | Lepidoptera | Jackai and Oghiakhe, 1989 |
| *Zea mays* | *Diabrotica virgifera* | Coleoptera | Hagan and Anderson, 1967 |
| | *Chilo zonellus (partellus)* | Lepidoptera | Kumar, 1992 |

A wild relative of pigeonpea, *Cajanus scarabaeoides* possesses twice the density of small, non-glandular trichomes (Type C) on pods as compared to pods of cultivated pigeonpea, *C. cajan* (Table 3). Fewer than 15% of neonate pod borers (*Helicoverpa armigera*) reach the feeding site on *C. scarabaeoides* (Peter and Shanower, unpublished). The high density of Types C and D non-glandular trichomes prevent neonate larvae from reaching the pod surface and they starve or desiccate before feeding. Larval survival was 22% on the pods of

*C. scarabaeoides* as compared to 89% on *C. cajan* pods (Shanower *et al.* 1997). Third instars avoid walking on pods and instead spend more time resting on and off the pod. Microscope observations confirmed that the trichomes in *C. scarabaeoides* irritate the larvae and prevent them from settling  on the pod surface. It has therefore been suggested that increasing the density of non-glandular trichomes on cultivated *C. cajan* pods could reduce damage and losses due to *H. armigera* (Shanower *et al.* 1997).

**Table 3.** Mean density (±SE) of trichome Types A-D on pods of three *Cajanus* spp. (Field sample, Winter season, 1994)

| *Cajanus* species | Type A | Type B | Type C | Type D |
|---|---|---|---|---|
| | | Number of trichomes/mm$^2$ | | |
| *C. cajan* | 2.6a | 0.09b | 71.4b | 2.4b |
| *C. platycarpus* | 2.5a | 0.12b | 0.5c | 0.7c |
| *C. scarabaeoides* | 0 | 7.23a | 155.6a | 5.4a |

Means within a column followed by the same letter are not significantly different at P=0.05; (Source: Romeis *et al.* 1999a).

Among the three *Cajanus* spp. studied, *C. scarabaeoides is* resistant to an array of insect pests (Table 4). Populations of *H. armigera*, the major pest of pigeonpea, were lowest on *C. scarabaeoides* followed by *C. platycarpus*. The pod wasp, *Tanaostigmoides cajaninae* was the only pest that infested pods of *C. scarabaeoides*. Female *T. cajaninae* insert eggs inside the pods and the larvae never come in contact with the pod trichomes. All of the immature stages of the pod fly occur inside the pod and trichomes are therefore not an effective resistance mechanism against this pest.

Trichomes have been used to develop soybean (*Glycine max*) accessions with resistance to several insect pests. Soybean leaves possess simple non-glandular trichomes and are classified as glabrous, normal or densely pubescent based on trichome density (Lambert and Kilen, 1989). Feeding and survival of *Trichoplusia*

**Table 4.** Major pests infesting pods of *Cajanus* spp. in relation to population abundance (Rainy season, 1995).

| Insect pests | *C. cajan* | *C. scarabaeoides* | *C. platycarpus* |
|---|---|---|---|
| *Helicoverpa armigera* | ++++ | + | ++ |
| *Adisura atkinsoni* | – | – | +++++ |
| *Lampedes boeticus* | ++++ | – | +++ |
| *Tanaostigmodes cajaninae* | + | +++++ | + |
| *Melanagromyza obtusa* | ++ | – | ++++ |
| *Exelastis atomosa* | ++ | – | ++++ |
| *Clavigralla spp.* | +++ | – | ++++ |

Percentage population: + 1–19%; ++ = 20-39%; +++ = 40=59%; ++++=60-79%; +++++=80–100%; (Source: Peter and Shanower, 1996).

*ni* and false melon beetle, *Atrychya menetriesi* on shaved and unshaved leaves demonstrated that leaf hairiness is correlated with resistance to these pests (Khan *et al.* 1986). Damage levels were similar between shaved leaves of trichome bearing accessions and unshaved leaves of trichomeless accessions. Plant hair are an important resistance mechanism against larvae of corn earworm, *Heliothis zea*, soybean looper, *Pseudoplusia includens* and velvetbean caterpillar, *Anticarsia gemmatalis*, though trichomes enhance oviposition by these Lepidopterans (Lambert *et al.* 1992). Gannon and Bach (1996) and Johnson (1975) reported higher larval mortality in Mexican bean beetle, *Epilachna verivestis*, on leaves with higher trichome density. Trichome orientation (Broersma *et al.* 1972) and length (Johnson, 1975; Turnipseed, 1977) are important resistant factors in soybean for potato leaf hopper, *Empoasca fabae* and springtail, *Deuterosmiathurus yumanensis*. The importance of this resistance mechanism also depends on insect body size. The population of the smaller-bodied *E. fabae* (body length=1.0-4.0 mm) decreased with increasing trichome length, regardless of trichome density, whereas *D. yumanensis* (body length = 0.2-0.4 mm) populations decreased with increasing trichome density (Turnipseed, 1977). The irregular shape produced by the highly pubescent accessions prevents normal attachment for feeding and oviposition of *E. fabae* (Lee *et al.* 1986) making the hirsute varieties more resistant.

Cotton (*Gossypium* spp.) is categorized as glabrous (smooth leaf, no hair) hirsute (medium length, normal hair density) or pilose (short, dense hairs) (Meyer, 1957). Among these, pilose genotypes confer resistance against several insect species (Wilson and George, 1986). Development of the cotton bollweevil (*Anthonomus grandis*) was slower on hairy buds compared to buds with less or no hair (Stephens and Lee, 1961). Reduced movement of 1st instar larvae of pink bollworm (*Pectinophora gossypiella*) has been observed on highly pilose versus glabrous leaves (Smith *et al.* 1975). Larvae pause regularly to swing their heads and sample the substrate while moving. This resulted in higher larval mortality due to starvation or desiccation. Ramalho *et al.* (1984) found that the larvae of *Heliothis virescense* were more likely to die from biotic (predators and parasites) and abiotic (high temperature and insecticides) factors because of slower movement on pubescent genotypes. Increased trichome density on lower leaf surface has been shown to provide greater resistance to cotton leaf worm, *Spodoptera littoralis* (Kamel, 1965). Trichomes also affect both leafhopper and whitefly feeding preferences on cotton. Resistance to sap sucking *Empoasca* spp. depends not only on hair density (Parnell *et al.* 1949), but also on hair length (Ambekar and Kalbhor, 1981). Leafhopper eggs are laid along the adaxial mid-rib. Since the Ovipositor length is shorter than the mid-vein hair length (Khan and Agarwal, 1984), egg laying is prevented or reduced. Similarly, leafhopper populations declined while whitefly populations increased with increase in the number of leaf trichomes. However, when trichome densities reached 70/13.7 mm² of leaf surface, whitefly and leafhopper populations declined (Butler *et al.* 1991). The incidence of cotton leaf curl bigeminivirus (CLCuV) was higher in S-12 accession as compared to NIAB-Krishma and

CIM-448 accessions. The density of leaf hair is low in S-12 that reduces the disease incidence caused by the vector, *Bemisia tabaci* (Ashraf *et al.* 1999).

The sorghum shootfly (*Antherigona soccata*) is an important insect pest infesting sorghum, *Sorghum bicolor*. Eggs are laid on the lower side of the leaves. After hatching, the larvae move down the leaf blade into the culm between the outer sheaths, before finally boring into the growing points (Chapman and Woodhead, 1985). Accessions with more trichomes hinder the movement of the larvae towards the target site, making them less susceptible (Gibson and Maiti, 1983; Raina, 1985). Similarly, wild sorghum species with high trichome density on the lower leaf surface are resistant to shootfly. These hair prevent larvae from penetrating and are considered a mechanical resistance mechanism developed by this species (Bapat and Mote, 1982; Blum, 1967).

The cereal leaf beetle, *Oulema melanopus* lays fewer eggs on wheat genotypes (*Triticum* spp.) with highly pubescent leaves compared to genotypes with glabrous leaves (Gallun *et al.* 1973). Both egg laying and larval survival were lower on cultivars with relatively longer and/or denser trichomes, though trichome length was of greater importance than trichome density. Papp *et al.* (1992) reported trichome density in seedlings and trichome length in flag leaves as a better indicator of resistance to *O. melanopus*. The Hessian fly, *Mayetiola destructor* and bird cherry oat aphid, *Rhopalosiphum padi* produce fewer eggs and have lower survivorship on pubescent compared to glabrous cultivars (Roberts and Foster, 1983).

## TRICHOMES AS A CHEMICAL BARRIER

Glandular trichomes confer resistance to several phytophagous insects and, because of the toxic and deterrent properties of their exudates, are often more effective than non-glandular trichomes in providing resistance to insect pests. Glandular trichomes secrete and/or accumulate a variety of terpenes, sucrose, glucose esters of fatty acids, nitrogenous compounds, phenolic compounds, fatty acids, proteins and polysaccharides (Kelsey *et al.* 1984). These generally act either as insect repellents or immobilize insects by entrapment. In addition to protecting plants by deterring herbivores, glandular trichomes may also attract pollinators, and these structures contribute to the flavor and aroma of many plants (McCaskill and Croteau, 1998). Fourteen volatile compounds have been identified in leaf trichomes of yellow squash. These compounds act as repellents for female pickleworm moths, *Diaphania nitidalis* (Peterson *et al.* 1994). The garden geranium, *Pelargonium* × *hortorum* resists mite colonization because of chemicals present in the glandular trichome exudates. Up to 90% of the trichome exudate from mite-resistant, *P. hortorum* inbreds contains anacardic acid with an unsaturated omega-5 alkyl chain. These omega 5-fatty acids (16:1 DELTA11 and 18:1 DELTA13) are present only in the mite resistant geraniums (Grazzini *et al.* 1999).

The commercial potato, *Solanum tuberosum* and its wild relatives, *Solanum* spp., possess two types of glandular trichomes. Type A is short and tetralobulate, and release the membrane bound chemicals (phenols) only after rupturing.

Type B is longer and has an ovoid gland at the tip, which continuously exudes a viscous fluid (Tingey and Laubengayer, 1981) that contains esters of carboxylic acids (Neal *et al.* 1989). After landing, an insect first encounters Type B exudates. In struggling to escape from the sticky coating, the insect disturbs Type A trichomes. Type A trichomes contain the polyphenol oxidase (PPO) enzymes that oxidize the phenols, hardening the sticky exudate. Thus, small bodied insects such as aphids and leafhoppers are trapped (Gibson and Turner, 1977) and starve to death (Gregory *et al.* 1986). The sucrose esters of carboxylic acids in Type B exudates also enhance the mortality rate and inhibit settling and probing behavior of *M. persicae* (Neal *et al.* 1990). These trichomes provide greater resistance to wild *Solanum* spp. than in the cultivated *S. tuberosum*. In general, wild species are more pubescent and have more glandular trichomes, and are hence, more resistant to insect pests than the cultivated relatives (Neal *et al.* 1990). Further, trichomes in *S. tuberosum* are resistant to detachment from their stalk and hence the exudate may not be released (Ave *et al.* 1987). The exudates of both trichomes also contain various sesquiterpenes (Ave and Tingey, 1986), including E-β-farnesene and β-caryophyllene of which the latter is known to act as an aphid alarm pheromone (Ave *et al.* 1987; Tingey, 1991). Ave *et al.* (1987) reported that other sesquiterpenes also alter aphid behavior.

Domesticated and wild tomato, *Lycopersicon* spp. possesses two types of glandular trichomes: Type IV, a multicellular stalk with a 2-4 celled glandular head and a monocellular base and Type VI, a very short unicellular stalk with a 4-8 celled glandular head. These trichomes confer resistance to a variety of arthropods. Wild relatives of *Lycopersicon* spp. possess a trichome-based defense, similar to that described for potato. When insects make contact with these glandular exudates, they are immobilized due to the browning and hardening of the exudates. Exudates of Type IV trichomes are responsible for such trappings (Isman and Duffey, 1982a,b). These exudates have both catecholic phenols (Ave and Tingey, 1986) and a viscous mixture of acylsugars (Goffreda *et al.* 1989). The presence of catecholic phenols and sucrose esters of fatty acids in the exudates also protect tomato, to some extent, from *H. zea* (Duffey, 1986) and *M. euphorbiae* (Goffreda *et al.* 1989). Liedl *et al.* (1995) have isolated acylsugars from trichomes of wild tomato, *Lycopersicon pennelli* that cause reduction in the settling of the adult silverleaf whiteflies, *Bemisia argentifolii*. Although, *L. esculentum* possesses Type VI trichomes, they secrete very low level of insecticidal ketones. Type VI trichomes also produce monoterpenes and sesquiterpenes including zingiberene which has been correlated with resistance of the Colorado potato beetle, *Leptinotarsa decemlineata* (Carter *et al.* 1988). Even the cultivated *L. hirsutum* showed a high resistance level to Eryophide mites, *Aculops lycopersica* and the moth, *Tuta absoluta* due to the high density of Type VI trichomes and consequently higher levels of tridecon-2-one in its leaves (Leite *et al.* 1999a,b).

The glandular trichomes of chickpea, *Cicer arietinum* secrete highly acidic (pH=1) exudates, containing primarily malic and oxalic acids (Rembold and Weigner, 1990). The quantity of malic acid in the exudate reportedly confers resistance to the leafminer, *Liriomyza cicerina* and the pod borer, *H. armigera*. However, Yoshida *et al.* (1995) observed an antibiotic effect of oxalic acid, but not malic acid, on *H. armigera* larvae.

## DISADVANTAGES

Plant trichomes, as an insect resistance factor, can also have an adverse effect on crop production because of negative impacts on natural enemies. Small, soft bodied entomophagous insects are generally more affected by these factors than larger ones (Obrycki, 1986). Natural enemy efficacy can be impaired by: 1) increasing search time, 2) decreasing residence time in the host/prey habitat, 3) chemical and/or physical entrapment, and 4) chemical repellents (David and Easwaramoorthy, 1988; Peter *et al.* 1995; Romeis *et al.* 1999b; Eigenbrode *et al.* 1999). These effects have been documented in predator/prey and parasitoid/host systems in cotton, chickpea, pigeonpea, potato and tomato.

The parasitization efficiency of the egg parasitoid, *Trichogramma pretiosum* and the predatory efficiency of the egg predator, *Chrysoperla* spp. are greatly influenced by trichomes in cotton. The lower efficacy is due to slower walking and searching speed, and adhesion or interference from exudates. Higher parasitism by *Trichogramma* spp. and predation by *Chrysoperla* spp. was recorded on smooth than on pilose cotton (Treacy *et al.* 1986).

The potato moth, *Phthorimaea operculella koehleri* lays more eggs on pubescent accessions and exploits pubescence to protect the eggs from parasitoids (Gurr, 1993). The egg parasitoid, *Copidosoma koehleri* is impaired and entrapped by the glandular trichomes of *L. esculentum* (Baggen and Gurr, 1995). Ten percent of host eggs were parasitized on the less hairy *Solanum chacoense* versus 0% on hairy *S. tuberosum* spp. *tuberosum* cv. *Bintje* and *L. esculentum*. The negative impact of trichomes on biological control is attributed to slower walking speed and reduced oviposition by the egg parasitoid, *Copidosoma koehleri* (Gooderham *et al.* 1998).

On pigeonpea, eggs of *H. armigera* are readily parasitized (>55%) by *Trichogramma chilonis* when placed on leaves but are almost unaffected (<1%) when placed on pigeonpea pods (Romeis *et al.* 1998). Parasitoids walked significantly faster on leaves than on pods, where their movement was inhibited by long trichomes. The higher density of glandular trichomes on pods compared to leaves resulted in the parasitoids being trapped frequently by glandular exudates (Romeis *et al.* 1998). Though, *H. armigera* is an important insect pest in chickpea, *Cicer arietinum*, its parasitoid, *T. chilonis* seldom parasitizes eggs on this host plant (Sithanantham *et al.* 1982). No parasitoid population was recorded in the field even after five releases of the parasitoid at weekly intervals (Romeis *et al.* 1999b). The highly acidic trichome exudates secreted by vegetative and reproductive parts of chickpea deterred *T. chilonis* and entrapped a small proportion (6.8%) of the parasitoids in the field (Romeis *et al.* 1999b). Other examples of *Trichogramma* spp. being trapped by sticky trichome exudates is reported from tomato and potato (Kauffman and Kennedy, 1989; Kashyap *et al.* 1991; Farrar *et al.* 1994).

The walking speed of larvae of several coccinellid predators are inversely related to trichome density (Obrycki, 1986). The predation rate of adult predatory mite, *Phytoseiulus persimilis* is also affected by trichome density, particularly when prey density is low. Predation rate is inversely proportional at prey densities of 1.3-2.5 eggs/cm² and relatively unaffected at 8 eggs/cm² (Krips *et al.* 1999). The small hooked trichomes in *Mentzelia pumila* leaves entrap and kill the predatory coccinellid beetle, *Hippodamia convergens* (Eisner *et al.* 1998). Similarly, the efficacy of two predators, the coccinellid, *Coleomegilla maculata* and the big-eyed bug, *Geocoris punctipes* is directly related to the abundance of methyl ketones detected in the glandular exudate in tomato (Barbour *et al.* 1993).

The negative effects of trichomes are not restricted to insect natural enemies. Phytophagous insects may also positively influenced by plant trichomes. The increased oviposition in *Lygus hesperus, Helicoverpa* spp., *Earias fabia* and *E. vitella* on hirsute over glabrous cotton cultivars are examples of this type of negative impact of trichomes (Mehta and Saxena, 1970; Lukefahr *et al.* 1975). A similar situation has also been reported in soybean for *Ophiomyia phaseoli* (Chiang and Norris, 1983). Therefore, it is essential that the full benefits and costs of plant surface wax and trichomes be known before determining if this approach will provide an overall benefit in terms of host plant resistance.

## BENEFIT AND COST RATIO

As insect herbivores and plants co-evolved, plants channelized resources, such as nitrogen, carbon and energy to the production and storage of secondary plant chemicals and structural changes to withstand herbivore feeding. The biochemical conversion of plant primary nutrients that are intended for growth and reproduction into secondary plant defense chemicals is a cost that must be recovered through enhanced survival and/or production (Simms, 1992). These structural and chemical changes make a plant either attractive or repulsive to insects. The trade-off between plant defenses and insect herbivory has not been fully explored. Several studies have focused on glandular trichomes as defense mechanisms. For example, the cost for producing glandular trichomes in potato is reduced fitness or lower productivity. The *Solanaceous* spp. with Type IV trichomes, produced smaller tubers, leaves and stolons than plants that did not producing Type IV trichomes (Kalazich and Plaisted, 1991). This may be due to competition for resources by a plant towards reproduction and trichome defense (Herms and Mattson, 1992), or due to genes that decrease plant fitness (Bergelson and Purrington, 1996). Initial attempts to measure the costs and benefits of glandular trichomes have not yet documented a net benefit for this resistance character (Elle *et al.* 1999).

Mauricio and Rausher (1997) used *Arabidopsis thaliana* to understand selection pressure strategies in plants. They planted 1728 *A. thaliana* in North Carolina and exposed half to natural enemies at natural densities and sprayed the other half regularly with pesticides to remove or reduce phytophagous insects and fungi. They measured the trichome density and glucosinolates concentration as indicators of selection, the total number of leaf damage holes and their diameter as indicators of herbivory, and the total number of fruit as an indicator of fitness. Plants sprayed with pesticides had lower trichome densities and glucosinolate concentrations than unsprayed plants. Natural enemies, however, modified the pattern of stabilizing selection on these two characters indicating the benefits of resistance and the costs in growth were resolvable.

*Datura wrightii* exists in two trichome phenotypes: one with >95% glandular trichomes (sticky genotype) and another with about 5% glandular trichomes (velvety genotype). These two phenotypes are under the control of a single gene, with the sticky allele dominant (van Dam *et al.* 1999). The overall herbivore fitness is reduced on sticky, as compared to velvety genotypes. Higher tolerance was recorded against herbivores when the sticky plant was watered. However, in the absence of herbivores, sticky plants produced 45% fewer viable seeds than the velvety genotypes. The production of viable seeds and the germination capabilities are also reduced in sticky genotypes. This indicates that the herbivore resistant characters can be costly (Ell *et al.* 1999). Moreover, glandular trichomes in several plants appear only in the juvenile stage and disappear in mature plants (Gibson, 1971; van Dam *et al.* 1999). These studies raise questions about how this adaptation is maintained in wild population. If truly undesirable and costly, natural selection should work against the maintenance of these characters. In contrast, many plants possess glandular trichomes on both vegetative and reproductive parts, and these confer resistance to one or more insect pests. Further investigation into the benefits and costs of these characters are needed to understand their role on plant growth, reproduction and defense.

## CONCLUSIONS

Plant surface morphology and chemistry play an important role in plant defense against an array of insect herbivores. The general message is that plant surface resistance mechanisms are specific, variable among species and differ for different pest species. They may reduce herbivore fitness, but in other well-documented cases, may increase fitness also. The physical and chemical changes in plants occurred either independently or were induced through pressure exerted by insect herbivores. This genetic plasticity has lead to the use of these structural and chemical characters as important host plant resistance factors. Their further use, in other crop plants and plant-insect systems, may provide additional benefits to human consumers.

## REFERENCES

Abd-El-Rahman, I. and Saleh, R. 1993. Preference for oviposition of *Chilo agamemnon* Bles. on several rice varieties and role of leaf pubescence in this preference. *Alexandria Journal of Agricultural Research*, **32**: 439–448.

Abernathy, C.O. and Thurston, R. 1969. Plant age in relation to the resistance of *Nicotiana* to the green peach aphid. *Journal of Economic Entomology*, **62**: 1356–1359.

Agarwal, R.A. 1969. Morphological characteristics of sugarcane and insect resistance. *Entomologia Experimentalis et Applicata*, **12**: 767–776.

Ambekar, J.S. and Kalbhor, S.E. 1981. Note on the plant characters associated with plant resistance to iassid, *Amrasca biguttula biguttula* Ishida in different varieties of cotton. *Indian Journal of Agricultural Science,* **51:** 816–817.

Ananthakrishnan, T.N. 1994. Viewpoints on the chemical ecology of insects. *Phytophaga,* **6:** 69–72.

Ananthakrishnan, T.N. 1996. Chemical signals, patterns of defence gene expression and molecular recognition in insect-plant interactions. *In·* Ananthakrishnan, T.N. (Ed.), *Biotechnological Perspectives in Chemical Ecology of Insects.* Oxford and IBH Publishing Co. Pvt Ltd., New Delhi pp. 4–14.

Ananthakrishnan, T.N. 1999. Induced responses signal diversity and plant defense: Implications in insect phytophagy. *Current Science,* **76(3):** 285–289.

Ashraf, M., Zafar, Z.U., McNeilly, T. and Veltkamp, C.J. 1999. Some morphoanatomical characteristics of cotton (*Gossypium hirsutum* L.) in relation to resistance to cotton leaf curl virus. *Angewandte-Botanik,* **73:** 3–4.

Ave, D.A., Gregory P. and Tingey, W.M. 1987. Aphid repellent sesquiterpenes in glandular trichomes of *Solanum berthaultii* and *Solanum tuberosum. Entomologia Experimentalis et Applicata,* **44:** 131–138.

Ave, D.A. and Tingey, W.M. 1986. Phenolic constituents of glandular trichomes in *Solanum berthaultii* and *Solanum polyadenium. American Potato Journal,* **63:** 473–480.

Baggen, L.R. and Gurr, G.M. 1995. Lethal effects of foliar pubescence of solanaceous plants on the biological control agent *Copidosoma koehleri* Blanchard (Hymenoptera: Encyrtidae). *Plant Protection Quartely,* **10:** 116–118.

Baldocchi, D., Verma, S.B., Rosenberg, N.J., Blad, B.L., Garay, A. and Specht, J.E. 1983. Leaf pubescence effects on the mass and energy exchange between soybean canopies and the atmosphere. *Agronomy Journal,* **75:** 537–542.

Balsdon, J.A., Espelie, K.E. and Braman, S.K. 1994. Cuticular lipids from azalea (*Rhododendron* spp.) and their potential role in host plant acceptance by azalea lace bug, *Stepphanitis pyrioides* (Scott). Biology, Chemistry, Systematic and Ecology.

Bapat, D.R. and Mote, U.N. 1982. Sources of shootfly resistance in sorghum. *Journal of Maharashtra Agricultural University,* **7:** 238–240.

Barbour, J.D., Farrar, R.R. and Kennedy, G.G. 1993. Interaction of *Manduca sexta* resistance in tomato with insect predators of *Helicoverpa zea. Entomologia Experimentalis et Applicata,* **68:** 143–155.

Barker, G.M. 1989. Grass host preferences of *Listronotus bonariensis* (Coleoptera: Curculionidae). *Journal of Economic Entomology,* **82:** 1807–1816.

Bergelson, J. and Purrington, C. 1996. Surveying patterns in the cost of resistance in plants. *Am. Nat.,* **148:** 536–558.

Bergman, D.K., Dillwith, J.W., Zaarrabi, A.A., Caddel, J.L. and Berberet, R.C. 1991. Epicuticular lipids of alfalfa relative to its susceptibility to spotted alfalfa aphids (Homoptera: Aphididae). *Environmental Entomology,* **20:** 781–785.

Bernays, E.A., Blaney, W.M., Chapman, R.F. and Cook, A.G. 1976. The ability of *Locusta migratoria* L. to perceive plant surface waxes. *In* T. Jermy (Ed.), Plenum Press *The Host Plant in Relation to Insect Behaviour and Reproduction,* New York pp. 35–40.

Bilderback, T.E. and Mattson, R.H. 1977. Whitefly host preference associated with selected biochemical and phenotypic characteristics of poinsettias. *Journal of American Society for Horticultural Science,* **102:** 327–331.

Blomquist, G.J. and Jackson, L.L. 1973. Incorporation of labelled dietary n-alkanes into cuticular lipids of the grasshopper *Melanoplus sanguinipes. Journal of Insect Physiology,* **19:** 1639–1647.

Blum, A. 1967. Varietal resistance of sorghum to the shootfly (*Atherigona varia var. soccata*). *Crop Science,* **7:** 461–462.

Bosabalidis, A.M. and Skoula, M. 1998. A comparative study of the glandular trichomes on the upper and lower leaf surfaces of *Origanum x intercedens* Rech. *Journal of Essential Oil Research,* **10(3):** 277–286.

Bodnaryk, R.P. 1992. Leaf epicuticular wax, an antixenotic factor in Brassicaceae that affects the rate and pattern of feeding of flea beetles *Phyllotreta cruciferae* (Goeze). *Canadian Journal of Plant Science*, **72**: 1295–1303.

Broersma, D.B., Bernard, R.L. and Luckmann, W.H. 1972. Some effects of soybean pubescence on populations of the potato leafhopper. *Journal of Economic Entomology*, **65**: 78–82.

Butler, G.D. Jr., Wilson, F.D. and Fisher, G. 1991. Cotton leaf trichomes on popybica and *Bemisia tabaci*. *Crop Protection*, **10**: 461–464.

Carter, M.R., Manglitz, G.R. and Sorensen, E.L. 1998. Resistance to the spotted alfalfa aphid (Homoptera: Aphididae) in simple-haired alfalfa plant introductions. *Journal of Economic Entomology*, **81**: 1760–1764.

Chapman, R.F 1982. Chemoreception: The significance of receptor populations. *Adv. Insect Physiol.*, **16**: 247–356.

Chapman, R.F. and Bernays, E.A. 1989. Insect behaviour at the leaf surface and learning behaviour at the leaf surface and learning as aspects of host plant selection. *Experientia*, **45**: 215–22.

Chapman, R.F. and Woodhead, S. 1985. insect behavior in sorghum resistance mechanisms. 137–147. *In* V. Kumble, ICRISAT, Patancheru, A.P., India. (Ed.) *Proceedings of the International Sorghum Entomology Workshop*, Texas A and M University, July 1984.

Chiang, H.S. and Singh, S.R. 1988. Pod hair as a factor in *Vigna vexillata* resistance to the pod-sucking bug, *Clavigralla tomentosicollis*. *Entomologia Experimentalis et Applicata*, **47**: 195–199.

Chiang, H.S. and Norris, D.M. 1983. Morphological and physiological parameters of soybean resistance to agromyzid beanflies. *Environmental Entomology*, **12**: 260–265.

Cole, R.A. and Riggal, W. 1992. Pleiotropic effects of genes in glossy *Brassica oleracea* resistance to *Brevicoryne brassicae*. *In* Menken, S.B.J., J.H. Visser and P. Harrewjin, Kluwer Academy, Dordrecht (Eds.). *Proceedings of 8th International Symposium on Insect-Plant Relationships*, pp. 313–315.

Dakshayani, J.S., Bentur, J.S. and Kalode, M.B. 1993. Nature of resistance in rice varieties against leaffolder *Cnaphalocrocis medinalis* (Guenee). *Insect Science Applicata*, **14**: 107–114.

David, H. and Easwaramoorthy, S. 1988. Physical resistance mechanisms in insect plant interactions. 45–70. *In* Ananthakrishnan, T.N. and Raman, A. (Eds). *Dynamics of insect-plant interactions: Recent advances and future trends*. Oxford and IBH Publishing Co., New Delhi, India.

Doss, R.P. and Shanks, C.H. 1988. The influence of leaf pubescence on the resistance of selected clones of beach strawberry (*Fragaria chiloensis* (L.) Duchensne) to adult black vine weevils (*Otiorhynchus sulcatus* F.). *Scientia Horticulturae*, **33**: 47–54.

Duffey, S.S. 1986. Plant glandular trichomes: Their partial role in defence against insects. 151–172. *In* B.E. Juniper and T.R.E. Southwood (Eds), *Insects and the plant surface*, Edward Arnold Publishers Ltd. London, UK.

Ehrlich, P.R. and Raven, P.H. 1964. Butterflies and plants a study in coevolution. *Evolution*, **18**: 586–604.

Eigenbrode, S.D. and Espelie, K.E. 1995. Effects of plant epicuticular lipids on insect herbivores *Annual Review of Entomology*, **40**: 171–94.

Eigenbrode, S.D. and Shelton, A.M. 1990. Behaviour of neonate diamondback moth larvae (Lepidopera: Plutellidae) on glossy-leafed resistant *Brassica oleracea* L. *Environmental Entomology*, **19**: 1566–1571.

Eigenbrode, S.D., Shelton, A.M. and Dickson, M.H. 1990. Two types of resistance to the diamondback moth (Lepidoptera: Plutellidae) in cabbage. *Environmental Entomology*, **19**: 1086–90.

Eigenbrode, S.D., Stoner, K.A., Shelton A.M. and Kain, W.C. 1991. Characteristics of leaf waxes of *Brassica oleracea* associated with resistance to diamondback moth. *Journal of Economic Entomology*, **83**: 1609–18.

Eigenbrode, S.D., Castagnola, T., Roux, M.B. and Steljes, L. 1996. Mobility of three generalist predators is greater on cabbage with glossy leaf wax than on cabbage with a wax bloom. *Entomologia Experimentalis et Applicata*, **81**: 335–43.

Eigenbrode, S.D., Kabalo, N.N. and Stoner, K.A. 1999. Predation, behaviour and attachment by *Chrysoperla plorabunda* larvae on *Brassica oleracea* with different surface waxblooms. *Entomologia Experimentalis et Applicata*, **90**: 225–35.

Eisner, T., Eisner, M. and Hoebeke, E.R. 1998. When defense backfires: Detrimental effect of a plant's protective trichomes on an insect beneficial to the plant. *Proceedings of National Academy of Science (U.S.A.)*, **95(8):** 4410–4414.

Elle, E., van Dam, N.M. and Hare, J.D. 1999. Cost of glandular trichomes a "resistance" character in *Datura wrightii* Regel (Solanaceae). *Evolution*, **53(1):** 22–35.

El Khidir, E. (1965) Bionomics on the cotton whitefly (*Bemisia tabaci* Genn.) in the Sudan and the effects of irrigation on population density of whiteflies. *Sudan Agriculture*, **1:** 8–22.

Elsey, K.D. and Wann, E.V. 1982. Differences in infestation of pubescent and glabrous forms of cucumber by pickleworms and melonworms. *Horticultural Science*, **17:** 253–254.

Espelie, K.E., Bernays, E.A. and Brown, J.J. 1991. Plant and insect cuticular lipids serve as behavioral cues for insects. *Arch. Insect Biochemistry and Physiology*, **17:** 223–33.

Farmer, E.E. and Ryan, C.A. 1990. Interplant communication: Airbrone methyl jasmonate induces synthesis of Proteinase inhibitors in plant leaves. *Proceedings of National Academy of Science (USA)*, **87:** 7713–16.

Farrar, R.R., Barbour, J.D. and Kennedy, G.G. 1994. Field evaluation of insect resistance in a wild tomato and its effects on insect parasitoids. *Entomologia Experimentalis et Applicata*, **71:** 211–226.

Feeny, P.P. 1975. Biochemical coevolution between insects and plants. *In* Gilbert, L.E. and P.H. Raven (Eds.), *Coevolution of Animals and Plants*, University of Texas, Austin, pp. 3-19.

Foster, S.P. and Harris, M.O. 1992. Foliar chemicals of wheat and related grasses influencing oviposition by Hessian fly, *Mayetiola destructor* (Say) (Diptera: Cecidomyiidae). *Journal of Chemical Ecology*, **18:** 1965–80.

Fox, L.R. 1981. Defense and dynamics in plant-herbivore systems , *American Zoology*, **24:** 853–64.

Gallun, R.,L., Roberts, J.J., Finney, R.E. and Patterson, F.L. 1973. Leaf pubescence of field grown wheat: A deterrent to oviposition by the cereal leaf beetle. *Journal of Environmental Quality*, **2:** 333–334.

Gannon, A.J. and Bach, C.E. 1996. Effects of soybean trichome density on Mexican bean beetle (Coleoptera, Coccinellidae) development and feeding preference. *Environmental Entomology*, **25:** 1077–82.

Gerhold, D.L., Craig, R. and Mummo, R.O. 1984. Analysis of trichome exudate from mite resistant geranium. *Journal of Chemical Ecology*, **10:** 713–722.

Gibson, P.T. and Maiti, R.K. 1983. Trichomes in segregating generations of sorghum matings. I. Inheritance of presence and density. *Crop Science*, **23:** 73–75.

Gibson, R.W. and Turner, R.H. 1977. Insect-trapping glandular hair on potato plants. *Pest Articles and News Summaries*, **23:** 272–277.

Gilbert, L.E. 1971. Butterfly-Plant Coevolution: Has *Passiflora adenopoda* won the selectional race with Heliconiine butterflies? *Science*, **172:** 585–586.

Goffreda, J.C., Mutschler, M.A., Ave, D.A., Tingey, W.M. and Steffens, J.C. 1989. Aphid deterrence by glucose esters in glandular trichome exudate of the wild tomato *Lycopersicon pennelli. Journal of Chemical Ecology*, **15:** 2135–2147.

Gooderham, J., Bailey, P.C.E., Gurr, G.M. and Baggen, L.R. 1998. Sub-lethal effects of foliar pubescence on the egg parasitoid *Copidosoma koehleri* and influence on parasitism of potato moth *Phthorimaea operculela. Entomologia Experimentalis et Applicata*, **87:** 115–118.

Grazzini, R.A., Paul, P.R., Hage, T., Cox-Foster, D.L., Medford, J.I., Craig, R. and Mumma, R.O. 1999. Tissue specific fatty acid composition of glandular trichomes of mite-resistant and— susciptible *Pelargonium x hortorum. Journal of Chemical Ecology*, **25(4):** 955–968.

Gregory, P., Tingey, W.M., Ave, D.A. and Bouthyette, P.Y. 1986. Potato glandular trichomes. A physicochemical defense mechanisms against insects. 160–167. *In* M.B. Green and P.A. Hedin (Eds), *Natural resistance of plants to pests: Role of allelochemicals*, American Chemical Society Symposium Series 296, Washington, DC, USA.

Gurr, G.M. 1993. Host-Plant resistance to the potato moth (*Phthorimaea operculella*). 117–118 (Vol 2). *In* B.C. Imrie and J.B. Hacker (Eds). *Proceedings of the Tenth Australian Plant Breeding Conference, Towards Responsible and Sustainable Agriculture*, Gold Coast, Australia.

Hagan, A.F. and Anderson, F.N., 1967. Nutrient imbalance and leaf pubescence in corn as factors influencing leaf injury by the adult western corn rootworm. *Journal of Economic Entomology*, **60:** 1071–1073.

Herms, D.A. and Mattson, W.J. 1992. The dilemma of plants to grow or defend. *Quart. Rev. Biol.*, **67:** 283:335.

Heupel, R.C. 1985. Varietal similarities and differences in the polycyclic isopentenoid composition of sorghum. *Phytochemistry*, **24:** 2929–37.

Hoffman, G.D. and McEvoy, P.B. 1985. The mechanism of trichome resistance in *Anaphalis margaritacea* to the meadow spittlebug, *Philaenus spumarius*. *Entomologia Experimentalis et Applicata*, **39:** 123–129.

Isman, M.B. and Duffey, S.S. 1982a. Toxicity of tomato phenolic compounds to the fruitworm, *Heliothis zea*. *Entomologia Experimentalis et Applicata*, **31:** 370–376.

Isman, M.B. and Duffey, S.S. 1982b. Phenolic compounds in the foliage of commercial tomato cultivars as growth inhibitors to the fruitworm, *Heliothis zea*. *Journal of the American Horticultural Society*, **107:** 167–170.

Jackai, L.E.N. and Oghiakhe, S. 1989. Pod wall brichomes and resistance of two wild cowpea, *Vigna vexillata* accessions to *Maruca* testulalis (Geyer) (Lepidoptera: Pyralidae) and *Clarigralla comentosicollis* stal. in *Bulletin of Entomological Research*, **79:** 595-605.

Jeffree, C.E. 1986. The cuticle, epicuticular waxes and trichomes of plants with reference to their structure, functions and evolution. 23–64. *In* B.E. Juniper and T.R.E. Southwood (Eds.), *Insects and the Plant Surface*, Edward Arnold Publishers Ltd., London, UK.

Johnson, H.B. 1975. Plant pubescence: An ecological perspective. *Botanical Review*, **41:** 233–258.

Johnson, A.W. and Severson, R.F. 1984. Leaf surface chemistry of tobacco budworm resistant tobacco. *J. Agric. Entomol.*, **1:** 23–43.

Kalazich, J.C. and Plaisted, R.L. 1991. Association between trichome characters and agronomic traits in *Solanum tuberosum* (L) × *S. berthaultii* (Hawkes) hybrids. *American Potato Journal*, **68:** 833–847.

Kamel, S.A. 1965. Relationship between leaf hairiness and resistance to cotton leaf worm. *Empire Cotton Growing Review*, **42:** 41–48.

Kashyap, R.K., Kennedy, G.G. and Farrar, R.R. 1991. Behavioral response of *Trichogramma pretiosum* Riley and *Telenomus sphingis* (Ashmead) to trichome/methyl ketone mediated resistance in tomato. *Journal of Chemical Ecology*, **17:** 543–556.

Kauffman, W.C. and Kennedy, G.G. 1989. Relationship between trichome density in tomato and parasitism of *Heliothis* spp. (Lepidoptera: Noctuidae) eggs by *Trichogramma* spp. (Hymenoptera: Trichogrammatidae). *Environmental Entomology*, **18:** 698–704.

Kelsey, R.G., Reynoids, G.W. and Rodriguez, E. 1984. *In* Rodriguez, E., Healy, P.L. and Mehta I. (Eds.). *The chemistry of biologically active constituents secreted and stored in plant glandular trichomes*, in Biology and Chemistry of Plant Trichomes, Plenum Press, New York. pp. 187–241.

Khan, Z.R. and Agarwal, R.A. 1984. Ovipositional preference of jassid, *Amrasca bigutula* Ishida on cotton. *Journal of Entomological Research*, **8:** 78–80.

Khan, Z.R., Ward, J.T. and Norris, D.M. 1986. Role of trichomes in soybean resistance to cabbage looper, *Trichoplusia ni*. *Entomologia Experimentalis et Applicata*, **42:** 109–117.

Klingauf, F., Nocker-Wenzeland, K., Rottger, U. 1978. Die Rolle peripherer Pflanzenwaches fur den Befall durch phytophage Insekten. *Z. Pflanzenkr. Pflanzensch.*, **85:** 228–37.

Kogan, M. and Ortman, E.E. 1978. Antixenosis—a new term proposed to replace Painters' "non-preference" modality of resistance, *ESA Bulletin*, **24.**

Kolattukudy, P.E. and Espelie, K.E. 1985. Biosynthesis of cutin, suberin, and associated waxes. *In* T. Higuchi (Ed.), *Biosynthesis and Biodegradation of Wood Components*, Academic Press, New York. pp. 161–207.

Krips, O.E., Kleijn, P.W., Williems, P.E.L., Gols, G.J.Z. and Dicke, M. 1999. Leaf hair influence searching efficiency and predation rate of the predatory mite *Phytoseiulus persimilis* (Acarina: Phytoseiidae). *Experimental and Applied Acarology*. **23(2):** 119–131.

Kumar, H. 1992. Inhibition of ovipositional response of *Chilo partellus* (Lepidoptera: Pyralidae) by the trichomes on the lower leaf surface of a maize cultivar. *Journal of Economic Entomology*, **85:** 1736–1739.

Lambert, L., Beach, R.M., Kilen, T.C. and Todd, J.W. 1992. Soybean pubescence and its influence on larval development and oviposition preference of lepidopterous insects. *Crop Science.* **32:** 463–466.

Lambert, L. and Kilen, T.C. 1989. Influence and performance of soybean lines isogenic for pubescence type on oviposition preference and egg distribution of corn earworm (Lepidoptera: Noctuidae). *Journal of Entomological Science*, **24:** 309–316.

Lampe, U. 1982. Examination of the influences of hooked epidermal hair of french beans (*Phaseolus vulgaris*) on the pea aphid *Acyrthosiphon pisum.* 419. J.H. Vosser and A.K. Minks (Eds.), In *Proceedings of the Fifth International Symposium on Insect-Plant Relationships*, Wageningen University, The Netherlands.

Lee, Y.I., Morgan, M. and Larsen, J.R. 1986. Attachment of the potato leaf hopper to soybean plant surfaces as affected by morphology of the pretarsus. *Entomologia Experimentalis et Applicata*, **42:** 101–107.

Leite, G.L.D., Picanco, M. and Guedes, R.N.G. 1999a. Influence of canopy height and fertilization levels on the resistance of *Lycopersicon hirsutum* to *Aculops lycopersici* (Acarina: Eriophyidae). *Experimental and Applied Acarology*, **23(8):** 633–642.

Leite, G.L.D., Picanco, M. and Lucia, T.M.C.D. and Moreira, M.D. 1999b. Role of canopy height in the resistance of *Lycopersicon hirsutum f.glabratum* to *Tuta absoluta* (Lepidoptera: Gelechiidae). *Journal of Applied Entomology*, **123(8):** 459–463.

Levin, D.A. 1973. The role of trichomes in plant defence. *Quarterly Review of Biology*, **48:** 3–15.

Liang, X., Dron, M., Cramer, C.L. Dixon, R.A. and Lamb, C.J. 1989. Differential regulation of phenylalanine ammonia lyase genes during development and by environmental cues. *Journal of Biology and Chemistry*, **264:** 144486–92.

Liedle, B.E., d.m. Lawson, K.K. White, J.A. Shapiro D.E. Cohen, W.G. Carson, J.T. Trumble and M.A. Matschler 1995. Acylsugars of wild tomato *Lycopersicon pennellii* alters settling and reduces ovipoisition of *Bemisia argentifolii*. *Journal of Economic Entomology*, **88:** 742–48.

Lukefahr, M.J. and Houghtaling, J.E. 1969. Resistance of cotton strains with high gossypol content to *Heliothis* spp. *Journal of Economic Entomology*, **62:** 588–91.

Lukefahr, M.J., Houghtaling, J.E. and Gruhm, D.G. 1975. Suppression of *Heliothis* spp. with cottons containing combinaitons of resistance characters. *Journal of Economic Entomology*, **68:** 743–746.

Lyman, J.M. and Cardona, C. 1982. Resistance in lima beans to a leafhopper, *Empoasca kraemeri*. *Journal of Economic Entomology*, **75:** 281–286.

Malausa, J.C., Daunay, M.C. and Bourgoin, T. 1988. Resistance of several varieties of egg plant, *Solanum melongena* L. to the greenhouse whitefly, *Trialeurodes vaporariorum* Westwood (Homoptera Aleyrodidae). *Agronomie*, **8:** 393–699.

Mauricio, R. and Rausher, M.D. 1997. Experimental manipulation of putative selective agents provides evidence for the role of natural enemies in the evolution of plant defense. *Evolution*, **51:** 1435–44.

McCaskill, D. and Croteau, R. 1998. Strategies for bioengineering the development and metabolism of glandular tissues in plants. *Nature Biotechnology*, **17:** 31–36.

Mehta, R.C. and Saxena, K.N. 1970. Ovipositional respsonses of the cotton spotted bollworm, *Earias fabia* (Lepidoptera: Noctuidae), in relation to its establishment on various plants. *Entomologia Experimentalis et Applicata*, **13:** 10–20.

Meyer, J.R. 1957. Origin and inheritance of $D_2$ smoothness in upland cotton. *Journal of Heredity* **68:** 249–250.

Neal J.J., Steffens, J.C. and Tingey, W.M. 1989. Glandular trichomes of *Solanum berthaultii* and its resistance to the Colorado potato beetle. *Entomologia Experimentalis et Applicata*, **51:** 133–140.

Neal, J.J., Tingey, W.M. and Steffens, J.C. 1990. Sucrose esters of carboxylic acids in glandular trichomes of *Solanum berthaultii* deter settling and feeding by the green peach aphid. *Journal of Chemical Ecology*, **16:** 1547–1555.

Ni, X., Quisenberry, S.S., Siegfried, B.D. and Lee, K.W. 1998. Influence of cereal leaf epicuticular wax on *Diuraphis noxia* probing behavior and nymphoposition *Entomologia Experimentalis et Applicata,* **89:** 111–118.

Obrycki, J.J. 1986. The influence of foliar pubescence on entomophagous species. 61–83. *In* D.J Boethal and R.D. Eikenbarry (Eds.). *Interactions of Plant Resistance and Parasitoids and Predators of Insects,* John Wiley and Sons, New York, USA.

Ogiakhe, S., Jackai, L.E.N., Makanjuola, W.A. and Hodyson, C.J. 1992. Morpho-distribution and the role of trichomes in cowpea (*Vigna unguiculata*) resistance to the legume pod borer, *Maruca testulalis* (Lepidoptera: Pyralidae). *Bulletin of Entomological Research,* **82:** 499–505.

Painter, R.H. 1951. Resistance of plants of insects, *Annual Review of entomology,* **3:** 267–90.

Papp, M., Kolarov, J. and Mesterhazy, A. 1992. Rleation between pubescence of seedling and flag leaves of winter wheat and its significance in breeding resistance to cereal laef beetle (Coleoptera: Chrysomelidae). *Environmental Entomology,* **21:** 700–705.

Parnell, F.R., King, H.E. and Ruston, D.F. 1949. Jassid resistance and hairiness of the cotton plant. *Bulletin of Entomological Research,* **39:** 539–575.

Patanakamjorn, S. and Pathak, M.D. 1967. Varietal resistance of rice to the Asiatic rice borer, *Chilo suppressalis* (Lepidoptera: Crambidae) and its association with various plant characters, *Annals of the Entomological Society of America,* **60:** 287–292.

Peregrine, W.T.H. and Catling, W.S. 1967. Studies on resistance in oats to the frit fly. *Plant Pathology,* **16:** 170–175.

Peter, A.J. and Shanower, T.G. 1996. Biochemical bases of host plant resistance to insects. pp. 111–123. *In:* T.A. Ananthakrishnan (Ed.) *Proccedings of Biochemical Bases of Host Plant resistance to insects,* National Academy of Agricultural Sciences, New Delhi.

Peter, A.J., Shanower, T.G. and Romeis, J. 1995. The role of plant trichomes in insect resistance: a selective review. *Phytophaga,* **7:** 41–64.

Peterson, J.K., Horvat, R.J. and Elsey, K.D. 1994. Squash leaf glandular trichome volatile: identification and influence on behavior of female pickleworm moth *Diaphania nitidalis. Journal of Chemical Ecology,* **20:** 2099–2109.

Petinez, M.T. 1986. Field screening of different systematic insecticides and varietal resistance against rice green leafhoppers (*Nephotettix impicticeps, Nephotettix apicalis*) and brown planthoppers (*Nilaparvata lugens*). *Scientific Journal,* **5(2)–6(1):** 155.

Popova, I. 1960. Chestnuts resistant to *Curculio elephas* (Gyllenhal). *Rastitelna Zashtita,* **8:** 69–71.

Powell, G., Maniar, S.P., Pickett, J.A. and Hardie, J. 1999. Aphid responses to non-host epicuticular lipids. *Entomologia Experiemntalis et Applicata,* **91:** 115–123.

Prokopy, R.J., Collier, R.H. and Finch, S. 1983. Leaf color used by cabbage root flies to distinguish among host plants. *Science,* **221:** 190–92.

Quiring, D.T., Timmins, P.R. and Park, S.J. 1992. Effect of variations in hooked trichome densities of *Phaseolus vulgaris* on longevity of *Liriomyza trifolii* (Diptera: Agromyzidae) adults. *Enprionmental Entomology,* **21:** 1357–1361.

Raina, A.K. 1985. Mechanisms of resistance to shootfly in sorghum: A review. 131–136. *In* V. Kumble (Ed.), ICRISAT, Patancheru, A.P. India. *Proceedings of the International Sorghum Entomology Workshop,* Taxes A and M University, July 1984.

Ramalho, F.S., Parrott, W.L., Jenkins, J.N. and McCarty, J.C. 1984. Effects of cotton leaf trichomes on the mobility of newly hatched tobacco budworms (Lepidoptera: Noctuidae). *Journal of Economic Entomology,* **77:** 619–621.

Raven, P.H., Evert, R.F. and Eichhorn, S.E. 1992. Biology of Plants, 5th edition, Worth, New York.

Rembold, H. and C. Weigner (1990) Chemical composition of chickpea, *Cicer arietinum* exudate. *Zeitschrift fuer Naturforschung,* **45:** 922–923.

Rensburg, Van., L., Peacock, J., Kruger, G.H.J. and van. Rensburg, L. 1999. Boundary layer, stomatal geometry and spacing, in relation to drought tolerance in four *Nicotiana tobacum* L. cultivars. *South African Journal of Plant and Soil.* **16(1):** 44–49.

Roberts, J.J. and Foster, J.E. 1983. Effect of leaf pubescence in wheat on the bird cherry oat aphid (Homoptera: Aphidae). *Journal of Economic Entomology,* **76:** 1320–1322.

Rogers, C.E., Gershenzon, J., Ohno, N., Mabry, T.J., Stipanovic, R.D. and Krietner, G.L. 1987. Terpenes of wild sunflowers (*Helianthus*): An effective mechanism against seed predation by larvae of the sunflower moth, *Homoeosoma electellum* (Lepidoptera: Pyralidae). *Environmental Entomology*, **16**: 586–592.

Romeis, J., Shanower, T.G. and Zebitz, C.P.W. 1998. Physical and chemical plant characters inhibiting the searching behaviour of *Trichgramma chilonis*. *Entomologia Experimentalis et Applicata*, **87**: 275–84.

Romeis, J., Shanower, T.G. and Peter, A.J. 1999a. Trichomes on pigeonpea (*Cajanus cajan* (L.) Millsp.) and two wild *Cajanus* spp. *Crop Science*, **39**: 564–69.

Romeis, J., Shanower, T.G. and Zebitz, C.P.W. 1999b. Why *Trichogramma* (Hymenoptera: Trichogramatidae) egg parasitoids of *Helicoverpa armigera* (Lepidoptera: Noctuidae) fail on chickpea. *Bulletin of Entomological Research*, **89**: 89–95.

Sambandam, C.N., Chelliah, S. and Natarajan, K. 1969. A note on Fox-face, *Solanum mammosum* L. a source of resistance to *Aphis gossypii* G. and *Epilachna vigintioctopunctata* F. *Annamalai Agriculture Research Annual*, **1**: 110–112.

Schreiner, I.H. 1990. Resistance and yield response to *Amrasca biguttula* (Homoptera: Cecadellidae) in eggplant (*Solanum melongena*). *Philippine Entomologist*, **8**: 661–669.

Seigler, S.D. 1983. Role of lipids in plant resistance to insects, In *Plant resistance to insects*, American Chemical Society Symposium series 208, U.S.A. pp. 303–327.

Shanower, T.G., Romeis, J. and Peter, A.J. 1996. Pigeonpea plant trichomes: Multiple trophic level interactions . pp 76–88. *In:* T.N. Anathakrishnan (Ed.), *Biotechnological Perspectives in Chemical Ecology of Insects*. Oxford & IBH Publishing Co., New Delhi.

Shanower, T.G., Yoshida, M. and Peter, A.J. 1997. Survival, Growth, Fecundity, and behavior of *Helicoverpa armigera* (Lepidoptera: Noctuidae) on pigeonpea and two wild *Cajanus* species. *Journal of Economic Entomology*, **90(3)**: 837–841.

Simms, E.L. 1992. Cost of plant resistance to herbivory. *In* Fritz, R.S. and E.L.Simms (Eds). *Plant Resistance to Herbivores and Pathogens: Ecology, Evolution, and Genetics*, University of Chicago Press, Chicago. pp. 392–405.

Sithanantham, S., Bhatnagar, V.S., Jadhav, D.R. and Reed, W. 1982. Some aspects of *Trichogramma* spp. in eggs of *Heliothis armigera* (Hb.) (Lepidoptera: Noctuidae). Paper presented at the international symposium on *Trichogramma* and other egg parasitoids, Antibes, France, 12 pp. (Limited distribution)

Smith, C.M. 1989. Plant resistance to insects-A Fundamental Approach. John Wiley & Sons, Inc. Canada. pp. 286.

Smith, D.H. and Webster, J.A. 1974. Leaf pubescence and cereal leaf beetle resistance to *Triticum* and *Avena* species. *Crop Science*, **14**: 241–243.

Smith, R.L., Wilson, R.L. and Wilson, F.D. 1975. Resistance of cotton plant hairs to mortality of first instar of the pink bollworm. *Journal* of *Economic Entomology*, **68**: 679–683.

Sosa, O. 1990. Oviposition preference by the sugarcane borer (Lepidoptera: Pyralidae). *Journal of Economic Entomology*, **83**: 866–868.

Southwood, R. 1986. Plant surfaces and insects—An overview. pp. 1–22. *In* B.E. Juniper and T.R.E. Southwood (Eds.), *Insects and the Plant Surface*. Edward Arnold Publishers Ltd. London, UK.

Spencer, K. 1988. Chemical Mediation of Coevolution. Academic press, New York.

Stadler, E. 1986. Oviposition and feeding stimuli in leaf surface waxes. *In* Juniper, B.E. and T.R.E. Southwood (Eds.), *Insects and the Plant Surface*, Edward Arnold, London. pp. 105–121.

Stephens, S.G. and Lee, H.S. 1961. Futher studies on the feeding and oviposition preferences of the boll weevil (*Anthonomus grandis*). *Journal of Economic Entomology*, **54**: 1085–1090.

Stoner, A.K. 1990. Glossy leaf wax and plant resistance to insects in *Brassica oleracea* under natural infestation. *Environmental Entomology*, **19**: 730–39.

Stoner, A.K. 1992. Density of imported cabbageworms (Lepidoptera: Pieridae), cabbage aphids (Homoptera: Aphididae) and fleabeetles (Coleoptera: Chrysomelidae) on glossy and trichome-bearing lines of *Brassica oleracea*. *Journal of Economic Entomology*, **85**: 1023–1030.

Stork, N.E. 1980. Role of waxblooms in preventing attachment to brassicas by the mustard bettle, *Phaedon cochleariae*. *Entomologia Experimentalis et Applicata*, **28**: 100–107.

Sutherst, R.W., Jones, R.J. and Schnitzerling, H.J. 1982. Tropical legumes of the genus *Stylosanthes* immobilize and kill cattle ticks. *Nature*, **295**: 320–321.

Sutherst, R.W. and Wilson, L.J., 1986. Tropical legumes and their ability to immobilize and kill cattle ticks. 185–194. *In* B.E. Juniper and T.R.E. Southwood (Eds), *Insects and the Plant Surface*, Edward Arnold Publishers Ltd., London, UK.

Talekar, N.S., Yang, H.C. and Lee, Y.H. 1988. Morphological and physiological traits associated with agromyzid (Diptera: Agromyzidae) resistance in mungbean. *Journal of Economic Entomology*, **81**: 1352–1358.

Thompson, K.E. 1963. Resistance to the cabbage aphid (*Brevicorne brassicae*) in *Brassica* plant. *Nature*, **198**: 209.

Thurston, R. 1970. Toxicity of trichome exudates of *Nicotiana* and *Petunia* species to tobacoo hornworm larvae. *Journal of Economic Entomology*, **63**: 272–274.

Thurston, R. and Webster, J.A. 1962. Toxicity of *Nicotiana gossei* Domin. to *Myzus persicae* (Sulzer) *Entomologia Experimentalis et Applicata*, **5**: 233–238.

Tingey, W.M. 1991. Potato glandular trichomes defensive activity against insect attack. 126–135. *In* P.A. Hedin (Ed) *Naturally Occuring Pest Bioregulators*, American Chemical Society Symposium Series 449, Washington, DC, USA.

Tingey, W.M. and Laubengayer, J.E. 1981. Defense against the green peach aphid and potato leafhopper by glandular trichomes of *Solanum berthaultii*. *Journal of Economic Entomology*, **74**: 721–725.

Treacy, M.F., Benedict, J.H., Segers, J.C., Morrison, R.K. and Lopez, J.D. 1986. Role of cotton trichome density in bollworm (Lepidoptera: Noctuidae) egg parasitism. *Environmental Entomology*, **15**: 365–368.

Turnipseed, S.G. 1977. Influence of trichome variations on populations of small phytophagous insects in soybean. *Environmental Entomology*, **6**: 815–817.

Uematsu, H. and Kawada, K. 1989. Possible role of cabbage leaf wax bloom in suppressing diamondback moth *Plutella xylostella* (Lepidoptera: Yponomeutidae) oviposition. *Appl. Entomol. Zool*, **24**: 253–57.

Uengprasertporn, U. 1985. Studies on the resistant characters of yard long bean, *Vigna unguiculata* (L.) Walp. sub. *sesquipedalis* (L.) Verde to the bean fly, *Ophiomyia phaseoli* (Tryon). Ph.D. thesis, Graduate School, Kasetsart University, Bangkok, Thailand, pp. 168.

Uthamasamy, S. 1985. Influence of leaf hairiness on the resistance of bhendi or lady's finger, *Abelmoschus esculentus* (L). Moench, to the leafhopper, *Amrasca devastans* (Dist.). *Tropical Pest Management*, **31**: 294–295.

van Dam, N.M. and Hare, J.D. 1998. Biological activity of *Datura wrightii* (Solanaceae) glandular trichome exudate against *Manduca sexta* (Lepidoptera: Sphingidae) larvae. *Journal of Chemical Ecology*, **24**: 1529–1549.

van Dam, N.M., Hare, J.D. and Elle, E. 1999. Inheritance and distribution of trichome phenotypes in *Datura wrightii*, *Journal of Heredity*, **90**: 220–27.

van Emden, 1991. The role of host plant resistance in insect pest mismanagement. *Bulletin of Entomological Research*, **81**: 123–126.

Verma, S.C. and Mathur, P.S. 1950. The epidermal characters of sugarcane leaf in relation to insect pests. *Indian Journal of Agricultural Science*, **20**: 387–389.

Visser, J.H. 1986. Host odor perception in phytophagous insects. *Annual Review of Entomology*, **31**: 121–44.

Walters, D.S., Harman, J., Craig, R. and Mumma, R.O. 1991. Effect of temperature on glandular trichome exudate composition and pest resistance in geraniums. *Entomologia Experimentals et Applicata*, **60**: 61–69.

Wellso, S.G. 1973. Cereal leaf beetle larval feeding, orientation, development and survival on four small grain cultivars in the laboratory. *Annals of the Entomological Society of America*, **66**: 1201–1208.

White, C. 1998. Effects of *Pisum sativum* surface waxbloom variation on herbivores and predators. M. Sc. Thesis, *Plant Soil and Entomological Sciences*, University of Idaho, Moscow, pp. 96.

Wilson, F.D. and George, B.W. 1986. Smoothleaf and hirsute cottons: Response to insect pests and yield in Arizona. *Journal of Economic Entomology*, **79**: 229–232.

Wogerek, W. and Dunajska, K. 1964. (The morphology and anatomy of lupin varieties resistant and nonresistant to the pea aphid *Macrosiphum pisum*.) (In Pol. Summary in En.) *Biuletyn Instytulu Ochrony Roslin*, **27**: 1–15.

Wolfenbarger, D.A. and Sleeman, J.P. 1963. Variation in susceptibility of soybean pubescent types, brand bean and runner bean varieties and plant introductions to potato leafhopper. *Journal of Economic Entomology*, **56**: 895–897.

Woodhead, S. 1983. Surface chemistry of Sorghum bicolor and its importance in feeding by *Locusta migratoria*. *Physiol. Entomol.*, **8**: 345–52.

Yoshida, M., Cowgill, S.E. and Wightman, J.A. 1995. Mechanism of resistance to *Helicoverpa armigera* (Lepidoptera: Noctuidae) in chickpea: Role of oxalic acid in leaf exudate as an antibiotic factor. *Journal of Economic Entomology*, **88**: 1783–1786.

Zvereva, L.E., Kozlov, V.M. and Niemela, P. 1998. Effects of leaf pubescence in Salix borealis on host-plant choice and feeding behaviour of the leaf beetle, *Melasoma lapponica*. *Entomologia Experimentalis et Applicata*, **89**: 297–303.

## Chapter 7

# Host Plant resistance to insects: Measurement, Mechanisms and Insect-Plant Environment Interactions

*H.C. Sharma, B.U. Singh and Rodomiro Ortiz*\*

### INTRODUCTION

Host plant resistance to insects is one of the most important methods of insect control. With the domestication of plants for agricultural purposes, farmers always selected the plants that were able to overcome adverse environmental factors, including insects and diseases. The plants that were susceptible to pests were generally eliminated, and only the resistant plants survived until the time of harvest. This process led to the natural selection of plant varieties resistant to insect pests. Because of this unintentional but continuous selection of plants over several hundreds of years, many landraces selected by the farmers evolved as having or accumulating genes conferring resistance to insects. The best examples of this process are: shootfly resistance in landraces of sorghum cultivated during the post-rainy season in India, sorghum midge resistance in genotypes originating from eastern Africa, and head bug resistance in *guineense* sorghums cultivated in western Africa.

Inspite of the importance of host-plant resistance (HPR) as a component of integrated pest management (IPM), breeding of plants resistant to insects has not been as rapidly accepted and developed as the breeding disease-resistant cultivars. This was partly due to the relative ease with which insect control is achieved with the use of insecticides. Another reason for the slow progress in developing insect-resistant cultivars has been the difficulties involved in ensuring adequate insect pressure for resistance screening. Insect-rearing programs are expensive, the development of technology requires several years, and may not produce the behavioral or metabolic equivalent of an insect population in

\* *International Crops Research Institute for the Semi-Arid Tropics (ICRISAT), Patancheru 502 324, Andhra Pradesh, India.* Email: ICRISAT@CGIAR.ORG.

nature. However, with the development of insect resistance to insecticides, adverse effects of insecticides on natural enemies and public awareness of environmental contamination and conservation, there has been a renewed interest in the development of insect-resistant cultivars. The establishment of international agricultural research centers, and the collection and evaluation of existing germplasm for insect resistance has given a renewed impetus to the identification and use of HPR in integrated pest management (IPM) worldwide.

"Resistance of plants to insects enables a plant to avoid or inhibit host selection, inhibit oviposition and feeding, and reduce insect survival and development, tolerate or recover from injury from insect populations that would cause greater damage to other plants of the same species under similar environmental conditions" (Smith, 1989). Resistance of plants to insects is the consequence of heritable plant characters that result in a plant being relatively less damaged than the plant without these characters. This property is generally derived from morphological and biochemical characteristics of the plants, which affect the behavior and biology of insects and influence the relative degree of damage caused by the insects. From an evolutionary point of view, resistance traits are pre-adaptive genetically inherited characters of the plant. Plants with such pre-adaptive genes resist the selective pressure of herbivore populations, and thus increase their chances of survival and production.

Plant resistance to insects is always relative, and the level of resistance is expressed in relation to the susceptible genotypes that are more severely damaged under similar conditions. This concept is important, since expression of resistance is dependent on environmental factors both in time and space. Pseudo-resistance or false resistance may occur in normally susceptible plants through avoidance of insect damage. Induced resistance may occur in plants because of variations in temperature, photoperiod, plant-water potential, and chemicals in the soil that induce the production and accumulation of secondary plant substances (phytoalexins) through increased activity of the phenylpropanoid metabolic pathway. Associate resistance occurs when susceptible plants grow in association with resistant plants and derive protection for insect damage from resistant plants. Associate resistance indicates that the diversion or delaying actions of mixtures of plant species can help in the slow development of an insect biotype that can overcome resistant cultivars. Various aspects of host-plant resistance to insects have been discussed by Painter (1951), Maxwell and Jennings (1980), Smith (1989), Smith *et al.* (1994) and Sharma *et al.* (1999c). The role of host plant resistance in pest management has been discussed earlier by Pathak (1970), Kogan (1975), Teetes (1985) Heinrichs (1988) and Sharma (1993). In this chapter, we summarize information on various aspects of insect-plant interactions in relation to research needs and the progress made in developing crop cultivars with resistance to insects for pest management.

## RESISTANCE SCREENING TECHNIQUES

Development and standardization of resistance screening techniques is the key for an effective resistance-breeding program. Knowledge concerning the periods of greatest insect density and 'hot-spots' is the first step to initiate work on resistance screening. Delayed plantings and the use of infester rows of a susceptible cultivar are other effective means of augmenting insect populations to screen for insect resistance (Sharma *et al.,* 1992; Smith *et al.,* 1994; Kalode and Sharma, 1995). Information on population peaks, hot-spots, and use of infester rows need to be accumulated and utilized to screen for resistance to a number of important crops pests.

Screening for insect resistance under natural conditions is a long-term process. Because of variation in insect populations in space and time, it is difficult to identify reliable and stable sources of resistance under natural infestation. To overcome these problems, it is important to develop multi-choice or no-choice screening techniques where the test cultivars can be subjected to an uniform insect pressure at the most susceptible stage of the crop. Such techniques have been developed for brown plant hopper (*Nilaparvata lugens*), gall midge (*Orseolea oryzae*) and stem borer (*Scirpophaga incertulas*) in rice (Kalode *et al.* 1989; Krishanaiah, 1995), sorghum shoot fly (*Atherigona soccata*), spotted stem borer (*Chilo partellus*), sorghum midge (*Stenodiplosis sorghicola*), and head bugs (*Calocoris angustatus*) in sorghum (Sharma *et al.,* 1992), Oriental armyworm (*Mythimna separata*) in pearl millet and sorghum (Sharma and Sullivan, 2000), leaf miner (*Aproarema modicella*), aphids (*Aphis craccivora*), jassids (*Empoasca kerri*), and tobacco caterpillar (*Spodoptera litura*) in groundnut (Wightman *et al.,* 1990), pod borer (*Helicoverpa armigera*) in chickpea and pigeonpea (Lateef, 1985; Reed and Lateef, 1990), spotted pod borer (*Maruca vitrata*) in pigeonpea and cowpea (Sharma *et al.,* 1999b), spotted stem borer (*C. partellus*) in maize (Swarup, 1987), pearl millet shoot fly (*Atherigona approximata*) (Sharma and Youm, 1999), and potato tuber moth (*Pthorimea opercullela*) in potato (Otriz *et al.,* 1990). While screening under artificial infestation is a necessity for most insect pests, it may not be an absolute requirement for insects that occur in epidemic proportions over many years or are endemic to particular areas, e.g., bollworms (*Pectinophora gossypiella, H. armigera* and *Earias vittella*) and jassids (*Amrasca biguttula biguttula*) in cotton, aphids (*Lipaphis erysimi* and *Brevicoryne brassicae*) in cruciferous crops, and pod borers (*H. armigera* and *M. vitrata*) in pigeonpea. Greater emphasis is needed on refining resistance screening techniques, and developing uniform and standard procedures for evaluating insect populations and damage. Entomologists should be critical and vigilant in their screening and selection procedures to become an effective instrument in breeding crop cultivars resistant to insect pests. Once we achieve this, it will be easier to develop insect-resistant cultivars for increasing and stabilizing crop production. The following techniques can be used to screen and select for resistance to insect pests in different crops.

## Manipulations of insect abundance

It is possible to manipulate insect abundance by field infestation, caging, artificial rearing and by evaluating insecticide-protected and unprotected plots.

**Field infestation.** Rarely is a researcher able to grow a group of plant genotypes and accurately evaluate insect damage. Without proper planning, either there will be insufficient insect numbers to cause adequate damage or insects occur at an inappropriate phenological stage of crop growth. Field infestations are nomally used to evaluate a large number of genotypes at an early stage of the resistance evaluation program. Unmanaged insect populations may be too low or unevenly distributed to expose all the genotypes to a unifrom level of insect density. Also, there are large differences in insect density over years and locations. Field evaluations are additionally influenced by nontarget insects, which may interfere with the damage caused by the target insect, which makes it difficult to achieve dependable screening of plant material for resistance to insects in the field. Managed or augmented insect density ensures a uniform distribution of insects, but the insects are subjected to mortality due to naturally occurring biological control agents and abiotic population regulation factors. The crop can be treated with selective insecticides before insect infestation takes place to avoid interference by natural enemies and nontarget pests (Sharma *et al.*, 1992). Insects on infested plants can also be protected from natural enemies by cages. The objective of all these approaches should be to have an optimum insect density to damage ratio that allows the researcher to observe maximum differences among the resistant and susceptible genotypes.

Several procedures can be employed to obtain adequate insect pressure for resistance screening. 'Hot-spots', where the insects are known to occur regularly in optimum numbers across the season, can be used efficiently for large-scale screening of the test material. Planting of mixed or uniform maturity susceptible cultivars as infester rows along the field borders or at regular intervals in the field helps to increase insect abundance. The infester rows may be planted in advance so that the insect can have sufficient time to multiply on the infester rows (Sharma *et al.*, 1988a; Smith *et al.*, 1994). The infester rows can be removed after infestation of the test materials has taken place. The test material can also be planted two to three times so that one of the plantings are exposed to adequate insect abundance, e.g., sorghum midge and head bugs (Sharma *et al.*, 1997). Such an approach also helps to reduce the chances of escape. Pest abundance can be augmented by placing nondestructive light, pheromone or kairomone traps. Indigenous insect populations can be collected from the surrounding areas and released in the test plots. In field screening under natural infestation, known resistant and susceptible genotypes should also be grown at regular intervals in the screening nursery (Ortiz *et al.*, 1995). For insects feeding on the reproductive parts of a plant (and if there are large differences at the flowering times of the test genotypes), resistant and susceptible genotypes of different maturities should be included. The test material flowering on the same day can also be tagged with different colored labels or marked with paint. This enables the comparison of the test material flowering on the same date with the resistant and susceptible controls of similar duration.

**Caging.** Caging insects with test plants is one of the most dependable methods of screening for insect resistance. In this method, considerable control is exercised to maintain uniform insect pressure on the test entries, and to infest the test plants at the same phenological stage. This prevents the insects from migrating away from the test plants. The cage also keeps natural enemies away from the insects. Such tests can be carried out under greenhouse and field conditions. Small cages can be developed to cover the plant parts to be tested or whole plants can be put under a cage (Smith, 1989; Sharma *et al.*, 1988b, 1992, 1997; Smith *et al.*, 1994). The size and shape of the cage is determined by the type and number of test plants needed for evaluation. For valid conclusions, resistant and susceptible checks should also be included and infested at the same time as the test genotypes.

**Supplementing with artificially reared insects.** Artificially reared insects can be made available throughout the year for screening tests. Artificial diets have been developed for several insect species (Singh and Moore, 1985). Egg masses or the first-instar larvae can then be spread uniformly in the test material by stapling the egg masses to the host plant or spreading eggs or larvae among the test plants by mixing them with an appropriate inert carrier (Taneja and Leuschner, 1985; Smith *et al.*, 1994; Sharma *et al.*, 1992). But, if it is not possible to rear insects on an artificial diet, insect colonies can be maintained on natural hosts (e.g., shoot fly, head bugs, and sorghum midge on the sorghum plants in the greenhouse) or the insects can be collected from the infested fields and released on to the test material in uniform numbers at the susceptible stage of the crop (Sharma *et al.*, 1992).

## Measurement of resistance

### *Direct measurements*

**Direct-feeding injury.** Measurements of insect damage to plants are often more useful than insect growth or development. Plant damage and the resulting reduction in yield or quality are important while establishing the goals of a crop improvement program. Often, measurements of yield reduction indicate direct insect feeding injury to plants. Plant damage can also be determined by measuring the incidence of tissue necrosis, fruit abscission and stem damage. The quality of produce can also be used to measure the effect of insect damage. Insect defoliation in plants is usually determined by rating scales that make use of visual estimates of plant damage based on percentages or numerical ratings. Several such rating scales have been developed to assess insect damage in crop plants. Direct measurements of leaf area are also used to measure insect damage. Indirect feeding injury measurements, such as plant growth, photosynthetic rates, transpiration rates, ehylene production and respiratory rates are also recorded. Feeding injury is measured as loss of yield under protected and unprotected conditions. Different levels of insect infestations are created using different spray regimes. Less affected genotypes with low regression coefficients are selected in comparison with the susceptible genotypes, which have high regression coefficients.

**Simulated feeding injury.** Insect feeding injury can be simulated by mechanical defoliation. However, plants respond somewhat differently to artificial defoliation than to actual insect feeding. Therefore, the relationship between artificial and natural insect feeding should be determined before results on artificial defoliation are accepted. Insect injury is also measured by injection of toxic insect secretions into plant tissues, e.g., application of crude extract of greenbug in sorghum.

**Correlation of plant factors with insect resistance.** Chemical or mechanical resistance is measured by concentrations of allelochemicals or the density or size of morphological structures present in the tissues of resistant plants. This permits the rapid determination of potentially resistant plant material. This also removes the variations associated with insect density and the effect of environmental factors on the expression of resistance to insects.

### Indirect measurements

**Sampling insect populations.** Insect abundance can be estimated by sampling at the plant site where damage has taken place and at the appropriate phenological plant stage and time. The population of immobile insects is measured visually, but this method is subjected to variations in colony size and pattern of insect distribution. Shaking the plants, use of sampling nets, use of traps  or actual counts are used to obtain an estimate of insect abundance.

**Measurements of insect feeding and development.** Insect development is monitored if antixenotic and antibiotic effects are exhibited by the resistant plants. Several measures of consumption and utilization of food by the insects are also used to determine the levels of plant resistance to insects (Waldbauer, 1968). Effect of plant resistance on insect feeding and development is measured in terms of the amount of food consumed per unit, body weight per day or leaf area consumed, duration of larval or pupal development, fecundity and insect survival. Antibiosis effects are expressed in terms of weight and size of insects, sex ratio and proportion of insects entering into diapause.

**Measurements of insect behavior.** Several techniques for studying insect behavior are used to quantify antixenosis mechanism of resistance. The responses of insects to volatile stimuli have been studied for several insects. Several designs of olfactometers have been used to observe insect behavior. Olfactory responses are also studied physiologically by electroantennograms, electroretinograms and by electronic feeding monitors.

## MECHANISMS OF RESISTANCE

### Antixenosis

Antixenosis is derived from a Greek word, *xeno*—meaning, "guest". It describes the inability of a plant to serve as a host to an insect herbivore. As a result, the insect is forced to change its host plant for feeding and oviposition. This term was proposed by Kogan and Ortman (1978) to replace the term nonpreference proposed earlier by Painter (1951). In some genotypes, antixenosis is the major component of resistance to insects (Sharma *et al.*, 1990b; Sharma and Nwanze, 1997). Antixenosis may be due to morphological or chemical plant factors that affect the insect behavior adversely, resulting in selection of an alternative host plant.

### Plant Defenses imparting antixenosis

**Olfaction.** This comprises chemical stimuli that affect the orientation of insects toward their host plants. These are perceived by sensilla basiconica. Chemicals that help in host identification and selection are called kairomones, and give an adaptive advantage to the insects. Chemicals that repel the insects away from their host plants are called allomones, and these give adaptive advantage to the plants. Plant odors that cause a long-range insect movement towards the plants are called attractants, while those eliciting a movement away from the plants are called repellents. Some plants also produce arrestants, which cause the insect movement to cease in close proximity of the host plant. Grasshoppers (Haskell *et al.*, 1962), Colorado potato beetle (de Wilde *et al.*, 1969), sorghum midge (Sharma *et al.*, 1990a), and sorghum head bugs (*C. angustatus* and *Eurystylus immaculatus*) (Sharma and Lopez, 1990; Sharma *et al.*, 1994) are attracted by the odors emanating from their host plants. Most plant species have a unique chemical profile of the volatile compounds in different plant parts, which play an important role in host plant selection by the insects. In some cases, volatile compounds play a crucial role in plant selection by the insects, e.g., sulphur compounds in onions (Pierce *et al.*, 1978), propanylbenzenes in carrots (Guerin *et al.*, 1983), and alcohols in potato (Visser *et al.*, 1979).

**Vision.** This involves the ability of insects to perceive spatial patterns using instinctive stimuli templates and their ability to detect differences in color, e.g., brightness, hue and the saturation of various wavelengths of light. Visual cues perceived by the insects during host selection are the result of the spectral quality of light, dimensions and the pattern or shape of objects (Prokopy and Owens, 1983). Host finding from a distance involves visual stimuli, e.g., aphids are attracted to yellow-white surfaces (Moericke, 1969), cotton white fly (*Bemisia tabaci*) to yellow-green and yellow-red (Hussain and Trehan, 1940), and sorghum midge (*S. sorghicola*) to white-yellow colors (Sharma *et al.*, 1990a). Acridids orient themselves towards vertical lines or objects (Williams, 1954; Mulkern, 1969), and some caterpillars and beetles are attracted towards vertical patterns (Heirholzer, 1950). Visual and chemical stimuli are perceived simultaneously during the orientation behavior of an insect to a host plant (Prokopy and Owens, 1983). During long-range orientation, an insect may use vision to recognize the shape of an object, and chemical stimuli to perceive the plant attractants. By genetically altering the color of the foliage, the antixenosis mechanism can be used to develop crop cultivars with resistance to insects, e.g., some cucurbit cultivars with silver colored leaves to aphids (Shifriss, 1981), red leaves in cotton to boll weevil (*Anthonomus grandis*) (Iseley, 1928) and spotted bollworm (*E. vittella*) (Sharma *et al.*, 1982), and birch leaves with lower spectrophotometric absorption rates to leaf miner (*Fenusa pusila* ) (Fiori and Craig, 1987).

**Thigmoreception.** Contact stimuli are perceived by trochoid sensilla on the insect body, tarsi, head and antenna. Such stimuli are received from leaf or stem trichomes, epidermal ridges and leaf margins. The plant morphological structures can act as feeding or oviposition stimuli. Some plant structures such as plant hairs can also act as feeding or oviposition deterrents, e.g. leaf hairs in cotton for resistance to jassids (*A. biguttula biguttula*) (Sharma and Agarwal, 1983a).

**Gustation.** Once an insect is in physical contact with a potential food source, it is in a position to test it thoroughly with its chemical sense. Insect species, with tarsal receptors, may detect chemicals on the surface of their substrate. Stimuli provided by the phytochemicals in vapor or liquid phase are electronically transmitted to the central nervous system. The response spectrum of the gustatory receptors depends on the range of chemicals within the host range of the insect in question. Chemicals eliciting continued feeding are called phagostimulants, and the stimuli that inhibit feeding are called phagodeterrents. The bite is smaller than the regular bite, and the food is chewed more carefully (Williams, 1954). Some apple aphids perceive a flavonoid, phloricin, after landing on the host plant (Klingauf, 1971), alkanes present on the leaf surface of *Vicia faba* exert arrestant effect on the pea aphid (*Acyrthosiphon pisum*) (Klingauf *et al.*, 1971), and tarsal receptors of the beetle (*Chrysolina brunsvicensis*) are simulated by hypericin, a quinone characteristic of *Hypericum* spp. (Reese, 1969). This beetle appears to be monophagous on *Hypericum* spp. Locusts appear to test their potential food source through labial and maxillary palpi (Blaney and Chapman, 1970). In sucking insects, contact chemostimulation takes place through the proboscis setae, which function as contact chemoreceptors (Schoonhoven, 1972). A series of compounds extracted from mulberry leaves have been associated with initial feeding, swallowing and continuous feeding by the larvae of *Bombyx mori* (Hamamura, 1970). These stimuli are perceived by sensilla styloconica, maxillary palpi and lateral gustatory receptors. Quantitative and qualitative differences in primary and secondary plant substances influence the gustatory processes that may, in turn, influence the host selection behavior of insects.

### Plant factors affecting insect behavior

**Trichomes.** Plant structures such as trichomes on the leaves of many crop plants affect the host selection by insects. Dense growth of erect trichomes deters feeding and oviposition by several insect species, e.g., leaf hairs on cotton leaves to cotton jassid (*A. biguttula biguttula*) (Sharma and Agarwal,

1983a), trichomes in soybean to *Trichoplusia ni* (Chiang and Norris, 1983; Khan *et al.*, 1986a), and trichomes on the under surface of sorghum leaves to sorghum shootfly (*A. soccata*) (Maiti and Bidinger, 1979; Maiti and Gibson, 1983). Glandular trichomes on the wild potato (*Solanum neocardenasii*) adversely affect the feeding behavior of the green peach aphid (*Myzus persicae*) (Lapointe and Tingey, 1986). However, in some cases, the plant pubescence can have an adverse effect on plant resistance to insects, e.g., pubescent pearl millet lines are preferred for oviposition and feeding by the fall armyworm (*Spodoptera frugiperda*) (Burton *et al.*, 1977). Pubescent cotton genotypes are also preferred for oviposition by the spotted bollworm (*E. vittella*), and, consequently suffer greater damage than the glabrous genotypes (Sharma and Agarwal, 1983b, c).

**Surface waxes.** Plant leaves are protected from pests by a layer of surface waxes over the epicuticle. When the sense organs on the insect tarsi and mouth receive negative chemical and tactile stimuli from the leaf surface, these stimuli play an important role in plant resistance to insects. Chemicals in the epicuticular wax of plants induce oviposition response in many phytophagous insects. In *Delia* (*brassicae*) *radicum*, oviposition on non-host plants can be induced in the leaves perfused with sinigrin (Staedler, 1978). Surface waxes from carrot leaves induce oviposition by *Psila rosea*. These have been identified as propenyl benzenes, furanocoumarins and polyacetylenes (Staedler and Buser, 1982). Klingauf *et al.* (1978) reported that oviposition by *Pegomya betae* is stimulated by the alkane fraction from the wax of *Beta vulgaris*.

**Tissue thickness.** Foliar toughness adversely affects the host selection by several insect species. Resistance in sorghum to shoot fly is related to thickened cells that surround the vascular bundles of leaves (Blum, 1968). In sugarcane, internode hardness confers resistance to the stem borer (*Diatraea saccharalis*) (Martin *et al.*, 1975). Cells thickened by increased layers of epidermal cells limit insect damage in rice, sugarcane and wheat (Wallace *et al.*, 1974)

### Antibiosis

Antibiosis includes the adverse effects of the physico-chemical characteristics of the plants on the biology of an insect attempting to use that plant as a host. Both chemical and morphological factors mediate antibiosis. Lethal effects may be acute, often affecting young larvae and eggs, and chronic effects lead to mortality of older larvae, pupae and adults. Individuals surviving the direct effects of antibiosis may have reduced body size and weight, prolonged period of development and reduced fecundity.

#### Plant Defenses imparting antibiosis

**Allelochemicals.** Allelochemicals not only effect the survival and feeding behavior, but may also have sub-chronic effects at low concentrations on growth rates, utilization of food, pupal development and fecundity (Reese and Beck, 1976; Sharma and Agarwal, 1982a; Sharma *et al.*, 1993a,c). Alkaloids, ketones, terpenoids, flavonoids and organic acids produced by the plants are toxic to insects. DIMBOA in maize, glycoalkaloids in potato, R-tomatine in tomato, gossypol in cotton and rutin and chlorogenic acid in tomato are also toxic to insects (Raman *et al.*, 1978; Sinden *et al.*, 1986; Elliger *et al.*, 1978; Sharma and Agarwal, 1982b; Sharma and Norris, 1991, 1994).

**Growth inhibitors.** Insect growth inhibition due to the presence of growth inhibitors or poor nutritional quality of the host plant are responsible for plant resistance to insects in several crops. Maysin in maize silks (Waiss *et al.*, 1979), coumesterol in soybean (Sharma and Norris, 1991) and terpenoids in pigment glands of cotton inhibit insect growth (Sharma and Agarwal, 1982a,b). Imbalanced ascorbic acid content in maize plants (Penny *et al.*, 1967), amino acid content in pea (Auclair *et al.*, 1957), low quantities of glutamic acid and asparagine in rice (Sogawa and Pathak, 1970) and low lysine content in sorghum (Singh and Jotwani, 1980) impart resistance to insects in these crops. The chronic effects of secondary plant substances affect the metabolism of insects that feed on the resistant plants.

**Morphological barriers.** Hypersensitive growth responses of plants, such as rapidly growing tissues of cotton bolls may kill bollworm larvae penetrating the bolls. Rapidly growing tissues of cotton bolls also kill the larvae of boll weevil (*Anthonomus grandis*) (Hinds, 1906), pink bollworm (*P. gossypiella*) (Adkisson, 1962) and spotted bollworm (*E. vittella*). Similar effects have been reported in long leaf pine trees of southern pine beetle (*Dendroctonus frontalis*) (Hodges *et al.*, 1979).

### Tolerance

The ability of plants to withstand or recover from damage caused by insect abundance, equivalent to that required to damage a susceptible cultivar, is termed as 'tolerance mechanism of resistance'. Expression of tolerance is determined by the inherent genetic capability to outgrow an insect infestation or to recover and add new plant growth after insect damage. From an agronomic perspective, the plants of a tolerant cultivar produce a greater yield than plants of a non-tolerant susceptible cultivar. But tolerance often occurs in combination with antixenosis and antibiosis components of resistance.

### Quantitative measurements of tolerance

Techniques used to measure tolerance to insects include increases in the size and growth-rates of leaves, stems, petioles, roots, and seeds or fruits (Wood, 1961; Panda and Hienrichs, 1983; Ortega *et al.*, 1980; Sharma and Nwanze, 1997). Seedling survival is a measure of tolerance in cereals. Productive tillers produced by the plants affected by the sorghum shoot fly and stem borer (Sharma and Nwanze, 1997), and increase in grain mass in midge-infested plants (Sharma *et al.*, 2000a), and less grain damage per unit number of head bugs on sorghum (Sharma and Lopez, 1993; Padma Kumari *et al.*, 2000) are measures of tolerance to insect feeding and damage.

### Factors affecting expression of tolerance

Environmental factors directly affect the expression of tolerance to insect damage in several crops. Temperature affects the tolerance to greenbug (*Schizaphis graminum*) in cereals (Schweissing and Wilde, 1978). High levels of nutrients affect the tolerance of the sorghum seedlings to greenbug (Schweissing and Wilde, 1979). Fertility conditions and moisture availability also affect tiller production in plants damaged by sorghum shoot fly and stem borers (Sharma, 1993).

## INSECT-PLANT INTERACTIONS

### Insect responses to plant defense

Plant defensive systems can be altered by biotic stress from insects. Breakdown of precursors, such as glucosinolates and cyanogens, to release toxic substances or the release of Tannins from vacuoles during tissue damage are typical plant defensive responses. Damage by the herbivores can also lead to decreased nutritional quality of food to other herbivores. Green and Ryan (1972) reported that mechanical injury or damage by Colorado potato beetle to tomato or potato leaves leads to a systemic increase in the concentration of Proteinase inhibitors in the plants. Oligosaccharides released from the tissues during wounding travel in the vascular system to initiate the accumulation of Proteinase inhibitors in the plants, and this might reduce subsequent herbivore attack. Similarly, the jassid damage in cotton leads to the production and accumulation of free phenols and Tannins (Sharma and Agarwal, 1983a). Grass bug (*Labops hesperius*) decreases the nutritional value of infested grasses by increasing fiber content (Todd and Kamm, 1974). Mechanically damaged leaves of birch (*Betula pubescens*) have higher levels of phenolics and phenolic trypsin inhibitor in the neighboring undamaged leaves (Haukioja and Niemela, 1977). Larvae of the authumnal moth (*Oporinia autumnata*) fed on birch, gypsy moth (*Lymantria dispar*) on gray birch (*Betula populifolia*) and black birch (*Quercus velutina*) (Wallner and Walton, 1979), and tent caterpillar (*Malacosoma californicum pulviale*) on alder (*Alnus rubra*) leaves from damaged trees showed adverse effects on survival, growth, pupation, pupal mass and fecundity. Insect or artificial defoliation of trees sometimes leads to increased attacks by secondary pests in oak (Stanley, 1965), and this is associated with decreased content of phenolics in the bark following defoliation (Parker, 1977).

### Offensive adaptation of insects to overcome induced plant defense system

Grasshopper (*Melanoplus differentialis*) feeds preferentially on wilted sunflower leaves as a result of stem girdling by the beetle (*Mecas inornata*) (Lewis, 1979). In some cases, the insects kill or severely damage the plant parts and feed on tissue far away from the site of damage. Salivary secretions of heteropterous bugs, aphids, psyllids and mites have been shown to contain plant growth controlling substances, including indole acetic acid (Miles, 1968; Osborne,

1973; Hori, 1976). Hemipterous saliva often contains polyphenoloxidase enzymes, which appear to neutralize plant defensive phenolic compounds or are involved in auxin like compounds *in situ* (Miles, 1968, 1978). Some Lepidopteran leaf miners retard the senescence of leaves on which they feed (Osborne, 1973). Control of plant metabolism by substances introduced by the herbivore into its host plant is exemplified by the gall forming arthropods that cause diversion of vascular system of the plant and produces profound developmental and structural changes in the surrounding areas (Miles, 1968) and usurp the defensive system of the plant for their own protection. Thus, plants defensively respond to attack by insects, and insects possess adaptations to exploit and induce changes in the nutritional quality of their food plants.

## Insect-plant-environment interactions

The most desirable form of insect resistance is the one that is stable across locations and seasons. However, climatic and edaphic factors influence the level and nature of resistance to insect pests (Kogan, 1975). Inherited characters, especially those involving physiological characteristics, are often influenced by environmental factors. Such an interaction in insect-plant relationship can be studied by testing a set of diverse sources of resistance across locations and seasons. Some of the factors that influence plant resistance are discussed below.

**Soil moisture.** Water stress influences the level of plant resistance. Populations of *Aphis fabae* have lower rates of reproduction on water stressed plants (McMurtry, 1962). High levels of water stress also reduces the damage by the sorghum shoot fly (*A. soccata*) (Soman *et al.*, 1994). However, water stressed plants of sorghum suffer greater damage by the sugarcane aphid (*Melanaphis sacchari*). Thus, moisture stress can alter the plant reaction to insect damage leading either to increased or lower susceptibility to insect damage. Atmospheric humidity also interferes with insect-plant interaction (Sharma *et al.*, 2000b). High humidity increases the ease of detection of odors, and thus, may influence host finding and non-preference mechanism of resistance to insects.

**Plant nutrition.** Nutrients have an ambivalent effect on plant resistance. In some instances, high levels of nutrients increase the level of insect resistance, and in others they increase the susceptibility. Unfertilized plots of sorghum suffered high shoot fly damage compared to those treated with normal dosage of fertilizers (Sharma, 1985). Application of nitrogenous fertilizers decreases the damage by shoot fly and spotted stem borer (*C. partellus*) in sorghum (Reddy and Narasimha Rao, 1975; Chand *et al.*, 1979); while Channabasavanna *et al.* (1969) reported a decrease in incidence of the sugarcane topborer (*Scirpophaga excerptalis*). However, high levels of nitrogen lead to greater damage by the cotton jassid (*A. biguttula biguttula*) (Purohit and Deshpande, 1992). Changes in nutrient supply also affects the resistance to greenbug (*S. graminum*) in sorghum (Schweissing and Wilde, 1979).

**Temperature.** Temperature is one of the most important physical factors of the environment affecting the behavioral and physiological interaction of insects and plants (Benedict and Hatfield, 1988). Temperature induced stress can cause changes in plant physiology and affect the expression of genetic resistance, resulting in changes in the levels of biochemical and morphological defenses or nutritional quality of the host-plant. Temperature not only affects plant growth, but also affects the biology, behavior and population dynamics (Tingey and Singh, 1980). In general, low temperatures have a negative effect on resistance (Kogan, 1975). Differences between resistant and susceptible genotypes of sorghum to greenbug become clear with an increase in temperature (Schweissing and Wilde, 1978). In alfalfa, the level of resistance to pea aphid and alfalfa aphid is enhanced at higher temperatures (Kogan, 1975). Sorghum midge-resistant genotypes suffer greater damage at lower temperatures (Sharma *et al.*, 2000b). There is still a great need to learn as to how temperature as a trophic factor affects the insect host-plant interactions.

**Photoperiod.** Photoperiod affects the development of both insects and the crop plants. Photoperiod also alters the physico-chemical characteristics of crop plants, and thus, influences the interaction between insects and crop plants. Failure or inability to grow certain crop plants during off-season may be largely because of the increased susceptibility to insects and diseases. Intensity and quality of light have been reported to influence biosynthesis of phenylpropanoids (Hahlbrock and Grisebach, 1979) and anthocyanins (Carew and Krueger, 1976). Prolonged exposure to high intensity light induces susceptibility in PI 227687 soybean plants (otherwise resistant) to cabbage looper (*Trichoplusia ni*) (Khan *et al.*, 1986b). Susceptibility in sorghum to midge (*S. sorghicola*) increases under longer daylength (Sharma *et al.*, 1999a, 2000b).

**Biotypes.** Biotypes are populations of insects capable of damaging and surviving on plants (varieties) known to be resistant to the other populations of the same species (Kogan, 1975). The term biotype has been used here for a group of insects primarily distinguishable on the basis of their interaction with relatively genetically stable varieties or clones of host plants. Most biotypes have been recorded in aphids. Because of parthenogenesis, even a single mutant aphid capable of feeding on a resistant genotype can build up into a new biotype (Pathak, 1970). In India, instances of emergence of new biotypes have been documented in case of brown planthopper and gall midge in rice (Bentur *et al.*, 1987). To overcome the problem of biotypes, diverse genotypes with moderate levels of resistance should be utilized in a breeding program to have stable resistance against a number of prevalent insect biotypes.

## PLANT RESISTANCE TO INSECTS IN CROP IMPROVEMENT

Newly developed cultivars fail to face the challenge of heterogenous pest populations in the field, and consequently are rejected by the farmers. The products of crop improvement programs have to be viable under existing field conditions while having the potential for increased yields under improved or high input situations. Sowing pest-resistant cultivars is especially useful under subsistence farming. Utilization of plant resistance as a control strategy has enormous practical relevance and additional emotional appeal (Davies, 1981). It is in this context that host-plant resistance assumes a central role in our efforts to increase the production and productivity of crops.

There is great amount of variability in: i) emphasis placed on plant resistance in crop improvement programs, ii) availability of effective resistance screening techniques, iii) progress made in identification and utilization of insect-resistant genotypes, iv) multilocational testing to understand genotype x environment interactions, v) evolution of insect biotypes, vi) emphasis given to insect resistance in identifying and releasing new crop cultivars, and vii) concerted efforts to spread and popularize insect-resistant genotypes in farmers' fields. Considerable progress has been made in identification and utilization of resistance for a number of insect pests. However, resistance-breeding programs are underway only for a few crop pests. This aspect needs greater emphasis and input from the entomologists, breeders and support from the research managers. Insect resistance should be one of the major components in the development and release of new crop cultivars. Several sources of resistance to insects have been identified, and resistance transferred into high yielding cultivars in several crops. Insect-resistant cultivars in desirable agronomic backgrounds have been developed in several crops and are being currently grown by the farmers (Table 1). Cultivars with multiple resistance to insects and diseases will be in greater demand in future and this requires a concerted effort from scientists involved in the crop improvement programs.

**Table 1.** Insect-resistant cultivars/sources of resistance identified/developed in different crops in India

| Common name | Scientific name | Cultivar(s) |
|---|---|---|
| Rice (*Oryza sativa* L.) (Kalode, 1987; Nadarajan and Skaria, 1992; Krishnaiah, 1995) | | |
| Gall midge | *Orseolia oryzae* Wood-Mason | IR 36*, Kakatiya*, Dhanya Lakshmi*, Surekha*, Phalguna*, Kunti*, Shakti, Shamlei*, Asha*, and Rajendradhan*. |
| Rice stem borer | *Scirpophaga incertulas* Walk. | Ratna*, Sasyasree*, Saket*, IET 3127, IET 2812, and MTU 5849*. |
| Brown planthopper | *Nilaparvata lugens* Stal. | Manasarowar*, Bhadra*, Jyoti *, Co 42*, IET 7575, IET 6315, and MTU 5249*. |
| Green leafhopper | *Nephotettix* spp. | IR 20*, IET 7301, IET 7303, IET 7302, and Vani*. |
| Whitebacked planthopper | *Sogatella furcifera* Horvath. | AR 133, IC 25687, Tangner*, and Amelbero*. |
| Stem borer, leaf folder, thrips, and blue beetle | | PTB 12, PTB 20, PT 321, and H4. |

contd.

**Table 1.** contd.

| Common name | Scientific name | Cultivar(s) |
|---|---|---|
| **Maize (*Zea mays* L.) (Swarup, 1987)[2]** | | |
| Spotted stem borer | *Chilo partellus* Swin. | Ganga 4, 5, 7, and 9*; Ganga Safed 2*; Deccan 101 and 103*; Him 123*; Ageti*; VL 54; C 1, 3, and 7*, Kanchan*, and Kundan*. |
| Pink stem borer | *Sesamia inferens* Walk. | Deccan 101 and 103*. |
| Maize shoot fly | *Atherigona* sp. | DMR 5, NCD, and VC 80. |
| **Sorghum (*Sorghum bicolor* (L.) Moench) (Sharma, 1993)** | | |
| Sorghum shoot fly | *Atherigona soccata* Rond. | M 35–1*, Swati*, SPV 491*, IS 18551,ICSV 700, and ICSV 705. |
| Spotted stem borer | *C. partellus* Swin. | E 302, E 303, IS 2205, ICSV 700, and ICSV 705. |
| Sorghum midge | *Stenodiplosis sorghicola* Coq. | DJ 6514, AF 28, ICSV 197, ICSV 745*, and ICSV 88013. |
| Head bug | *Calocoris angustatus* Leth. | IS 17610, IS 17645, CSM 388, and Chencholam*. |
| **Pigeonpea (*Cajanas cajan* (L.) Millsp.) (Reed and Lateef, 1990)** | | |
| Pod borer | *Helicoverpa armigera* Hb. | ICPL 332*, PPE 45–2 (ICP 1964), MA 2, and ICPL 84060. |
| Pod fly | *Melanagromyza obtusa* Malloch | ICP 10531-EI, ICP 7941 E1, ICP 7946-E1, and ICP 7176-5. |
| **Chickpea (*Cicer arietinum* L.) (Lateef, 1985)** | | |
| Pod borer | *H. armigera* Hb. | ICC 506, ICCV 7*, ICC 6663, PDE 2, G 645, Dulia*, ICC 10667, and ICC 5264. |
| **Pea (*Pisum sativum* L.) (Lal, 1987)** | | |
| Pod borer | *Etiella zinkenella* Triet. | EC 33860, Bonville*, T 6113*, PS 410, 2S 21, 172M, and PS 410. |
| Leaf miner | *Phytomyza horticola* Goureau | P 402, PS 41–6, T 6113, PS 40, KMPR 9, P 402, and P 200. |
| **Cowpea (*Vigna unguiculata* L.) (Lal, 1987)** | | |
| Pod borer | *Maruca vitrata* (Geyer) | Banswara*, G 20, C 55, CR 2–55, P 1461, and G7. |
| Jassid | *Empoasca kerri* Pruthi | JG 10–72, NS 19–4–1, C 152, and 3–779 (1159). |
| Aphid | *Aphis craccivora* Koch. | P 1473, P 1476, and MS 9369. |
| Galerucid beetle | *Maduracia obscurella* Jacoby | 5269. |
| **Urd (*Vigna mungo* (L.) Hepper) (Lal, 1987)** | | |
| Pod borer | *H. armigera* Hb. | Kalai*, 338–3, Krishna*, and Co 3*, 4* and 5*. |
| Jassid | *E. kerri* Pruthi | Sinkheda 1*, Krishna*, H 70–3, and UPBI*. |
| Stem fly | *Ophiomyia centrosematis* de Merjere | Karaillet*, Killikullam*, 338/3, P 58, Co 3*, Co 4*, and Co 5*. |
| Galerucid beetle | *M. obscurella* Jacoby | Pusa 1*, KG 3, Krishna*, T 9, T 27, G 1*, and H 11*. |

contd.

**Table 1.** contd.

| Common name | Scientific name | Cultivar(s) |
|---|---|---|

**Mung (*Vigna radiata* (L.) Wlczek) (Lal, 1987).**

| | | |
|---|---|---|
| Pod borer | *M vitrata* | J1, LM 11, P 526, and P 336. |
| White fly | *Bemisia tabaci* Guen. | ML 337, ML 5, MH 85–61, and ML 325. |
| Stem fly | *O. centrosematis* | Co 3. |
| Galerucid bettle | *M. obscurella* | Jawahar 45*, PIMS 3, Gujarat 1*, R 12–16–3, S 9, ML 4, ML 6, PIMS 3, and PIMS 4*. |

**Groundnut (*Arachis hypogaea* L.) (Wightman *et al.* 1990)**

| | | |
|---|---|---|
| Leaf miner | *Aproaerema modicella* Duv. | ICGV 86031, ICGS 156 (M 13)*, FDRS 10, and ICG 57, 156, 541, 7016, 7404, and 9883. |
| Tobacco caterpillar | *Spodoptera litura* F. | ICGV 86031 and FDRS 10*. |
| Jassid | *E. kerri* | NcAc 2230, M 13, ICG 5043, and ICG 5049. |
| Thrips | *Thrips palmi* Karny | M 13*, Robut 33–1*, ICG 5043, and ICG 5044. |

**Soybean (*Glycine max* Merrill) (Sundaram and Sundara Babu, 1992)**

| | | |
|---|---|---|
| Leaf miner | *A. modicella* | PI 227687, Nimsoy*, PL 507, J 9 75-1, and C 18687. |

**Rape and mustard (*Brassica* spp.) (Bakhetia, 1987)**

| | | |
|---|---|---|
| Mustard aphid | *Lipaphis erysimi* Kalt. | C 294, Laha 101*, Pusa Kalyani*, RLM 84, 185, 195 and 185–59, 15, 34 and 52* in *B. juncea* Hooks f.& Thoms. Guilliver*, GSA, Regent*, GSV 2, V 3, 6, 8 and 13, GS 47, 86, 123, 139, and 391, and V 3, 6, 8, and 13 in *B. napus* L., Tora*, CDA-Span*, and Sariahi* in *B. campestris* Hooks f. & Thoms. |

**Cotton (*Gossypium* spp.) (Reddy and Rosaiah, 1987; Sundramurthy and Chitra, 1992)**

| | | |
|---|---|---|
| Pink bollworm | *Pectinophora gossypiella* Saund. | G27, LD 135*, Lohit, Abadhita , MCU 7, Sujata, Digvijay, and Saguineum* |
| Spotted bollworm | *Earias vittella* Fab. | L 1245*, JK 119–25–54, BCS 10, BSC 10–75, FBRN 2–6, HAO 66–107–1/1, G 27, and Sanguineum* |
| Cotton jassid | *Amrasca biguttula biguttula* Ishida | Khandwa 2, Badnawar, MCU 5*, Krishna*, Mahalaxmi*, Sujay*, Sanguineum*, and Eknath* |
| Whitefly | *B. tabaci* | LK 861*, Amravathi*, Kanchan*, Supriya*, and LPS 141*. |

**Sugarcane (*Saccharum officinarum* L.) (David, 1987).**

| | | |
|---|---|---|
| Stkla borer | *Chilo auricilius* Dudge | Co 7302* and CoS 767. |
| Top borer | *Scirpophage exerptalis* Wlk. | Co 7224, Co 6, and Co 1158. |
| Internode borer | *Chilo sacchariphagus indicus* (K.) | Co 6806, Co 6217, Co 975, and Co 77–1. |
| Sugarcane | *Melanaspis glomerata* Green | CoS 671, Co 8014, Co 6217, Co 1132, Co 611, and Co 6907. |

contd.

**Table 1.** contd.

| Common name | Scientific name | Cultivar(s) |
|---|---|---|
| White fly | *Aleurolobus barodensis* Mask. | Co 671. |
| White grub | *Holotrichia* spp. | Co 6304, Co 1158, and Co 5510. |

**Tobacco (*Nicotiana* spp.) (Chari, 1987)**

| | | |
|---|---|---|
| Tobacco leaf caterpillar | *Spodoptera litura* Fab. | GT 4* and DWFC*. |
| Aphid | *Myzus persicae* Sulz. | Jamaica*, Cuban*, Fransons*, Little rittendant*, and Sumatra*. |
| Stem borer | *Scrobipalpa heliopa* Low | SBR 1* and SBR 2*. |

**Brinjal (*Solanum melongena* L.) (Jiyani *et al.*, 1992)**

| | | |
|---|---|---|
| Shoot/ fruit borer | *Leucinodes orbonalis* G. | Chaklasi doli*, Doli 5*, Pusa purple*, SM 67, and SM 68. |
| Jassid | *A. biguttula biguttula* | Chaklasi Doli*, Doli 5*, and Pusa purple*. |
| White fly | *B. tabaci* | Pusa purple*. |

**Tomato (*Lycopersicum esculentum* Mill.) (Mishra and Mishra, 1992)**

| | | |
|---|---|---|
| Fruit borer | *H. armigera* | BT 1, T 32*, and T 27*. |

**Lady's finger (*Abelmoschus esculentus* (L) Moench (Ahmad and Rizvi, 1992)**

| | | |
|---|---|---|
| Shoot/ fruit borer | *E. vittella* | AE 57, PMS 8, Parkins long green*, PKX 9275, and Karnual special*. |

**Potato (*Solanum tuberosum* L.)**

| | | |
|---|---|---|
| Potato tuber moth | *Phthormoea opercullela* Zeller | QB 1A 21–29. |

* Released for cultivation.

## INFLUENCE OF PLANT RESISTANCE ON POPULATION DYNAMICS OF INSECTS

Plant resistance has been used to suppress pest populations for many years. However, only a few pest species can be controlled by the use of resistant varieties alone. Adequate levels of resistance are present against a few pests only, e.g., sorghum midge and green bug in sorghum; leafhoppers and gall midge in rice; jassids in cotton; jassids and thrips in groundnut; and stem borers, scale insects, and whitefly in sugarcane. Cultivars with low to moderate levels of resistance can be extremely useful for population suppression of the target pests over a period of time. The impact of resistant varieties on pest populations can be demonstrated by making use of simple insect models of Knipling (1964) as adopted by Adkisson and Dyck (1980), Teetes (1985) and Sharma (1993). The impact of growing a resistant, a moderately-resistant, and a susceptible variety on insect populations over a period of time has been explained for sorghum midge (*S. sorghicola*) in Fig. 1. The effect of resistant cultivars would be similar for other insects depending on the level of resistance and the mechanisms of resistance. The rate of multiplication of sorghum midge is six times on a susceptible cultivar (CSH 1), and three times on a moderately-resistant cultivar IS (12664C) as compared with the resistant cultivar (DJ 6514) (Sharma, 1985). Based on the rates of insect multiplication, there would be 1296 and 81 times as many insects in areas planted to CSH 1 and IS 12664C, respectively, as compared with areas cropped with DJ 6514, in which the insect population would remain constant. Thus, plant resistance has a great influence on insect populations, which is cumulative overtime. These models can also explain the situations where minor pests become very serious  with the introduction of newly developed high-yielding susceptible cultivars.

The levels of resistance quite often are not adequate, and in such situations, host-plant resistance in combination with insecticides or natural enemies can reduce pest densities over time. One application or insecticides (with 90% mortality) will lead to 810 insects in an area planted to a moderately-resistant cultivar (IS 12664C) as compared with 12,960 insects in an area planted with a susceptible cultivar (CSH 1). Similarly, moderate levels of plant resistance in combination with natural enemies (e.g., *Tetrastichus diplosidis* on sorghum midge resulting in 50% parasitization) can reduce the insect numbers to 506 on IS 12664C and 8,100 on CSH 1 (Sharma, 1985).

## INFLUENCE OF PLANT RESISTANCE ON ECONOMIC THRESHOLD LEVELS (ETLs)

Insect resistant cultivars not only decrease the pest populations, but also delay the time required by the pest to attain the ETL depending on the nature of resistance and the criterion on which it is based (Fig. 1). If the ETL is based on damage (e.g., percentage of deadhearts for sorghum shoot fly and stem borer, number of leaves damaged by aphids or percentage leaf area consumed by the armyworms), and the insect population increases over the season, then a susceptible cultivar will suffer economic loss in July, a moderately-resistant cultivar in August, and a resistant cultivar can withstand pest densities until the end of the season (Fig. 1a) (Sharma, 1993). In case of insects in which the damage is limited to a particular stage and a short span of time (e.g., deadheart formation due to sorghum shoot fly and stem borers), a cultivar can be planted up to a period when insect density is below ETL. If ETL is based on adults, which is a non-damaging stage of the insect (e.g., sorghum midge adults or number of moths caught in pheromone or light traps), the ETL will increase with an increase in the level of insect resistance  (Fig. 1b). In the case of sorghum midge, the ETL is 1 adult per panicle on a susceptible cultivar (CSHI), 25 adults per panicle on a moderately-resistant cultivar (ICSV 745), and 50 adult per panicle on a highly-resistant cultivar (ICSV 197) (Sharma *et al.*, 1993b). If the ETL is based on adults, which also cause damage (e.g., head bugs), and the resistance is based on non-preference and antibiosis (which will

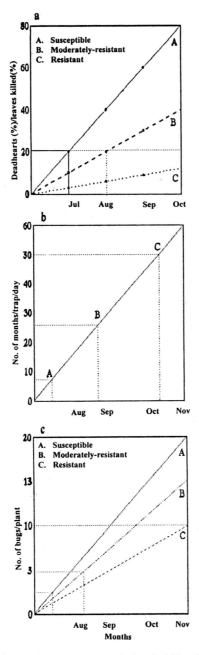

**Fig. 1.** Effect of host plant resistance on economic threshold levels (ETL) when the ETL is based on damage (a), non-damaging adult stage (b), and damaging adults on genotypes with non-preference and antibiosis mechanisms (c). A, B, and C are the susceptible, moderately-resistant and resistance genoypes, respectively.

decrease the rate of population increase), then the time taken by the insect to attain ETL will also be extended by the resistant cultivars (Fig. 1c) (Sharma, 1993).

## CONCLUSIONS

Considerable progress has been made in developing standard techniques to screen for resistance under adequate and uniform insect pressure. There is a need to establish insect rearing and greenhouse facilities to undertake screening and breeding for resistance to insects. The evaluation and selection criteria need to be standardized for many insects. Multilocational testing of the identified sources and breeding material need to be strengthened to identify stable and diverse source of resistance, and establish the presence of new insect biotypes. Resistance to insects should be given as much emphasis as yield to identify new varieties and hybrids for cultivation by the farmers. Insect-resistant varieties exercise a constant and cumulative effect on insect populations over time and space, have no adverse effects on the environment, reduces the need to use pesticides, has no extra-cost to the farmers and does not require inputs and application skills by the farmers. However, the development of a pest resistant variety can take several years (5-10 years). The presence or selection of an insect biotype capable of infesting the resistant variety can limit the effectiveness of resistant varieties. Introduction of varieties with resistance to insects will form the backbone of pest management programs in future for sustainable crop production and environment conservation.

## REFERENCES

Adkisson, P.L. 1962. *Cotton Stocks Screened for Resistance to Pink Bollworm, 1961–1962.* Texas Agriculture Experiment Station, Miscellaneous Publication 606. Texas, U.S.A.

Adkisson, P.L. and Dyck, V.A. 1980. Resistant varieties in pest management systems. *In:* F.G. Maxwell and P.R. Jennings (Eds.). *Breeding Plants Resistant to Insects.* John Wiley & Sons, New York, USA. pp. 234–251.

Ahmad, H. and Rizvi, S.M.A. 1992. Varietal resistance of *Abelmoschus esculentus* (L.) Moench to the Okra shoot and fruit borer, *Earias vittella* (Fabr.). *In:* National Seminar on Changing Scenario in Pests and Pest Management in India, 31 Jan—1Feb 1992. Plant Protection Association of India, Rajendranagar, Hyderabad, Andhra Pradesh, India.

Auclair, J.L., Maltais, J.B. and Cartier, J.J. 1957. Factors in resistance of peas to the pea aphid, *Acyrthosiphon pisum* (Harris) (Homoptera: Aphididae). II. Amino acids. *Canadian Entomolgist* 10: 457–464.

Bakhetia, D.R.C. 1987. Insect pests of rapeseed-mustard and their management. *In:* M. Veerabhadra Rao, and S. Sithanantham (Eds.). *Plant Protection in Field Crops.* Plant Protection Association of Inida, Rajendranagar, Hyderabad Andhra Pradesh, India. pp. 249–260.

Benedict, J.H. and Hatfield, J.L. 1988. Influence of temperature induced stress on host plant suitability to insects. *In* E.A. Heinrichs (Ed.) *Plant stress-Insect Interactions.* John Wiley and Sons, New York, USA. pp. 139–165.

Bentur, J.S., Srininvasan, T.E. and Kalode, M.B. 1987. Occurrence of a virulent rice gall midge (GM) *Orseolia oryzae* Wood-Mason biotype (?) in Andhra Pradesh, India. *International Rice Research Newsletter,* 12: 33–34.

Blaney, W.M. and Chapman, R.F. 1970. The function of the maxillary palps of Acrididae (Orthoptera). *Entomologia Experimentalis et Applicalis* 13: 363–376.

Blum, A. 1968. Anatomical phenomena in seedlings of sorghum varieties resistant to the sorghum shootfly (*Atherigona varia soccata*). *Crop Science* **8**: 388–390.

Burton, G.W., Hanna, W.W., Johnson Jr, J.C., Leuck, D.B., Monson, W.G., Powell, J.B., Wells, H.D. and Widstrom, N.W. 1977. Pleiotropic effects of the trichomless gene in pearl millet transpiration, forage quality and pest resistance, *Crop Science* **17**: 613–616.

Carew, D.P. and Krueger, J. 1976. Anthocyanidins of *Catharanthus* callus cultures. *Phytochemistry* **15**: 442.

Chand, P., Sinha, M.P. and Kumar, A. 1979. Nitrogen fertilizer reduces shootfly incidence in sorghum. *Science and Culture* **45**: 61–62.

Channabasavanna, G.P., Venkat Rao, B.V. and Rajagopal, G.K. 1969. Preliminary studies on the effect of incremental levels of phosphatic fertilizer on the incidence of jowar shootfly. *Mysore Journal of Agriculture Sciences* **3**: 253–255.

Chari, M.S. 1987. Insect pests of tobacco and their management. *In:* M. Veerabhadra Rao and S. Sithanantham (Eds.). *Plant Protection in Field Crops.* Plant Protection Association of India, Rajendranagar, Hyderabad, Andhra Pradesh, India. pp. 349–368.

Chiang, H.C. and Norris, D.M. 1983. Morphological and physiological parameters of soybean resistance to agromyzid bean flies. *Environmental Entomology* **12**: 260–265.

David, H. 1987. Insect pests of sugarcane and their management. *In* Veerabhadra Rao, M. and Sithanantham, S. (Eds.). *Plant Protection in Field Crops.* Plant Protection Association of India, Rajendranagar, Hyderabad, Aṇdhra Pradesh, India. pp. 319–388.

Davies, J.C. 1981. Pest losses and control of damage on sorghum in developing countries. The realities and myths. *In* House, L.R., Mughogo, L.K. and Peacock, J.M (Eds.). *Sorghum in the Eighties.* International Crops Research Institute for the Semi-Arid Tropics (ICRISAT), Patancheru, Andhra Pradesh, India. pp. 215–224.

de Wilde, J., Lambers-Suverkropp, K., Hills, R. and van Tol, A. 1969. Responses to air flow and airbrone plant odour in the Colorado potato beetle. *Netherlands Journal of Plant Pathology* **75**: 53–57.

Elliger, C.A., Chan B.G. and Waiss Jr, A.C. 1978. Relative toxicity of minor cotton terpenoids compared to gossypol. *Journal of Economic Entomology* **71**: 161–164.

Fiori, B.J. and Craig, D.C.W. 1987. Relationship between color intensity of leaf supernatants from resistant and susceptibe birch trees and rate of oviposition by the birch leaf miner (Hymenoptera: Tenthredinidae). *Journal of Economic of Entomology* **80**: 1331–1333.

Green, T.R. and Ryan, C.A. 1972. Wound-induced proteinase inhibitor in plant leaves: A possible defense mechanism against insects. *Science* **175**: 776–777.

Guerin, P.M., Stadler, E. and Buser, H.R. 1983. Identification of host plant attractants for the carrot fly, *Psila rosae. Journal of Chemical Ecology* **9**: 843–861.

Hahlbrock, K. and Grisebach, H. 1979. Enzymatic controls in the biosynthesis of lignins and flavonoids. *Annual Review of Plant Physiology* **30**: 105–130.

Hamamura, Y. 1970. The substances that control the feeding behavior and the growth of the silkworm *Bombyx mori* L. *In* Wood, D.L., Silverstein, R.M. and Nakajima. M. (Eds.). *Control of Insect Behaviour by Natural products.* Academic Press, New York, USA. pp. 55–80.

Haskell, P.T., Paskin, M.W.J. and Moorehouse, J.E. 1962. Laboratory observations on factors affecting the movements of hoppers of the desert locust. *Journal of Insect Physiology* **8**: 53–78.

Haukioja E. and Niemela, P. 1977. Retarded growth of the geometrid larva after mechanical damage to the leaves of its host tree. *Annals of Zoologic Fennici* **14**: 48–52.

Heinrichs, E.A. 1988. Role of insect-resistant varieties in rice IPM systems. *In:* P.S. Teng, and K.L. Heong. (Eds.). *Pesticide Management and Integrated Pest Management in Southeast Asia.* Consortium for International Crop Protection, Maryland, USA.

Hierholzer, 1950. Ein beitrag zur frage der orientierung von *Ips curvidens* Germ. *Zeitschrift ful Tier Psychologie* **7**: 588–620.

Hinds, W.E. 1906. Proliferation as a factor in the natural control of the Mexican cotton boll weevil. *U.S. Department of Agriculture Bureau, Entomology Bulletin* **59**: 45 pp.

Hodges, J.D., Elam, W.W., Watson, W.F. and Nebeker, T.B. 1979. Oleoresin characteristics and susceptibility of four southern pines to southern pine beetle (Coleoptera, Scolytidae) attacks. *Canadian Entomologist* **111**: 889–896.

Hori, K. 1976. Plant-growth regulating factor in the salivary gland of several heteropterous insects. *Comparative Biochemistry and Physiology* **23**: 1075–1080.

Hussain, M.A. and Trehan, K.N. 1940. Final report on the scheme of investigation on the whitefly of cotton in the Punjab. *Indian Journal Agricultural Sciences* **10**: 101–109.

Iseley, D. 1928. The relation of leaf color and leaf size to boll weevil infestation. *Journal of Economic Entomology* **21**: 553–559.

Jiyani, D.B., Patel, N.C., Ratnapara, H.C., Patel, J.R. and Borad, P.K. 1992. Varietal resistance in brinjal to insect pests and diseases. *In: National Seminar on Changing Scenario in Pests and Pest Management in India.* 31 Jan—1 Feb 1992. Plant Protection Association of India, Rajendranager, Hyderabad, Andhra Pradesh, India.

Kalode, M.B. 1987. Insects pests of rice and their management. *In* Veerabhadra Rao, M. and Sithanantham, S. (Eds.). *Plant Protection in Field Crops.* Plant Protection Association of India, Rajendranagar, Hyderabad, Andhra Pradesh, India. pp. 61–74.

Kalode, M.B. and Sharma, H.C. 1985. Host plant resistance to insects: Progress, Problems, and future needs. *In* Sharma, H.C. and Veerbhadra Rao, M. (Eds.). *Pests and Pest Management in India: the Changing Scenario.* Plant Protection Association of India, Rajendranagar, Hyderabad, Andhra Pradesh, India. pp. 229–243.

Kalode, M.B., Bentur, J.S. and Srinivasan, T.E. 1989. Screening and breeding rice for stem borer resistance. *In* Nwanze, K.F. (Ed.). *International Workshop on Sorghum Stem Borers.* International Crops Research Institute for the Semi-Arid Tropics (ICRISAT), Patancheru 502 324, Andhra Pradesh, India. pp. 153–157.

Khan, Z.R., Norris, D.M., Chiang, H.S. and Oosterwyk, A.S. 1986a. Light induced susceptibility in soybean to cabbage looper, *Trichoplusia ni* (Lepidoptera: Noctuidae). *Environmental Entomology* **15**: 803–808.

Khan, Z.R., Ward, J.T. and Norris, D.M. 1986b. Role of trichomes in soybean resistance to cabbage looper, *Trichoplusia ni. Entomologica Experimentalis et Applicata* **42**: 109–117.

Klingauf, F. 1971. Die wirkung des glucosids phlorizin auf das wirtswahlverhalten von *Rhopalosiphum insertum* (Walk) und *Aphis pomi* De Geer (Homoptera: Aphididae). *Zeitschrift fur Angewandte Entomologie* **68**: 41–55.

Klingauf, F., Nocker-Wenzel, K., und Klein, W. 1971. Einfluss einger Wachskomponenten von *Vicia faba* L. auf das Wirtswahlverhalten von *Acythosiphon pisum* (Harris) (Homoptera: Aphididae) *Zeitschrift fur Pflanzen Pathologie und Pflanzenschutz* **78**: 641–648.

Klingauf, F., Nocker-Wenzel, K., und Rottger, U. 1978. Die rolle peripherer pflanzenwachsefur den befall durch phytophage inseketen. *Zeitschrift fur Pflanzenkrankheiten und Pflanzenschutz* **85**: 228–237.

Knipling, E.F. 1964. *The Potential Role of Sterility Method for Insect Population Control with Special Reference to Combining this Method with Conventional Methods.* United States Department of Agriculture, Agriculture Research Service Bulletin No. 3388. 54 pp.

Kogan, M. 1975. Plant resistance in pest management. *In* Metcalf, R.L. and Luckmann, W.H. (Eds.). *Introduction to Insect Pest Management.* John Wiley and Sons, New York, USA. pp. 103–146.

Krishanaiah, K. 1995. Changing scenario in rice insect pest problems. *In:* Sharma, H.C. and Veerbhadra Rao, M. (Eds.). *Pests and Pest Management in India: The Changing Scenario.* Plant Protection Association of India, Rajendranagar, Hyderabad, Andhra Pradesh, India. pp. 11–18.

Kogan, M. and Ortman, E.E. 1978. Antixenosis: A new term proposed to replace Painter's "nonpreference" modality of resistance. *Bulletin of Entomological Society of America* **24**: 175–176.

Lal, S.S. 1987. Insect pests of mung, urid, cowpea and pea and their management. *In* Veerabhadra Rao, M. and Sithanantham, S. (Eds.). *Plant Protection in Field Crops.* Plant Protection Association of India, Rajendranagar, Hyderabad, Andhra Pradesh, India. pp. 185–202.

Lapointe, C. and Tingey, W.M. 1986. Glandular trichomes of *Solanum neocardensii* confer resistance to green peach aphid (Homoptera: Aphididae). *Journal of Economic Entomology* **79**: 1264–1268.

Lateef, S.S. 1985. Gram podborer (*Heliothis armigera*) (Hub.) resistance in chickpeas. *Agriculture, Ecosystems & Environment* **14**: 95–102.

Lewis, A.C. 1979. Feeding preference for diseased and wilted sunflower in the grasshopper, *Melanoplus differentialis*. *Entomologia Experimentalis et Applicata* **26**: 202–207.

Maiti, R.K. and Bidinger, F.R. 1979. A. simple approach to identification of shootfly tolerance in sorghum. *Indian Journal of Plant Protection* **7**: 135–140.

Maiti, R.K. and Gibson, R.W. 1983. Trichomes in segregating generations of sorghum matings, II. Association with shoot fly resistance. *Crop Science* **23**: 76–79.

Martin, G.A., Richard, C.A. and Hensley, S.D. 1975. Host resistance to *Diatraea saccharalis* (F.): Relationship of sugarcane node hardness to larval damage. *Environmental Entomology* **4**: 687–688.

Maxwell, F.G. and Jennings, P.R. (Eds.) 1980. *Breeding Plants Resistant to Insects*. John Wiley and Sons, New York, USA. 683 pp.

McMurtry, J.A. 1962. Resistance of alfalfa to spotted alfalfa aphid in relation to environmental factors. *Hilgardia* **32**: 501–539.

Miles, P.W. 1968. Insect secretion in plants. *Annual Review of Phytopathology* **43**: 137–164.

Miles, P.W. 1978. Redox reactions of hemipterous saliva in plant tissues. *Entomologia Experimentalis et Applicata* **24**: 534–539.

Misra, N.C. and Misra, S.N. 1992. Varietal performance of tomatoes in wilt and fruit borer prone north-eastern ghat zone of Orissa. *In:* National Seminar on Changing Scenario in Pests and Pest Management in India, 31 Jan—1 Feb 1992. Plant Protection Association of India, Rajendranagar, Hyderabad, Andhra Pradesh, India.

Moericke, V. 1969. Host plant specific color behavior by *Hyalopterus pruni* (Aphididae). *Entomologia Experimentalis et Applicata*. **12**: 524–534.

Mulkern. G.B. 1969. Behavioral influences on food selection in grasshoppers (Orthoptera: Acrididae). *Entomologia Experimentalis et Applicata* **12**: 509–523.

Nadarajan, L. and Skaria, B.P. 1992. Host plant resistance to insect pests in rice in Kerala. *In:* National Seminar on Changing Scenario in Pests and Pest Management in India, 31 Jan—1Feb 1992. Plant Protection Association of India, Rajendranagar, Hyderabad, Andhra Pradesh, India.

Ortega, A., Vasal, S.K., Mihm, J.A. and Hershey, C. 1980. Breeding for insect resistance in maize. *In* F.G. Maxwell, and P.R. Jennings (Eds.). *Breeding Plants Resistant to Insects*. Wiley, New York, USA. pp. 372–419.

Ortiz, R., Raman, K.V., Wanaga, M. and Palacios, M. 1990. Breeding for resistance to potato tuber moth, *Pthorimea opercullela* (Zeller) in diploid potatoes. *Euphytica* **50**: 119–125.

Ortiz, R., Vuylsteke, D., Ferris, R.S.B. and Dumpe, B. 1995. Banana weevil resistance and corn hardness in banana germplasm. *Euphytica* **86**: 95–102.

Osborne, D.J. 1973. Mutual regulation of growth and development in plants and insects. *In:* Van Emden, H.F. (Ed.). *Insect Plant Relationship*. Blackwell, Oxford, UK. pp. 33–42.

Padma Kumari, A.P., Sharma, H.C. and Reddy, D.D.R. 2000. Components of resistance to sorghum head bug, *Calocoris angustatus*. *Crop Protection*, **19**: 385-392.

Painter, R.H. 1951. *Insect Resistance in Crop Plants*. MacMillan, New York, USA. 520 pp.

Panda, N. and Heinrichs, E.A. 1983. Levels of tolerance and antibiosis in rice varieties having moderate resistance to the brown planthopper, *Nilaparvata lugens* (Stal) (Hemiptera: Delphacidae). *Environmental Entomology* **12**: 1204–1214.

Parker, J. 1977. Phenolics in black oak bark and leaves. *Journal of Chemical Ecology* **3**: 489–496.

Pathak, M.D. 1970. Genetics of plants in pest management. *In* Rabb, R.L. and Guthrie, F.E. (Eds.). *Concepts of Pest Management*. North Carolina State University, Raleigh, North Carolina, USA. pp. 138–157.

Penny, L.H., Scott, G.E. and Guthrie, W.D. 1967. Recurrent selection for European corn borer resistance in maize. *Crop Science* **7**: 407–409.

Pierce, H.D., Vernon, R.S., Borden, J.H. and Oehlschlager, A.C. 1978. Host selection by *Hylemya antiqua* (Meigen): Identification of three new attractants and oviposition stimulants. *Journal of Chemical Ecology* **4**: 65–72.

Prokopy, R.J. and Owens, E.D. 1983. Visual detection of plants by herbivorous insects. *Annual Review of Entomology* **28:** 337–364.

Purohit, M. and Deshpande, A.D. 1992. Effect of nitrogenous fertilizer application on cotton leafhopper, *Amrasca biguttula biguttula* (Ishida). *In:* National Seminar on Changing Scenario in Pests and Pest Management in India, 31 Jan—1 Feb 1992. Plant Protection Association of India, Rajendranagar, Hyderabad, Andhra Pradesh, India.

Raman, K.V., Tingey, W.M. and Gregory, P. 1978. Potato glycoalkaloids: effects on survival and feeding behaviour of the potato leafhopper. *Journal of Economic Entomology* **72:** 337–341.

Reddy, A.S.N. and Rosaiah, B. 1987. Insect pest management in cotton. *In:* Veerbhadra Rao, M. and Sithanantham, S. (Eds.). *Plant Protection in Field Crops.* Plant Protection Association of India, Rajendranagar, Hyderabad, Andhra Pradesh, India. pp. 393–300.

Reddy, K.S. and Narasimha Rao, D.V. 1975. Effect of nitrogen application on shootfly incidence and grain maturity in sorghum. *Sorghum Newsletter* **18:** 23–24.

Reed, W. and Lateef, S.S. 1990. Pigeonpea: pest management. *In* Nene, Y.L., Hall, S.D. and Shiela V.K. (Eds.). *The Pigeonpea.* C.A.B. International, Wallingford, Oxon, UK. pp. 349–374.

Reese C.J.C. 1969. Chemoreceptor specificity associated with choice of feeding site by the beetle, *Chrysolina brunsvicensis* on its food plant, *Hypericum hirsutum. Entomologia Experimentalis et Applicata* **12:** 565–583.

Reese, J.C. and Beck, S.D. 1976. Effects of certain allelochemics on the black cutworm, *Agrotis ipsilon:* effects of catechol, dopamine, and chlorogenic acid on larval growth, development and utilization of food. *Annals of the Entomological Society of America* **69:** 68–72.

Schoonhoven, L.M. 1972. Some aspects of host selection and feeding in phytophagous insects. *In* Rodriguez, J.G. (Ed.). *Insect and Mite Nutrition.* North-Holland, Amsterdam, The Netherlands. pp. 557–566.

Schweissing, F.C. and Wilde, G. 1978. Temperature influence on greenbug resistance of crops in the seedling stage. *Environmental Entomology* **7:** 831–834.

Schweissing, F.C. and Wilde, G. 1979. Temperature and plant nutrient effects on resistance of seedling sorghum to the greenbug. *Journal of Economic Entomology* **72:** 20–23.

Sharma, H.C. 1985. Strategies for pest control in sorghum in India. *Tropical Pest Management* **31:** 167–185.

Sharma, H.C. 1993. Host plant resistance to insects in sorghum and its role in integrated pest management. *Crop Protection* **12:** 11–34.

Sharma, H.C., Abraham, C.V. and Stenhouse, J.W. 2000a. Inheritance of compensation in grain mass and volume in sorghum and their association with resistance to sorghum midge, *Stenodiplosis sorghicola. Euphytica* (Submitted).

Sharma, H.C. and Agarwal, R.A. 1982a. Consumption and utilization of bolls of different cotton genotypes by larvae of *Earias vittella* F. and effect of gossypol and Tannins on food utilization. *Zietschrift fur Angewandte Zoologie* **68:** 13–38.

Sharma, H.C. and Agarwal, R.A. 1982b. Effect of some antibiotic compounds in *Gossypium* on the post-embryonic development of spotted bollworm (*Earias vittella* F.). *Entomologia Experimentalis et Applicata* **31:** 225–228.

Sharma, H.C. and Agarwal, R.A. 1983a. Role of some chemical components and leaf hairs in varietal resistance in cotton to jassid, *Amrasca biguttula biguttula* Ishida. *Journal of Entomological Research* **7:** 145–149.

Sharma, H.C. and Agarwal, R.A. 1983b. ovipositional behavior of spotted bollworm, *Earias vittella* Fab. on some cotton genotypes. *Insect Science and its Application* **4:** 373–376.

Sharma, H.C. and Agarwal, R.A. 1983c. Factors affecting genotypic susceptibility to spotted bollworm (*Earias vittella* Fab.) in cotton. *Insect Science and its Application* **4:** 363–372.

Sharma, H.C., Agarwal, R.A. and Singh, M. 1982. Effect of some antibiotic compounds in cotton on post-embryonic development of spotted bollworm (*Earias vittella* F.) and the mechanism of resistance in *Gossypium arboreum. Proceedings of Indian Academy of Sciences (B)* **91:** 67–77.

Sharma, H.C., Doumbia, Y.O., Haidara, M., Scheuring, J.F., Ramaiah, K.V., and Beninati, N.F. 1994. Sources and mechanisms of resistance to sorghum head bug, *Eurystylus immaculatus* Odh in West Africa. *Insect Science and its Application* **15:** 39–48.

Sharma, H.C., Leuschner, K. and Vidyasagar, P. 1990a. Factors influencing oviposition behavior of sorghum midge, *Contarinia sorghicola* Coq. *Annals of Applied Biology* 116: 431–439.

Sharma, H.C. and Lopez, V.F. 1990. Mechanisms of resistance in sorghum to head bug, *Calocoris angustatus*. *Entomologia Experimentalis et Applicata* 57: 285–294.

Sharma, H.C, and Lopez, V.F. 1993. Survival of *Calocoris angustatus* (Hemiptera: Miridae) nymphs on diverse sorghum genotypes. *Journal of Economic Entomology* 86: 607–613.

Sharma, H.C., Lopez, V.F. and Nwanze, K.F. 1993a. Genotypic effects of sorghum accessions on fecundity of sorghum head bug, *Calocoris angustatus* Lethiery. *Euphytica* 65: 167–175.

Sharma, H.C., Mukuru, S.Z., Manyasa, E. and Were, J. 1999a. Breakdown of resistance to sorghum midge, *Stenodiplosis sorghicola*. *Euphytica* 109: 131–140.

Sharma, H.C. and Norris, D.M. 1991. Chemical basis of resistance in soybean to cabbage looper, *Trichoplusia ni*. *Journal of Science of Food and Agriculture* 55: 353–364.

Sharma, H.C. and Norris, D. M. 1994. Biochemical mechanisms of resistance to insects in soybean: Extraction and fractionation of antifeedants. *Insect Science and its Application*, 15: 31–38.

Sharma, H.C. and Nwanze, K.F. 1997. *Mechanisms of Resistance to Insects in Sorghum and Their Usefulness in Crop Improvement*. Information Bulletin No 45. International Crops Research Institute for the Semi-Arid tropics (ICRISAT), Patancheru 502 324, Andhra Pradesh, India. 51 pp.

Sharma, H.C., Saxena, K.B. and Bhagwat, V.R. 1999b. *Legume Pod Borer*, Maruca vitrata: *Bionomics and Management Information Bulletin No. 55*. International Crops Research Institute for the Semi-Arid Tropics (ICRISAT), Patancheru 502 324, Andhra Pradesh, India. 35 pp.

Sharma, H.C., Singh, F. and Nwanze, K.F. (Eds.) 1997. *Plant resistance to insects in Sorghum*. International Crops Research Institute for the Semi-Arid Tropics (ICRISAT), Patancheru 502 324, Andhra Pradesh, India. 216 pp.

Sharma, H.C., Singh, B.U., Hariprasad, K.V. and Bramel-Cox, P.J. 1999c. Host plant resistance to insects in integrated pest management for a safer environment. *Proceedings, Academy of Environmental Biology* 8: 113–136.

Sharma, H.C., Taneja, S.L., Leuschner, K. and Nwanze, K.F. 1992. *Techniques to Screen Sorghums for Resistance to Insects*. Information Bulletin no. 32. International Crops Research Institute for the Semi-Arid Tropics (ICRISAT), Patancheru 502 324, Andhra Pradesh, India. 48 pp.

Sharma. H.C. and Sullivan, D.J. 2000. Screening for plant resistance to the Oriental armyworm, *Mythimna separata* (Lepidoptera: Noctuidae) in pearl millet, *Pennisetum glaucum*. *Journal of Agricultural and Urban Entomology* (in press).

Sharma, H.C., Venkateswarlu, G. and Sharma, A. 2000b. Genotype x environment interactions for expression of resistance to sorghum midge, *Stenodiplosis sorghicola*. *Euphytica* (submitted).

Sharma, H.C., Vidyasagar, P. and Leuschner, K. 1988a. Field screening for resistance to sorghum midge (Diptera: Cecidomyiidae). *Journal of Economic Entomology* 81: 327–334.

Sharma, H.C.,Vidyasagar, P. and Leuschner, K. 1988b. No-choice cage technique to screen for resistance to sorghum midge (Diptera: Cecidomyiidae). *Journal of Economic Entomology* 81: 415–422.

Sharma, H.C., Vidyasagar, P. and Leuschner, K. 1990b. Components of resistance to the sorghum midge, *Contarinia sorghicola*. *Annals of Applied Biology* 116: 327–333.

Sharma, H.C., Vidyasagar, P. and Nwanze, K.F. 1993b. Effect of host-plant resistance on economic injury levels for the sorghum midge, *Contarinia sorghicola*. *International Journal of Pest Management* 39: 435–444.

Sharma, H.C., Vidyasagar, P. and Subramanian, V. 1993c. Antibiosis component of resistance in sorghum to sorghum midge, *Contarinia sorghicola*. *Annals of Applied Biology* 123: 469–483.

Sharma, H.C. and Youm, O. 1999. Integrated pest management in pearl millet with special reference to host plant resistance to insects. *In* Khairwal, I.S., Rai, K.N. andrews, D.J. and Harinarayana, G. (Eds.). *Pearl Millet Improvement*. Oxford and IBH Publishing Company Pvt. Ltd., New Delhi, India. pp. 381–425.

Shifriss, O. 1981. Do *Cucurbita* plants with silvery leaves escape virus infection? *Cucurbit General Cooperative Report* 4: 42–45.

Sinden, S.L., Sanford, L.L., Cantelo, W.W. and Deahl, K.L. 1986. Leptine glycoalkaloids and resistance to the Colorado pototo beetle (Coleoptera: Chrysomelidae) in *Solanum chacoense*. *Environmental Entomology* 15: 1057–1062.

Singh, S.P. and Jotwani, M.G. 1980. Mechanism of resistance in sorghum to shootfly. III. Biochemical basis of resistance. *Indian Journal of Entomology* 42: 551–566.

Singh, P. and Moore, R.F. (Eds.) 1985. *Handbook of Insect Rearing*. Vol. I and II. Elsevier, New York, USA.

Smith, C.M. 1989. *Plant resistance to insects*. John Wiley and Sons, New York, USA. 286 pp.

Smith, C.M., Khan Z.R. and Pathak, M.D. 1994. *Techniques for Evaluating Insect Resistance in Crop Plants*. CRC Press, Inc., Boca Raton, Florida, USA. 320 pp.

Sogawa, K. and Pathak, M.D. 1970. Mechanisms of brown planthopper resistance in Mudgo variety of rice (Hemiptera: Delphacidae). *Applied Entomology and Zoology* 5: 145–158.

Soman, P., Nwanze K.F., Laryea, K.B., Butler, D.R. and Reddy, Y.V.R. 1994. Leaf surface wetness in sorghum and resistance to shootfly, *Atherigona soccata*: role of soil and plant water potentials. *Annals of Applied Biology* 124: 97–108.

Staedler, E. 1978. Chemo-reception of host plant chemicals by ovipositing females of *Delia* (*Hylemya*) *brassicae*. *Entomologia Experimentalis et Applicata* 24: 710–720.

Staedler, E. and Buser, H.R. 1982. Oviposition stimulants for the carrot fly in the surface wax of carrot leaves. *In* Visser, J.H. and A.K. Minks (Eds.). *Proceedings of the 5th International Symposium on Insect Host Plant Relationships*. PUDOC, Wageningen, The Netherlands.

Stanley, J.M. 1965. Decline and mortality of red and scarlet oaks. *Forestry Science* 11: 2–17.

Sundaram, M.K. and Sundara Babu, P.C. 1992. Screening varieties for resistance to soybean leaf miner, *Aproaerema modicella* (Deventer) (Lepidoptera: Gelechiidae) in Tamil Nadu. *In:* National Seminar on Changing Scenario in Pests and Pest Management in India, 31 Jan—1 Feb 1992. Plant Protection Association of India. Rajendranagar, Hyderabad, Andhra Pradesh, India.

Sundaramurthy, V.T. and Chitra, K.L. 1992. Integrated pest management in cotton. *Indian Journal of Plant Protection* 20: 1–17.

Swarup, P. 1987. Insect pest management in maize. *In* Veerbhadra Rao, M. and Sithanantham, S. (Eds.) *Plant Protection in Field Crops*. Plant Protection Association of India, Rajendranagar, Hyderabad, Andhra Pradesh, India. pp. 105–112.

Taneja, S.L. and Leuschner, K. 1985. Methods of rearing, infestation, and evaluation for *Chilo partellus* in sorghum. *In* Proceedings, International Sorghum Entomology Workshop, 15—21 July 1984, Texas A&M University, Texas, USA. International Crops Research Institute for the Semi-Arid Tropics (ICRISAT), Patancheru 502 324, Andhra Pradesh, India. pp. 175–188.

Teetes, G.L. 1985. Insect resistant sorghums in pest management. *Insect Science and its Application* 6: 443–451.

Tingey, W.M. and Singh, S.R. 1980. Environmental factors affecting the magnitude and expression of resistance. *In* Maxwell, F.G. and Jennings, P.R. (Eds.). *Breeding Plants Resistant to Insects*. John Wiley and Sons, New York, USA. pp. 87–113.

Todd, J.G. and Kamm, J.A. 1974. Biology and impact of a grass bug: *Labops hesperius* Uhler in Oregon rangeland. *Journal of Range Management* 27: 453–458.

Visser, J.H., Van Straten, S. and Maarse, T. 1979. Isolation and identification of volatiles in the foliage of potato, *Solanum tuberosum*, a host plant of the Colorado potato beetle, *Leptinotarsa decemlineata*. *Journal of Chemical Ecology* 5: 11–23.

Waiss Jr., A.C., Chan, B.G., Elliger, C.A., Wiseman, B.R., McMillian, W.W., Widstrom. N.W., Zuber, M.S. and Keaster, A.J. 1979. Maysin, a flavone glycoside from corn silks with antibiotic activity toward corn earworm. *Journal of Economic Entomology* 72: 256–258.

Waldbauer, G.P. 1968. The consumption and utilization of food by insects. *Advances in Insect Physiology* 5: 229–288.

Wallace, L.E., McNeal, F.H. and Berg, M.A. 1974. Resistance to both *Oulema melanoplus* and *Cephus cinctus* in pubescent-leaved and solid stemmed wheat selections. *Journal of Economic Entomology* 67: 105–107.

Wallner, W.E. and Walton, G.S. 1979. Host defoliation. A possible determinant of gypsy moth population quality. *Annals of the Entomological Society of America* 72: 62–67.

Wightman, J.A., Dick, K.M., Ranga Rao, G.V., Shanower, T.G. and Gold, C.G. 1990. Pests of groundnut in the semi-arid tropics. *In* S.R. Singh (Ed.). *Insect Pests of Legumes*. Longman and Sons Ltd., New York, USA. pp. 243–322.

Williams, L.H. 1954. The feeding habits and food preferences of Acrididae and the factors which determine them. *Transactions of the Royal Entomological Society* (London) **105**: 423–454.

Wood Jr., E.A. 1961. Description and results of a new greenhouse technique for evaluating tolerance of small grains to the greenbug. *Journal of Economic Entomology* **54**: 303–305.

# Chapter 8

# Defensive Tactics of Caterpillars against Predators and Parasitoids

*Brent A. Salazar and Douglas W. Whitman**

## INTRODUCTION

The fate of most animals is a quick crushing death between predator jaws, or a slow wasting away from internal or external parasites. Predators and parasites exert strong selective forces on prey populations, culling from each generation those individuals whose genes code for poor defenses. In response, prey have evolved an impressive diversity of anti-predator and anti-parasite defenses, many of which serve as classical examples of evolution. The larvae of Lepidoptera are no exception. In this chapter, we review the varied ways that this group of insects—the caterpillars—defend themselves from predators and parasitoids.

The Order Lepidoptera (butterflies and moths) is the second largest group of living organisms, surpassed only by the beetles in number of known species. Estimates vary, but most agree that worldwide there are approximately 200,000 species of Lepidoptera, about 15,000 of which are butterflies, the rest being moths (Scott, 1986). Lepidoptera are found in almost every terrestrial environment, from subarctic to tropical, alpine to desert, and are often conspicuous and abundant. Most caterpillars feed on plants thus, areas with high plant diversity generally contain high butterfly and moth diversity. It is mainly for this reason that more Lepidopteran species are found in the tropics than elsewhere. By far, the American tropics have the richest Lepidoptera fauna followed by tropical Southeast Asia and the East Indies (Scott, 1986).

At first impression, caterpillars seem to be relatively vulnerable prey because they lack defenses common to other animals. They are nearly blind. They lack cuticular armor. They possess short legs and a hydrostatic skeleton that limits their mobility. They cannot fly, jump, or run quickly. In fact, they tend to be

*4120 Biological Sciences, Illinois State University, Normal, IL 61790, USA. dwwhitm@ilstu.edu

rather sluggish. Most feed on foliage, where they often sit exposed and vulnerable to searching predators.

Caterpillars may be constrained from evolving certain defenses because of their body plan, life history needs and conflicting selective forces. The primary function of the larval stage of butterflies and moths is to feed and assimilate nutrients so that enough resources can be acquired to complete metamorphosis and adult reproduction (Heinrich, 1979). It is, therefore, advantageous for caterpillars to be efficient feeders, and indeed they are. Caterpillars are designed foremost as feeding machines—they are basically a mouth, a gut, and an anus; defense appears to be of secondary importance. Similar tradeoffs between competing needs are seen in the time spent feeding. For example, to quickly obtain nutrients and to maximize growth and development, caterpillars should spend as much time foraging as possible (Slansky, 1993). They could feed twenty-four hours a day; but most do not. Many feed only during the night, probably because feeding exposes caterpillars to natural enemies, and there are relatively fewer enemies at night (Heinrich, 1993). Likewise, caterpillars are poilkilotherms and could greatly increase their feeding, assimilation and metabolic rates, and shorten their developmental time by thermomaximizing (increasing their body temperature by feeding in sunlight) (Casey, 1993). However, few do so, presumably because such conspicuous behavior would increase predation. It appears that caterpillars, like all organisms, are trapped between competing needs.

Despite the morphological and life history limits place on them, caterpillars have still managed to evolve a very impressive diversity of defensive capabilities. This chapter explores some of these defensive strategies, but it is hardly all encompassing. Completely covering the enormous diversity of caterpillar defenses would be an almost impossible task. It is our hope to simply illustrate some of the varied and astounding strategies that caterpillars have evolved to thwart the deprivations of predators and parasitoids.

## THE NATURAL ENEMIES OF CATERPILLARS

A great diversity of predators, parasitoids and microbes attack caterpillars; this chapter emphasizes the former two groups. Included in the predator group are many species of spiders, scorpions, centipedes, frogs, toad s, lizards, marsupials, shrews, moles, skunks, raccoons, rodents, armadillos, bats and primates. However, the greatest predatory groups on caterpillars are undoubtedly insectivorous birds and other insects.

A great number of bird species take caterpillars, and this group is considered to be the primary force driving the evolution of many caterpillar defenses (Greenberg and Gradwohl, 1980; Pasteels *et al.* 1983; Brower 1984; Rothschild, 1985; Heinrich, 1993). Indeed, caterpillars comprise the main food for some birds in certain seasons (Solomon *et al.* 1977; Holmes *et al.* 1986). For example, in a heavily forested area of southern Illinois during the spring, populations of migrant warblers were correlated with populations of lepidopterous larvae.

The birds appeared to feed entirely on these larvae and chose either the smaller size classes (under 15 mm), or the smaller species (especially leaf rollers) (Graber and Graber 1983). The warblers ingested 1.2-1.7 times their own weight in caterpillars per day. Some authors suggest that bird predation may account for most of the long-term mortality of caterpillars in northeastern deciduous woodlands (Holmes *et al* 1979).

With their excellent color vision, birds are especially good at locating even cryptic caterpillars. Their high metabolism requires that they forage throughout the day, and their wings and small size allows them to search any part of plant foliage, from ground level to the canopy.

Predatory insects are also major natural enemies of caterpillars, and include some mantids, katydids, earwigs, minute pirate bugs, ambush bugs, assassin bugs, damsel bugs, stink bugs, snakeflies, mantidflies, lacewings, ground beetles, rove beetles, lady beetles, vespid wasps, thread-waisted wasps and others (Balduf 1935, 1939; Clausen 1940; Brues 1946; New 1991; Montllor and Bernays 1993). Some caterpillars are cannibalistic, and a few species are even predatory (Montgomery 1982; Whitman *et al.* 1994). However, the most important insect predators of caterpillars are ants (McNeil *et al.* 1978; Laine and Niemelä 1980; Jones 1987). These active hunters are abundant in most terrestrial ecosystems. Their social nature allows them to build large populations of small workers that forage individually (Hölldobler and Wilson 1990). Thus, a single colony with relatively low biomass can forage over a large area.

Insect parasitoids also are great natural enemies of caterpillars (Fig. 1) (Weseloh 1993). Included are wasps in families Ichneumonidae, Brachonidae, Pteromalidae, Encyrtidae and the superfamily Chalcidoidea, as well as others, and flies of the families Tachinidae and Sarcophagidae (Askew 1971; Waage and Greathead 1986). These parasitoids typically lay their eggs in, on, or near caterpillars (Fig. 2); the resulting parasitoid larva consumes the caterpillar, killing it. Although parasitization rates in caterpillars vary according to species, year, season and locality, rates from 30% to 95% are common (Wilkinson 1966; Askew 1968; Embree 1971; Huffaker *et al.* 1971; Mayer and Beirne 1974; Mumma and Zettle 1977; Kato 1984).

The different groups of caterpillar natural enemies use different senses and hunting behaviors to locate their prey. Birds use vision and respond to shape, color, motion, etc. (Schuler 1990). In contrast, parasitoids often orient chemically and respond to caterpillar body and habitat odors (Hendry *et al* 1973; Schmidt 1974; Vinson 1976). For example, the odor of caterpillar feces and the odor of caterpillar host plants, especially damage plants, are highly attractive to some parasitoids (Vinson *et al.* 1975; Roth *et al.* 1978; Turlings *et al.* 1990; Whitman and Nordlund 1994). Although ants and spiders also use visual and chemical stimuli, many respond to prey tactile or vibratory cues (Hölldobler and Wilson 1990). Gleaning bats, on the other hand, hunt by ultrasound and occasionally pluck caterpillars off leaves.

The use of different sensory modalities by different groups of predators allows them to hunt at different times. Birds use vision, and are, thus, largely

**Fig. 1.** The ichneumonid parasitoid, *Netelia heroica*, locates and stings its host, temporarily paralyzing the caterpillar until its egg can be inserted.

**Fig. 2.** *Ceratomia catalpae*, with parasitoid pupae attached to the outside of its body.

restricted to diurnal foraging. Many insectivorous mammals hunt at night or are crepuscular. Likewise, reliance on non-visual stimuli allows bats and some parasitoids, ants and spiders to hunt at night.

Different natural enemies also use different methods to overcome their caterpillar prey. Large predators like birds and mice use brute mechanical force, whereas some smaller predators such as parasitoids or predatory bugs rely on stealth, speed or venoms (Askew 1971; New 1991).

Lepidophagous carnivores also differ in their learning capacity. Birds and mammals are highly adaptive and flexible. They can develop "search images," whereby they quickly learn to identify and locate a particular prey species, and forage preferentially for that specific prey item (Schuler 1990). Likewise, vertebrates easily learn to avoid certain prey after an unpleasant experience with that prey. In some cases, only a single encounter with a toxic or venomous prey is required to establish long-lasting food aversion. In contrast, short-lived invertebrates primarily hunt by instinct and are less capable of learning and remembering (Theodoratus and Bowers 1999). However, even invertebrates can alter their search and attack behavior following encounters with rewarding or punishing prey (Berenbaum and Miliczky 1984; Hölldobler and Wilson 1990: Vet *et al.* 1995; Honda *et al.* 1998).

## CATERPILLAR DEFENSES

Given the great diversity in the types of natural enemies that attack caterpillars, the different learning capabilities of these enemies, and the varied sensory modalities and behaviors used, caterpillars are faced with a daunting task. They must defend themselves from both diurnal and nocturnal predators. Caterpillars must dissuade large and small, solitary and social carnivores, deter visually, chemically and tactially oriented predators, and confront intelligent predators with the ability to learn. Additionally, caterpillars face different sets of natural enemies as they grow, and often must switch defensive strategies during their relatively short development. Defending against all these different threats is almost an impossible task. A defense that acts against birds (warning colors, toxins, spines) may have no value against ants. Likewise, a behavior that reduces parasitoid attack, such as moving away from odorous feeding damage, may attract sharp-eyed birds. Nocturnal activity may lower bird predation but increase predation by nocturnal hunters. Clearly, caterpillars are caught between conflicting defensive pressures.

Caterpillars have evolved numerous tactics to defend themselves from the onslaught of predators and parasitoids. These defensive tactics can be morphological (color, tough cuticles, spines, hairs etc.), mutualistic (ant attendance), chemical (internal toxins, glands, regurgitation, etc.), physiological (encapsulation), or behavioral (shelter building, thrashing, feeding at night, removing frass, etc.), or more than likely, some combination therein. This chapter outlines these defenses and, where possible, gives some indication of the efficacy of these defenses against predators and parasitoids.

## MORPHOLOGICAL DEFENSES

### Crypsis

Crypsis is considered to be the most common morphological defense used by caterpillars; however, it usually consists of an ensemble of various morphological and behavioral traits, which, in combination, form an effective defense (Cott 1940; Edmunds 1974). Crypsis can be divided into two types, eucrypsis and mimesis, which grade into one another (Lederhouse 1990). In eucrypsis, the organism blends into its general background (Robinson 1969), whereas in mimesis the organism resembles an inedible object such as a twig, leaf, or animal excrement (Pasteur 1982). In eucrypsis, the model is the surrounding substrate and matching it in terms of color, pattern, texture, shadow and orientation is essential, but size and shape are less important. In mimesis, the model is a specific inedible object and matching the overall background is not as significant; mimetic animals often contrast with their substrate (Endler 1981). However, to be effective, mimetic animals should approximate the size and shape of the model.

The individual components that contribute to Crypsis are well studied (Edmunds 1974; Stamp and Wilkens 1993). Countershading eliminates shadows and caterpillars that feed right side up often have dark dorsums and light venters (Ruiter 1955). In contrast, caterpillars that feed upside down, such as the tobacco hornworm, *Manduca sexta*, have light dorsums and dark venters. Bold disruptive patterns draw attention away from the body outline, making identification more difficult. In general, the color, size and spacing of cryptic patterns match those of the substrate. Hence, caterpillars feeding or resting on small-leafed or fine-patterned plants (such as *Acacia* or juniper) have fine patterns that match the plant (Fig. 3). Those that feed on coarse-patterned plants, such as mesquite, possess larger patterns that approximate those of the host plant. Caterpillars feeding on large leaves are often a single color with almost no pattern (e.g. *Nadata gibbosa*). The saturniid, *Sphingicampa hubbardi*, feeds on mesquite and lives in the Chihuahuan desert of North America. It possesses silver-colored patches that reflect light, causing it to resemble the compound leaves of its food plant with bright light shining through from between the individual leaflets. Morphology and texture are critical components of Crypsis. Tapered heads and tails allow an inconspicuous transition from caterpillar to substrate. Caterpillars that rest against twigs or branches are often concave on their undersides, with a fringe of hairs that break up the shadow between the body edge and the bark (e.g.*Acronicta* spp).

Cryptic caterpillars can be of any color but they generally match the substrate they rest or feed on. Hence, caterpillars are often brown or green. In general, brown caterpillars (e.g *Eumorpha typhon, Sphecodina abbotti*) rest during the day on twigs, bark, or in leaf litter on the ground, and only move onto foliage to feed at night (Heinrich 1979). Green cryptically colored caterpillars (e.g. *Amphiphyra pyramidoides, Macrurocampa marthesia)* rest on foliage and may feed both during the day and night (Heinrich 1979). Other colors can be considered

**Fig. 3.** A geometrid caterpillar demonstrating how well it matches the scale pattern of its host plant. Photo by Erick Greene.

cryptic. For example, a yellow caterpillar is cryptic when it feeds on a yellow colored flower.

Crypsis can be an effective anti-predator defense (Ruiter 1952; Clarke *et al.* 1963; Herrebout *et al.* 1963; Curio 1970). For example, when great tits (*Parus major*) were offered green (cryptic) versus yellow (non-cryptic) pine looper caterpillars (*Bupalus piniarus*) on green pine needles, the birds found and consumed significantly more of the conspicuous yellow morph (Boer 1971). The problem with Crypsis, however, is that it can fail as a defense once the predator discovers the ruse. When Boer (1971) first offered his birds 20 green caterpillars, then gave them a choice between a yellow or green caterpillar, they took significantly more of the green, presumably because they had developed a search image for the latter. This may be why some cryptic prey exhibit high color polymorphism (and sometimes polyethism); predators may be less able to establish a specific search image against highly variable prey (Harvey and Greenwood 1978; Curio 1970; Boer 1971; Orsak and Whitman 1987).

Prey not only demonstrate visual Crypsis, they can also employ chemical, acoustic or tactile Crypsis. For example, *Vespula vulgaris* is a predatory wasp of caterpillars that primarily hunts by olfaction. By smearing twigs with the hemolymph from green or yellow caterpillars of *Bupalus piniarus,* Boer (1971) showed that significantly more wasps landed on twigs with green caterpillar hemolymph due to some unidentified chemical difference. Boer suggested that these two forces (predation by birds and wasps) seemed to balance each other out, since both color morphs were maintained in the caterpillar population.

Several families of caterpillars resemble twigs, but none do it quite as well as the Geometridae (Fig. 4). Many have fleshy projections that resemble buds or leaf scars (Fig. 5) and often there is an astounding match between the appearance of the caterpillar and the twig of its host plant. Caterpillars that rest on pubescent twigs appear pubescent, and caterpillars that rest on smooth twigs are smooth. Geometrids often remain motionless during the day with the anterior portion of their bodies fully extended away from the twig, and only feed at night when it is safer for them to move onto green foliage. Some suspend the anterior portion of their body with a silken thread, presumably to help them remain motionless and support the weight of their body (Ruiter 1952; Heinrich 1979).

Some early instar caterpillars, in the genus *Papilio*, resemble animal excrement (bird droppings). They, like the twig-mimicking Geometrids, remain motionless during the day and only move to feed under the cover of darkness. Their resting position, however, is quite different from the Geometrids. During the day, they sit conspicuously on the surfaces of leaves as if to advertise that they

**Fig. 4.**   The geometrid, *Selene kentaria*, resembles a broken twig of its host plant. Note the way in which the head and thoracic legs are held, completing the disguise. Photo by David Wagner.

**Fig. 5.** Geometrid caterpillars often have projections that resemble the buds or leaf scars of their host plant.

have just been dropped from above. As mentioned previously, mimics should match the size of their models. This may explain the "transformational" mimicry (a type of developmental polymorphism) observed in many *Papilio* species (Matthews and Matthews 1978). Early instars resemble bird droppings (Klots 1951), and rest on top of leaves, whereas the later, larger instars are green and rest on the stems and undersides of leaves (Edmunds 1974). Fecal mimicry can be quite exacting: *Limenitis archippus* (which resembles animal excrement even up to the final instar) possess a glossy coating making them appear to be freshly laid (Fig. 6).

A different type of developmental cryptic polymorphism is seen in the geometrid *Nemoria arizonaria* (Greene 1989). Caterpillars that emerge in the spring feed on and resemble the scaly flower clusters (catkins) of oaks, whereas those that hatch in the summer, when catkins are absent, resemble oak twigs. Diet may control this polymorphism; caterpillars fed catkins, which are low in Tannins, developed into catkin mimics, whereas those fed Tannin-rich leaves developed into twig mimics (Greene 1989).

Ruiter (1952) studied mimesis, and found that jays *(Garrulus glandarius)* and chaffinches *(Fringilla coelebs)* could not distinguish twigs from the very twig-like Geometrids of the genera *Ennomos* and *Biston*. The caterpillars were only attacked if they moved or were accidentally stepped on. After the caterpillars

**Fig. 6.**    All five instars of *Papilio cresphontes* are bird dropping mimics. Here, the caterpillar (with the U-shaped white band running across the middle of its body) is pictured with starling excrement. Photo by Brent Salazar.

were discovered, however, the birds quickly developed a search image and were able to find other caterpillars more frequently.

### Thick and tough cuticles

Some caterpillars derive a defensive benefit from a tough cuticle. For example, earlier instars and newly molted fifth instar *Manduca sexta* caterpillars have softer, thinner cuticles and are successfully attacked by *Apanteles congreatus* wasps. However, fully tanned fifth instar *M. sexta* larvae are seldom parasitized by this species because the Ovipositor of the wasp is unable to pierce the caterpillar's thick cuticle (Beckage and Riddiford 1978). Likewise, the parasitoid *Campoletis sonorensis* is deterred by the sclerotized integument and thicker subcuticular layers of fat of the larger caterpillars of *Helicoverpa zea* (Schmidt 1974).

Aposematic, chemically defended insects have long been thought to have tough and flexible cuticles, which allows them to survive experimental testing by predators without suffering serious injury (Poulton in Fisher 1930). This idea was questioned because it was thought that conspicuous insects were usually killed during sampling by predators who had not yet learned to avoid them. Hence, conspicuous coloration or chemical defense could not evolve by individual selection, but only via kin or group selection. However, Järvi *et al.* (1981) noted that the distastefulness and tough cuticle of *Papilio machaon* enabled them to be handled by naïve great tits *(Parus major)* without suffering ill effects. The authors suggested that the cost of being aposematic, and thus conspicuous to predators, can actually be quite small and that aposematic coloration could evolve through individual selection and not require kin selection. These conclusions were supported by similar results for three other aposematic caterpillars, all of which survived predator attacks (Wiklund and Järvi 1982).

### Waxy cuticles

Some caterpillars, such as *Attacus atlas* and *Samia cynthia ricini* produce copious amounts of cuticular wax that may act as a mechanical defense against small carnivores (Bowers and Thompson 1965; Jones *et al.* 1982). Presumably, natural enemies or their eggs have difficulty grasping or adhering to the slippery, loose, waxy covering. Likewise, some predators may be distracted or deterred by the waxes; however, to our knowledge, these hypotheses have never been tested.

### Hairs and Spines

Many caterpillars possess spines or hairs that act as mechanical deterrents to predators or parasitoids. These cuticular structures vary from long, thin, flexible hairs to stout, sharp bristles. Some spines are branched, forming tight, entangling masses. Others are sharp, barbed, or brittle, and act like tiny needles, penetrating the mouths, eyes, or other soft tissues of predators. Some break easily, remaining embedded in the hapless predator where they provide a continuing reminder of the consequence of attacking such prey. It is not surprising that many birds (and other predators) simply refuse to attack spiny or hairy caterpillars. Heinrich and Collins (1983) found that black-capped chickadees, *Parus atricapillus*, would not even peck at spiny *Nymphalis antiopa* or *Arisota rubicunda* larvae, or hairy *Hemerocampa leucostigma* or *Halysidota maculata*. Heinrich (1979) also observed that *Malacosoma disstria* were extremely common at his field site in Minnesota, USA, yet no birds fed on them, probably due to their hairiness. He also found that a pair of red-eyed vireos, *Vireo olivaceus*, brought back 27 caterpillars to feed their young, of which only one was hairy, and that the bird spent several minutes batting this one caterpillar on branches, in an attempt to knock off the hairs. However, no defense is effective against all predators, and for virtually every defense there are certain predators that have evolved or learned to overcome that defense (Yosef and Whitman 1992). Indeed, some birds, such as yellow-billed cuckoos, are known to accept hairy caterpillars quite readily (Knapp and Casey 1986).

Thick coverings of long hairs or spines can also act as a physical barrier to arthropod predators or parasitoids. Stamp (1982) found that the tubercle-spine length of third instar *Euphydryas phaeton* caterpillars was similar to the Ovipositor length of two major parasitoids of these caterpillars: *Apanteles euphydryidis* and

*Benjaminia euphydryadis.* The tubercle-spine length of the fourth instars is greater than the Ovipositor length of *A. euphydryidis,* and may deter parasitism by preventing Ovipositor penetration. Likewise, Bardwell and Averill (1996) found that no *Lymantria dispar* larvae were killed during two of their experiments, and that the long hairs of the caterpillar provided a potent defense against spider attack. Similarly, Weseloh (1976) observed that successful parasitization of *Lymantria dispar* larvae by *Cotesia melanoscela* wasps fell from 57% with first instar larvae to 8% with fourth instar larvae, because long hairs and thrashing movements of older larvae prevented parasitoid access. However, hairs and spines are clearly only partially effective mechanical deterrents against arthropods, as demonstrated by the fact that virtually all species of caterpillars, be they hairy, spiny, or smooth are attacked by parasitoids.

## MUTUALISM

### Ant attendance

Some caterpillars in the families Lycaenidae and Riodinidae gain defensive benefits from mutualistic associations with ants (Downey 1962; Malicky 1970; Harvey and Webb 1980; Douglas 1983; Cottrell 1984; Pierce 1987; Hölldobler and Wilson 1990; De Vries 1991a). The interaction is usually mediated by a honeydew-like secretion produced by specialized exocrine glands found in various locations on the bodies of the caterpillars (Pierce 1983). The secretions appease or reward the ants (Kitching 1983; Pierce 1983), and consist largely of carbohydrates (e.g., fructose, glucose, sucrose) and amino acids (e.g. methionine, serine). The latter is metabolically costly for the caterpillars to synthesize since it requires diverting proteins from growth and development to defense (Baylis and Pierce 1993). The benefits of this interaction for the ants are not so clear-cut, because, in some cases, the caterpillars are parasitic and consume the ant brood (Cottrell 1984). However, the caterpillars gain protection from predators and parasitoids that might otherwise attack them if the ants were not present (Malicky 1970; Fiedler and Maschwitz 1988, 1989a, b; Savignano 1990; DeVries 1991b).

Sometimes, caterpillars are actually carried by the ants into the safe interior of the ant's nest for the night, then returned to the plant at dawn. More commonly, however, the ants tend the caterpillars as they feed on foliage and attack and drive away small predators or parasitoids (Pierce and Easteal 1986; Pierce *et al.* 1987). When *Formica* ants were excluded from host plants containing *Glaucopsyche lygdamus* caterpillars, rates of parasitism by four different parasitoid species doubled (Pierce and Mead 1981; Pierce and Easteal 1986). The only exception was parasitism by *Apanteles cyaniridis,* which showed no treatment effects the second year of the study (Pierce and Easteal 1986). Likewise, no *Jalmenus evagoras* prepupae were parasitized by *Brachymeria regina* wasps when *Iridomyrmex anceps* ants were present in Australian *Acacia* trees, whereas 88% of prepupae were parasitized when ants were excluded (Pierce *et al.* 1987). Again, however, no defense is perfect, and ant attendance did not protect young larvae from attack by one *Apanteles* species.

Interestingly, nutritional versus defensive trade-offs may sometimes occur in lycaenid-ant mutualisms. Adult lycaenids like *Ogyris amaryllis* have been shown to oviposit on *Amyema maidenii,* a nutritionally inferior food plant over *Amyema preissii,* a nutritionally superior food plant, if attendant ants are present (Atsatt 1981).

## CHEMICAL DEFENSES

### Internal toxins

Many Lepidopteran larvae synthesize toxic defensive compounds (Owen 1971; Davis and Nahrstedt 1984; Marsh *et al.* 1984; Rothschild 1985; Witthohn and Naumann 1987). The Zygaenidae release cyanide from all stages of their life cycle including the egg (Jones *et al.* 1962), but it is the larvae and adults that synthesize the cyanoglucosides that are the source of the HCN (Nahrstedt and Davis 1983). The nymphalid genera *Heliconius* and *Acraea* also autogenously synthesize cyanoglucosides (Owen 1970; Nahrstedt and Davis 1981). Both zygaenids and the HCN-producing nymphalids *(Heliconius* and *Acraea)* are rejected by birds and other predators (Pocock 1911; Swynnerton 1915; Lane 1959; Frazer and Rothschild 1961; Brower *et al.* 1963; Marsh and Rothschild 1974). Fung *et al.* (1988) found that *Yponomeuta cagnagellus* (Yponomeutidae) sequestered isosiphonodin and siphonodin from its host plant *Euonymus europaeus,* but also found isosiphonodin in six other species of *Yponomeuta* that did not feed on *E. europaeus.* They suggest that, for these six species, this compound is most likely synthesized and may serve an anti-predatory function.

Other caterpillars do not synthesize defensive compounds, but instead sequester them from host plants (Brower and Brower 1964a,b; Duffey 1980; Blum 1983; Brower 1984; Rothschild 1985; Brattsten 1986; Bowers 1990, 1992). These plant-derived compounds are quite varied and include alkaloids in *Arctia, Battus, Creatonotos, Pyrrharctia, Spilosoma, Spodoptera, Uresiphita* and *Utetheisa* (Kelly *et al.* 1987; Boppré 1990; Montllor *et al.* 1990; Urizula *et al.* 1987; Fordyce, 2000), azoxyglycosides in *Eumaeus* and *Seirarctia* (Bowers and Larin 1989), cardenolides in *Danaus* (Brower 1984), and glucosinolates in *Pieris* (Marsh and Rothschild 1974; Aplin *et al.* 1975). Usually, the toxic plant compounds obtained by the larvae are retained by the pupae, adult, and, in some cases, even the subsequent eggs (Aplin *et al.* 1975). However, in other species, such as the catalpa sphinx, *Ceratomia catalpae,* and the buckeye, *Junonia coenia,* sequestered iridoid glycosides appear only in the caterpillars (Bowers and Puttick 1986). Interestingly, the larvae of the catalpa sphinx are warningly colored and gregarious whereas the adults are cryptic, solitary, and palatable (Bowers and Farley 1990).

The most famous example of the defensive Sequestration of poisonous plant compounds is the Monarch butterfly, *Danaus plexippus.* Monarch caterpillars obtain cardiac glycosides (cardenolides) from their host plants, which include various species of milkweed (Asclepiadaceae). The cardiac glycosides, which act as potent vertebrate heart toxins and emetics, are transferred through the pupal stage, into the adult. Brower *et al.* (1968) found that blue jays *(Cyanocitta cristata bromia)* repeatedly vomited after consuming Monarch butterflies and, subsequently, rejected further Monarchs on sight, a clear example of food aversion learing. Monarchs also deter many other predators (Bowers 1980; Bowers and Larin 1989; Montllor *et al.* 1991). However, a surprising number of natural ememies have been able to break through or circumvent Monarch defenses (Whitman 1988). For example, Monarch caterpillars are successfully attacked by a number of parasitoids (Reichstein *et al.* 1968). In addition, mice, orioles and grosbeaks consume hundreds of thousands of overwintering adult Monarchs in Mexico each year (Calvert *et al.* 1979; Brower and Calvert 1985; Brower *et al.* 1985). The mice and grosbeaks have evolved physiological tolerances to cardenolides (Fink and Brower 1981; Glendinning *et al.* 1988), whereas the orioles consume only those Monarchs with low cardenolide levels (Fink and Brower 1981).

### Defense Glands

Some caterpillars possess defensive glands consisting of an internal, nonreversible, cuticle-lined reservoir in which defensive secretions are stored (Detwiler 1922; Carpenter 1938; Dethier 1939; Hintze 1969; Pavan and Dazzini-Valcurone 1976; Weatherston *et al.* 1979; Rothschild 1985; Whitman *et al.* 1990). When attacked by predators, the caterpillar expels the secretions.

Most caterpillars of the family Notodontidae are well known for emitting a foul smelling liquid from their ventral prothoracic gland when disturbed, including *Dicranura vinula, Schizura leptinoides, Schizura concinna* and *Heterocampa manteo* (Poulton 1887; Poulton 1888; Monro *et al.* 1962; Eisner *et al.* 1972a). The secretion typically contains formic acid and ketones, and, in some cases, can be sprayed several centimeters at attackers. Targeting is accomplished by moving the forebody, but also by muscles that direct the aim of the gland's nozzle (Fig. 7) (Alsop 1970; Whitman *et. al.* 1990*).* Eisner *et al.* (1972a) found that *Heterocampa manteo* could aim its secretion with high accuracy. When sprayed at lycosid spiders, the spiders immediately stopped their attack and engaged in intensive cleaning activities. The formic acid spray of *Schizura leptinoides* is a deterrent to birds *(Cyanocitta cristata)*, lizards *(Anolis equestris)*, and toad s *(Bufo americanus)* (Eisner *et al.* 1972a).

Some saturniids spray an irritating secretion from dorsal thoracic glands when attacked (Jones *et al.* 1982*),* and a few caterpillars of the family Cossidae, *Cossus cossus* and *Zeuzera pyrina*, secrete polyphenols from their mandibular glands *(Pavan and Dazzini-Valcurone 1976).* Similar phenolics have not been found in the defensive secretions of any other Lepidopteran larvae. In addition, Pocock (1911) found that *Cossus cossus* larvae were distasteful to birds.

### Osmeteria

Osmeteria are eversible glands found in Papilionid larvae. They are located middorsally just behind the head, and are extruded via hydrostatic pressure, releasing volatile chemicals (Fig. 8) (Young *et al.* 1986). In *Papilio machaon* the

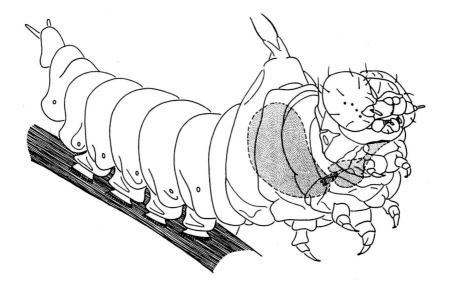

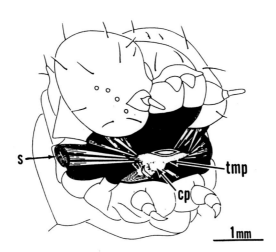

**Fig. 7.**  (Top) Basal and apical chambers of the ventral prothoracic gland of *Schizuranicornis*. (Bottom) A cutaway of the neck region of *S. unicornis* showing the muscles involved in directing the formic acid spray. Note the positions of the spiracle(s), terminal membranous pocket of the gland (tmp), and a small pocket of cells (cp) extending downward from the junction of the ejaculatory duct. (Figure from Alsop 1970).

secretion consists of isobutyric and 2-methylbutyric acid, and that of *Troides amphtysus* contained sequestered aristoloic acid. (Eisner and Meinwald 1965; Nishida *et al.* 1993). Osmeterial secretions repel certain predators (Eisner *et al.*

**Fig. 8.**   Papilionid larvae, including *Papilio polyxenes* seen here, evert their Y-shaped Osmeteria
when disturbed, releasing volatile chemicals. Photo by Erick Greene.

1970; Honda 1980a, b, 1981; Wiklund and Sillén-Tullberg 1985), especially
invertebrates such as ants, spiders (Honda 1983; Damman 1986), and praying
mantids (Chow and Tsai 1989). However, Osmeteria are not particularly effective
against *Polistes* wasps or larger vertebrate predators such as birds (Järvi *et al.*
1981; Honda 1983).

### Regurgitation and Defecation

The simplest and perhaps most common form of caterpillar chemical defense
is regurgitation of gut contents (Eisner *et al.* 1972b, 1974; Blum 1981; Stamp
1982, 1984, 1986; Nishio 1983; Brower 1984; Watherston *et al.* 1986; Cornell *et al.*
1987; Whitman *et al.* 1990). Caterpillars feed on plants, and plants contain a
great diversity of repellent, irritating or toxic compounds. It is not surprising
then that the first line of defense for some caterpillars is to regurgitate a sticky
brew of digestive acids, enzymes, and recently consumed plant chemicals onto
attacking carnivores. Peterson *et al.* (1987) found that the regurgitated fluid of
*Malacosoma americanum* contained hydrogen cyanide and benzaldehyde obtained
from the leaves of its host plant, *Prunus serotina*, and that the fluid effectively
repelled ants.

   Defecation in response to disturbance has been documented for only a few
species of Lepidoptera (Whitman *et al.* 1990; Theodoratus and Bowers 1999).

Monarch feces contain a substantial amount of cardenolides (Roeske *et al.* 1976) and the larvae release an additional anal fluid with unknown defensive attributes when severely disturbed (Brower 1984). Brower (1984) also observed accumulation of frass on milkweed leaves below feeding larvae and suggested that this may serve as a visual or chemical aposematic display towards diurnal or nocturnal predators. However, larval defecation upon disturbance has never been proven to be a defense against predators.

## Hairs and Spines

Many caterpillars possess urticating or stinging hairs and spines (Fig. 9) (Stamp 1982; Rothschild 1985; Whitman *et al.* 1990). They can be found in the families Arctiidae, Lasiocampidae, Limacodidae, Lymantriidae, Megalopygidae, Noctuidae, Notodontidae, Nymphalidae, Saturniidae, Zygaenidae and others (Beard 1963; Frazer 1965; Pesce and Delgado 1971; Picarelli and Valle 1971; Jong and Bleumink 1977; Quiroz 1978; Kawamoto and Kumada 1984; Novak *et al.* 1987). These hairs and spines are designed to pierce the skin or cuticle of predators and release their toxic chemicals. Filled or coated with irritating or painful venoms, these sharp and brittle structures often break off and remain embedded in animal tissues. Some urticating hairs are lightweight and easily dislodged during predator attack. They float through the air into the eyes or

**Fig. 9.** The puss moth caterpillars (Family: Megalopygidae) are often suited with long setae and can inflict a painful sting.

respiratory systems of predators, distracting or disabling them. The venoms often contain histamine, or histamine-like substances (Kawamoto and Kumada 1984). Venomous hairs and spines can effectively deter vertebrate predators such as rodents, lizards and birds (Root 1966; Bellows *et al*.1982) and humans have died from allergic reactions following envenomization from caterpillar spines.

### Chemical Camouflage

Some caterpillars may protect themselves from predators by hiding or masking their body odors, or by smelling and tasting like their host plant. An example of the latter case are *Mechanitis polymnia* (Ithomiinae), from Brazil, which are not attacked by *Camponotus* ants when feeding on *Solanum tabacifolium*, but are attacked and killed when placed on leaves of other plant species. The cuticular lipid profiles of the caterpillar and *S. tabacifolium* match closely, suggesting that the predatory ants simply do not recognize the caterpillars as prey against the similar chemical background of their host plant (Portugal, 2001).

### APOSEMATISM

Chemically defended insects often possess vivid colors or patterns that are thought to serve as a visual warning to potential predators. These aposematic (or warning) color patterns, usually consist of red, orange or yellow alternating with black; however, blues and purples sometimes serve as warning colors. Predators often learn to avoid such brightly-colored insects after only one or two unpleasant experiences. This food aversion conditioning is facilitated if the prey species is highly memorable. Hence, well-defended prey benefit from being conspicuous and from having an appearance different from edible prey. Odor, sound and behavior can also serve a warning function, and thus, well-defended prey may employ visual, chemical, acoustic or behavioral aposematism.

Both the genetic makeup and learning capabilities of predators influence avoidance of aposematic prey. Individual predators learn by trial and error which prey should be avoided, and in some cases only a single unpleasant encounter with a poisonous or venomous prey is required to establish a life-long avoidance of that prey. In addition, selection has acted to alter the population genetic structure so that many predators are born with an innate fear of aposematic prey, and reject them the first time encountered (Coppinger 1970). Neophobia (fear of new) also plays a role in avoidance of aposematic prey and predators that encounter strange, new, brightly colored prey are often hesitant to attack (Coppinger 1970; Schuler and Hesse 1985).

Many caterpillars are aposematically colored, and many of these are probably chemically defended. Examples include caterpillars in the families of Arctiidae, Danaidae, Papilionidae and Zygaenidae.

Chemically defended insects often possess a suite of characteristics that in combination provides a greater defense. Referred to as the Chemical Defense Syndrome (CDS), this ensemble of morphological, behavioral and physiological traits often includes chemical defense, aposematic coloration, warning scents, gregariousness, large size, sluggishness and diurnal activity (Pasteels *et al.* 1983; Whitman *et al.* 1985). Each individual component of the CDS contributes to the overall defense of the insect. The chemical defense, be it stinging hair, defensive glands or synthesized or sequestered toxins make the prey unpalatable to predators. The aposematic pattern or odor warns both experienced and naïve predators of the dangers of attacking that prey. The aggregation presents a larger, more striking, more memorable and more threatening image to potential predators. Aggregations may also be more effective at repelling small arthropod natural enemies through group defenses such as communal thrashing (Cornell *et al.* 1987). In addition, because most predators will typically sample only one or two individuals from an aggregation of toxic prey, a prey individual can lower its chance of being eaten by remaining in a large group. Large caterpillars may more effectively repel small arthropod predators, and sluggish caterpillars may fail to elicit attack responses from motion-oriented predators (see Hatle and Whitman, this volume).

Additional benefits accrue from the CDS. For example, a release from predation (especially from birds) allows CDS insects to feed on top of leaves in direct sunlight, and, thus, maintain high body temperatures and rapid development. In addition, reduced predation from vertebrates releases some chemically defended insects from size constraints (edible insects need to be small and inconspicuous in order to hide from sharp-eyed birds). Thus, toxic caterpillars that are avoided by birds and other vertebrates may have the luxury of evolving a larger body size. Large size may benefit adult dispersal, fecundity, physiological homeostasis and defense against small arthropod predators. Large caterpillars may more easily chew through tough or thick foliage, and aggregated caterpillars may more easily overcome various plant defenses. Slow, sluggish behavior may have metabolic benefits. Hence, the type of defense used by a species can greatly influence other life history traits.

## BATESIAN AND MULLERIAN MIMICRY

Although mimicry is well studied in adult Lepidoptera, it is less known in caterpillars. In Batesian mimicry, a palatable mimic resembles an unpalatable model of a different species, and Mullerian mimicry is the resemblance of two or more unpalatable species to each other. In Batesian mimicry the mimic gains a selective advantage by resembling an unpalatable species that is avoided by predators. Hence, Batesian mimics "parasitize" the relationship between the model and its predators, presumably to the disadvantage of the model. In Mullerian mimicry, all participating species gain an advantage because once a predator has sampled one unpalatable species, it learns to avoid others with a similar pattern.

Batesian mimicry in caterpillars is considered by some investigators to be nonexistent (Sillen-Tullberg 1988). There are, however, a few potential cases where Batesian mimicry may occur. One involves the nymphalids *Chlosyne harrisii* and *Euphydryas phaeton*. *Chlosyne harrisii* feeds on *Aster umbellatus* and closely resembles the larvae of *E. phaeton*, which are unpalatable and emetic due to the Sequestration of iridoid glycosides from their host plant, *Chelone glabra* (Bowers 1980, 1983, 1988, 1993). Larvae of the two species are virtually identical, and both are found in the northeastern United States where they share a similar habitat (wet meadows) and phenology (Bowers 1983). Although the palatability of *C. harrisii* has not been tested, the adults are readily consumed by blue jays suggesting that the larvae may also be edible (Bowers 1983).

Mullerian mimicry may also exist among Lepidopteran caterpillars. The larvae of two Geometrids, *Meris alticola* and *Neoterpes graefiaria*, both feed on *Penstemon* species, closely resemble each other, are sympatric and sequester iridoid glycosides, leading some to suggest that they are Mullerian mimics (Poole 1970; Bowers 1980; Stermitz *et al.* 1988). However, as with *C. harrisii* (above), feeding experiments with potential predators need to be performed in order to verify this hypothesis.

## PHYSIOLOGICAL DEFENSES

### Encapsulation

Several caterpillars are able to vanquish parasitoids by encapsulation after the parasitoid has successfully deposited its eggs internally (Salt 1955, 1963, 1968; Sato 1976). Encapsulation generally involves the aggregation of the host's blood cells around the parasitoid egg or larvae. The blood cells adhere to the parasitoid in large numbers, flattening themselves against the surface of the parasitoid and, later, against the blood cells already surrounding the parasitoid (Askew 1971). Encapsulated parasitoids are believed to die from suffocation, however, sometimes the capsule is incomplete or the parasitoid is able to wiggle free (Askew 1971). In a two-year field study, *Apanteles glomeratus* successfully parasitized only 3 out of 549 *Pieris melete* larvae examined because of the caterpillar's ability to encapsulate the developing parasitoid larvae (Ohsaki and Sato 1990).

Encapsulation is thought to be a general response against most objects invading the insect hemocoel. However, for almost every caterpillar species, there are parasitoids that have apparently evolved the biochemical machinery to hide from, disrupt or otherwise circumvent the host's encapsulation response. Hence, most caterpillar species are successfully parasitized by at least one, if not many, species of parasitoids.

## BEHAVIORAL DEFENSES

### Shelters

Many caterpillars construct shelters, and these shelters are highly variable. Some caterpillars roll individual leaves, others tie together several leaves at a time, binding them with silken threads (Rawlins 1984; Feichtinger and Reavey 1989). Others, such as *Malacosoma americanum* and *Euphydryas phaeton*, construct large colonial webs or "tents" (Fitzgerald 1980; Stamp 1982, 1984; Fitzgerald *et al.* 1988). Some caterpillars construct sequential shelters as they mature, whereas others simply add to existing shelters. Ruehlmann *et al.* (1988) describes how *Herpetogramma aeglealis* sequentially inhabits approximately five shelters of three distinct types: a bundle, fiddlehead and globe shelter. One problem with shelters is that they are often stationary and sometimes caterpillars must leave them in order to feed. Bag moths, *Phereoeca* spp., have solved this problem by constructing a silken bag around the body to which is attached sticks, leaves or other debris. The tube is carried about while feeding (Aiello 1979). When alarmed, the caterpillar withdraws into its bag and pinches the aperture closed, but remains attached to the plant by a silken thread (Edmunds 1974).

Leaf tying and wriggling by *Sparganothis sulfureana* larvae was highly effective in deterring Salticid predation (Bardwell and Averill 1996), and Leaf tying and gregarious feeding reduced the impact of natural enemies on *Omphalocera munroei* (Damman 1987). Leaf ties built of old leaves were more effective at reducing predation than were those built of young leaves, probably because old leaves maintained their shape better. If given a choice, larvae preferred nutritionally poor old leaves arranged to form a shelter over nutritionally superior young leaves without a shelter, suggesting that protection from enemies was more important than nutrition to the larvae. Shelter building can even protect caterpillars from ants. Vasconcelos (1991) studied the ant-plant *Maieta guianensis*, and found that the attendant ants effectively deterred most herbivores. However, two shelter building caterpillars successfully blocked ant attack and, thus, were able to feed on the plant: *Acrospila gastralis* constructed ant-proof leaf cylinders, and *Stenoma charitarca* made ant-proof tunnels of silk and feces.

## Feeding Inside Plant Tissues

Many caterpillars bore inside various plant tissues, mine inside individual leaves, or feed in plant galls. Price *et al.* (1987) suggests that an advantage of such feeding modes is to place a barrier of plant tissue between the larvae and its potential predators. In addition, leaf characteristics such as waxiness, spines, trichomes or plant defensive chemicals may deter some predators (Owen 1975). Harcourt (1966) found that the earlier instars of *Pieris rapae* feed exposed on outer cabbage leaves, but that later instars moved to the center of the plant presumably as an evolutionary response to parasitoids and avian and invertebrate predators. However, as a countervailing selective force, the likelihood of death by pathogens increased from 0%, while on outer leaves, to 34% within the plant. Baker (1970) also studied *P. rapae*, but attributed the movement of later instars into the center of the plant to feed as an adaptive response only to avian predation. Both studies, however, indicate that the selection pressure driving these caterpillars to feed within the plant must be great since exposure to pathogens increases quite significantly once inside.

Lepidopteran leaf miners are necessarily small and flat, and generally spend their entire larval life inside one or a few leaves (Needham *et al.* 1928). Hiding inside plants can be an effective defense against generalist predators such as ants, mantids, predaceous bugs, lizards, etc. that do not invade plant tissue when hunting. For example, less than 10% of the 6078 identifiable caterpillars gathered by ants in Canadian forests were leaf miners as opposed to external feeding caterpillars (McNeil *et al.* 1978).

Although internal feeding protects caterpillars from some natural enemies, it clearly is not effective against all, because almost all boring, mining and galling caterpillar species are attacked by specialized parasitoids (Askew 1968; Kato 1984). Predators invade tunnels, birds pick apart galls and leaf rolls to obtain insects, woodpeckers extract borers hidden in wood, and parasitoids simply drill through the plant tissue with their long Ovipositors (Robinson and Holmes 1982). Again, no defense is perfect.

*Phytomyza lonicerae*, a leaf miner of honeysuckle, may have found a way around the problem of parasitoids drilling through their host plant to oviposit. Its leaf mining pattern often branches or crosses within a leaf, which may lead a parasitoid in the wrong direction long enough for it to give up searching that leaf (Kato 1984). Parasitoids may require a relatively longer time to locate endophytic versus exophytic prey and to bore through the plant tissue to reach such prey. This additional searching and handling time could make parasitoids more susceptible to their own natural enemies. Interestingly, Sato and Higashi (1987) found that leaf miners were less parasitized when ants were present, possibly because the ants drove off the parasitic wasps.

### Biting

Some caterpillars, when attacked by predators or parasitoids, fight back by biting (Salt 1938; Stamp 1984). *Helicoverpa zea* is extremely agile and, when touched by a parasitoid, quickly turns and bites, sometimes wounding or killing the parasitoid with its sharp and powerful mandibles.

### Thrashing

Thrashing is a common caterpillar response to molestation by parasitoids or predators (Jones *et al.* 1982). Hopper (1986) found that *Microplitis croceipes* exhibited a longer handling time for less preferred instars of *Heliothis virescens* because of a reluctance to parasitize and difficulty in ovipositing into small, thrashing first instars. Similar results were observed for *Campoletis sonorensis* attacking *Helicoverpa zea* (Schmidt 1974). *Plodia interpunctella* caterpillars thrashed so strongly in response to attack by *Venturia canescens* wasps that many eggs were dropped or failed to adhere to the host (Ridout 1981). Thrashing may be especially effective in grouped or aggregated larvae (Cornell *et al.* 1987).

Thrashing is often combined with biting, regurgitation and head or tail flicking (Morris 1963; Iwao and Wellington 1970; Frank 1971; Suzuki *et al.* 1980; Stamp 1986; Martin *et al.* 1989). In combination, these behaviors can dissuade, wound or kill small arthropods. Danks (1975) observed that head flicking and biting by *Helicoverpa zea* and *Heliothis virescens* greatly hindered parasitoid oviposition. *Heliothis punctiger* larvae avoided capture by juvenile and adult

*Oechalia schellenbergii* with a combination of defensive behaviors that included head flicking, biting, tail flicking, oral spitting, rolling over, dropping off the plant and ceasing movement (Awan 1985). In response to parasitoid attack, *Trichoplusia ni* flicks its head, bites at and eats the eggs and burrowing larvae of parasites (Brubaker 1968).

### Molting

Many parasitoids attach their eggs to the cuticles of caterpillars, and some parasitoid larvae adhere to the cuticle and feed externally (Askew 1971). In some cases, caterpillars can rid themselves of these external parasitoids by simply Molting (Salt 1938). Molting, as a defense, works best when performed frequently. However, frequent Molting requires rapid growth, which requires either more time spent feeding or food with a higher nutritional content. Interestingly, Danks (1975) found that for final instar *Helicoverpa zea*, between 25% and 54% of tachinid eggs were molted away on the most favorable food plants, but only 3% to 21% on the least favorable. For penultimate instar caterpillars, close to 100% of tachinid eggs were shed.

### Integumental Coverings

For polyphagous cryptic caterpillars, matching the immediate host plant may be problematic, if the cryptic pattern is genetically determined. *Synchlora aerata* (Geometridae) may have solved this dilemma. The larvae feed on many species of plants in the family Asteraceae, and excise and attach petals and other plant fragments to their bodies (Figs 10-11). Hence, their appearance changes to match the species of plant they are feeding on, and they apparently attach additional material regularly, so that their disguise is always fresh (Wagner, in press).

### Startle displays

Startle displays are more commonly associated with adult butterflies and moths than with their larvae; however, larvae can also execute very convincing Startle displays. In some cases, the caterpillars are relatively cryptic, but when attacked, they change their form and position to expose eyespots, or horns. Some caterpillars have even come to resemble reptiles (Curio 1965). Sphingids, such as *Panacra* and *Leucorampha* spp., are particularly well known for this type of display. These caterpillars rest upside-down beneath a leaf or branch, and when disturbed, raise and inflate their head (Edmunds 1974). In *Panacra*, the dorsal part of the head bears false eyes and resembles the head of a common sympatric snake (Morrell 1969). In *Leucorampha*, the ventral surface of the head contains the false eyes (Moss 1920). The death head caterpillar, *Acherontia atropos*, possess two eyespots and when disturbed hunches its thorax up and gnashes its mandibles to make a rapid clicking sound (Edmunds 1974). *Proserpinus terlooii*, from southwestern USA, has numerous false eyespots running laterally down its body and very convincingly resembles a lizard in appearance (Fig. 12).

**Fig. 10.** *Synchlora aerata*, a polyphagous feeder on members of the sunflower family (Asteraceae), attaches petals and other plant fragments to its body in order to disguise itself. By doing so, the caterpillar is able to match the species of plant it is feeding on. Photo by David Wagner.

In North America, several late instar caterpillars of the genus *Papilio* resemble snakes (Scott 1986). When disturbed, they inflate their false head and thrash it about vigorously. In *Papilio troilus*, the resemblance to a snake may be emphasized when the caterpillar rests within its leaf roll with only the false head protruding; the abdomen, which does not resemble a snake, remains hidden in the leaf roll.

### Hiding Evidence of Feeding

Some predators can locate caterpillars by their feeding damage. For example, *Parus atricapillus*, the black-capped chickadee, uses leaf damage as a cue even when prey are not visually conspicuous (Heinrich and Collins 1983). Real *et al.* (1984) noted that blue jays, *Cyanocitta cristata*, quickly differentiated between projected photographic images of whole leaves versus caterpillar damaged leaves. Thurston and Prachuabmoh (1971) found that grackles (*Quisculos quiscula*) took more fourth and fifth instar *Manduca sexta* larvae than earlier instars. Because the larvae fed from the undersurface of the leaves and were not initially visible, they concluded that the birds were attracted first to the

**Fig. 11.** *Synchlora aerata* on *Rudbeckia* (Asteraceae). Note how its appearance changes (see Fig. 10) to match the species of plant it is feeding on. Photo by David Wagner.

damaged leaf, and then to the insect. In addition, the larger instars fed more on the tips and edges of the leaves, making the damage more apparent.

Heinrich (1979) noted that palatable caterpillars behaved as if they were attempting to hide leaf damage from visually oriented predators. For example, *Phaeosia rimosa* always fed on a leaf until it was completely consumed, so that no feeding damage was evident. *Sphecodina abbotti* fed on the edges of grape leaves and would usually chew off lobes so that the damage was not readily apparent. Most interestingly, *Catocola cerogama* and several others performed petiole clipping: after feeding, they chewed through the thick petiole of the partially consumed leaf (Fig. 13), letting it fall to the ground. After petiole clipping, the caterpillar moved to a resting spot on a twig. Feeding in this manner leaves no evidence of leaf damage except for a small portion of the petiole where it attaches to the twig. When *Papilio polyxenes* feeds on its umbellifer hosts, it produces smooth edges, which also give little evidence of feeding damage (Codella and Lederhouse 1984).

One of the most exquisite examples of a caterpillar hiding its feeding damage comes from the family Notodontidae. Many species of this family feed on the edges of leaves and actually place their bodies into the space where they previously fed, causing the edge of the leaf to appear smooth. To complete the ruse, the larvae often exhibit color patterns that resemble leaf edges. One

**Fig. 12.** Many Sphingids look like reptiles, including *Proserpinus terlooii,* which resembles a
lizard. Photo by Brent Salazar.

*Heterocampa* species, from Arizona, USA, has a white line with reddish blotches
running dorsally along the length of its body, which resembles the dried edge
of the leaf of its oak food plant. When it slips its body into the groove made by
previous feeding, it is almost unnoticeable.

### Disperse Feeding Throughout a Plant

Caterpillars often forage for only short periods of time, and some make many
small and widely spaced feeding injuries. For example, *Amphiphyra pyramidoides*
takes short meals, then quickly moves up to a meter away from the damaged
leaf (Heinrich 1979). Likewise, *Sphecodina abbotti* moves widely on its host
plant, never staying to finish any one leaf. Similar foraging patterns have been
described by Young (1972) and Schultz (1983) and are usually considered
strategies to confuse visually hunting birds (Heinrich and Collins 1983).

Dispersing damage throughout the plant may also confuse natural enemies
that use odors to locate their prey (Hassell 1968; Odell and Godwin 1984;
Whitman and Eller 1990). It is well documented that parasitoids, and some
predators, orient to the odors of caterpillar-damaged plants (Whitman 1988;
Whitman and Nordlund 1994). In some cases, the plants may have evolved to
release specific parasitoid-attracting volatile compounds when attacked by
caterpillars, but not when artificially damaged (Turlings *et al.* 1990, 1993;

**Fig. 13.** To hide evidence of their feeding damage, some caterpillars, including *Antheraea polyphemus*, chew through the petiole of the leaf they were feeding on, letting the partially consumed leaf fall to the ground. Photo by Bernd Heinrich.

Turlings and Tumlinson 1992; Kainoh *et al.* 1999). Hence, plants and parasitoids may function jointly against leaf-damaging herbivores. Some caterpillars may have evolved to counter such plant-to-parasitoid communication by dispersing their feeding, and, thus, leaving numerous odorous sites throughout the plant, which confuse and frustrate searching natural enemies.

## Nocturnal Feeding

Many caterpillars feed only at night under the cover of darkness, presumably because there are fewer nocturnal predators (Campbell *et al.* 1975; Heinrich 1979). In many habitats, birds are probably the primary diurnal predators and arthropods the primary nocturnal predators. Predators typically attack only certain sized prey, and birds and terrestrial arthropods differ greatly in size. Hence, tiny caterpillars might escape bird predation, but not arthropod predation, whereas large caterpillars might be better able to fend off arthropod predation (e.g. ants), but would be more visible (and a greater reward) to birds. Birds may not only suppress the diurnal activity of caterpillars, but also of entomophagous arthropods. This predicts that small, early instar caterpillars should feed during the day, whereas large, late instar caterpillars should feed at night. Leonard (1970) found a shift in feeding schedule, with the first three instars of *Lymantria dispar* preferring to feed during the day, and the last three instars only feeding at night. Johnson (1984) also found that some *Catocala* spp. feed both day and night in the first instar and rest on leaves between feeding bouts, but that later instars feed only at night and between feedings rest on twigs. Dempster (1967) estimated that arthropod predation accounted for between 50 and 60 per cent of the mortality of young instars of *Pieris rapae*, and that it occurred chiefly at night.

### Movement away from Feeding Sites

Caterpillars often rest considerable distances away from feeding sites (Herrebout *et al.* 1963; Young 1972; Heinrich 1979; Heinrich and Collins 1983; Johnson 1984). Curio (1970) examined the four color morphs of *Erinnyis ello* and found that brown morphs rested for many hours during the day on the base of the trunk, whereas the other color morphs typically rested in foliage. The larvae were preyed upon during the day by a wasp (*Polistes crinitus*) and to a lesser extent by an anole (*Anolis lineatopus*), but because the wasps only searched for prey in the foliage of the host plant, the brown form suffered less predation. Interestingly, in this example, the caterpillars were both polymorphic and polyethic, and the behavior matched the specific color.

At low population densities, about two-thirds of fourth, fifth, and sixth instar *Lymantria dispar* larvae rested during the day in bark fissures or under bark flaps on trees, and about one-third rested in leaf litter (Campbell *et al.* 1975). Regardless of where they rested, all the caterpillars ascended trees to feed at night (Leonard 1970). When these caterpillars rested in bark fissures or under bark flaps during the day, they gained protection from generalist predators (Weseloh 1988).

Some caterpillars rest completely off of their host plant between feeding bouts. For example, Damman (1986) observed third, fourth, and fifth instar *Eurytides marcellus* caterpillars moving a short distance from their host plant to rest in litter when not feeding. Young (1972) occasionally observed *Morpho peleides limpida* larvae resting on branches of other plants in intimate contact with its host plant. A few caterpillars may alter their behavior in response to predator density. For example, in the presence of stinkbugs, *Junonia coenia* caterpillars tended to move more between host plants than caterpillars foraging in the absence of these predators (Stamp and Bowers 1991).

### Pupate away from Host Plant

Caterpillars that finish their larval development need to select a safe site to pupate. A lower density of pupae in a given area may reduce the formation of search images by generalist predators (West and Hazel 1982). Caterpillars can travel far distances to pupate, effectively separating themselves from their host plant (West and Hazel 1979). *Nymphalis antiopa* larvae can travel well over

100 m (Besemer and Meeuse 1938). Young *et al.* (1986) observed that *Papilio anchisiades* pupated on the trunk of its host plant, on nearby buildings, or on other substrates near the host plant, and suggested that dispersal away from the plant reduced mortality from natural enemies.

## Dropping

Many caterpillars simply drop off of their host plant when attacked by predators or parasitoids (Baker 1970; Myers and Campbell 1976; Awan 1985; Heads and Lawton 1985; Cornell *et al.* 1987). This is a risky strategy, especially for younger caterpillars who may starve, desiccate, or fall prey to ground predators before relocating their host plant (Dethier 1959; Rausher 1979). However, some caterpillars have found a solution to this problem; they fall off, but remain attached to the leaf by a silken thread produced by a spinneret near their labial palps (Allen *et al.* 1970; Dempster 1971). This can be an effective way to avoid predation and parasitism, but not always. When *Plathypena scabra* drops from a silken thread, its parasitoid, *Diolcogaster facetosa*, locates the thread and grasps it with her tarsi. She then slides, head first, slowly down the thread towards the suspended larvae. When the wasp is within a few millimeters of the host, she quickly thrusts her abdomen forward to oviposit. Oviposition is very rapid, and usually requires less than a second (Yeargan and Braman 1986). When the hyperparasitoid *Mesochorus discitergus* locates a small (second instar) *Plathypena scabra* suspended by a thread, it hangs head downward from the edge of the leaf, suspended by its hind tarsi. It then uses its fore legs to gather up the caterpillar's silken thread, thus lifting the larvae slowly toward it (Yeargan and Braman 1989). When the caterpillar is within grasping distance, the wasp grabs the larva, and both fall to the ground or remain suspended on the caterpillar's silken thread while the hyperparasitoid parasitizes the primary parasitoid within the caterpillar.

## Playing Dead

Playing dead (thanatosis) is uncommon among Lepidopteran larvae. Rawlins (1984), however, did find a few external feeding, cryptically colored, mycophagous caterpillars that became immobile when disturbed. Whether or not this behavior is an effective means of defense against predators or parasitoids has not been demonstrated. However, these caterpillars may gain protection by looking, smelling and tasting like fungi. Hence, they may chemically and visually mimic fungus, a mimicry reinforced by immobility.

## Group Defense

Many caterpillars are thought to derive defensive benefits from aggregation. Colonial tent caterpillars build silken structures that sometimes deter birds and other generalist predators. Many chemically defended species are gregarious, which allows them to pool their defenses. This includes species with urticating hairs such as the western grape leaf skeletonizer, *Harrisina brillians*, and species with internal toxins such as *Zygaena lonicerae*. Gregarious caterpillars often perform group displays in response to predators or parasitoids (Morris 1963; Hougue 1972; Myers and Smith 1978; Stamp 1984). In *Hemileuca lucina*, the entire aggregation head flicks for up to 20 minutes in response to tachinid flies. This behavior appeared to often prevent fly contact with the larvae (Stamp and Bowers 1990). Disturbance by parasitoids resulted in simultaneous head flicking by the colonial, web-making caterpillars of *Euphydryas phaeton* (Stamp 1982). The behavior lasted up to 14 minutes and occasionally forced the parasitoid to move to an unoccupied portion of the web or an adjacent leaf.

### Frass Chains

Some nymphalid caterpillars, such as *Epiphile adrasta adrasta*, *Temenis laothoe liberia* and *Pseudonica flavilla canthara*, construct chains of frass (Muyshondt 1973 a,b,c), which apparently serve a defensive function (Muyshondt 1974, 1976; Casagrande and Mielke 1985; DeVries 1987). Soon after emerging, the larvae move to the leaf edge, nibble around the end of a vein and affix to it pellets of frass stuck together with silk, until the vein seems to project beyond the leaf limits. As the caterpillars feed and defecate, new frass pellets are added to the chain. First and second instar larvae remain on the frass chain, except when feeding.

Freitas and Oliveira (1992) characterized this behavior for the nymphalid, *Eunica bechina*, whose adults lay their eggs singly on plants that bear extrafloral nectaries. They suggested that the behavior of constructing Frass Chains was related to defense against flightless predators, especially ants, which would not cross over the feces to attack the caterpillars. They, then, demonstrated this by placing live Termites on leaves and on the ends of Frass Chains. Termites placed on leaves were attacked by foraging ants significanlty more than those placed on Frass Chains (Freitas and Oliveira 1996).

### Removal of Frass

Some caterpillars perform behaviors that serve to distance themselves from their feces. The function for such behaviors is unknown; however, several hypotheses have been proposed. The first and, probably, most obvious is that frass may harbor pathogenic microbes or parasites (Tanada 1953a,b,c, 1955a,b, 1956; Biever and Wilkinson 1978). Separation from such pathogens or parasites could confer a selective avantage to caterpillars that remove frass from their immediate vicinity. For shelter building caterpillars, such as the Hesperiidae, Tortricidae, Gelechiidae and Pyralidae, the build-up of frass within a shelter may necessitate the building of a new shelter. Building new shelters is energetically costly, and although most of these species build more than one shelter during their development (see previous section), having to build additional shelters because of compiling frass would be detrimental. In addition, since most shelter building caterpillars feed and construct shelters at night, additional shelter building would only further reduce the time available for feeding. Frass also serves as a chemical beacon to parasitoids that use olfaction in searching for their host (Lewis *et al.* 1976; Vinson 1976), and may serve as a visual marker for some predators with acute vision such as birds. Frass removal would reduce the searching efficiency of such natural enemies.

There are two behaviors used by caterpillars to remove frass from their general vicinity. In projectile defecation (fecal firing), the feces is forcibly ejected from the anus like a cannonball. In some cases, the pellet travels quite far. Frohawk (1913), observed that some hesperiids could eject their feces up to one meter. Since then, projectile defecation has been anecdotally reported by several authors (Rawlins 1984; Friedlander 1986). When ready to defecate, *Achalarus casica* (Hesperiidae) moves so that its anus is clear of the top leaf of its shelter, and fires its feces up to 103 cm away. It then swings its head back and forth near the leaf edge, apparently to make sure the pellet is, in fact, gone. When a fresh piece of its own frass was placed in its shelter, immediately after firing, the caterpillar nudged the pellet off the leaf with its head (Salazar, personal observations).

Caveney *et al.* (1998) characterized the mechanism behind fecal firing for the hesperiid, *Calpodes ethlius* (Fig. 14). Situated above the anus of many shelter building caterpillars, including *C. ethlius*, is an anal comb, which is attached to the lower surface of the anal plate. Rather than acting as a lever and flicking fecal pellets away, the anal comb serves as a latch to prevent the premature distortion of the lower wall of the anal plate until the anal haemocoel compartment is fully pressurized. The anal comb is swung into position during pellet extrusion by retractor muscles and held in place by a catch formed by a blood-swollen torus of everted rectal wall. When the caterpillar is ready to defecate, it contracts its anal prolegs, which raises the hemostatic pressure in the anal compartment and causes the comb to slip over the toral catch. This causes the underside of the anal plate to move rapidly backwards as the blood pressure is released, projecting the pellet resting against it through the air.

Housecleaning is the second behavior used by caterpillars to remove feces, and, again, is observed primarily in shelter building larvae. In this behavior, fecal pellets are either picked up with mandibles or thoracic legs and thrown off of the host plant, or pushed off of the host plant by the head (Fig. 15) (Clarke 1971; Aiello 1979; Poirier and Borden 1995; Salazar, unpublished data). In *Pieris rapae*, larvae pick up frass pellets in their mandibles and throw them off the plant by swinging the head and thorax sideways (Usher 1984). McFarland (1988) described this behavior for an Australian geometrid, *Heliomystis electrica*. The caterpillar arched the anterior portion of its body backwards, grabbed the frass pellet resting on its anus with its thoracic legs, and then sprang its body back into place, sending the pellet flying through the air.

Although visually hunting animals, such as birds, may use frass as a means of locating prey, to our knowledge, this has never been demonstrated. Parasitoids, however, are well known to use kairomones from frass to locate their hosts (Fig. 16) (Lewis and Jones 1971; Hendry *et al.* 1973; Nettles and Burks 1975; Vinson *et al.* 1975; Roth *et al.* 1978; Sato 1979; Mohyuddin *et al.* 1981;

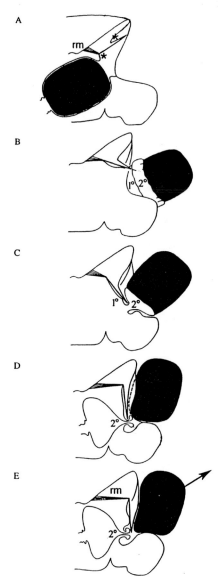

**Fig. 14.** The sequence of positions of the pressure plate and anal comb during preparation for pellet discharge. (A) Comb in a resting position just before the fecal pellet is voided (rm), retractor muscles. (B) After the pellet is half way out of the rectum, the comb is extended over the primary torus (1°) and hooked beneath the dorsal margin of the secondary torus (2°). This is accomplished by the contraction of its retractor muscles. (C) As it arches forwards, the anal comb remains tightly wedged between the primary and secondary tori. (D) The anal comb in fully flexed state. The comb acts as a "latch" that trips when it is forced backwards over a "catch" formed by the secondary torus. (E) Position of the pressure plate and anal comb at the moment of pellet discharge. (Figure and legend from Caveney *et al.* 1998; Company of Biologist Ltd.).

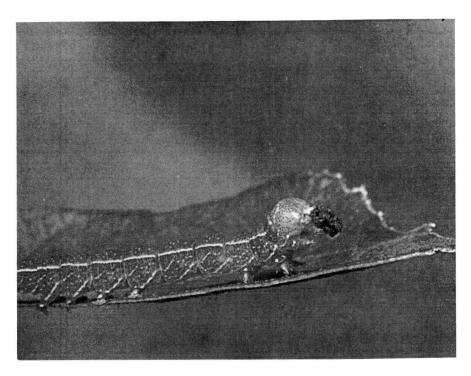

**Fig. 15.** *Panopoda rufimargo* removing its feces from its host plant leaf.

**Fig. 16.** *Microplitis croceipes* inspecting feces from its caterpillar host, *Helicoverpa zea.*

Nettles 1982; Nordlund *et al.* 1988; Takabayashi and Takahashi 1989; Alborn *et al.* 1995; Mattiacci and Dicke 1995). Jones *et al.* (1971) was able to isolate and identify from the frass of *Helicoverpa zea* 13-Methylhentriacontane, the host-seeking stimulant for *Microplitis croceipes*.

Using the mouth to move frass seems counter to the idea that Housecleaning is a means of avoiding infection by pathogens. Pathogens are often transmitted orally, and the spores of some bacterial and viral pathogens are found in frass (Tanada 1953b; Weiser 1961). Shelter builders may indeed houseclean or fecal fire in order to conserve space within their shelter, but this also has not been demonstrated experimentally. However, Usher (1984) showed that frass removed by *Pieris rapae* caterpillars conferred a selective advantage against *Cotesia glomerata* wasps. Therefore, the hypothesis that Housecleaning might have evolved in response to the host seeking behavior of parasitoids has found some experimental support. Clearly, projectile defecation and Housecleaning must be beneficial to have evolved and persisted. However, demonstration of these benefits awaits further experimentation.

**CONCLUSION**

Throughout this chapter, we have emphasized a number of broad concepts. First, caterpillars are attacked by a great diversity of predators and parasitoids and these natural enemies use different strategies to locate and subdue their prey. Within its short life, a caterpillar may need to defend itself from predators and parasitoids, nocturnal and diurnal enemies, visually oriented and chemically oriented enemies, large and small predators, and solitary and social predators. No caterpillar could possibly evolve defenses against all these potential natural enemies—there are simply too many of them. A defense against one predator may fail against another, or, worse, may even make the caterpillar more susceptible to another. Thus, there are tradeoffs. A caterpillar may be very successful at deterring one particular type of natural enemy or one type of hunting strategy, but less successful against another. There are also tradeoffs between defense and other life history requirements; in many cases, defense directly conflicts with the need to feed and develop rapidly.

Another broad theme is that defensive traits do not exist in isolation, but are usually linked together in defensive ensembles consisting of numerous morphological, behavioral, and physiological traits, which, together, form a comprehensive defense. Two such comprehensive strategies are Crypsis and the Chemical Defense Syndrome (CDS), both of which involve numerous specific linked traits. Crypsis usually includes color, pattern, body shape, immobility, nocturnal activity and a dispersed and low population density. The CDS joins together a different set of traits, including aposematism, aggregation, diurnal behavior, large size, sluggishness and chemical defense.

We have stressed an evolutionary approach. We believe that caterpillar defenses have evolved in direct response to selection by natural enemies. Furthermore, we suggest that there has been at least a diffuse co-evolution between caterpillars and these natural enemies such that defensive adaptations in caterpillars have driven the evolution of hunting adaptations in predators and parasitoids in a reciprocal fashion. Plants, too, are under selective pressure and may have evolved mechanisms to alert natural enemies to caterpillar presence by emitting parasitoid-attractive odors from caterpillar-damaged leaves. Parasitoids have evolved the capability to detect and respond to these plant volatiles at extremely low concentrations, and caterpillars appear to have evolved various behaviors to counteract this plant-to-parasitoid "communication". Hence, plants, caterpillars, and natural enemies appear to be linked together as co-evolving units, each evolving their own strategies in response to the adaptations of the others.

Finally, we have stressed that no defense is perfect. Not only are there simply too many different types of natural enemies and hunting strategies to defend against, but under an evolutionary scenario, we would always expect some successful predation. This is because once a defense nears being completely effective against natural enemies, adding resources to improve it is wasteful. Additionally, there would be little selective pressure to improve a defense that is already highly effective. In contrast, specialized natural enemies would be under great selective pressure to improve their ability to overcome that defense. Hence, for every caterpillar defense, there should always be some natural enemies that can breach that defense.

## ACKNOWLEDGEMENTS

We would like to dedicate this chapter to Demetrio and Verna Salazar for their patience and generosity. We thank D. Alsop, S. Caveney, E. Greene, B. Heinrich and D. Wagner for contributing figures, J. Armstrong for his help in the use of QuickScan35, and the Advanced Entomology Group at Illinois State University for editorial suggestions.

## REFERENCES

Aiello, A. 1979. Life history and behavior of the case-bearer *Phereoeca allutella* (Lepidoptera: Tineidae). *Psyche*, **86**: 125–136.

Alborn, H.T., Lewis, W.J. and Tumlinson, J.H. 1995. Host-specific recognition kairomone for the parasitoid *Microplitis croceipes* (Cresson). *J. Chem Ecol.*, **21**: 1697–1708.

Allen, D.C., Knight, F.B. and Foltz, J.L. 1970. Invertebrate predators of the jackpine budworm, *Choristoneura pinus*, in Michigan. *Ann. Entomol. Soc. Am.*, **63**: 59–64.

Alsop, D.W. 1970. Defensive glands of arthropods: comparative morphology of selected types. Ph. D. thesis, Cornell University, Ithaca.

Aplin, R.T., D'Arcy Ward, T. and Rothschild, M. 1975. Examination of the large white and small white butterflies (*Pieris* spp.) for the presence of mustard oil glycosides. *J. Entomol. A*, **50**: 73–78.

Askew, R.R. 1968. A survey of leaf-miners and their parasites on *Laburnum*. *Trans. R. Entomol. Soc. London*, **120**: 1–37.

Askew, R.R. 1971. *Parasitic Insects*. London: Heinemann Educational Books Ltd.

Atsatt, P.R. 1981. Ant-dependent food plant selection by the mistletoe butterfly *Ogyris amaryllis* (Lycaenidae). *Oecologia*, **48**: 60–63.

Awan, M.S. 1985. Anti-predator ploys of *Heliothis punctiger* (Lepidoptera: Noctuidae) caterpillars against the predator *Oechalia schellenbergii* (Hemiptera: Pentatomidae). *Aust. J. Zool.*, **33**: 885–890.

Baker, R.R. 1970. Bird predation as a selective pressure on immature stages of the cabbage butterflies, *Pieris rapae* and *P. brassicae*. *J. Zool. Lond.*, **162**: 43–59.

Balduf, W.V. 1935. *The Bionomics of Entomophagous Coleoptera*. New York, NY: Swift.

Balduf, W.V. 1939. *The Bionomics of Entomophagous Insects, Part II*. Guilford, England: John Swift.

Bardwell, C.J. and Averill, A.L. 1996. Effectiveness of larval defenses against spider predation in cranberry ecosystems. *Environ. Entomol.*, **25**: 1083–1091.

Baylis, M. and Pierce, N.E. 1993 The effects of ant mutualism on the foraging and diet of lycaenid caterpillars. *In* N.E. Stamp and T.M. Casey (Eds.), *Caterpillars: Ecological and Evolutionary Constraints on Foraging*. New York, NY: Chapman and Hall, Inc. pp. 404–421.

Beard, R.L. 1963. Insect toxins and venoms. *Annu. Rev. Entomol.*, **8**: 1–18.

Beckage, N.E. and Riddiford, L.M. 1978. Development interactions between the tobacco hornworm *Manduca sexta* and its braconid parasite *Apanteles congregatus*. *Entomol. Exp. Appl.*, **23**: 139–151.

Bellows, T.S.Jr., Owens, J.C. and Huddleston, E.W. 1982. Predation of range caterpillar, *Hemileuca oliviae* (Lepidoptera: Saturniidae) at various stages of development by different species of rodents in New Mexico during 1980. *Environ. Entomol.*, **11**: 1211–1215.

Berenbaum, M.R. and Miliczky, E. 1984. Mantids and milkweed bugs: Efficacy of aposematic coloration against invertebrate predators. *Am. Midl. Nat.*, **111**: 64–68.

Besemer, A.F.H. and Meeuse, B.J.D. 1938 Rouwmantels. *Levende Nat.*, **43**: 1–12.

Biever, K.D. and Wilkinson, J.D. 1978. A stress-induced granulosis virus of *Pieris rapae*. *Environ. Entomol.*, **7**: 572–573.

Blum, M.S. 1981. *Chemical Defenses of Arthropods*. New York: Academic Press.

Blum, M.S. 1983. Detoxication, deactivation, and utilization of plant compounds by insects. *In* P.A. Hedin (Ed) *Plant resistance to insects*. ACS Symposium Series 208. Washington, D.C.: American Chemical Society. pp. 265–275.

Boer, M.H. den 1971. A colour polymorphism in caterpillars of *Bupalus piniarius* (L.) (Lepidoptera: Geometridae). *Neth.J.Zool.*, **21**: 61–116.

Boppré, M. 1990 Lepidoptera and pyrrolizidine alkaloids. *J. Chem Ecol.*, **16**: 165–186.

Bowers, M.D. 1980. Unpalatability as a defense strategy of *Euphydryas phaeton* (Lepidoptera: Nymphalidae). *Evolution*, **34**: 586–600

Bowers, M.D. 1983. Mimicry in North American checkerspot butterflies: *Euphydryas phaeton* and *Chlosyne harrisii* (Nymphalidae). *Ecol. Entomol.*, **8**: 1–8

Bowers, M.D. 1988. Plant allelochemistry and mimicry, *In* P. Barbosa and D. Letourneau (Eds.), *Novel Aspects of Insect-Plant Interactions*. Wiley, New York. pp. 273–311.

Bowers, M.D. 1990. Recycling plant natural products for insect defense. *In* D.L. Evans and J.O. Schmidt (Eds) *Insect Defenses: Adaptive Mechanisms and Strategies of Prey and Predators*. Albany, NY: State University of New York Press. pp. 353–386.

Bowers, M.D. 1992. Unpalatability and the cost of chemical defense in insects. *In* B. Roitberg and M.B. Isman (Eds), *Chemical Ecology of Insects: An Evolutionary Approach*. Chapman and Hall. New York: pp. 216–244.

Bowers, M.D. 1993. Aposematic caterpillars: Life-styles of the warningly colored and unpalatable. *In* N.E. Stamp and T.M. Casey (Eds.), *Caterpillars: Ecological and Evolutionary Constraints on Foraging*. Chapman and Hall, Inc. New York, NY: pp. 331–371.

Bowers, M.D. and Farley, S. 1990 The behaviour of gray jays (*Perisoreus canadensis*) toward palatable and unpalatable Lepidoptera. *Anim. Behav.*, **39**: 699–705.

Bowers, M.D. and Larin, Z. 1989. Acquired chemical defense in the lycaenid butterfly *Eumaeus atala*. *J.Chem. Ecol.*, **15**: 1133–1146.

Bowers, M.D. and Puttick, G.M. 1986. The fate of ingested iridoid glycosides in Lepidopteran herbivores. *J. Chem. Ecol.*, **12**: 169–178.

Bowers, W.S. and Thompson, M.J. 1965. Identification of the major constituents of the crystalline powder covering the larval cuticle of *Samia cynthia ricini* (Jones). *J. Insect Physiol.*, **11**: 1003–1011.

Brattsten, L.B. 1986. Fate of ingested plant allelochemicals in herbivorous insects. *In* L.B. Brattsten and S. Ahmad (Eds.), *Molecular Aspects of Insect Plant Associations*. New York: Plenum. pp. 211–255.

Brower, L.P. 1984. Chemical defense in butterflies. *In* R.I. Vane-Wright and P.R. Ackery (Eds), *The Biology of Butterflies*. New York: Academic Press. pp. 109–134.

Brower, L.P. and Brower, J.V.Z. 1964a. Birds, butterflies, and plant poisons: A study in ecological chemistry. *Zoologica*, **48**: 65–84.

Brower, L.P. and Brower J.V.Z. 1964b. Birds, butterflies, and plant poisons: A study in ecological chemistry. *Zoologica*, **49**: 137–159.

Brower, L.P. and Calvert, W.H. 1985. Foraging dynamics of bird predators on overwintering Monarch butterflies in Mexico. *Evolution* **39**: 852–868.

Brower, L.P., Brower, J.V.Z. and Collins, C.T. 1963. Experimental studies of mimicry-7. Relative palatability and Müllerian mimicry among neoteropical butterflies of the sub-family *Heliconiinae*. *Zoologica*, **48**: 65–84.

Brower, L.P., Ryerson, W.N., Coppinger, L.L. and Glazier, S.C. 1968. Ecological chemistry and the palatability spectrum. *Science,* **161**: 1349–1351.

Brower, L.P., Horner, B.E., Marty, M.A., Moffitt, C.M. and Villa, R.B. 1985. Mice (*Peromyscus maniculatus, P. spicilegus,* and *Microtus mexicanus*) as predators of overwintering Monarch butterflies (*Danaus plexippus*) in Mexico. *Biotropica*, **17**: 89–99.

Brubaker, R.W. 1968. Seasonal occurrence of *Voria ruralis*, a parasite of the cabbage looper, in Arizona, and its behavior and development in laboratory culture, *J. Econ. Entomol.*, **61**: 306–309.

Brues, C.T. 1946. *Insect Dietary*. Cambridge, Massachusetts: Harvard University Press.

Calvert, W.H., Hedrick, L.E. and Brower, L.P. 1979. Mortality of the Monarch butterfly (*Danaus plexippus* L.): avian predation at five overwintering sites in Mexico. *Science*, **204**: 847–851.

Campbell, R.W., Hubbard, D.L. and Sloan, R.J. 1975. Patterns of gypsy moth occurrence within a sparse and numerically stable population. *Environ. Entomol.*, **4**: 535–542.

Carpenter, G.D.H. 1938. Audible emission of defensive forth by insects. *Proc. Zool. Soc. Lond. A*, **108**: 243–252.

Casagrande, M.M. and Mielke, O.H.H. 1985. Estágios imaturos de *Agrias claudina claudianus* Staudinger (Lepidoptera, Nymphalidae, Charaxinae). *Rev. Brasil Entomol.*, **29**: 139–142.

Casey, T.M. 1993. Effects of temperature on foraging of caterpillars. *In* N.E. Stamp and T.M. Casey (Eds.), *Caterpillars: Ecological and Evolutionary Constraints on Foraging*. Chapman and Hall, Inc. New York, NY: pp. 5–28.

Caveney, S., McLean, H. and Surry, D. 1998. Faecal firing in a skipper caterpillar is pressure-driven. *J. Exp. Biol.*, **201**: 121–133.

Chow, Y.S. and Tsai, R.S. 1989. Protective chemicals in caterpillar survival. *Experientia*, **45**: 390–392.

Clarke, C.A., Dickson, C.G.C. and Sheppard, P.M. 1963. Larval color pattern in *Papilio demodocus*. *Evolution*, **17**: 130–137.

Clarke, J.F. 1971. The Lepidoptera of Rapa Island. *Smith Cont. Zool.*, **56**: 1–282.

Clausen, C. 1940. *Entomophagous Insects*. New York: McGraw-Hill.

Codella, S.G. and Lederhouse, R.C. 1984. Foraging strategy and leaf damage patterns of black swallowtail caterpillars, *Papilio polyxenes*. *Bull. New Jersey Acad. Sci.*, **29**: 37.

Coppinger, R.P. 1970. The effects of experience and novelty on avian feeding behavior with reference to the evolution of warning coloration in butterflies. II. Reactions of naïve birds to novel insects. *Amer. Nat.*, **104**: 323–336.

Cornell, J.C., Stamp. N.E. and Bowers, M.D. 1987. Developmental change in aggregation, defense and escape behavior of buckmoth caterpillars, *Hemileuca lucina* (Saturniidae). *Behav. Ecol. Sociobiol.*, **20**: 383–388.

Cott, H.B. 1940. *Adaptive Coloration in Animals*. London: Methuen.

Cottrell, C.B. 1984. Aphytophagy in butterflies: Its relationship to myrmecophily. *Zool. J. Linn. Soc.*, **80**: 1–57.

Curio, E. 1965. Die schlangenmimikry einer sudamerikanischen schwarmerraupe. *Natur und Museum*, **95**: 207–211.

Curio, E. 1970. Validity of the selective coefficient of a behaviour trait in hawkmoth larvae. *Nature*, **228**: 382.

Damman, H. 1986. The osmaterial glands of the swallowtail butterfly *Eurytides marcellus* as a defense against natural enemies. *Ecol. Entomol.*, **11**: 261–265.

Damman, H. 1987. Leaf quality and enemy avoidance by the larvae of a pyralid moth. *Ecology*, **68**: 88–97.

Danks, H.V. 1975. Factors determining levels of parasitism by *Winthemia rufopicta* (Diptera: Tachinidae), with particular reference to *Heliothis* spp. (Lepidoptera: Noctuidae) as hosts. *Can. Entomol.*, **107**: 655–684.

Davis, R.H. and Nahrstedt, A. 1984. Cyanogenesis in insects. *In* G.A. Kerkut and L.I. Gilbert (Eds.), *Comprehensive Insect Physiology, Biochemistry and Pharmacology*, Vol. 11. Pergamon Press. Oxford: pp. 635-654.

Dempster, J.P. 1967. The control of *Pieris rapae* with DDT. I. The natural mortality of the young stages of *Pieris*. *J. Appl. Ecol.*, **4**: 485–500.

Dempster, J.P. 1971. The population ecology of the cinnabar moth *Tyria jacobaeae* L. (Lepidoptera, Arctiidae). *Oecologia*, **7**: 26–67.

Dethier, V.G. 1939. Prothoracic glands of adult Lepidoptera. *J.N.Y. Entomol. Soc.*, **47**: 131–144.

Dethier, V.G. 1959. Food-plant distribution and density and larval dispersal as factors affecting insect populations. *Can. Entomol.*, **91**: 581–596.

Detwiler, J.D. 1922. The ventral prothoracic gland of the red-humped apple caterpillar (*Schizura concinna* Smith & Abbot). *Can. Entomol.*, **54**: 175–191.

DeVries, P.J. 1987. *The Butterflies of Costa Rica and Their Natural History*. Princeton, New Jersey: Princeton University Press.

DeVries, P.J. 1991a. Evolutionary and ecological patterns in myrmecophilous riodinid butterflies. *In* C.R. Huxley and D.F. Cutler (Eds) *Ant-Plant Interactions*. Oxford University Press. Oxford: pp. 143–156.

DeVries, P.J. 1991b. Mutualism between *Thisbe irenea* larvae and ants, and the role of ant ecology in the evolution of myrmecophilous butterflies. *Biol. J. Linn. Soc.*, **43**: 179–195.

Douglas, M.M. 1983. Defense of bracken fern by arthropods attracted to axillary nectaries. *Psyche*, **90**: 313–320.

Downey, J.C. 1962. Myrmecophily in *Plebejus* (*Icaricia*) *icarioides* ( Lepidop.: Lycaenidae). *Entomol. News*, **73**: 57–66.

Duffey, S.S. 1980. Sequestration of plant natural products by insects. *Annu. Rev. Entomol.*, **25**: 447–477.

Edmunds, M. 1974. *Defense in Animals: A Survey of Anti-predator Defenses*. Essex: Longman.

Eisner, T. and Meinwald, Y.C. 1965. Defensive secretion of a caterpillar (*Papilio*). *Science*, **150**: 1733–1735.

Eisner, T., Pliske, T.E., Ikeda, M., Owen, D.F., Vazquez, L., Perez, H., Franclemont, J.G. and Meinwald, J. 1970. Defense mechanisms of arthropods XXVII. Osmeterial secretion of Papilionid caterpillars (*Baronia, Papilio, Eurytides*). *Ann Entomol. Soc. Am.*, **63**: 914–915.

Eisner, T., Kluge, A.F., Carrel, J.C. and Meinwald, J. 1972a. Defense mechanisms of arthropods. XXXIV. Formic acid and acyclic ketones in the spray of a caterpillar. *Ann. Entomol. Soc. Am.*, **65**: 765–766.

Eisner, T., Jutro, P., Aneshansley, D.J. and Niedhauk, R. 1972b. Defense against ants in a caterpillar that feeds on ant-guarded scale insects. *Ann. Entomol. Soc. Am.*, **65**: 987–988.

Eisner, T., Johnessee, J.S., Carrel, J., Hendry, L.B. and Meinwald, J. 1974. Defensive use by an insect of a plant resin. *Science*, **184**: 996–999.

Embree, D.G. 1971. The biological control of the winter moth in eastern Canada by introduced parasites. *In* C.B. Huffaker (Ed.), *Biological Control*. Plenum Press. New York pp. 217–268.

Endler, J.A. 1981. An overview of the relationships between mimicry and Crypsis. *Biol. J. Linn. Soc.*, **16**: 25–31.

Feichtinger, V.E. and Reavey, D. 1989. Changes in movement, tying and feeding patterns as caterpillars grow: The case of the yellow horned moth. *Ecol. Entomol.*, **14**: 471–474.

Fiedler, K. and Maschwitz, U. 1988. Functional analysis of the myrmecophilous relationships between ants (Hymenoptera: Formicidae) and lycaenids (Lepidoptera:Lycaenidae) II. Lycaenid larvae as trophobiotic partners of ants-a quantitative approach. *Oecologia*, **75**: 204–206.

Fiedler, K. and Maschwitz, U. 1989a. Functional analysis of the myrmecophilous relationships between ants (Hymenoptera: Formicidae) and lycaenids (Lepidoptera: Lycaenidae). I. Release of food recruitment in ants by lycaenid larvae and pupae. *Ethology*, **80**: 71–80.

Fiedler, K. and Maschwitz, U. 1989b. The symbiosis between the weaver ant, *Oecophylla smaragdina*, and *Anthene emolus*, an obligate myrmecophilous lycaenid butterfly. *J. Nat. Hist.*, **23**: 833–846.

Fink, L.S. and Brower, L.P. 1981. Birds can overcome the cardenolide defence of Monarch butterflies in Mexico. *Nature*, **291**: 67–70.

Fisher, R.A. 1930. *The Genetical Theory of Natural Selection*. Oxford: Clarendon Press.

Fitzgerald, T.D. 1980. An analysis of the daily foraging patterns of laboratory colonies of the eastern tent caterpillar, *Malacosoma americanum* (Lepidoptera:Lasiocampidae) recorded photoelectronically. *Can. Entomol.*, **112**: 731–738.

Fitzgerald, T.D., Casey, T.M. and Joos, B. 1988. Daily foraging schedule of field colonies of the eastern tent caterpillar *Malacosoma americanum*. *Oecologia*, **76**: 574–578.

Fordyce, J.A. 2000. A model without a mimic: aristolochic acids from the California pipevine swallowtail, *Philenot hirsuta*, and its host plant, *Aristolochia California*. *J. Chem. Ecol.*, **26**: 2567-2578.

Frank, J.H. 1971. Carabidae (Coleoptera) as predators of the red-backed cutworm (Lepidoptera:Noctuidae) in central Alberta. *Can Entomol.*, **103**: 1039–1044.

Frazer, J.F.D. 1965. The cause of urtication produced by larval hairs of *Arctia caja* (L.) (Lepidoptera:Arctiidae). *Proc. Roy. Entomol. Soc. Lond.*, **40**: 96–100.

Frazer, J.F.D. and Rothschild, M. 1961. Defence mechanisms in warningly-coloured moths and other insects. *Proc. 11th Int. Congr. Entomol. Vienna 1960*. **1**: 249–256.

Freitas, A.V.L. and Oliveria, P.S. 1992. Biology and behavior of the netotropical butterfly *Eunica bechina* (Nymphalidae) with special reference to larval defense against ant predation. *J. Res. Lepid.*, **31**: 1–11.

Freitas, A.V.L. and Oliveira, P.S. 1996. ants as selective agents on herbivore biology: Effects on the behaviour of a non-myrmecophilous butterfly. *J. Anim. Ecol.*, **65**: 205–210.

Friedlander, T. 1986. Taxonomy, phylogeny, and biogeography of *Asterocampa* Rober 1916 (Lepidoptera, Nymphalidae, Apaturinae). *J. Res. Lepid.*, **25**: 215–338.

Frohawk, F.W. 1913. Fecal ejection in hesperids, *Entomologist*, **49**: 201–202.

Fung, S.Y., Herrebout, W.M., Verpoorte, R. and Fischer, F.C. 1988. Butenolides in small ermine moths, *Yponomeuta* spp. (Lepidoptera:Yponomeutidea), and spindle-tree, *Euonymus europaeus* (Celastraceae). *J.Chem. Ecol.*, **14**: 1099–1111.

Glendinning, J.I., Alonso Mieja, A. and Brower, L.P. 1988. Behavioral and ecological interactions of foraging mice (*Peromyscus melanotis*) with overwintering Monarch butterflies (*Danaus plexippus*) in Mexico. *Oecologia*, **75**: 222–227.

Graber, J.W and Graber, R.R. 1983. Feeding rates of warblers in spring. *Condor*, **85**: 139–150.

Greenberg, R. and Gradwohl, J. 1980. Leaf surface specializing birds and arthropods in a Panamanian forest. *Oecologia*, **46**: 114–124.

Greene, E. 1989. A diet-induced developmental polymorphism in a caterpillar. *Science*, **243**: 643–646.

Harcourt, D.G. 1966. Major factors in survival of the immature stages of *Pieris rapae* L. *Can. Entomol.*, **98**: 653–662.

Harvey, D.J. and Webb, T.A. 1980. ants associated with *Harkenclenus titus, Glaucopsyche lygdamus*, and *Celestrina argiolus* (Lycaenidae). *J. Lepid. Soc.*, **34**: 371–372.

Harvey, P.H. and Greenwood, P.J. 1978. Anti-predator defence startegies: some evolutionary problems. *In* J.R. Krebs and N.B. Davies (Eds.), *Behavioural Ecology: An Evolutionary Approach*. Blackwell Scientific Publications. Oxford, London: pp. 129–151.

Hassell, M.P. 1968. The behavioral response of a tachinid fly (*Cyzenis albicans*) (Fall), to its host, the winter moth (*Operophtera brumata* (L.)). *J. Anim. Ecol.*, **37**: 627–639.

Heads, P.A. and Lawton, J.H. 1985. Bracken, ants and extrafloral nectaries. III. How insect herbivores avoid ant predation. *Ecol. Entomol.*, **10**: 29–42.

Heinrich, B. 1979. Foraging strategy of caterpillars: Leaf damage and possible predator avoidance strategies. *Oecologia*, **42**: 325–337.

Heinrich, B. 1993. How avian predators constrain caterpillar foraging. *In* N.E. Stamp and T.M. Casey (Eds.), *Caterpillars: Ecological and Evolutionary Constraints on Foraging.* Chapman and Hall, Inc. New York, NY: pp. 224–247.

Heinrich, B. and Collins, S.L. 1983. Caterpillar leaf damage, and the game of hide-and-seek with birds. *Ecology,* **64:** 592–602.

Hendry, L.B., Greany. P.D. and Gill, R.J. 1973. Kairomone mediated host-finding behavior in the parasitic wasp *Orgilus lepidus. Entomol. Exp. Appl.,* **16:** 471–477.

Herrebout, W.M., Kuyten, P.J. and de Ruiter, L. 1963. Observations on colour patterns and behaviour of caterpillars feeding on Scots pine. *Arch. Neerl. Zool.,* **15:** 315–357.

Hintze, C. 1969. Histologische Untersuchungen am Wehrsekretbeutel von *Cerura vinula* L. und *Notodonta anceps* Goeze (Notodontidae, Lepidoptera). *Z. Morph. Tiere,* **64:** 1–8.

Hogue, C.L. 1972. Protective function of sound perception and gregariousness in *Hylesia* larvae (Saturniidae: Hemileucinae). *J. Lepid. Soc.,* **26:** 33–34.

Hölldobler, B. and Wilson, E.O. 1990. *The ants.* Cambridge, Mass.: Belknap Press.

Holmes, R.T., Schultz, J.C. and Nothnagle, P. 1979. Bird predation on forest insects: An exclosure experiment. *Science,* **206:** 462–463.

Holmes, R.T., Sherry, T.W. and Sturges, F.W. 1986. Bird community dynamics in a temperate deciduous forest: long-term trends at Hubbard Brook. *Ecol. Monogr.,* **56:** 202–220.

Honda, K. 1980a. Volatile constituents of larval Osmeterial secretions in *Papilio protenor demetrius. J. Insect Physiol.,* **26:** 39–45.

Honda, K. 1980b. Osmeterial secretions of Papilionid larvae in the genera *Luehdorfia, Graphium* and *Atrophaneura* (Lepidoptera). *Insect Biochem.,* **10:** 583–588.

Honda, K. 1981. Larval Osmeterial secretions of the swallowtails (*Papilio*). *J. Chem. Ecol.,* **7:** 1089–1113.

Honda, K. 1983. Defensive potential of components of the larval Osmeterial secretion of Papilionid butterflies against ants. *Physiol. Entomol.,* **8:** 173–179.

Honda, T., Kainoh, Y. and Honda, H. 1998. Enhancement of learned response to plant chemicals by the egg-larval parasitoid, *Ascogaster reticulatus* Watanabe (Hymenoptera:Braconidae). *Appl. Entomol. Zool.,* **33:** 271–276.

Hopper, K.R. 1986. Preference, acceptance, and fitness components of *Microplitis croceipes* (Hymenoptera: Braconidae) attacking various instars of *Heliothis virescens* (Lepidoptera: Noctuidae). *Environ. Entomol.,* **15:** 274–280.

Huffaker, C.B., Messenger, P.S. and DeBach, P. 1971. The natural enemy component in natural control and the theory of biological control. *In* C.B. Huffaker (Ed.), *Biological Control.* Plenum Press. New York: pp. 16–67.

Iwao, S. and Wellington, W.G. 1970. The influence of behavioral differences among tent-caterpillar larvae on predation by a pentatomid bug. *Can. J. Zool.,* **48:** 896–898.

Järvi, T., Sillén-Tullberg, B. and Wiklund, C. 1981. The cost of being aposematic. An experimental study of predation on larvae of *Papilio machaon* by the great tit *Parus major. Oikos,* **36:** 267–272.

Johnson, J.W. 1984. The immature stages of six California *Catocala* (Lepidoptera: Noctuidae). *J. Res. Lepid.,* **23:** 303–327.

Jones, C.G., Young, A.M., Jones, T.H. and Blum, M.S. 1982. Chemistry and possible roles of cuticular alcohols of the larval *Atlas* moth. *Comp. Biochem. Physiol. B,* **73:** 797–801.

Jones, D., Parsons, J. and Rothschild, M. 1962. Release of hydrocyanic acid from crushed tissues of all stages in the life-cycle of species of the *Zygaeninae* (Lepidoptera). *Nature,* **193:** 52–53.

Jones, R.E. 1987. ants, parasitoids, and the cabbage butterfly *Pieris rapae. J. Anim. Ecol.,* **56:** 739–749.

Jones, R.L., Lewis, W.J., Bowman, M.C., Beroza, M. and Bierl, B.A. 1971. Host-seeking stimulant for parasite of corn earworm: isolation, identification, and synthesis. *Science,* **173:** 842–843.

Jong, M.C.J.M. de and Bleumink, K. 1977. Investigative studies of the dermatitus caused by the larvae of the brown tail moth, *Euproctis chrysorrhoea* L. (Lepidoptera, Lymantridae). IV. Further characterization of skin reactive substances. *Arch. Dermatol. Res.,* **259:** 263–281.

Kainoh, Y., Tanaka, C. and Nakamura, S. 1999. Odor from herbivore-damaged plant attracts the parasitoid fly *Exorista japonica* Townsend (Diptera: Tachinidae). *Appl. Entomol. Zool.,* **34:** 463–467.

Kato, M. 1984. Mining pattern of the honeysuckle leaf-miner *Phytomyza lonicerae*. *Res. Popul. Ecol.*, **26**: 84–96.

Kawamoto, F. and Kumada, N. 1984. Biology and venoms of Lepidoptera. *In* A.T. Tu (Ed.), *Handbook of Natural Toxins 2*. Marcel Dekker. New York: pp. 291–330.

Kelly, R.B., Seiber, J.N., Segall, D.D. and Brower, L.P. 1987. Pyrrolizidine alkaloids in overwintering Monarch butterflies (*Danaus plexippus*) from Mexico. *Experientia*. **43**: 943–946.

Kitching, R.L. 1983. Myrmecophilous organs of the larvae and pupae of the lycaenid butterfly *Jalmenus evagoras* (Donovan). *J. Nat. Hist.*, **17**: 471–481.

Klots, A.B. 1951. *A Field Guide to the Butterflies*. Boston: Houghton Mifflin.

Knapp, R. and Casey, T.M. 1986. Thermal ecology, behavior, and growth of gypsy moth and Eastern tent caterpillars. *Ecology*, **67**: 598–608.

Laine, K.J. and Niemelä, P. 1980. The influence of ants on the survival of mountain birches during an *Oporinia autumnata* (Lep., Geometridae) outbreak. *Oecologia*, **47**: 39–42.

Lane, C.D. 1959. A very toxic moth. The five spot Burnet (*Zygaena trifolii* Esp.) *Entomol. Month Mag.*, **95**: 93–94.

Lederhouse, R.C. 1990. Avoiding the hunt: Primary defenses of Lepidopteran caterpillars. *In* D.L. Evans and J.O. Schmidt (Eds), *Insect Defenses: Adaptive Mechanisms and Strategies of Prey and Predators*. Albany, New York: State University of New York Press. pp. 175–189.

Leonard, D.E. 1970. Feeding rhythm of the gypsy moth. *J. Econ. Entomol.*, **63**: 1454–1457.

Lewis, W.J. and Jones, R.L. 1971. Substance that stimulates host-seeking by *Microplitis croceipes* (Hymenoptera: Braconidae), a parasite of *Heliothis* species. *Ann. Entomol. Soc. Am.*, **64**: 471–473.

Lewis, W.J., Jones, R.L., Gross, H.R. (Jr.) and Nordlund, D.A. 1976. The role of kairomones and other behavioral chemicals in host finding by parasitic insects. *Behav. Biol.* **16**: 267–289.

Malicky, H. 1970. New aspects on the association between lycaenid larvae (Lycaenidae) and ants (Formicidae, Hymenoptera). *J. Lepid. Soc.*, **24**: 190–202.

Marsh, N. and Rothschild, M. 1974. Aposematic and cryptic Lepidoptera tested on the mouse. *J. Zool. Lond.*, **174**: 89–122.

Marsh, N., Rothschild, M. and Evans, F. 1984. A new look at butterfly toxins. In *The Biology of Butterflies*. R.I. Vane-Wright and P.R. Ackery (Eds.), Academic Press. New York: pp. 135–139.

Martin, W.R. (Jr.), Nordland, D.A. and Nettles, W.C. Jr. 1989. Ovipositional behavior of the parasitoid *Palexorista laxa* (Diptera: Tachinidae) on *Heliothis zea* (Lepidoptera: Noctuidae) larvae. *J. Entomol. Sci.*, **24**: 460–464.

Matthews, R.W. and Matthews, J.R. 1978. *Insect Behavior*. New York: John Wiley.

Mattiacci, L. and Dicke, M. 1995. The parasitoid *Cotesia glomerata* (Hymenoptera:Braconidae) discriminates between first and fifth larval instars of its host *Pieris brassicae*, on the basis of contact cues from frass, silk, and herbivore-damaged leaf tissue. *J. Insect Behav.*, **8**: 485–498.

Mayer, D.F. and Beirne, B.P. 1974. Aspects of the ecology of apple leaf rollers (Lepidoptera: Tortricidae) in the Okanagan Valley, British Columbia. *Can Entomol.*, **106**: 349–352.

McFarland, A.N. 1988. *Portraits of South Australian Geometrid Moths*. Lawrence, Kansas: Allen Press.

McNeil, J.N., Delisle, J. and Finnegan, R.J. 1978. Seasonal predatory activity of the introduced red wood ant, *Formica lugubris* (Hymenoptera: Formicidae) at Valcartier, Quebec, in 1976. *Can. Entomol.*, **110**: 85–90.

Mohyuddin, A.I., Inayatullah, C. and King, E.G. 1981. Host selection and strain occurrence in *Apanteles flavipes* (Cameron) (Hymenoptera: Braconidae) and its bearing on biological control of graminaceous stem-borers (Lepidoptera: Pyralidae). *Bull. Entomol. Res.*, **71**: 575–581.

Monro, A., Meinwald, J. and Eisner, T. 1962. Work cited in L.M. Roth and T. Eisner. chemical defenses of arthropods. *Annu. Rev. Entomol.*, **7**: 107–137.

Montgomery, S.L. 1982. Biogeography of the moth genus *Eupithecia* in Oceania and the evolution of ambush predation in Hawaiian caterpillars (Lepidoptera: Geometridae). *Entomol. Gen.*, **8**: 27–34.

Montllor, C.B. and Bernays, E.A. 1993. Invertebrate predators and caterpillar foraging. *In* N.E. Stamp and T.M. Casey (Eds.), *Caterpillars: Ecological and Evolutionary Constraints on Foraging*. Chapman and Hall, Inc. New York, NY: pp. 170–202.

Montllor, C.B., Bernays, E.A. and Barbehenn. R.V. 1990. Importance of quinolizidine alkaloids in the relationship between larvae of *Uresiphita reversalis* (Lepidoptera: Pyralidae) and a host plant, *Genista monspessulana*, *J. Chem. Ecol* **16**: 1853–1865.

Montllor, C.B., Bernays, E.A. and Cornelius, M.L. 1991. Responses of two hymenopteran predators to surface chemistry of their prey: significance for an alkaloid-sequestering caterpillar. *J. Chem. Ecol.*, **17**: 391–399.

Morrell, R. 1969. Play snake for safety. *Animals,* **12**: 154–155.

Morris, R.F. 1963. The effect of predator age in prey defense on the functional response of *Podisus maculiventris* Say to the density of *Hyphantria-cunea* Drury. *Can. Entomol.*, **95**: 1009–1020.

Moss, A.M. 1920. Sphingidae of Para, Brazil. *Novit. Zool.*, **27**: 333–424.

Mumma, R.O. and Zettle, A.S. 1977. Larval and pupal parasites of the oak leafroller, *Archips semiferanus*. *Environ. Entomol.*, **6**: 601–605.

Muyshondt, A. 1973a. Notes on the life cycle and natural history of butterflies of EI Salvador. II A. *-Epiphile adrasta adrasta* (Nymphalidae-Catonephelinae). *J.N.Y. Entomol. Soc.*, **81**: 214–223.

Muyshondt, A. 1973b. Notes on the life cycle and natural history of butterflies of EI Salvador. III A. *-Temenis laothoe liberia* (Nymphalidae-Catonephelinae). *J.N.Y. Entomol. Soc.*, **81**: 224–233.

Muyshondt, A. 1973c. Notes on the life cycle and natural history of butterflies of EI Salvador. IV A. *-Pseudonica flavilla canthara* (Nymphalidae-Catonephelinae). *J.N.Y. Entomol. Soc.*, **81**: 234–242.

Muyshondt, A. 1974. Notes on the life cycle and natural history of butterflies of EI Salvador. III-*Anaea (Consul) fabius* (Nymphalidae). *J. Lepid. Soc.*, **28**: 81–89.

Muyshondt, A. 1976. Notes on the life cycle and natural history of butterflies of EI Salvador. VII. *Archaeoprepona demophon centralis* (Nymphalidae). *J. Lepid. Soc.*, 30: 23–32.

Myers, J.H. and Campbell, B.J. 1976. Predation by carpenter ants: A deterrent to the spread of cinnabar moth. *J. Entomol. Soc. Brit. Col.*, **73**: 7–9.

Myers, J.H. and Smith, J.N.M. 1978. Head flicking by tent caterpillars: A defensive response to parasite sounds. *Can. J. Zool.*, **56**: 1628–1631.

Nahrstedt, A. and Davis, R.H. 1981. The occurrence of the cyanoglucosides, linamarin and lotaustralin, in *Acraea* and *Heliconius* butterflies. *Comp. Biochem. Physoil. B*, **68**: 575–577.

Nahrstedt, A. and Davis, R.H. 1983. Occurrence, variation and biosynthesis of the cyanogenic glucosides Linamarin and Lotustralin in species of the Heliconiini (Insecta: Lepidoptera). *Comp. Biochem. Physiol.B*, **75**: 65–73.

Needham, J.G., Frost, S.W. and Tothill, B.H. 1928. *Leaf-Mining Insects*. Baltimore: Williams & Wilkens.

Nettles, W.C. Jr. 1982. Contact stimulants from *Heliothis virescens* that influence the behavior of females of the tachinid, *Eucelatoria bryani*, *J. Chem. Ecol.*, **8**: 1183–1191.

Nettles, W.C.Jr. and Burks, M.L. 1975. A substance from *Heliothis virescens* larvae stimulating larviposition by females of the tachinid, *Archytas marmoratus*. *J. Insect Physiol.*, **21**: 965–978.

New, T.R. 1991. *Insects as Predators*. Kensington, Australia: New South Wales University Press.

Nishida, R., Weintraub, J.D., Feeny, P. and Fukami, H. 1993. Aristolochic acids from *Thottea* spp. (Aristolochiaceae) and the Osmeterial secretions of *Thottea* feading trodine swallowtail larvae (Papilionidae). *J. Chem. Ecol.*, **19**: 1587-1594.

Nishio, S. 1983. The fates and adaptive significance of cardenolides sequestered by larvae of *Danaus plexippus* (L.) and *Cycnia inopinatus* (Hy. Edwards). Ph.D. thesis. University Micorfilms, University of Georgia, Athens, GA.

Nordlund, D.A., Lewis, W.J. and Altieri, M.A. 1988. Influences of plant-produced allelochemicals on the host/prey selection behavior of entomophagous insects. In *Novel Aspects of Insect Plant Interactions*. P. Barbosa and D.K. Letourneau (Eds.) New York: Wiley. pp. 65–90.

Novak, F., Pelissou, V. and Lamy, M. 1987. Comparative morphological, anatomical and biochemical studies of the urticating apparatus and urticating hairs of some Lepidoptera: *Thaumetopoea pityocampa* Schiff., *Th. processionea* L. (Lepidoptera, Thaumetopoeidae) and *Hylesia metabus* Cramer (Lepidoptera, Saturniidae). *Comp. Biochem. Physiol. A.* **88**: 141–146.

ODell, T.M. and Godwin, P.A. 1984. Host selection by *Blepharipa pratensis* (Meigen) a tachinid parasite of the gypsy moth, *Lymantria dispar* L. *J. Chem. Ecol.*, **10**: 311–320.

Ohsaki, N. and Sato, Y. 1990. Avoidance mechanisms of three *Pieris* butterfly species against the parasitoid wasp *Apanteles glomeratus*. *Ecol. Entomol.*, **15**: 169–176.

Orsak, L. and Whitman, D.W. 1987. Chromatic polymorphism in *Callophrys mossii bayensis* larvae (Lycaenidae): spectral characterization, short-term color shifts, and natural morph frequencies. *J. Res. Lepid.*, **25**: 188–201.

Owen, D.F. 1970. Mimetic polymorphism and the palatability spectrum. *Oikos*, **21**: 333–336.

Owen, D.F. 1971. *Tropical Butterflies*. Oxford: Clarendon Press.

Owen, D.F. 1975. The efficiency of blue tits *Parus caeruleus* preying on larvae of *Phytomyza ilicis*. *Ibis*, **117**: 515–516.

Pasteels, J.M., Gregoire, J. and Rowell-Rahier, M.1983. The chemical ecology of defense in arthropods. *Annu. Rev Ent.*, **28**: 263–289.

Pasteur, G. 1982. A classificatory review of mimicry systems. *Ann. Rev. Ecol. Syst.*, **13**: 169–199.

Pavan, M. and Dazzini-Valcurone, M. 1976. Sostanze di difesa dei lepidotteri. *Pubbl. 1ˢᵗ Entomol. Agr. Univ. Pavia*, **3**: 3–23.

Pesce, H. and Delgado, A. 1971. Poisoning from adult moths and caterpillars. *In* W. Bücherl and E.E. Buckley (Eds.), *Venomous Animals and Their Venoms III*. Academic Press. New York: pp. 119–156.

Peterson, S.C., Johnson, N.D. and LeGuyader, J.L. 1987. Defensive regurgitation of allelochemicals derived from host cyanogenesis by eastern tent caterpillars. *Ecology*, **68**: 1268–1272.

Picarelli, Z.P. and Valle, J.R. 1971. Pharmacological studies on caterpillar venoms. *In* W. Bücherl and E.E. Buckley (Eds), *Venomous Animals and Their Venoms III*. New York: Academic Press. pp. 103–118.

Pierce, N.E. 1983. The ecology and evolution of symbioses between lycaenid butterflies and ants. Ph. D. thesis, Harvard University, Cambridge.

Pierce, N.E. 1987. The evolution and biogeography of associations between lycaenid butterflies and ants. *Oxford Surv. Evol. Biol.*, **4**: 89–116.

Pierce, N.E. and Easteal, S. 1986. The selective advantage of attendant ants for the larvae of a lycaenid butterfly, *Glaucopsyche lygdamus*. *J. Anim. Ecol.*, **55**: 451–462.

Pierce, N.E. and Mead, P.S. 1981. Parasitoids as selective agents in the symbiosis between lycaenid butterfly larvae and ants. *Science*, **211**: 1185–1187.

Pierce, N.E., Kitching, R.L., Buckley, R.C., Taylor, M.F.J. and Benbow, K.F. 1987. The costs and benefits of cooperation between the Australian lycaenid butterfly, *Jalmenus evagoras*, and its attendant ants. *Behav. Ecol. Sociobiol.*, **21**: 237–248.

Pocock, R.T. 1911. On the palatability of some British insects, with notes on the significance of mimetic resemblance. (With notes on the experiments by E.B. Poulton). *Proc. Zool. Soc. Lond.*, **1911**: 809–868.

Poirier, L.M. and Borden, J.H. 1995. Oral exudate as a mediator of behavior in larval eastern and western spruce budworms (Lepidoptera: Torticidae). *J. Insect Behav.*, **8**: 801–811.

Poole, R.W. 1970. Convergent evolution in the larvae of two *Penstemon*-feeding Geometrids (Lepidoptera: Geometridae). *J. Kan. Entomol. Soc.*, **43**: 292–297.

Portugal, A.H.A. 2001. Defesas quimicas em larvas da borboleta *Mechanitis polymnia* (Nymphalidae: Ithomiinae). MSc Thesis, Universidade Estadual de Campinas. pp. 176.

Poulton, E.B. 1887. The secretion of pure aqueous formic acid by lepidopterous larvae for the purpose of defence. *Br. Assoc. Adv. Sci. Rept.*, **57**: 765–766.

Poulton, E.B. 1888. The secretion of pure formic acid by lepidopterous larvae for the purpose of defence. *Br. Assoc. Adv. Sci Rept.*, **5**: 765–766.

Price, P.W., Fernandes, G.W. and Waring, G.L. 1987. Adaptive nature of gall insects. *Environ. Enotomol.*, **16**: 15–24.

Quiroz, A.D. 1978. Venoms of Lepidoptera. *In* S. Bettini (Ed.), *Arthropod Venoms*. Springer-Verlag. Berlin: pp. 555–611.

Rausher, M.D. 1979. Egg recognition: Its advantage to a butterfly. *Anim. Behav.*, **27**: 1034–1040.

Rawlins, J.E. 1984. Mycophagy in Lepidoptera. *In* Q. Wheeler and M. Blackwell (Eds) *Fungus-Insect Relationships*. Columbia University Press. New York: pp. 382–423.

Real, R.G. Ianazzi, R., Kamil, A.C. and Heinrich, B. 1984. Discrimination and generalization of leaf damage by blue jays (*Cyanocitta cristata*). *Anim, Learn. Behav.*, **12:** 202–208.

Reichstein, T., Euw, J. von, Parsons, J.A. and Rothschild, M. 1968. Heart poisons in the Monarch butterfly. *Science*, **161:** 861–868.

Ridout, L.M. 1981. Mutual interference: Behavioural consequences of encounters between adults of the parasitoid wasp *Venturia canescens* (Hymenoptera: Ichneumonidae). *Anim Behav.*, **29:** 897–903.

Robinson, M.H. 1969. Defenses against visually hunting predators. *Evol. Biol.*, **3:** 225–259.

Robinson, S.K. and Holmes, R.T. 1982. Foraging behavior of forest birds: The relationships among search tactics, diet, and habitat structure. *Ecology*, **63:** 1918–1931.

Roeske, C.N. Seiber, J.S., Brower, L.P. and Moffitt, C.M. 1976. Milkweed cardenolides and their comparative processing by Monarch butterflies (*Danaus plexippus*). *Recent Adv. Phytochem.*, **10:** 93–167.

Root, R.B. 1966. The avian response to a population outbreak of the tent caterpillar, *Malacosoma constrictum* (Stretch) (Lepidoptera: Lasiocampidae). *Pan-Pac. Entomol.*, **42:** 48–52.

Roth, J.P., King. E.G. and Thompson, A.C. 1978. Host location behavior by the tachinid, *Lixophaga diatraeae*. *Environ. Entomol.*, **7:** 794–798.

Rothschild, M. 1985. British aposematic Lepidoptera. *In* J.H. Heath and A.M. Emmet (Eds), *The Moths and Butterflies of Great Britain and Ireland*, Vol. 2. Essex: B.H. and A. Harley Ltd. pp. 9-62.

Ruehlmann, T.E., Matthews, R.W. and Matthews, J.R. 1988. Roles for structural and temporal shelter-changing by fern-feeding Lepidopteran larvae. *Oecologia*, **75:** 228–232.

Ruiter, L. de. 1952. Some experiments on the camouflage of stick caterpillars. *Behaviour*, **4:** 222–232.

Ruiter, L. de. 1955. Countershading in caterpillars. *Arch Neerl. Zool.*, **11:** 1–57.

Salt, G. 1938. Experimental studies in insect parasitism. VI. Host suitability. *Bull Entomol. Res.*, **29:** 224–246.

Salt, G. 1955. Experimental studies in insect parasitism. VIII. Host reactions following artificial parasitization. *Proc. Royal Soc. B*, **144:** 380–398.

Salt, G. 1963. The defence reactions of insect metazoan parasites. *Parasitology*, **53:** 527–642.

Salt, G. 1968. The resistance of insect parasitoids to the defence reactions of their hosts. *Biol. Rev.*, **43:** 200–232.

Sato, Y. 1976. Experimental studies on parasitization by *Apanteles glomeratus* L. (Hymenoptera: Braconidae). I. Parasitization to different species of genus *Pieris*. *Appl. Entomol. Zool.*, **11:** 165–175.

Sato, Y. 1979. Experimental studies on parasitization by *Apanteles glomeratus*. IV. Factors leading a female to the host. *Physiol. Entomol.*, **4:** 63–70.

Sato, H. and Higashi, S. 1987. Bionomics of *Phyllonorycter* (Lepidoptera, Gracillariidae, on *Quercus*. II. Effects of ants. *Ecol. Res.*, **2:** 53–60.

Savignano, D. 1990. Associations between ants and larvae of the Karner blue. Ph. D. thesis, University of Texas, Austin.

Schmidt, G.T. 1974. Host-acceptance behavior of *Campoletis sonorensis* toward *Heliothis zea*. *Ann. Entomol. Soc. Am.*, **67:** 835–844.

Schuler, W. 1990. Avian predatory behavior and prey distribution. *In* D.L. Evans and J.O. Schmidt (Eds.) *Insect Defenses: Adaptive Mechanisms and Strategies of Prey and Predators*. State University of New York Press. Albany, New York: pp. 151–171.

Schuler, W. and Hesse, E. 1985. On the function of warning coloration: A black and yellow pattern inhibits prey-attack by naïve domestic chicks. *Behav. Ecol. Sociobiol.*, **16:** 249–255.

Schultz, J.C. 1983. Habitat selection and foraging tactics of caterpillars in heterogeneous trees. *In* R.F. Denno and M.S. McClure (Eds.), *Variable Plants and Herbivores in Natural and Managed Systems*. Academic Press. New York: pp. 61–90.

Scott, J.A. 1986. *The Butterflies of North America*. Stanford, California: Stanford University Press.

Sillén-Tullberg, B. 1988. Evolution of gregariousness in aposematic butterfly larvae: A phylogenetic analysis. *Evolution*, **42:** 293–305.

Slansky, F.Jr. 1993. Nutritional ecology: the fundamental quest for nutrients. *In* N.E. Stamp and T.M. Casey (Eds.), *Caterpillars Ecological and Evolutionary Constraints on Foraging.* Chapman and Hall, Inc. New York, NY pp. 29–91.

Solomon, M.E., Glen, D.M., Kendall, D.A. and Milsom, N.F. 1977. Predation of overwintering larvae of codling moth (*Cydia pomonella* (L.) by birds. *J. Appl. Ecol.,* **13:** 341–352.

Stamp, N.E. 1982. Behavioral interactions of parasitoids and Baltimore checkerspot caterpillars (*Euphydryas phaeton*). *Environ. Entomol.,* **11:** 100–104.

Stamp, N.E. 1984 . Interactions of parasitoids and checkerspot caterpillars *Euphydryas* spp. (Nymphalidae). *J. Res. Lepid.,* **23:** 2–18.

Stamp, N.E. 1986. Physical constraints of defense and response to invertebrate predators by pipevine caterpillars (*Battus philenor:*Papilionidae). *J. Lepid. Soc.* **40:** 191–205.

Stamp, N.E. and Bowers, M.D. 1990. Parasitism of New England buckmoth caterpillars (*Hemileuca lucina:* Saturniidae) by tachinid flies. *J. Lepid. Soc.,* **44:** 199–200.

Stamp, N.E. and Bowers, M.D. 1991. Indirect effect on survivorship of caterpillars due to presence of invertebrate predators. *Oecologia,* **88:** 325–330.

Stamp, N.E. and Wilkens, R.T. 1993. On the cryptic side of life: being unapparent to enemies and the consequences for foraging and growth of caterpillars. In N.E. Stamp and T.M. Casey (Eds), *Caterpillars: Ecological and Evolutionary Constraints on Foraging.* Chapman and Hall, Inc. New York, NY: pp. 283–330.

Stermitz, F.R., Gardner, D.R. and McFarland, N. 1988. Iridoid glycoside Sequestration by two aposematic *Penstemon*-feeding geometrid larvae. *J. Chem. Ecol.,* **14:** 435–441.

Suzuki, N., Kunimi, Y., Uematsu, S. and Kobayashi, K. 1980. Changes in spatial distribution pattern during the larval stage of the fall webworm, *Hyphantria cunea* Drury (Lepidoptera: Arctiidae). *Res. Popul. Ecol.,* **22:** 273–283.

Swynnerton, C.F. 1915. Birds in relation to their prey: Experiments on Wood Hoopoes, small Hornbills and a babbler. *J. So. Afr. Orn. Un.,* **11:** 32–108.

Takabayashi, J. and Takahashi, S. 1989. Effects of host fecal pellet and synthetic kairomone on host-searching and postoviposition behavior of *Apanteles kariyai,* a parasitoid of *Pseudaletia separata. Entomol. Exp. Appl.,* **52:** 221–227.

Tanada, Y. 1953a. Susceptibility of the imported cabbageworm to *Bacillus thuringiensis* Berliner. *Proc. Hawaiian Entomol. Soc.,* **15:** 159–166.

Tanada, Y. 1953b. A microsporidian parasite of the imported cabbageworm in Hawaii. *Proc. Hawaiian Entomol. Soc.,* **15:** 167–175.

Tanada, Y. 1953c. Description and characterisitics of a granulosis virus of the imported cabbageworm. *Proc. Hawaiian Entomol. Soc.,* **15:** 235–261.

Tanada, Y. 1955a. Field observations on a microsporidian parasite of *Pieris rapae* (L.) and *Apanteles glomeratus* (L.) *Proc. Hawaiian Entomol. Soc.,* **15:** 609–616.

Tanada, Y. 1955b. Susceptibility of the imported cabbageworm to fungi: *Beauveria* spp. *Proc. Hawaiian Entomol. Soc.,* **15:** 617–622.

Tanada, Y. 1956. Microbial control of some lepidopterous pests of Crucifers. *J. Econ Entomol.,* **49:** 320–329.

Theodoratus, D.H. and Bowers, M.D. 1999. Effects of sequestered iridoid glycosides on prey choice of the prairie wolf spider, *Lycosa carolinensis. J. Chem. Ecol.,* **25:** 283–291.

Thurston, R. and Prachuabmoh, O. 1971. Predation by birds on tobacco hornworm larvae infesting tobacco. *J. Econ. Entomol.,* **64:** 1548–1549.

Turlings, T.C. and Tumlinson, J.H. 1992. Systemic release of chemical signals by herbivore-injured corn. *Proc. Natl. Acad. Sci.,* **89:** 8399–8402 .

Turlings, T.C., Tumlinson, J.H. and Lewis, W.J. 1990. Exploitation of herbivore-induced plant odors by host-seeking parasitic wasps. *Science,* **250:** 1251–1253.

Turilings, T.C., McCall, P.J., Alborn, H.T. and Tumlinson, J.H. 1993. An elicitor in caterpillar oral secretions that induces corn seedlings to emit chemical signals attractive to parasitic wasps. *J. Chem. Ecol.,* **19:** 411–425.

Urzua, A., Rodiriguez, R. and Cassel, B. 1987. Fate of ingested aristolochic acids in *Battus archidamus. Biochem. Syst. Ecol.,* **15:** 687-690.

Usher, B.F. 1984. Housecleaning behavior of an herbivorous caterpillar: Selective and behavioral implications of frass-throwing by *Pieris rapae* larvae. Ph.D. thesis, Cornell University, Ithaca.

Vasconcelos, H.L. 1991. Mutualism between *Maieta guianesis* Aubl., a myrmecophytic melastome, and one of its ant inhabitants: Ant protection against insect herbivores. *Oecologia*, **87**: 295–298.

Vet, L.E.M., Lewis, W.J. and Cardé, R.T. 1995. Parasitoid foraging and learning. *In* R.T. Cardé and W.J. Bell (Eds). *Chemical Ecology of Insects 2*. Chapman and Hall Inc. New York, NY: pp. 65–101.

Vinson, S.B. 1976. Host selection by insect parasitoids. *Ann. Rev. Entomol.*, **21**: 109–133.

Vinson, S.B. Jones, R.L., Sonnet, P., Beirl. B.A. and Beroza, M. 1975. Isolation, identification and synthesis of host-seeking stimulants for *Cardiochiles nigriceps*, a parasitoid of the tobacco budworm. *Entomol. Exp. App.*, **18**: 443–450.

Waage, J. and Greathead, D. 1986. *Insect Parasitoids*. London, England: Academic Press.

Wagner, D.L., Ferguson, D.C., McCabe, T.L. and Reardon, R.C. *Geometroid Caterpillars of Northeastern and Appalachian Forests*. USFS Technology Transfer Bulletin, in press.

Weatherston, J., Percy, J.E., MacDonald, L.M. and MacDonald, J.A. 1979. Morphology of the prothoracic defensive gland of *Schizura concinna* (J.E. Smith) (Lepidoptera: Notodontidae) and the nature of its secretion. *J. Chem. Ecol.* **5**: 165–177.

Weatherston, J., MacDonald, J.A., Miller, D., Riere, G., Percy-Cunningham. J.E. and Benn, M.H. 1986. Ultrastructure of exocrine prothoracic gland of *Datana ministra* (Drury) (Lepidoptera:Notodontidae) and the nature of its secretion. *J. Chem. Ecol.*, **12**: 2039–2050.

Weiser, J. 1961. Die Mikrosporidien als Parasiten der Insekten. *Monographien zur Angew Entomologie*, **17**: 149 pp.

Weseloh, R.M. 1976. Reduced effectiveness of the gypsy moth parasite, *Apanteles melanoscelus*, in Connecticut due to poor seasonal synchronization with its host. *Environ. Entomol.*, **5**: 743–746.

Weseloh, R.M. 1988. Effects of microhabitat, time of day, and weather on predation of gypsy moth larvae. *Oecologia*, **77**: 250–254.

Weseloh, R.M. 1993. Potential effects of parasitoids on the evolution of caterpillar foraging behavior. *In* N.E. Stamp and T.M. Casey (Eds.). *Caterpillars: Ecological and Evolutionary Constraints or Foraging*. Chapman and Hall, Inc. New York, NY: pp 203–223.

West, D.A. and Hazel, W.N. 1979. Natural pupation sites of swallowtail butterflies (Lepidoptera:Papilionidae): *Papilio polyxenes* Fabr., *P. glaucus* L. and *Battus philenor* (L.) *Ecol. Entomol.*, **4**: 387–392.

West, D.A. and Hazel, W.N. 1982. An experimental test of natural selection for pupation site in swallowtail butterflies. *Evolution*, **36**: 152–159.

Whitman, D.W. 1988. Plant natural products as parasitoid cuing agents. *In* H.G. Cuttler (Ed.), *Biologically Active Natural products: Potential Use in Agriculture*. ASC Symp. Ser. 380. Washington D.C.: American Chemical Society. pp. 386–396.

Whitman, D.W. and Eller, F.J. 1990. Parasitic wasps orient to green leaf volatiles. *Chemoecology*, **1**: 69–76.

Whitman, D.W. and Nordlund, D.A. 1994. Plant chemicals and the location of herbivorous arthropods by their natural enemies. *In* T.N. Ananthakrishnan (Ed.), *Functional Dynamics of Phytophagous Insects*. Oxford & IBH Publishing Co. New Delhi: pp. 133–159.

Whitman, D.W., Blum, M.S. and Jones, C. 1985. Chemical defense in *Taeniopoda eques* (Orthoptera:Acrididae): role of the metathoracic secretion. *Ann. Ent. Soc. Amer.*, **78**: 451–455.

Whitman, D.W., Blum, M.S. and Alsop, D.W. 1990. Allomones: chemicals for defense. *In* D.L. Evans and J.O. Schmidt (Eds.), *Insect Defenses: Adaptive Mechanisms and Strategies of Prey and Predators*. State University of New York Press. Albany, New York: pp. 289–351.

Whitman, D.W., Blum, M.S. and Slansky, F.Jr. 1994. Carnivory in phytophagous insects. *In* T.N. Ananthakrishnan (Ed), *Functional Dynamics of Phytophagous Insects*. Oxford & IBH Publishing Co. New Delhi: pp. 161–205.

Wiklund, C. and Järvi, T. 1982. Survival of distasteful insects after being attacked by naive birds: A reappraisal of the theory of aposematic coloration evolving through individual selection. *Evolution*, **36**: 998–1002.

Wiklund, C. and Sillén-Tullberg, B. 1985. Why distasteful butterflies have aposematic larvae and adults, but cryptic pupae: Evidence from predation experiments on the Monarch and the European swallowtail. *Evolution*, **39**: 1155–1158.

Wilkinson, A.T.S. 1996. *Apanteles rubecula* Marsh, and other parasites of *Pieris rapae* in British Columbia. *J. Econ. Entomol.,* **59**: 1012–1018.

Witthohn, K. and Naumann, C.M. 1987. Cyanogenesis-a general phenomenon in the Lepidoptera? *J. Chem. Ecol.,* **13**: 1789–1809.

Yeargan, K.V. and Braman, S.K. 1986. Life history of the parasite *Diolcogaster facetosa* (Weed) (Hymenoptera:Braconidae) and its behavioral adaptation to the defensive response of a Lepidopteran host. *Ann. Entomol. Soc. Am.,* **79**: 1029–1033.

Yeargan, K.V. and Braman, S.K. 1989. Life history of the hyperparasitoid *Mesochorus discitergus* (Hymenoptera: Ichneumonidae) and tactics used to overcome the defensive behavior of the green cloverworm (Lepidoptera: Noctuidae). *Ann. Entomol. Soc. Am.,* **82**: 393–398.

Yosef, R. and Whitman, D.W. 1992. Predator exaptations and defensive adaptations in evolutionary balance: no defense is prefect. *Evol. Ecol.,* **6**: 527–536.

Young, A.M. 1972. Adaptive strategies of feeding and predator-avoidance in the larvae of the neotropical butterfly, *Morpho peleides limpida* (Lepidoptera:Morphidae). *J.N.Y. Entomol. Soc.,* **80**: 66–82.

Young, A.M., Blum, M.S., Fales, H.H. and Bian, Z. 1986. Natural history and ecological chemistry of the neotropical butterfly *Papilio anchisiades* (Papilionidae). *J. Lepid Soc.,* **40**: 36–53.

# Chapter 9

# Sluggish Movement of Conspicuous Insects as a Defense Mechanism against Motion-oriented Predators

*John D. Hatle and Douglas W. Whitman*

## INTRODUCTION

Movement is an important element of insect defense ensembles. Perhaps the two most common insect defense strategies involve motion (Edmunds, 1974). At one extreme of the motion continuum are insects that remain motionless to avoid predator detection (Evans, 1984). Some insects 'freeze' when they detect predators (Fig. 1), while others simply pass long periods without moving (Fig. 2) or move extremely slowly (Fig. 3). Insects that use lack of motion as a defense tend to be cryptic and/or camouflaged. Insect Crypsis has been well studied and recently reviewed (Edmunds, 1990; Stamp and Wilkens, 1993).

At the opposite end of the motion continuum are insects that use rapid movement to flee predators. Examples include dragonflies, houseflies and some butterflies, all of which are active, conspicuous and difficult to capture. Again, many examples of prey fleeing predators exist in the literature (e.g. Edmunds, 1974; Ydenberg and Dill, 1986).

Remaining motionless (and undetected) and moving away rapidly are common and effective defense strategies in insects. The adaptive advantages of remaining motionless and fleeing are clear. In contrast, exhibiting intermediate levels of movement when in the presence of predators (neither avoiding detection nor fleeing) would seem to be maladaptive. Such prey would be both easily detected and easily captured, and, therefore, strong selection pressures against intermediate levels of movement would seem likely. Despite this, intermediate levels of movement are common. Indeed, the association of

Illinois State University, Department of Biological Sciences, Behavior, Ecology, Evolution, and Systematics Section, Campus Box 4120, Normal, IL, USA 61790, jhatle@ilstu.edu, dwwhitm@ilstu.edu

**Fig. 1.** *Phrynotettix robustus* Bruner, a cryptic grasshopper from southeast Arizona, USA. This ground-dwelling species closely resembles the color and texture of the pebbles of its desert habitat. When approached by predators, these 'toad lubbers' freeze, making them virtually undetectable by vision. Photo by D. Lightfoot.

**Fig. 2.** Most moths, like this noctuid from Arizona, USA, are cryptic and remain motionless and undetected during daytime.

**Fig. 3.** Most walking sticks, like this Australian species, reduce detection from predators by cryptic coloration and sluggish movement.

intermediate movement, aposematic coloration, chemical defense and conspicuous behavior in insects (Pasteels *et al.* 1983) is so common, it has been termed the Chemical Defense Syndrome (CDS; Whitman *et al.* 1985). In addition to many anecdotal descriptions, intermediate levels of movement in aposematic insects have been quantitatively described at least twice (Chai and Srygley, 1990; Hatle and Faragher, 1998).

How do we reconcile the frequent existence of this seemingly maladaptive trait? The existence of intermediate movement in aposematic insects would be somewhat resolved if it could be shown that this behavior is actually beneficial. In fact, we have recently demonstrated that intermediate levels of movement can be adaptive in aposematic insects when the movement is sufficiently slow to fail to release the attack response of motion-oriented predators (Hatle and Faragher, 1998; see below). In this chapter, we refer to this type of movement as 'sluggish'.

The notion that sluggish movement can serve as a defense engenders several new theoretical questions and directions for research. To address this need, in this chapter, we: 1) summarize the few papers on sluggish movement as a defense in aposematic insects; 2) list representative aposematic prey that exhibit sluggish movement; 3) list representative motion-oriented predators from which these prey might escape; 4) propose a definition of sluggish movement as a defense; 5) discuss possible scenarios for the evolution of sluggish movement as a defense; and 6) suggest future work which we believe to be important for determining the impact of sluggish movement in prey evolution.

## SLUGGISH MOVEMENT CAN ACT AS A DEFENSE AGAINST MOTION-ORIENTED PREDATORS

Sluggish movement has long been anecdotally noted in aposematic prey (e.g.

Bates, 1862.) To our knowledge, the first quantitative demonstration of slow movement in unpalatable prey, in comparison with palatable prey, was Chai and Srygley (1990), who showed that unpalatable Costa Rican butterflies flew more slowly and less erratically than palatable butterflies. They discussed the implications of these results for the evolution of butterfly body morphology; palatable butterflies have stouter bodies and greater wing musculature, presumably to facilitate a burst of speed to escape predators (Srygley and Chai, 1990; Marden and Chai, 1991). This implicitly suggests that unpalatable butterflies may not need well-developed wing musculature because they do not need to flee predators. However, Chai and colleagues (Chai and Srygley, 1990; Srygley and Chai, 1990; Marden and Chai, 1991) did not discuss the possibility that the slower movement of unpalatable butterflies might discourage motion-oriented predators.

More recently, we have shown that sluggish movement in aposematic insects can actually be beneficial in encounters with motion-oriented predators. To examine sluggish movement as a defense, we used lubber grasshoppers, *Romalea microptera*, as prey (Hatle and Faragher, 1998). Lubber grasshoppers are sluggish, aposematic, chemically defended, gregarious and conspicuous (Fig. 4; Whitman, 1988; 1990). In our experiments, lubber grasshoppers moved approximately four times more slowly than crickets, regardless of the presence of frogs (Hatle and Faragher, 1998). This suggests that lubber grasshoppers move more slowly than an insect that readily uses fleeing as a secondary defense. In our experiments, we used Northern leopard frogs, *Rana pipiens*, as predators. Northern leopard frogs are sit-and-wait, motion-oriented predators (i.e. ambush predators; Anderson, 1993).

**Fig. 4.** An aggregation of chemically defended, aposematic, Eastern lubber grasshoppers, *Romalea microptera*, from Georgia, USA. Early instars form tight aggregations from which individuals slowly move a few centimeters when feeding.

Because sluggish movement frequently co-occurs with chemical defense, one of the difficulties in examining the putative defensive role of sluggish movement is separating the effects of motion from chemical defense. If a slow moving prey is chemically defended from a motion-oriented predator, this could obscure the benefits of sluggish movement. Fortunately for us, lubber grasshoppers obtain part of their chemical defense from their diet; lubbers fed toxic plants are better-defended, whereas lubbers fed innocuous plants are more poorly defended (Jones *et al.* 1989; Blum *et al.* 1990; Hatle and Spring, 1998). Therefore, to minimize the chemical defence of the grasshoppers, while retaining their aposematic coloration and sluggish behavior, we fed them solely Romaine lettuce and oats, which are low in defensive chemicals. We offered these lettuce-fed grasshoppers to the frogs. In a five-day experiment, the frogs did not learn an aversion to the grasshoppers and, in fact, tended to attack more lubbers as the experiment continued. Hence, the lettuce-fed lubbers did not appear to deter the frogs. We now had an experimental system with a sluggish-moving, aposematic prey that was not chemically defended from a motion-oriented predator. Thus, we could directly test the role of sluggish movement in defense (Hatle and Faragher, 1998).

We manipulated the movement of the grasshoppers by tethering them and then either inducing them with motion or allowing them to exhibit their natural sluggish movement (shown as a choice test in Fig. 5). The grasshoppers were offered to frogs in a balanced experimental design. Of 23 frogs, seven preferentially ate motion-induced grasshoppers whereas only one preferentially ate sluggish-moving grasshoppers (P=0.0273). In addition, the frogs oriented toward (P=0.0078) and snapped at (P=0.0067) motion-induced grasshoppers significantly sooner than sluggish-moving grasshoppers. These data imply that the natural sluggish movement of the grasshoppers can delay or fail to release the attack response of frogs. Therefore, sluggish movement could be beneficial in encounters with motion-oriented predators in nature (Hatle and Faragher, 1998). Importantly, these experiments were conducted with a

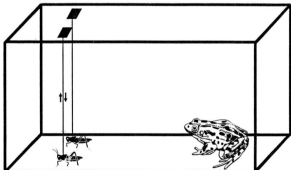

**Fig. 5.** We manipulated the movement of grasshoppers by tethering them and then either inducing them with motion or allowing them to exhibit their natural sluggish movement. The 'fast-moving' grasshopper was jigged by an electric motor that plucked the tether thread about once every second. The 'sluggish-moving' grasshopper was allowed to move about on the tether without further manipulation (Hatle and Faragher, 1998).

conspicuous insect, which does not appear to be avoiding predator detection in nature, and almost certainly was not avoiding predator detection in our laboratory trials. We interpret our results as suggesting that lubber grasshoppers were not avoiding detection but instead dissuading attack.

This defensive strategy (permit detection, but do not release attack) predicts that the prey will not attempt to flee. Fleeing would require rapid movement and is likely to release the attack response of the motion-oriented predator. This is a new twist on Ydenberg and Dill's (1986) hypothesis on the costs and benefits of when prey should flee approaching predators. Their model predicts that the distance at which potential prey will begin to flee an approaching predator will decrease with increasing risk of fleeing. Ydenberg and Dill (1986) envisioned this risk as either encountering another predator or the cost of halting foraging. For sluggish prey, the risk of fleeing becomes the release of the predator's attack response. Young lubber grasshoppers will not typically flee small predators, such as birds and mice, until the predators are within 10 cm (Whitman, unpublished data). Furthermore, most lubbers will not release their defensive secretion until actually seized (Whitman *et al.* 1991). This fits well with Ydenberg and Dill's (1986) predictions. We hypothesize that sluggish-moving prey will not flee until approaching predators are attacking (i.e. rapidly moving toward their prey). This strategy allows sluggish-movement to act as a defensive trait as long as possible, at which time it becomes more profitable to attempt to flee.

Sluggish motion may have particular significance for aggregated prey. Motion-oriented predators preferentially attack moving prey. Therefore, in gregarious prey, it may be beneficial to move more sluggishly than your neighbors. In this context, instead of sluggish movement being seen as an advantage rapid or saltatory movement in gregarious insects may be a disadvantage. We hypothesize that sluggish movement might be advantageous when motion-oriented predators encounter gregarious arthropod prey, because the predator will cull the fastest-moving (and hence, most stimulating) individuals from the aggregation. We have begun to examine this hypothesis experimentally. We tethered two grasshoppers, induced one with motion, and allowed the second to move sluggishly (Fig. 5). To mimic an aggregation, we released five additional untethered grasshoppers between the two tethered grasshoppers (aggregation not shown in Fig 5.) We then released a frog and recorded the first grasshopper to be attacked. In these trials, only one sluggish grasshopper was attacked, whereas 19 motion-induced grasshoppers were attacked (Hatle and Whitman, unpublished data). These data suggest that sluggish movement may be especially important in gregarious insects, because motion-oriented predators are most likely to attack the faster-moving insects in the group.

## FREQUENCY OF SLUGGISH MOVEMENT IN APOSEMATIC PREY

Sluggish movement is common in aposematic prey. For example, slow movement is predominant in caterpillars (Heinrich, 1993; Stamp and Wilkens,

1993), and many are aposematic. To demonstrate the frequency and phylogenetic distribution of this defensive ensemble, Table 1 lists examples of slow-moving aposematic prey that might use sluggish movement as a defense.

## MOTION-ORIENTED PREDATORS AND PREY MOTION

Predators of insects respond to many types of stimuli to subdue their prey. Many predators are primarily visual, and visual motion-orientation is undoubtedly of great importance for many entomophagous carnivores. For this paper, we define a motion-oriented predator (i.e. sit-and-wait or ambush predator) as one that has been shown to preferentially attack moving prey over non-moving prey. Importantly, we only include predators that attack, as opposed to detect, prey in response to motion. Table 2 lists representative examples of predators of insects that have been shown to attack in response to prey motion. Many others predators likely attack prey in response to prey movement, including dragonflies, assassin bugs, mantid flies, tiger beetles and predaceous diving beetles.

For these predators, the rate of motion is often an important factor in triggering predatory response. According to Roth (1986), the common assumption that amphibians (which are the classic vertebrate motion-oriented predator) attack everything that is the right size and moves, is incorrect. Instead, motion-oriented predators can be quite selective and refuse to attack slow prey. For many predators, attack rate is positively correlated with the rate of prey movement, up to a certain limit (Barth, 1985; Roth, 1986). Hence, it may not be surprising that prey, even conspicuous prey, might reduce predator attack by moving slowly. For example, Freed (1984) offered five different types of insect prey to the jumping spider *Phidippus audax*. Prey mortality mirrored prey velocity; slower-moving prey were killed less often than faster-moving prey (Fig. 6). These results suggested that a certain level of prey velocity (2-5 cm/s) was the best releaser of spider attacks. Freed (1984) did not control the size, shape and color of the prey.

Predators have evolved to attack the jerky and saltatory movements characteristic of their insect prey. For example, the praying mantis (*Parastagmatoptera*) preferentially attacked prey with jerky movement over prey with a constant rate of movement (Rilling *et al.* 1959). Also, the common toad (*Bufo bufo*) has been shown to preferentially attack prey that move in a jerky fashion at 1-2 Hz (Fig. 7; adapted from Borchers *et al.* 1978).

The optic-neural bases for selective responses to prey motion have been studied in several motion-oriented predators. In frogs, a variety of retinal ganglion cells exist in the retina, and different ganglion cells respond to different visual stimuli. One cell type responds preferentially to moving objects and has been called the "bug detector" (Maturana *et al.* 1960). Prey recognition is further defined in the optic tectum region of the brain, where specialized ganglion cells respond preferentially to visual stimuli moving at higher velocities (Fig. 8; Roth, 1986). In jumping spiders, the optimotor response increases with increased image speed (Land, 1969).

**Table 1.** A representative list of conspicuous, aposematic prey that are sluggish-moving. We only include terrestrial prey in our list because, whether or not bright coloration in marine animals such as coral reef fishes is aposematic, is unknown (Guilford, 1990). This Table is not meant to be comprehensive, but is instead intended to demonstrate that many prey are both sluggish and conspicuous.

| Order | Species | Reference | Description of movement and coloration |
|---|---|---|---|
| Araneae | *Argiope* spp. argiope spiders | M.C. Crowley, pers. comm. | argiopes are large, conspicuous spiders that rest on webs and rarely move |
| Orthoptera | *Romalea microptera* Eastern lubber grasshopper | Hatle and Faragher 1998 | lubbers moved significantly more slowly than crickets aposematic....black and yellow |
| | *Taeniopoda eques* Western lubber grasshopper | Whitman *et al.* 1985 | sluggish, flightless aposematic.... with bright red hind wings |
| | *Dactylotum variegatum* rainbow grasshopper | Neal *et al.* 1994 | flightless....delays initiation of escape responses bright orange and yellow-green marks on dark blue |
| Hemiptera | *Oncopeltus fasciatus* milkweed bugs | personal observation Isman *et al.* 1977 | sluggish moving and gregarious brightly colored |
| Coleoptera | *Oreina gloriosa* alpine beetle | Eggenberger & Rowell-Rahier 1992 | slow-moving, brightly colored, and they often aggregate |
| Lepidoptera | unpalatable adults from Costa Rica | Chai and Srygley 1990 Marden and Chai 1991 | quantifiably slower than palatable, sympatric adults |
| | *Danaus plexippus* Monarch butterfly | personal observation Brower 1971 | slow flying bright orange, black, and white bespeckled |
| | *Euphydryas phaeton* checkerspot butterflies | Bowers 1980 | slow-flying and sluggish; non-evasive behaviour black with red and yellow markings |
| | Notodontidae larvae notodontid caterpillars | Heinrich 1993 | slow, often motionless green background with sharp black and colorful markings |
| Gastropoda | *Ariolimax columbianus* banana slugs | personal observation | crawl slowly on the forest floor, esp. on hiking trails bright yellow |

contd.

**Table 1.** contd.

| Order | Species | Reference | Description of movement and coloration |
|-------|---------|-----------|----------------------------------------|
| Caudata | *Notophthalmus* spp. red eft | Petranka, 1998 | aggregates of 20-46 newts move slowly about pond bottoms bright coloration of the efts functions as warning coloration |
| | *Triturus* spp. terrestrial salamander | R.G. Jaeger, personal comm. | move about on the surface of the ground during the day, and none of them quickly |
| | | Petranka, 1998 | the bright color appears to have an aposematic function |
| Anura | *Dendrobates* spp. poison dart frogs | Duellman and Trueb, 1986 | aposematic prey with a sit-and-wait hunting mode |

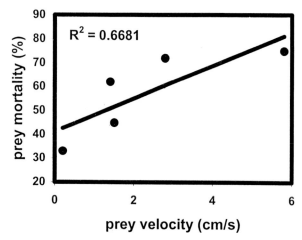

**Fig. 6.**  The relationship between prey velocity and prey mortality for a jumping spider (*Phidippus audax*) attacking five insect prey that differed in rate of movement. Prey velocity followed the order: caterpillar<flea beetle <stink bug <soldier beetles < housefly (adapted from Freed, 1984).

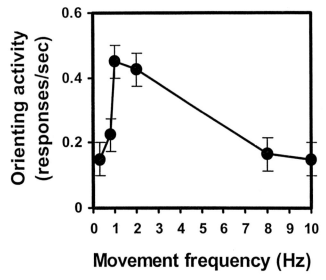

**Fig. 7.**  The attack response of the common toad (*Bufo bufo*) toward square prey dummies is most strongly released by movement that cycles at 1-2 Hz. Toad s will not attack unmoving prey (adapted from Borchers *et al.* 1978). Conspicuous prey with sluggish movement typically exhibit fluid movement and therefore would have a movement frequency of less than 1 Hz.

Sluggish movement may also deter predators that respond to vibrations created by prey. A visually oriented hemipteran (*Velia caprai*) has been shown to attack prey only if they create vibrations (Meyer, 1971). It may be that

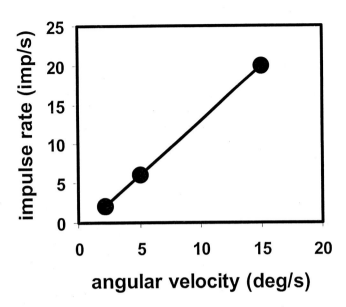

**Fig. 8.** Relationship between image velocity and nerve impulse rate for one type of ganglia in the optic tectum region of the frog brain (adapted from Roth, 1986). The response of four other optic tectum ganglia types were similar, with all responding more intensely to faster-moving images.

sluggish-moving prey do not produce sufficient vibrations to release attack from this type of predator (J.O. Schmidt, personal communication).

## TOWARDS A DEFINITION OF SLUGGISH MOVEMENT AS A DEFENSE

We previously defined sluggish movement as a movement that was "neither (1) running away from nor (2) apparently avoiding visual detection by the predator" (Hatle and Faragher, 1998). Here, we suggest a new term and definition for this behavior and elaborate on our conception of this behavior. We recommend that this behavior be called "conspicuous sluggish movement" and be defined as sluggish movement in well-defended, conspicuous prey.

Conspicuous sluggish movement is a movement behavior that inhibits predator attack because it is insufficiently rapid or jerky to release the attack response of the predator. Conspicuous sluggish movement is not avoiding detection or fleeing the predator. We further suggest that conspicuous sluggish movement only be used with aposematic prey to distinguish it from slowed movement in cryptic prey that is presumably used to deter detection (e.g. Sih, 1986).

From studies on the release of attacks by motion-oriented predators, we presently understand what *is not* conspicuous sluggish movement better than we understand what *is* conspicuous sluggish movement. Roth (1986) defines the functional unit of neural prey recognition as a 'recognition module' (similar to the 'bug detector' ganglion cells in frog eyes of Maturana *et al.* 1960). In an ecological context, the recognition module describes the prey movement and shape parameters that trigger the neural cascade that results in the predator attacking the prey. We hypothesize that conspicuous sluggish movement does not meet the requirements of the predator's recognition module. Sluggish prey may have evolved a movement behavior that falls outside the bounds of their predator's recognition module, and hence can escape the predator's attack, even if detected. For many predators, the behaviors that the predator will not attack (e.g. conspicuous sluggish movement) could be determined by video manipulation of the movement of a prey image. Release of predatory behaviors in response to video images has been observed in several predators, including frogs and toad s (Roster *et al.* 1995), jumping spiders (Clark and Uetz, 1990), and birds (Evans and Marler, 1991). This type of experiment could clarify our conception of conspicuous sluggish movement.

Sluggish movement is defensive when it fails to release the attack response of motion-oriented predators. This requires that the movements be fluid and not jerky or saltatory, which would tend to more strongly release predator attack responses (Luthardt and Roth, 1979; Ewert, 1976). A prey could move slowly from point A to point B, but if its movement is saltatory (e.g. run a few cm, stop, run a few cm, etc.), it is not likely to avoid predator attack. Further, even jerky movement of the appendages, without moving from point A to point B, can release orientation in praying mantids (Rilling *et al.* 1959). Hence, prey that exhibit this type of behavior cannot be described as sluggish-moving. We believe the term 'sluggish' portrays the lack of jerky movement better than the term 'slow'.

We prefer to classify conspicuous sluggish movement as a primary defense mechanism. Primary defenses are traditionally defined as those mechanisms that are displayed regardless of the presence of the predator. Our observations of conspicuous sluggish movement in the lubber grasshopper largely fit this requirement (Whitman, 1988; Hatle and Faragher, 1998). Sluggish movement in lubber grasshoppers is exhibited until an approaching predator nearly touches a grasshopper, at which time the grasshopper flees. It should be noted that a prey could begin to exhibit sluggish movement only after the prey detects the predator (e.g. Sih, 1986) and still deter attack. We know of no such descriptions for aposematic prey in literature. Finally, Edmunds (1974) requires that a primary defense mechanism inhibits predator detection of prey. Clearly, sluggish movement in aposematic insects does not meet this component of Edmund's (1974) definition of primary defense.

## ON THE EVOLUTION OF SLUGGISH MOVEMENT

Because sluggish movement in conspicuous prey is seemingly maladaptive, its existence is of special interest. At least five hypotheses, that are not mutually exclusive, could contribute to the explanation of the evolution of conspicuous sluggish movement. Most of these hypotheses rely on the prior evolution of another defense mechanism, usually chemical defense.

First, our data suggest that conspicuous sluggish movement would be more likely to evolve in chemically defended prey (Hatle *et al.* in press). To test this, we manipulated the movement and odor (=chemical defense) of lubber grasshoppers to produce prey that were: 1) sluggish-moving and had strong odor; 2) sluggish-moving and faint odor; 3) fast-moving and strong odor; and 4) fast-moving and faint odor. We then offered these prey to frogs. In two independent experiments, frogs attacked sluggish-moving and strong odor lubbers significantly later than they attacked all three other prey types. This is the first demonstration that sluggish movement can act together with chemical defense to produce a better-defended prey. The preponderance of sluggish movement in chemically defended insects (Table 1) may be due to the fact that conspicuous sluggish movement is more effective when in combination with chemical defense.

Second, sluggish movement may have evolved in well-defended prey because fast movement was not needed to flee from predators and sluggish movement saved energy. Unpalatable Costa Rican butterflies fly more slowly and less erratically than palatable, sympatric butterflies (Chai and Srygley, 1990; Marden and Chai, 1991). The unpalatable butterflies have less mass devoted to flight musculature than palatable butterflies, saving the energy needed to maintain muscle tissue. Marden and Chai (1991) imply that slower movement may have evolved simply because these insects did not need to move fast to escape predators. In this light, sluggish movement might have secondarily become a defense mechanism (as an exaptation) against motion-oriented predators.

Third, sluggish-moving aposematic prey are often gregarious (Sillén-Tullberg, 1988; Sillén-Tullberg and Leimar, 1988; Vulinec, 1990; Gamberale and Tullberg, 1998; Lindström *et al.* 1999). Chance encounters between the gregarious group and motion-oriented predators could result in the predators culling the fastest-moving prey from the aggregation. This is, perhaps, more easily conceived as a disadvantage for fast-moving, gregarious prey than an advantage for sluggish-moving prey. Nonetheless, it would serve the same purpose. This would especially benefit sluggish-moving individuals if the prey was unpalatable, and the predator terminated attacking after consuming only a few prey (the fastest-moving prey). It would also be more likely to evolve if predators became satiated and ate only the fastest-moving prey before leaving the gregarious group. In addition, now that aposematism has been shown to be potentially density-dependent (Sword, 1999), we hypothesize that sluggish movement is also density dependent.

Fourth, sluggish-movement may be especially beneficial and therefore likely to evolve in aposematic prey. Vertebrate predators often shun conspicuous, aposematic prey, either innately (Coppinger, 1970; Schuler and Hesse, 1985) or as a result of learning (Brower and Brower, 1962; Brower *et al.* 1960; Gittleman and Harvey, 1980; Roper and Wistow, 1986). This attack deterrence is facilitated if the warning signal is unambiguous. This may be problematic with motion-oriented predators that have evolved a rapid feeding reflex. The short interval between detection and attack may not allow predators time to process the aposematic signal. In addition, warning patterns would presumably be more difficult to detect in a rapidly moving prey. Hence, sluggish and aposematic prey may deliver a strong warning signal by: 1) delaying attack and allowing more time for the signal to be processed; and 2) being more easily recognized than fast-moving prey. We hypothesize that the defensive benefits of sluggish movements and aposematic coloration interact positively. In other words, we predict that sluggish-moving, aposematic prey will be attacked less than (or later than) sluggish-moving, cryptic prey.

Fifth, conspicuous sluggish movement seems more likely to evolve in monophagous prey. Specialist herbivores are often better defended than generalists (Bernays and Cornelius, 1989; Jones *et al.* 1989; Dyer, 1995; Sword, 1999) because these individuals acquire higher concentrations of a specific defensive compound. Concomitantly, monophagy could be more likely to evolve in sluggish individuals, simply because they would not move to new host plants. In addition, if they ingest only a chemically defended plant, these monophagous individuals might become sick, creating a proximate mechanism for sluggish movement. One or a combination of these evolutionary forces (or others) may have cooperated to produce slow movement in CDS insects.

## SUGGESTIONS FOR FUTURE WORK

Much work remains on the importance of conspicuous sluggish movement. We suggest four directions for future research that promise to be fruitful; 1) determining the generality of the phenomena in laboratory experiments; 2) determining the frequency of sluggish movement acting as a defense in nature; 3) investigating possible pathways of the evolution of conspicuous sluggish movement; and 4) determining the relative importance of conspicuous sluggish movement within the broader defense ensemble of the CDS.

First, and we feel most important, is studying the generality of conspicuous sluggish movement. In other words, is this a common or rare phenomenon? How many prey gain protection from conspicuous sluggish movement, and how many predators are deterred by conspicuous sluggish movement? Certainly, there are many sluggish aposematic prey (Table 1) and many motion-oriented predators (Table 2). We suggest testing predator-prey pairs that would be likely to come in contact in nature. For example, both lubber grasshoppers and frogs thrive in the wetlands in the SE United States (but they might not be active at the same time of day). We feel that pairing true bugs or caterpillars with frogs, toad s, lizards and spiders from the same microhabitats would be particularly profitable.

Perhaps of more interest than the responses of motion-oriented predators would be to test the survivorship of sluggish prey with seek-and-find predators (e.g. birds). Many motion-oriented predators employ a sit-and-wait strategy (e.g. frogs). These predators are less likely to contact an aggregation of sluggish prey than seek-and-find predators that are constantly moving in search of prey. Although seek-and-find predators may chose prey more by shape and color than by motion, they are likely to also be attracted to prey motion, and more

**Table 2.** A representative list of primarily motion-oriented, terrestrial predators. These predators could potentially be deterred by the conspicuous sluggish-moving prey that are listed in Table 1.

| Order | Species | Reference | Description of factors that release attack |
|---|---|---|---|
| Araneae | *Phidippus audax* jumping spider | Freed, 1984 | prey with a higher velocity also had higher mortality |
| | *Rabidosa rabida* wolf (lycosid) spiders | Rovner, 1993 | movement elicits attack |
| Orthoptera | *Parastagmatoptera* spp. and *Hierodula* spp. praying mantids | Rilling *et al.* 1959 (in New, 1991) | strike is best released when the prey's appendages make rapid jerking movements; smooth movement of the prey are relatively ineffective |
| Hempitera | *Velia caprai* | Meyer, 1971 (in New, 1991) | reacts to visual stimuli only in concert with vibrations |
| Coleoptera | *Notiophilus biguttatus* European carabid groundbeetle | Ernsting and Jansen, 1978 (in New, 1991) | actively searching predator that prefers an active springtail species as prey over a relative inactive springtail species |
| Scorpions | scorpions | M.C. Crowley, pers. comm. | pectines (tactile organs that sense vibrations) |
| Caudata | *Ambystoma tigrinum* tiger salamander | Lindquist and Bachmann, 1982 | primarily visual and secondarily olfactory |
| | *Hydromantes italicus* European plethodontid | Roth, 1976 | stimulus movement is an important but not decisive factor in the release of tongue projection |
| | *Salamandra salamandra* fire salamader | Luthardt and Roth, 1979 | prey catching behavior is best released by saltatory movement with stimulus parallel to the movement |

contd.

**Table 2.** contd.

| Order | Species | Reference | Description of factors that release attack |
|---|---|---|---|
| Anura | *Bufo bufo* common toad | Borchers *et al.* 1978; Ewert, 1976 | prey movement pattern may have strong effects on the release of attack |
| | *Rana pipiens* Northern leopard frog | Anderson, 1993 | regardless of prey type, *R. pipiens* does not strike prey unless it is moving |
| Squamata | Viperida; *Agkistrondon contortrix* copperhead; *Crotalus* spp. | Mattison, 1999; Mattison, 1999 | viperids are especially motion-oriented |
| | Iguanidae and Agamidae; *Sceloperous* spp. fence lizards | Cooper, 1989 | diurnal sit-and-wait predators that appear to rely largely on visual cues for identification of prey |
| | *Anolis carolinensis* green anoles; *Eumeces fasciatus* five-lined skinks | Burghardt, 1964 | prey movement is important in releasing attack response |
| Class Aves | *Tyto alba* barn owls | Kaufman, 1974 | attacked live mice first 27 times and attacked dead mice first 0 times |
| | *Otus asio* screech owls | Kaufman, 1974 | attacked live mice first 14 times and attacked dead mice first once |
| Mammalia | *Onychomys torridus* grasshopper mice | Langely and Knapp, 1982 | attacked crickets carrying an odor associated with conditioned aversion because motion overcame aversion |

likely to contact gregarious prey. Whether seek-and-find predators, after detecting prey, preferentially attack and seize prey that are moving in comparison to motionless prey, is unclear.

To date, sluggish movement has only been shown to function as a defense in the laboratory. The demonstration of conspicuous sluggish movement acting as a defense in nature would be an important step, validating that this phenomenon is important in natural systems. Perhaps the most difficult step in this research would be finding an appropriate experimental system. Ideally, this would involve discovering morphologically identical populations of a species that differ only in movement rate, and then quantifying the survivorship of the two populations in response to a motion-oriented predator. However, we know of no prey species that has both sluggish-moving and fast-moving 'morphs'. One simple alternative would be to compare morphologically similar species that differ in movement rates. A better, but more demanding, alternative would be to artificially breed sluggish-moving and fast-moving strains of a conspicuous insect that are morphologically identical. The survivorship of these strains in mesocosms with a motion-oriented predator could be compared.

Placing conspicuous sluggish movement into an evolutionary context is also important for understanding the importance of this trait in ecological time. We believe that conspicuous sluggish movement was unlikely to evolve in species lacking other defensive traits. Hence, the evolution of conspicuous sluggish movement can be addressed by pairing sluggish movement with other defenses. We have already initiated this direction of research, pairing sluggish movement with chemical defense (Hatle *et al.* in press) and gregarious behavior (Hatle and Whitman, unpublished data). Sluggish movement could also be paired with aposematic coloration. We hypothesize that sluggish movement will interact positively with all these defenses. In other words, we predict that conspicuous sluggish movement will increase the protective value of chemical defense, gregarious behavior and aposematic coloration. Such a result would provide an evolutionary context for the origin of conspicuous sluggish movement, because sluggish movement otherwise seems unlikely to be defensive.

Finally, prey typically do not rely on single defenses, but on ensembles of interacting defense elements (Pearson, 1989; Marples *et al.* 1994). A study of all the defenses of a chemically defended species would provide information on the relative contribution of sluggish movement to defense of chemically defended prey. These prey often exhibit aposematic coloration, gregarious behavior and sluggish movement (Pasteels *et al.* 1983). The relative importance of each of these defenses in deterring predators is unclear. Our understanding of the interactions of these defenses is also unclear. This experiment would involve offering prey with all possible combinations of these four defenses (allomones, aposematic coloration, gregarious behavior and sluggish movement) to both seek-and-find and sit-and-wait predators. The data could be statistically tested using a path analysis, which would reveal the relative importance of each defensive trait at each step of the predatory response. Such an experiment would not only further our understanding of the importance of sluggish movement as a defense, but would also clarify the role of each trait in the defense of chemically defended prey.

## ACKNOWLEDGEMENTS

We thank the ISU Advanced Entomology group for critically reading the manuscript and Amy J. Mitchell of the Illinois State Museum for creating Fig. 5.

## REFERENCES

Anderson, G.W. 1993. The modulation of feeding behavior in response to prey type in the frog *Rana pipiens. J. Exp. Biol.*, **197:** 1–12.

Barth, F.G. 1985. *Neurobiology of Arachnids.* Springer-Verlag: Berlin.

Bates, H.W. 1862. Contributions to an insect fauna of the Amazon Valley, Lepidoptera: Heliconidae *Trans. Linnaean Soc., Zool.,* **XXIII:** 495–566.

Bernays, E.A. and Cornelius, M.L. 1989. Generalist caterpillar prey are more palatable than specialists for the generalist predator *Iridomyrmex humilis. Oecologica*, **79:** 427–430.

Blum, M.S., Severson, R.F., Arrendale, R.F., Whitman, D.W., Escoubas, P., Adeyeye, O. and Jones, C.G. 1990. A generalist herbivore in a specialist mode: metabolic, sequestrative, and defensive consequences. *J. Chem. Ecol.*, **16:** 223–244.

Borchers, H-W., Burghagen, H., Ewert, J-P. 1978. Key stimuli of prey for toad s (*Bufo bufo* L.): configuration and movements patterns. *J. Comp. Physiol.*, **125:** 189–192.

Bowers, M.D. 1980. Unpalatability as a defense strategy of *Euphydryas phaeton* (Lepidoptera: Nymphalidae). *Evolution*, **34:** 586–600.

Brower, L.P. 1971. Prey coloration and predator behavior. *In: V.G. Dethier (Ed.) Topics in Animal Behavior.* Harper & Row, Publ., Inc. New York: pp. 66–76.

Brower, L.P., Brower, J.V.Z. and Westcott, P.W. 1960. Experimental studies of mimicry 5. The reactions of toad s (*Bufo terrestris*) to bumble bees (*Bombus americanus*) and their robberfly mimics (*Mallophora bomboides*), with a discussion of aggressive mimicry. *Am. Nat.*, **94:** 343–356.

Brower, J.V.Z. and Brower, L.P. 1962. Experimental studies of mimicry. 6. the reaction of toad s (*Bufo terrestris*) to honeybees (*Apis mellifera*) and other dronefly mimics (*Eristalis vinetorum*). *Am. Nat.*, **96:** 297–307.

Burghardt, G.M. 1964. Effects of prey size and movements on feeding behavior of the lizards *Anolis carolinensis* and *Eumeces fasciatus. Copeia*, **3:** 576–578.

Chai, P. and Srygley, R.B. 1990. Predation and the flight, morphology, and temperature of neotropical rainforest butterflies. *Am. Nat.*, **135:** 748–765.

Clark, D.L. and Uetz, G.W. 1990. Video image recognition by the jumping spider, *Maevia inclemens* (Araneae: Salticidae). *Anim. Behav.*, **40:** 884–980.

Cooper, W.E. Jr. 1989. Absence of prey odor discrimination by iguanid and agamid lizards in applicator tests. *Copeia*, **1989:** 472–478.

Coppinger, R.P. 1970. The effect of experience and novelty on avian feeding behavior with reference to the evolution of warning coloration in butterflies. II. Reactions of naïve birds to novel insects *Am. Nat.*, **104:** 323–335.

Duellman, W.E. and Trueb, L. 1986. Biology of Amphibians. McGraw-Hill Book Company, New York: p. 670.

Dyer, L.A. 1995. Tasty generalists and nasty specialists? Antipredator mechanisms in tropical Lepidopteran larvae. *Ecology,* **76:** 1483–1496.

Edmunds, M. 1974. Defence in Animals: A Survey of Anti-predator Defences. Harlow, Longman Group, Essex: p. 357.

Edmunds, M. 1990. The evolution of cryptic coloration. *In* D.L. Evans, J.O. Schmidt (Eds.) *Insect Defenses: Adaptive Mechanisms and Strategies of Prey and Predators.* State University of New York Press, Albany: pp. 3–22.

Eggenberger, F. and Rowell-Rahier, M. 1992. Genetic component of variation in chemical defense of *Oreina gloriosa* (Coleoptera: Chrysomelidae). *J. Chem. Ecol.*, **18**: 1375–1387.

Ernsting, G. and Jansen, J.W. 1978. Interspecific and intraspecific selection by the predator *Notiophilus biguttatus* F. (Carabidae) concerning two collembolan prey species. *Oecologia*, **33**: 173–183.

Evans, H.E. 1984. *Insect biology.* Addison-Wesley Publ. Co., Reading, MA: p. 436.

Evans, C.S. and Marler, P. 1991. On the use of video images as social stimulus in birds: audience effects on alarm calling. *Anim. Behav.*, **41**: 17–26.

Ewert, J-P. 1976. The visual system of the toad : behavioral and physiological studies on a pattern recognition system. *In* K.V. Fite, (Ed.) *The Amphibian Visual System: a Multidisciplinary Approach.* Academic Press, New York: pp. 141–202.

Freed, A.N. 1984. Foraging behavior in the jumping spider Phidippus audax: bases for selectivity, *J. Zool. Lond.*, **203**: 49–61.

Gamberale, G. and Tullberg, B.S. 1998. Aposematism and gregariousness: the combined effect of group size and coloration on signal repellence. *Proc. R. Soc. Lond. B.*, **265**: 889–894.

Gittleman, J.L. and Harvey, P.H. 1980. Why are distasteful prey not cryptic? *Nature*, **286**: 149-160.

Guilford, T. 1990. The evolution of aposematism. *In* D.L. Evans, J.O. Schmidt (Eds.) *Insect Defenses: Adaptive Mechanisms and Strategies of Prey and Predators.* State University of New York Press, Albany: pp. 23–62.

Hatle, J.D. and Faragher, S.G. 1998. Slow movement increases the survivorship of a chemically defended grasshopper in predatory encounters. *Oecologia*, **115**: 260–267.

Hatle, J.D. and Spring, J.H. 1998. Inter-individual variation in Sequestration (as measured by energy dispersive spectroscopy) predicts efficacy of defensive secretion in lubber grasshoppers. *Chemoecology*, **8**: 85–90.

Hatle, J.D., Salazar, B.A. and Whitman, D.W. In press. Slow movement and repugnant odor are positively interacting insect defensive traits in encounters with frogs. *J. Insect Behavior.*

Heinrich, B. 1993. How avian predators constrain caterpillar foraging. *In* N.E. Stamp and T.M. Casey (Eds.) *Caterpillars: Ecological and Evolutionary Constraints on Foraging.* Chapman & Hall. New York: pp. 283–330.

Jones, C.G., Whitman, D.W., Compton, S.J., Silk, P.J. and Blum, M.S. 1989. Reduction in diet breadth results in Sequestration of plant chemicals and increases efficacy of chemical defense in a generalist grasshopper. *J. Chem. Ecol.*, **15**: 1811–1822.

Kaufman. D.W. 1974. Differential predation on active and inactive prey by owls. *Auk*, **91**: 172–173.

Land, M.F. 1969. Movements of the retinae of jumping spiders (Salticidae: Dendryphantinae) in response to visual stimuli. *J. Exp. Biol.*, **51**: 471–493.

Langely, W.L. and Knapp, K. 1982. Importance of olfaction to suppression of attack response through conditioned taste aversion in the grasshopper mouse. *Behav. Neural Biol.*, **35**: 368–378.

Lindquist, S.B. and Bachmann, M.D. 1982. The role of visual and olfactory cues in the prey catching behavior of the tiger salamander, *Ambystoma tigrinum. Copiea*, **1982**: 81–90.

Lindström, L., Alatalo, R.V. and Mappes, J. 1999. Reactions of hand-reared and wild-caught predators toward warningly colored, gregarious, and conspicuous prey, *Behav. Ecol.*, **10**: 317–322.

Luthardt, G. and Roth, G. 1979. The relationship between stimulus orientation and stimulus movement pattern in the prey catching behavior of *Salamandra salamadra. Copiea*, **1979**: 442–447.

Marden, J.H and Chai, P. 1991. Aerial predation and butterfly design: how palatability, mimicry, and the need for evasive flight constrain mass allocation. *Am. Nat.*, **138**: 15–36.

Marples, N.M., van Veelen, W. and Brakefield, P.M. 1994. The relative importance of colour, taste and smell in the protection of an aposematic insect *Coccinella septempunctata. Anim. Behav.*, **48**: 967–974.

Mattison, C. 1999. Snake. DK Publishing Inc., New York: pp. 192.

Maturana, H.R., McCulloch, W.S., Lettvin, J. Y. and Pitts, W.H. 1960. Anatomy and physiology of vision in the frog (*Rana pipiens). J. Gen. Physiol. (Suppl.)*, **43**: 129–175.

Meyer, H.W. 1971. Visuelle Schlusselreize fur die Anslosung der Beutefanghandlung bem

Bachwasserlaufer Velia caprai (Hemiptera, Heteroptera) 1. Untersuchung der raumlichen und zeitlicher Reizparameter mit formverschiedenen Attrappen. *Z. vergl. Physiol,* **72**: 260–297.

Neal, P.R., Stromberg, M.K. and Jepson-Innes, K.A. 1994. Aposematic coloration in *Dactylotum variegatum* (Orthopera: Acrididae): support from vertebrate feeding trials. *Southwestern Naturalist,* **39**: 21–25.

New, T.R. 1991. *Insects as Predators.* New South Wales Press, Kensington: pp: 178.

Pasteels, J.M. Gregoire, J.-C. and Rowell-Rahier, M. 1983. The chemical ecology of defense in arthropods. *Ann. Rev. Entomol.,* **28**: 263–289.

Pearson, D.L. 1989. What is the adaptive significance of multicomponent defensive repertoires? *Oikos,* **54**: 251–253.

Petranka, J.W. 1998. *Salamanders of the United States and Canada.* Smithsonian Institution Press, Washington. pp: 587.

Rilling, S., Mittelstaedt, H. and Roeder, K.D. 1959. Prey recognition in the praying mantis. *Behaviour,* **14**: 164–184.

Roper, R.J. and Wistow, R. 1986. Aposematic colouration and avoidance learning in chicks. *Quarterly J. Exp. Psychol.,* **38**: 141–149.

Roster, N.O., Clark, D.L. and Gillingham, J.C. 1995. Prey catching behavior in frogs and toad s using video-simulated prey. *Copiea,* **1995**: 496–498.

Roth, G. 1976. Experimental analysis of the prey catching behavior of *Hydromantes italicus* Dunn (Amphibia, Plethodontidae). *J. Comp. Physiol.,* **109**: 47–58.

Roth, G. 1986. Neural mechanisms of prey recognition: an example in amphibians. *In:* M.E. Feder, G.V. Lauder (Eds.) *Predator-Prey Relationships: Perspective and Approaches from the Lower Vertebrates.* Univ. Chicago Press, Chicago: pp. 198.

Rovner, J.S. 1993. Visually mediated responses in the lycosid spider *Rabidosa rabida:* the roles of different pairs of eyes. *Mem. Queensland Mus.,* **33**: 635–638.

Schuler, W. and Hesse, E. 1985. On the function of warning coloration: black and yellow pattern inhibits prey-attack by naïve domestic chicks. *Oecologica,* **16**: 249–255.

Sih, A. 1986. Antipredator responses and the perception of danger by mosquito larvae. *Ecology,* **67**: 434–441.

Sillén-Tullberg, B. 1988. Evolution of gregariousness in aposematic butterfly larvae: a phylogenetic analysis. *Evolution,* **42**: 293–305.

Sillén-Tullberg, B. and Leimar, O. 1988. The evolution of gregariousness in distasteful insects as a defense against predators. *Am. Nat.,* **132**: 723–734.

Srygley, R.B. and Chai, P. 1990. Predation and the elevation of thoracic temperature in brightly colored neotropical butterflies, *Am. Nat.,* **135**: 766-787.

Stamp, N.E. and Wilkens, R.T. 1993. On the cryptic side of life: being unapparent to enemies and the consequences for foraging and growth of caterpillars. *In:* N.E. Stamp, T.M Casey (Eds.) *Caterpillars: Ecological and Evolutionary Constraints on Foraging.* Chapman & Hall, New York: pp: 283 -330.

Sword, G.A. l999. Density-dependent warning coloration. *Nature,* **397**: 217.

Vulinec, K. 1990. Collective security: aggregation by insects as a defense. *In* D.L. Evans, J.O. Schmidt (Eds.) *Insect Defenses: Adaptive Mechanisms and Strategies of Prey and Predators.* State University of New York Press, Albany: pp: 251–288.

Whitman, D.W. 1988. Allelochemical interactions among plants, herbivores, and their predators. *In* P. Barbosa, D. Letourneau (Eds.) *Novel Aspects of Insect-plant Interactions.* John Wiley & Sons, Inc., New York: pp: 11-64.

Whitman, D.W. 1990. Grasshopper chemical communication. *In:* R. Chapman, A. Joern (Eds.) *Biology of Grasshoppers.* John Wiley & Sons, New York: pp. 357–391.

Whitman, D.W., Blum, M.S. and Jones, D.G. 1985. Chemical defense in *Taeniopoda eques* (Orthoptera: Acrididae): role of the metathoracic secretion. *Ann. Entomol. Soc. Am.,* **78**: 451–455.

Whitman, D.W, Billen, J.P.J., Alsop, D. and Blum, M.S. 1991. Anatomy, ultrastructure, and functional morphology of the metathoracic tracheal defensive glands of the grasshopper *Romalea guttata.* *Can. J. Zool.,* **69**: 2100–2108.

Ydenberg, R.C. and Dill, L.M. 1986. The economics of fleeing from predators. *Adv. Study Behav.* **16**: 229-249.

# Chapter 10

# Molecular Approaches to Host Plant Defences against Insects

*T.N. Ananthakrishnan*

## INTRODUCTION

Traditional breeding techniques generally depend on crosses with closely related plants, either subspecies or wild ancestral species of the crop. In recent years, genetic engineering has opened up an exciting frontier in agriculture. Molecular engineering techniques have overcome incompatibility barriers so that genes could be inserted from closely or distantly related species envisaging production of crops with a variety of added resistance factors against pests. Of added significance is that genetically engineered pest resistant cultivars could help limit the use of insecticides, so that pest resistant genes become an important natural resource. While traditional methods involve several crosses before the right chance combination of genes to result in offspring with the desired combination of traits, molecular biological methods allow manipulation of one gene at a time instead of depending on the recombination of a large number of genes. Today, it is possible to insert individual genes for specific traits directly into an established genome, with facilities to control the way these genes express themselves in the new variety of crops. By focussing specifically on the desired trait, molecular gene transfer can shorten the time required to develop new varieties. Genetic engineering, besides involving the ability to develop a particular gene for a desired trait, also enables isolation of the gene, understand its function and regulation, modify the gene and reintroduce it into the natural host. It is the ability to isolate and clone individual genes that have played a significant role in the development of biotechnology. Insect resistant and genetically modified crops like maize, potato and cotton are commercially grown in several countries. Vectors used to transfer genes include viruses, plasmids and mobile segments of DNA, the transposon elements. Genetic engineering also requires the ability to manipulate individual cells as recipients of isolated genes. Besides enabling production of clones, cell culture allows for regeneration of somatoclonal variants or plants with altered genetic traits, which can prove useful as new improved crops.

Genetic engineering offers a wide scope to transfer derived genes that synthesize, store and degrade secondary products to other plants without the co-transfer of undesirable traits. This results in the acceleration of the development of new varieties, besides allowing the transfer of genes across incompatibility barriers. Further, genes can also be introduced from sources that are totally unavailable to conventional plant breeding. In short, the strategy in genetic engineering is to assemble gene packages in host plants, and, through this approach, a multimechanistic form of resistance can be tailored to different crops. It is possible to establish multiline mixtures in an otherwise individual genetic background by creating new isogenic lines containing different resistance genes. The use of multilines forms part of the IPM strategy to improve the durability of resistant varieties and thwart the development of biotypes in insects (Chelliah, 1994).

Ever since the first transgenic tobacco plants expressing foreign proteins were obtained, progress in genetic engineering to develop insect resistant plants has been rapid. Most of the early successes have been generated through the mediation of *Agrobacterium tumefaciens*, a soil phytopathogen. Transgenic plants are generated in nature by *A. tumefaciens* through the transfer of some of its own DNA into susceptible plants. The genes are carried on a plasmid, which can be manipulated for the transfer of specific segments of foreign DNA. By moving them into vulnerable plants, their increased protection against insects has been demonstrated (Schell, 1987). In subsequent years, many microorganisms colonizing plant surfaces have been recognized. *Pseudomonas, Rhizobium, Erwinia*, etc. have the potential to function as delivery systems for plant protection.

## BASIC TECHNIQUES IN MOLECULAR STUDIES

Today it is possible to identify, isolate, cut and splice genes and transfer them from one species to another; these four steps are made possible by enzymes obtained mainly from bacteria. The development of techniques to purify DNA and cut it from different organisms at specific sites through the development of restrictive endonucleases, and joining them together in a plasmid, have now become common. Amplification of DNA pieces in a test tube using polymerase chain reaction (PCR) so as to result in sufficient quantity for making DNA analysis is well known. Detection of specific fragments of DNA is possible through the RFLP technique. (Restriction fragment length polymorphism), wherein DNA is first extracted from the tissues and different restriction DNA enzymes are added to different DNA sample. The RAPD (Rapid amplified polymorphic DNA) is used to characterize and detect genes, species and polymorphism. Gene identification, cloning and expression to produce a host of substances now offer an effective tool in biological control and molecular approaches also attempt to use Recombinant DNA technique to produce more virulent strains to pathogens.

The mass production of clones of crop varieties is achieved through tissue culture, which is the most important means by which agriculture, horticulture and ornamental species are propagated (micropropagation). Microinjection used to deliver genetic material into plant cells, including individual chromosomes or cellular organelles could result in improved cultivars with new traits.

Transfer of foreign genes in plant has also been made possible by the Ti-plasmid found in the soil bacterium *Agrobacterium tumefaciens*. The Ti-plasmid is adapted to transfer foreign genes with plants to obtain stable and heritable expression of genes (Moses, 1987).

## DEFENCE RELATED RESPONSES

Defence genes are activated, which may function as deterrents or antifeedants or genes may encode for enzymes such as phenylalanine ammonia lyase (PAL) which synthesizes defence-related products such as phytoalexins, caryophyllene, polygodial, warburganal, etc. Different defence genes are activated simultaneously within the same cell and this activation may involve encoding a set of enzymes that function within a specific metabolic pathway (Ayers, 1992). Genes encoding for structural and functionally diverse products are also activated. Different classes of genes relating to resistance strategy are expressed sequentially. Some of the more important defence related proteins involve hydroxyproline rich glycoproteins (HRGP), glycine rich glycoproteins (GRP), which act directly as deterrents, and the proteins functioning as enzyme inhibitors—amylase and Proteinase inhibitors. Other proteins can be toxic proteins such as lectins and thionins; hydrolyses such as $\beta$, 1-3, glucanases, proteinases and antimicrobial compounds like phytoalexins. Others, like PR proteins or pathogenesis related proteins, are of unknown function being products of genes that accumulate in wound response (Bennet and Wallsgrove, 1994).

The sequence of events leading from defence responses, such as the simultaneous synthesis of diverse proteins and the nature of their expression at various loci, is basic to further research on their significance in terms of insect-plant interactions (Bowles, 1990). Defence depends on a battery of mechanisms, some Constitutive, becoming enhanced later, and induced ones occurring after the event. Another gene marker exercise relates to the induction of ehylene forming enzymes. They regulate ehylene biosynthesis during a wound response, which in turn triggers the activity of phenolic enzymes and phenols. Similarly, Proteinase inhibitors are known antifeedants and have been shown to exhibit toxic effects when genes encoding them have been expressed in transgenic plants. Even intracellular signalling events are recognized, linking recognition of oligosaccharides at the cell surface to changes in gene activity. Injury leads to the release of cell wall fragments and the impact of these on ion transport may lead to systemic signalling effects (Bowles, 1990, 1992).

## ROLE OF SIGNALLING CHEMICALS

Recent researches have shown that lipoxygenases catalyse the oxidation of polyunsaturated fatty acids, like linolenic acid by molecular oxygen, giving rise to the formation of hydroperoxidases, which can be further converted enzymatically into other compounds (Ayers, 1992). One of them is jasmonic acid or methyl jasmonate, which has hormone-like actions eliciting proteinase gene inhibitor expression and is exhibited in interplant signalling. Methyl jasmonate, a common secondary plant substance, induces synthesis of Proteinase inhibitors in the treated and nearby plants (Farmer and Ryan, 1990; Ananthakrishnan, 1994).

Investigations on defence-related lectins, that have carbohydrate binding properties, and specificity towards foreign glycan structures, have indicated their protective potential. Lectins are the only class of plant proteins capable of recognizing and binding glycoconjugates present on the surface of the intestinal tract. Lectins are known to cause deleterious effects and a study of the effect of several commercially available plant lectins on the cowpea weevil has indicated that only those from peanut, wheat germ, potato and thorn apple had a significant inhibitory effect on the development of weevil larvae. Of all lectins, only ricin from castor was 100 per cent lethal to the larvae of *Diabrotica undecimpunctata*, when applied topically to the artificial diet. An increased insect resistance has also been reported for tobacco plants expressing the pea lectin gene. The effect of pea lectin was additive to that of the cowpea trypsin inhibitors in tobacco plants expressing the genes of both the lectin and protease inhibitors (Peumans and Van damme, 1995). More recently, mannose-binding agglutinins from garlic have shown anti-insect properties on sucking insects (Gatehouse, 1993). It is likely that toxicity of lectins is based on highly specific binding to receptor molecules that are located in the gut of insects. Lectins may act synergistically with the protective proteins like protease inhibitors, *alpha*-amylase inhibitors, ribosome inactivating proteins and thionins, establishing a highly efficient defence system against insects.

Further, plants being a source of natural resistance mechanisms, which have the potential of being translated into new insect control chemical products, provided the genetic elements of resistance are suitable for incorporation into desirable cultivars (Bowers, 1988).

## PROTEINASE INHIBITORS

Of the many natural defence mechanisms involved in plants against herbivorous insects, the synthesis of Proteinase inhibitors is an important one (Green and Ryan, 1972). In contrast to the *Bt* endotoxins, these proteins have anti-metabolic activity against a wide range of insects. They are present in the tissue of some plants at relatively high concentrations, participating in complex defence responses with other molecules produced by the plant. The introduction of specific Proteinase inhibitors into plants, which do not produce these molecules, is an alternative approach for obtaining crops that are resistant to insect attack. Four classes, serine, thiol, metallo and aspartyl-Proteinase inhibitors, have been identified (Brunke and Meeusen, 1991). Presence of trypsin inhibitor in the diet of insects reduces the effective concentration of trypsin available for digestion. This triggers a series of events, including the inability to obtain amino acids from ingested food.

The advantages of using Proteinase inhibitors as insect-control agents are that they are active against a wide range of insects, besides their use as a second mechanism to help prevent development of insects that are resistant to *Bt* endotoxin. The need for high levels of protein for insect killing and the potential need to regulate protein expression to specific plant organs are major disadvantages.

The first gene of plant origin that was successfully transferred to another plant species through genetic engineering, resulting in enhanced insect resistance, was that isolated from cowpea, encoding a trypsin inhibitor. The trypsin inhibitor is an effective antimetabolic against a number of Lepidopteran, coleopteran and orthopteran field and storage pests. The coding sequence of cowpea trypsin inhibitor (CPTI) has been successfully inserted into tobacco plants and such plants afforded protection against *H. zea, Spodoptera litura* and *Manduca sexta*. Besides, the genes encoding protease inhibitors isolated from cowpea conferred resistance when expressed in tobacco. The tomato inhibitor II gene, when expressed in tobacco, has also been shown to confer insect resistance. The tomato and potato inhibitor II gene encodes a trypsin inhibitor and expression of the gene in tobacco results in increased levels of protection against the tobacco hornworm, *M. sexta*.

Ever since the first transgenic tobacco plants expressing foreign protein virus were obtained, progress in genetic engineering to develop resistant plants has been considerable. Most of the engineered plants have been generated through *A. tumefaciens*-mediated gene transfer, since it is very efficient and versatile vector for the stable introduction of genes into plants. Currently, around 25 different plant species have been transformed using different systems (Gasser and Fraley, 1992).

## TRANSGENIC CROPS

Recent advances in molecular genetics have shown that genes regulating basic development processes have been modified to control responses to biotic stresses. Genes coding for phenylalanine ammonia lyase are developmentally regulated and are normally expressed at particular phases in the developmental of healthy plants (Liang *et al.* 1989). It seems to be possible that genes regulating resistance could be manipulated to increase or intensify their expression, to the advantage of man (Ayers, 1992).

Transgenic plant technology has emerged as a useful tool in producing pest resistant crops by introducing entirely new genes into plant genes. Inserting genes into crop plants involved initially the soil bacterium. *A. tumefasciens*, as vector, wherein during infection transfer of foreign DNA occurs along with portion of its own plasmid DNA with a plant cell. The discovery that the aerobic, grampositive spore-forming bacterium, *Bacillus thuringiensis* (*Bt*), which produces a toxic protein during sporulation, has introduced a new era of a genetically engineered biocontrol agent. Moving *Bt* insecticide genes from the original bacterium to plants, through the mediation of a vector, *Pseudomonas flourescens*, allows crop protection from insect attack. Another class of proteins, the Proteinase inhibitors, are highly effective against certain insects when combined with *Bt* entotoxins. The advantages of transgenic plants involve improved targeting of toxins, season-long protection in all plants without the need for expensive equIPMent and less toxicity exposure, advantages that are both economic and environmental. The toxin can be delivered to the pest directly because the active ingredients are expressed within the plant (Dilawari and Dhaliwal, 1996).

Transgenesis is aimed at altering or improving the product derived from crop plants. Modification of metabolic pathways by the introduction of different enzymes from other species can result in a different product. Plants genetically modified to improve fitness advantage in some environments such as tolerance to environmental stress and pest resistance have been produced (Rogers and Parker, 1995). The ability to move bacterial genes, encoding natural insecticidal proteins into plants, enables the development of crops intrinsically resistant to insect attack, with advantage over conventional chemicals (Brunke and Meeusen, 1991).

More than 30 crop species, including some major field crops-corn, soybean and rice have now been successfully transformed with insect crystal protein (ICP) genes to become insect resistant. Receptor binding accounts to a greater extent each individual type of ICP. Binding studies have shown that one insect may be susceptible to different ICPs, depending upon the presence of different receptors on the midgut wall. For example, Cry1A (b) and Cry 1B recognized different receptor glycoproteins. Crystal proteins and their genes have been classified based on their structure and activities spectrum, into Cry I (Lepidopteran species), Cry II (Lepidoptera and Diptera), Cry III (Coleoptera) and Cry IV (Diptera). Each of these are functionally divided into several toxin types. This heterogeneity in toxin production is responsible for some of the diversity in activity spectrum among strains. There is also a high degree of heterogenicity among binding sites in some species, suggesting that some sites may bind a single toxin whereas others may bind two or more toxins (McGaughey and Whalon, 1992). *Gossypium hirsutum* has been genetically modified to incorporate Recombinant DNA form *Bacillus thuringiensis kurstakii* (*Btk*) which codes for delta-endotoxin proteins. Transgenic *Btk* cotton expressing Cry IA (b) and Cry IA (c) insecticidal proteins have shown significant protection against *Helicoverpa zea* and *Heliothis virescens* (Halcomb *et al.* 1996). While insect resistant plants with ICP genes are expected to be commercialized, concern is expressed that a large-scale introduction of crops containing a monogenic resistance trait could rapidly lead to the development of resistance within an insect population. This is true of field resistance of *Plutella xylostella* to Bt (Van Rie, 1991).

Some desirable areas of development for genetic engineering techniques that have great potential to benefit sustainable agriculture relate to enhancing crop resistance to insects, drought resistance and salt tolerance in crops, nitrogen fixation in cereal and other crops, development of perennial grain crops and improved botanical pesticides (Paoletti and Pimental, 1996). While traits for resistance to different insect pests already exist in some cultured crops, their occurrence already existing in related wild varieties provide an enormous gene pool for development of host plant resistance. With the need for increased input of cereals and pulses, genetic engineering to develop them to provide their own nitrogen by bacterial symbiosis has been on the cards. Untapped genetic sources of beneficial soil microbes such as *Rhizobium*, living in the nodules of roots of leguminous plants, have been utilized, since *Rhizobium* releases ammonia into the soil where other bacteria convert it into nitrates, acting as a natural fertilizer for other plants. Genes enhancing nitrogen fixing process as well as those involved in the mode of attachment to nodules have been identified, besides boosting the efficiency of these genes. These root bacteria are also used as vectors for introducing other genetic characters into a crop, such as resistance to root disease. Whether genes from these bacteria could be transferred into non-leguminous crops, such as wheat or corn, is a moot question. A growing range of agricultural biotechnology programs exist and commercialization of genetic engineered products relate to resistance to *Bt* toxin in cotton, enhanced nitrogen fixation to increase yield in alfalfa and resistance to herbicide in soybeans and in weed control.

## CONCLUSIONS

The take off stage in genetic engineering or biotechnology in insect pest management has been initiated, but increased efforts have to be made towards a better understanding of the biological regulation of plant-gene action in order to produce ecologically accepted crops. Emphasis should be on the genetics of plant-herbivore interaction in natural and agricultural systems and on the genetics of plant secondary metabolism (Gould, 1988). Genetically engineered, pest resistant cultivars, would tend to limit the use of synthetic chemicals, making pest resistance genes as an important natural resource. While the challenge of moving toxin-coding DNA sequences into several crops has been met, there is a need to move these toxin-coding groups into more crops enabling these genes to be expressed at higher levels. However, the production of crops that express resistance coding DNA sequences should be undertaken only when needed. Considering the rapid pace at which genetic engineering techniques are emerging, the development of plants the produce toxins in specific tissues during the growing season, to control insects deserves high consideration.

Resistance management should involve tactics, which are coordinated against pests, in particular transgenic plants, where a combination with *Bt* or other allelochemical mortality factors could be most useful. The increasing attention given to biotechnology may enable more effective production of predators and parasites, and genetic manipulation may produce more competitive organisms. The incorporation of genes conferring pesticides resistance into predators and parasites would facilitate their integration with the use of pesticides. Concern has also been expressed regarding the environmental side effects of the expressed toxin and suggestion is made that the selective advantage of the engineered transgenic plants might increase their natural level of invasiveness. Besides threats to non-target species posed by toxins cannot be ignored (Thacker, 1993, 1994).

## REFERENCES

Ananthakrishnan, T.N. 1994. View point in the chemical ecology of insects. *Phytophaga* **6**: 69–72.

Ayers, P.G. (ed.) 1992. *Pests and Pathogens*. Bios Scientific Publishers: Oxford, UK.

Bennett, R. and Wallsgrove, R.M. 1994. Secondary metabolites in plant defence mechanisms. *New Phytol.*, **127**: 617–633.

Bowers, W.S. 1988. Allelochemicals in insect management. *In:* D.W. Roberts and R.R., Granados (Eds). *Biochemistry, Biological Pesticides and Novel Plant-Pest Resistance for Insect Pest Management.* Boyce Thompson Institute for plant Research at Cornell University, Ithaca, New York, USA, pp. 123–131.

Bowles, D. 1990. Defence related protein in higher plants. *A Rev. Biochem*, **59**: 873–907.

Bowles, D. 1992. Signals in the wounded plant. *In* P.G. Ayers (Ed.) *Pests and Pathogens*. Bios Scientific Publishers, Oxford, UK, pp. 33–38.

Brunke, K.J. and Meeusen, R.L. 1991. Insect control in the genetically engineered crops. *Tibtech.* **9**: 197–200.

Chelliah, S., 1994. Current developments in genetic manipulation of crops for insect resistance *Phytophaga*, **6**: 89–93.

Dilawari, V.K. and Dhaliwal, G.S. 1996. Biotechnology and host plant resistance to insect: Opportunities and challenges. *In* T.N. Ananthakrishnan (Ed.) *Biotechnological Perspectives in Chemical Ecology of Insects*. Oxford & IBH Publishing Co. Pvt. Ltd.: New Delhi, India. pp. 36–53.

Farmer, E.E and Ryan, C.A. 1990. Interplant communications: Airborne methyl jasmonate induces synthesis of Proteinase inhibitors in plant leaves. *Proc. Natl. Acad. Sci. USA*, **87**: 7713–7716.

Gasser, C.S. and Fraley, R.T. 1992. Transgenic crops. *Scient. Am.*, **266** (6): 34–39.

Gatehouse, J.A. 1993. Approaches to insect resistance using transgenic plants. *Phil. Trans. Royal Soc. London*, B**342**: 279–286.

Gould, F. 1988. Evolution of biology and genetically engineered crops. *Bioscience*, **38**: 26–33.

Green, T.R. and Ryan, C.A. 1972. Wound-induced Proteinase inhibitors in plant leaves: A possible defence mechanism in insects. *Science*, **175**: 776–777.

Halcomb, J.L., Benedict, J.H. Cock, B. and Ring, D.R. 1996. Survival and growth of bollworm and tobacco budworm on transgene and transgenic cotton expressing a Cry IA insecticidal proteins (Lepidoptera: Noctuidae). *Envir. Ent.*, **25**: 251–255.

Liang, X., Dron, M., Gramer, C.L., Dixin, R.A. and Lamb, C.J. 1989. Differential regulation of phenylalanine ammonia lyase genes during development and by environmental clues. *J. Biol. Chem.*, **264**: 14486–14492.

McGaughey, W.H. and Whalon, M.E. 1992. Managing insect resistance to *Bacillus thuringiensis* Toxins Sci., **258**: 1451–1455.

Moses, P.B. 1987. Gene transfer methods applicable to agricultural organisms. *In Agricultural Biotechnology*. Natural Academy Press: 149–185.

Paoletti, M.G. and Pimental, D. 1996. Genetic engineering in agriculture and the environment. *Bioscience*, **46**: 665–673.

Peumans, W.J. and Van damme, E.J.M. 1995. The role of lectins in plants defence. *Hist. J.*, **2**: 253–271.

Rogers, H.J. and Parker, H.C. 1995. Transgenic plants and the environment. *E. Eypt. Biol.*, **46**: 467–488.

Schell, J.S. 1987. Transgenic plants as tools to study the molecular organisation of plant genes. *Science*, **237**: 1176–1183.

Thacker, J.R.M. 1994. Transgenic plants *Trends Ecol. Evol.*, **9**: 486.

Thacker, J.R.M. 1993. Transgenic crop plants and pest control. *Sci. Prog.*, **776**: 207–219.

Van Rie, J. 1991. Insect control with transgenic plants: Resistance proof? *Tibtech.*, **9**: 177–179.

# Chapter 11

# Chemical Ecology: The Future

*T.N. Ananthakrishnan*

## INTRODUCTION

Do plants recognize the difference between mechanical damage and insect damage? (Ananthakrishnan 1988). In the last decade, assumptions and hypotheses have turned into realities. Besides recognition and acquired immunology, limited evidence is currently available to indicate that healthy plants can detect and respond defensively to volatiles emitted by plants in the neighbourhood (Dicke 1994). While the foundation for the study of insect-plant interaction was laid by the adaptation of insects to plant secondary defence chemistry, recent advances relating to stress-induced metabolism have revealed the dynamics of such interactions, with the induction of a whole range of biosynthetic pathways. This has led to the initiation of defence metabolites in other parts of the plant, not to mention of the volatiles released, which besides enabling detection of herbivores to plants, also enable natural enemies to detect their hosts. While the preceding chapters of this book have essentially dealt with various aspects of insects and plant defence dynamics, an overall assessment of their implications for the future is outlined here.

Chemical signals within species play a critical role, whereby females of insect parasitoids use chemical cues to locate, identify, and exploit their hosts (Tumlinson *et al.* 1992, Turlings 1994). With the growth of the host insects and plants, the chemical cues emanating from the host plants tend to change, resulting in a variety of such signals whose nature and reliability vary. The searching parasitoid, therefore, 'utilizes a dynamic continuum of semiochemicals to locate and exploit the hosts'. The introduction of trap crops of intercropping capitalizing on the principle of masking odours of one plant with that of another is now utilized in agricultural practice. Such trap crops attract insects enabling target crops to escape pest attack (Landis *et al.* 2000).

## INDUCED DEFENSE STRATEGIES

With a better appreciation of the distinction between Constitutive and induced defenses, it has become possible to decipher the chemical signals produced by damaged plants. Physiological changes following feeding by herbivores reduce the quality of plant tissue for subsequent feeding. Different plant species and genotypes use different signals and consequently different physiological and biochemical processes involving diverse metabolic products result in induced resistance. Such signals—generated rapidly at the site of insect damage in the plants—travel rapidly through the plant eliciting induced responses (Baldwin 1994, Karban and Baldwin 1997). Knowledge of such cues is useful in experimental induction of resistance in crop plants. The inductive process initiates several changes that provide great variability in the levels of multiple defenses (Stout *et al.* 1996). Such chemical defenses afford greater flexibility allowing plants to switch chemical defense options more rapidly and irreversibly than many other forms of defense (Berenbaum and Zangerl 1995). Induced response vary spatially that they can remain localized or can spread within a plant as a systematic response or may even spread from a damaged plant to an undamaged neighbour (Farmer and Ryan 1990). While increasing plant fitness, induced defence responses confer resistance against herbivores so that inducibility can be viewed as an 'effective strategy if it minimizes costs or maximizes benefits to the plant for a given level of investment in defense' (Agrawal and Karban 1998). Recent studies have emphasized the usefulness of induced plant allelochemicals to natural enemies, with Host plant location being greatly enhanced by the volatiles released by the plants following feeding by the herbivore (Ananthakrishnan 1988, Dicke 1994, Dicke *et al.* 1993, Rose *et al.* 1998, Turlings 1994, Turlings *et al.* 1998). However, the need for a better understanding related to the effects of inducible plant responses on the behavioural and developmental aspects of parasitoids and herbivores on insect and plant coevolution cannot be ignored, since the important role of herbivore-induced volatiles in parasitoid-host location offers exciting opportunities in the biological control of insects (Poppy 1997, Havill and Raffa 2000). There can be no better example than jasmonate-induced plant defenses and increased parasitism (Thaler 1999). While herbivore location is greatly enhanced by volatile chemicals released by plants following insect feeding, the composition and timing of these volatile releases, while indicating an active role in plant defense also suggests a coevolutionary relationship between plants and parasitoids (Mattiacci *et al.* 1994, Peer and Tumlinson 1997, Havill and Raffa 2000).

## DIVERSITY OF PLANT SIGNALS AND ELICITOR MECHANISMS

Activation of defensive genes in crop plants by the action of a herbivore or a pathogen can result from the action of a variety of signalling molecules that are released in complex temporal patterns following initial invasion of tissues. These signals are transported locally by diffusion through intra-or extracellular fluids that permeate towards the site of attack or systemically through the translocatory system of the plant (Farmer and Ryan 1990). For example, a lipid-derived molecule, methyl jasmonate acts as a volatile signal and induces accumulation of proteinase-inhibitor proteins. These inhibitor genes are induced by oligosaccharide fragments (oligouronides) from plant cell walls. Jasmonic acid occurs naturally in plants and its methyl ester (methyl jasmonate) regulates gene expression in plants. Synthesis of Proteinase inhibitors I and II, for example, in members of Solanceae, can be triggered by cell wall oligosaccharides, besides inducing the synthesis of systemiñ, an 18 amino-acid-polypeptide and abscisic acid (Farmer and Ryan 1992, Ryan and Pearce 1998, Raymond *et al.* 1995). Higher concentrations of the defense-signalling molecule, jasmonic acid is known to result from the damage induced by the hornworm caterpillar (McCloud and Baldwin 1998).

The key to emission of plant signals for foraging success of parasitoids are substances from the oral secretions of the herbivores. Two oral secretions which augment the release of plant volatiles have so far been identified from chewing insects. *Pieris brassicae* caterpillar regurgitant contains β-glucosidase which triggers the emission of isothiocyanates in cabbage (Mattiacci *et al.* 1995). Enzyme activity in the regurgitant is retained enabling cleaving of sugars coupled with organic compounds which then become more volatile and get released. Volicitin, a low-molecular weight fatty acid derivative in the oral secretions of the beet army worm caterpillar induces corn seedlings to release volatile chemical signals (Alborn *et al.* 1997). A precursor of the volatile compound is obtained from the host plants and the bioactive product is formed within the caterpillar. Structurally, Volicitin is an octadeca-trienoate conjugated to an amino acid suggesting that the elicitor molecule interacts with the octadecenoid pathway in the herbivore-damaged plant (Peer *et al.* 1998). In the octadecenoid-derived plant signals, the $C_{12}$ compounds are derived from the $C_{18}$ unsaturated fatty acids (e.g. linolenic and linoleic acids), resulting in intracellular signalling molecules such as jasmonic acid (Farmer and Ryan 1992).

The need to identify more regulators which control the synthesis and release of plant volatiles is becoming critically imperative. Though several analyses for volatiles in arthropod oral secretions have been made, their biochemical pathways are not yet clear. It is equally important to recognize that distinct responses are induced by elicitors of different molecular structures from diverse herbivore species (Peer and Tumlinson 1999).

## ANALYSIS OF CHEMICAL SIGNALS

Mechanism of odour perception, modes of channelling and integration of messages, and their behaviour in response to the messages are important in interpreting arthropod relationships to host plants. Complexity of the nervous system and the neural inputs within the central nervous system reflect this. Steps in the host-recognition behaviour of phytophagous insects and that of their natural enemies are critical in IPM strategies, since different plants use different signals or a single plant species may use more than one. Amplification of signals is a necessary event and an understanding of the physiology of glomeruli in the brain where processing of sensory inputs takes place, becomes essential. The stimulatory plant odour induces the production of action potentials; their firing rates increase with the intensity of the odour. Several examples demonstrate that specific olfactory sense cells respond to compounds like allyl-isothiocyanate as well as the existence of cells with a number of acceptor sites which respond to a range of compounds (Pickett *et al.* 1999). To serve the requirements of a chemical communication system crucial for reproductive success, there is a need to understand how chemical signals are processed by the olfactory receptor cells of the behaviour of the concerned insect. In short, the steps that constitute the olfactory information pathway need to be worked out (Hildebrandt 1995). Since insects respond to a variety of semiochemicals, the quality and quantity of odoriferous compounds present in the natural environment are encoded in the patterns of activity. These messages are decoded and integrated in the olfactory centres of the central nervous system, thus regulating insect behaviour.

## TRANSGENICS AND INSECT RESISTANCE

Individual plants varying in their level of resistance against insects have been a source for plant breeders to increase resistance in crop cultivars. This exercise, though in trial process, has been supplemented if not totally replaced by insect-resistance transgenes. When introduced into plants they increase the level of resistance, so that genetically-engineered crop resistance to insects pests not only widens the potential pool of useful genes, but also permits introduction of a number of different desirable genes at a single event (Hilder and Boulter 1999). The significant initiative in developing pest resistant Transgenics has been with the incorporation and expression of endotoxins of *Bacillus thuringiensis* *(Bt)* in plants, the toxin binding to glycoprotein receptors in the brush border membrane of the gut epithelium of insects. As a result of binding, the toxin rapidly percolates into the cell membrane forming a pore and leading to the lysis of epithelial cells. With separate strains of *Bt* producing a variety of crystal proteins with distinct host range, it has been possible to transfer and express *Bt* toxin in several plant species. With expression levels from native bacterial genes being too low, the use of strong promoters and enhancers became necessary. *Bt* genes have been expressed successfully in such crops as cotton, rice, tobacco, potato, egg plant, maize, peanut, chick pea, and soy bean (Schuler *et al.* 1998).

Two other major groups of plant-derived genes used to transfer insect resistance are proteinase and amylase inhibitors and lectins. Different proteinases predominate in different insects, such as serine-lyase in Lepidoptera; serine and cysteine-Proteinase inhibitors mostly in Lepidopteran and coleopteran species. These inhibitors have been obtained from species of Fabaceae, Solanaceae, and Poaceae, the active inhibitor identified being the cowpea trypsin inhibitor against bollworms. Amylase inhibitors have been essentially from *Phaseolus vulgaris* inhibiting $\alpha$-amylase in the midguts of stored-products beetles, blocking larval development. Lectins are carbohydrate-binding proteins useful against Coleoptera, Lepidoptera, Homoptera, and Diptera. While further investigations are in progress, interest is also being evinced in the expression of plant secondary metabolites which are usually produced through multigene pathways. To increase the range of pests affected the possibility of combing different genes in plants is also being explored (Schuler *et al.* 1998).

## CONCLUSION

Some basic aspects relating to chemical ecology as a whole, notably detailed aspects on the effects of different herbivores on the induction of plant volatiles and the degree of specificity of the responses deserve further attention, besides the nature of different elicitors in insects and their specificity. A better understanding of different signal pathways and of the functional organization of olfactory pathways involved in information processing appear feasible. Such a knowledge may also lead to "the development of new plant varieties with enhanced chemical defenses or to methods of inoculating plants with elicitors to increase their resistance to insect pests" (Paré and Tumlinson 1998).

## REFERENCES

Agrawal, A. A. and Karban, R. 1998. Why induced defenses may be favored over Constitutive strategies of plants? *In* R. Tollrian and C. D. Harvell (Eds) *The Ecology and Evolution of Induced Defenses*, pages 45-61. Princeton University Press, Princeton, USA.

Alborn, H., T. M Turlings, T. C., Jones, T. H., Stenhagen, G., Longhrin, J. H. and Tumlinson, J. H. 1997. An elicitor of plant volatiles from beet armyworm oral secretion. *Science*, **276**: 945-949.

Anantharishnan, T. N. 1988. Insect-plant interaction: Future outlook. *In* T. N. Ananthakrishnan and A. Raman (Eds) *Dynamics of Insect-Plant Interaction*, pages 219-223. Oxford & IBH Publishing Company, New Delhi, India.

Baldwin, I. 1994. Chemical changes rapidly induced by folivory. *In* E. Bernays (Ed) *Insect-Plant Interaction V*, pages 1-23. CRC Press, Boca Raton, Florida, USA.

Berenbaum, M. and Zangerl, A. R. 1995

Dicke, M. 1994. Local and systemic production of volatile herbivore-induced terpenoids: Their role in plant-carnivore mutualism. *Jour. Pl. Phisiol.*, **143**: 465-472.

Dicke, M., van Barlen, P., Wessels, R. and Dijkman, H. 1993. Herbivory induces systemic production of plant volatiles that attract predators of the herbivore: extraction of the endogenous elicitor. *Jour. Chem. Ecol.*, **19**: 581-599.

Farmer, E. E. and Ryan, C. A. 1990. Interplant communication: airborne methyl jasmonate induces synthesis of Proteinase inhibitors in plant leaves. *Proc. Nat. Acad. Sci. USA*, **87**: 7713-7716.

Farmer, E. E. and Ryan, C. A. 1992. Octadecenoid jasmonate precursors activate the synthesis of wound-inducible Proteinase inhibitors. *Pl. Cell*, **4**: 129-134.

Havill, N. P. and Raffa, K. F. 2000. Compound effects of induced plant responses on insect herbivores and parasitoids: implications for tritrophic interactions. *Ecol. Ent.*, **25**: 171-179.

Hildebrandt, J.G. 1995. Analysis of chemical signals by nervous system. *Proc. Nat. Acad. Sci. USA*, **92**: 67-74.

Hilder, V. A. and Boulter, D. 1999. Genetic engineering of crop plants for insect resistance—a critical review. *Crop Prot.*, **18**: 177-191.

Karban, R. and Baldwin, I.T. 1997. *Induced Responses to herbivory.* University of Chicago Press, Chicago, USA. 319 pp.

Landis, D. A., Wratten, W. D. and Gurr, G. M. 2000. Habitat management of conserve natural enemies of arthropod pests in agriculture. *Annu. Rev. Entomol.*, **45**: 175-201.

Mattiacci, L., Dicke, M. and Posthumus, M. A. 1995. β-glucosidase: An elicitor of herbivore-induced plant odor that attracts host searching parasitic wasps. *Proc. Nat. Acad. Sci. USA*, **92**: 2036-2040.

McCloud, E. S. and Baldwin, I.T. 1998. Herbivory and caterpillar regurgitants amplify the wound-induced increases in jasmonic acid, but not nicotine in *Nicotiana sylvestris*. *Planta*, **203**: 430-435.

Paré, P. W., Alborn, H. J. and Tumlinson, J. H. 1998. Concerted biosynthesis of an insect elicitor of plant volatiles. *Proc. Nat. Acad. Sci. USA*, **95**: 13971-13975.

Paré, P. W. and Tumlinson, J. H. 1998. Plant volatiles as a defense against insect herbivores. *Plant Physiol.*, **121**: 325-331.

Pickett, J. A., Wadham, L. J. and Woodcock, C. M. 1998. Insect supersense: mate and host location by insect as model system for explaining olfactory interaction. *The Biochemist*, **20**: 8-13.

Pickett, J. A., Smiley, W. M. and Woodcock, C.M. 1999. Secondary metabolites in plant-insect interaction: dynamic system of induced and adaptive responses. *Adv. Bot. Res.*, **30**: 91-115.

Poppy, E. M. 1997. Tritrophic interaction: Improving ecological understanding and biological control? *Endeavour*, **21**: 61-65.

Reymond, P., Grünberger, S., Paul. K., Müller, M. and Farmer, E. E. 1995. Oligogalactouronide defense signals in plants: Large fragments interact with the plasma membrane *in vitro*. *Proc. Nat. Acad. Sci. USA*, **92**: 4145-4149.

Rose, U.S., Manukian, A., Heath R. R. and Tumlinson, J. H. 1996. Volatile semiochemicals released from undamaged cotton leaves: A systemic response of living plants to caterpillar damage. *Plant Physiol.*, **111**: 487-495.

Ryan, C. A. and Pearce, G. 1998. Systemin: A polypeptide signal for plant defense genes. *Annu. Rev. Cell and Develop. Biol.*, **14**: 1-17.

Schuler, T.H., Poppy, G.M., Kerry, B.Q. and Denholm, I. 1998. Insect resistant transgenic plants. *Trends in Biotech.*, **16**: 168-175.

Stout, M. J., Workman, K. V., Bostock, R. M. and Duffey, S. S. 1995. Specificity of induced resistance in the tomato, *Lycopersicon esculentum*. *Oecologia*, **113**: 74-81.

Tumlinson, J. H. Turlings, T. C. J. and Lewis, W. J. 1993. Semiochemically mediated foraging behaviour in beneficial parasitic insects. *Arch. Insect Biochem. Physiol.*, **22**: 385-391.

Tumlinson, J. H. Turlings, T. C. J. and Lewis, W. J. 1992. The semichemical complexes that mediate parasitoid foraging. *Agricult. Zool. Rev.*, **5**: 221-252.

Thaler, J. S. 1999. Jasmonate-inducible plant defences cause increased parasitism of herbivores. *Nature*, **399**: 386-388.

Turlings, T. C. J. 1994. The active role of plants in the foraging successes of entomophagous insects. *Norweg. Jour. Agricult. Sci.*, **16**: 211-219.

Turlings T. C. J., Loughtin, J. H., McCall, P. R., Rose, U. S., Lewis, W. J. and Tumlinson, J. H. 1995. How caterpillar-damaged plants protect themselves by attracting parasitic wasps. *Proc. Nat. Acad. Sci. USA*, **92**: 4169-4174.

# Index

D0686980

# James I

# James I

## SCOTLAND'S KING *of* ENGLAND

### JOHN MATUSIAK

The History Press

Look not to find the softness of a down pillow in a crown,
but remember that it is a thorny piece of stuff and full of continual cares.

*James I, Meditations of Matthew 27*

*For my father*

First published 2015

The History Press
The Mill, Brimscombe Port
Stroud, Gloucestershire, GL5 2QG
www.thehistorypress.co.uk

British Library Cataloguing in Publication Data.
A catalogue record for this book is available from the British Library.

ISBN 978 0 7509 5562 1

Typesetting and origination by The History Press
Printed in Malta by Melita Press

# Contents

# 1 ⚕ Heir to Scotland's Woe

No more tears now. I will think upon revenge.

*Words attributed to Mary Queen of Scots by Claude Nau de la Boiselliere,*
*her confidential secretary from 1575–86*

In the mid-morning of 19 June 1566, Mary Queen of Scots was gratefully delivered of her first and only live-born child in a tiny closet tightly lodged in the south-east wing of Edinburgh's ancient castle. Her labour had, it seems, been long and arduous, and ten days earlier, plainly fearing the worst, she had written her will. At that time, too, she had sent to Dunfermline Abbey for a sacred reliquary containing the skull of St Margaret, set in silver-gilt and 'enriched with several pearls and precious stones', which she intended to sustain her throughout the ordeal to come. Accordingly, as Mary endured the torment within her chamber's sombre panelled walls, the remains of the saint – a Catholic queen of Scotland like herself – duly loomed above her, along with the arms of the House of Stuart and a series of embossed crowns and thistles adorning the ceiling overhead. Beside her all the while stood Margaret Asteane, her midwife, specially garbed for the occasion in a brand new gown of black velvet, not far from the royal cradle, which was likewise draped in finest fabric.

By 11 a.m., however, the midwife's task was ended. For the queen was 'lighter of a bonny son' whom, she promptly predicted, 'shall first unite the two kingdoms of Scotland and England'. The boy had entered the world, like Napoleon after him, with a fine 'caul', or birth membrane, covering his head – an augury, it was said, of future greatness – and his mother's lofty hopes seemed far from fanciful, since her royal cousin across the Border was, of her own admission, 'but a barren stock'. If, therefore, Elizabeth I should now die childless, or if any plot against her life were to run its fatal course, Scotland's queen was not only the obvious female successor in her own right, but, much more importantly still, the bearer of a healthy male heir. And the blood of Mary's son, directly drawn from Henry VII through both his parents, was of plainly purer stock than any other rival.

All, then, was swiftly set for outward rejoicing throughout the northern kingdom, though not before Mary's secret messenger, Sir James Melville, was safely past the Border at Berwick on route hotfoot to London. Thereafter, nobles, officers of state and common folk alike gave solemn thanks in Edinburgh's Great Kirk, as the castle's mighty guns – long a stirring symbol of national pride – boomed their glad approval. Deputations and messages of goodwill arrived from far and wide, further couriers were dispatched to France and Savoy, and loyal toasts were heartily raised to Scotland's fledgling 'Solomon'. Later that night, 500 bonfires would blaze on Scottish hillsides, as all the due and proper customs associated with any royal birth were studiously observed.

But the mask and show of celebration was mainly sham, since Mary Queen of Scots was also Scotland's woe. It was not for nothing that she had shunned the comfort of Holyroodhouse as her birthing place and made instead for Edinburgh and the security it afforded. Nor were all the salutations she now received by any means sincere. Indeed, for most of the vested interests in her restless kingdom, the newborn child represented little more than a fresh and unwelcome complication of a political and religious situation already critically dangerous. Powerful sections of the nobility had hoped, for their own self-interested motives, that he might never be born, and the stilted congratulations of John Spottiswoode, the Lothian superintendent of the Protestant Kirk of Scotland's General Assembly, could not conceal his misgivings that the new heir would inevitably be baptised a Roman Catholic, with all that this entailed for the reformed religion that had made such rapid strides since its apparent triumph only six years earlier. Even the child's father, Henry, Lord Darnley, had already done his feckless yet malignant best to prevent the birth of the son who shattered his best chance of seizing the throne for himself.

It was Darnley, moreover, who had sedulously propagated the rumour that his wife's new son was merely the bastard offspring of David Riccio, her Italian secretary and musician, whom he had helped to murder in her very presence just four months previous. Jaundiced, jealous, vain and volatile – resembling 'more a woman than a man' and stricken by inner demons of his own devising, which he could neither tame by infidelity nor dowse with drink – the queen's husband was now a pox-ridden parody of the dashing blonde-haired lover who had first dazzled his bride only two years earlier 'as the properest and best proportioned long man that she had ever seen'. Both Mary and Darnley knew, furthermore, that she too had been 'struck with great dread' and in 'extreme fear' for her life when Riccio met his end, even though, within hours of the new birth, the queen's abject husband was once again reminding his wife of her subsequent promise to 'forgive and forget all'.

But while the queen might dutifully forgive, forgetting was another matter. 'What if Fawsdonsyd's pistol had shot?' she had asked her husband, recalling that fateful night when a gun, which had allegedly 'refused to give fyr', had been pressed to her own breast by one of Darnley's accomplices. 'What wold have become of him [the child] and me both?' Nor could she ever entirely quash those spiteful rumours propagated by her husband that would continue to shadow her son's legitimacy. It was vital to Mary, of course, that Darnley should swiftly undo as much of the harm he had already wrought with his loose and ill-intentioned tongue, and he was soon compelled to acknowledge the child in the presence of the queen's half-brother, the Earl of Moray, as well as the earls of Mar, Atholl and Argyll, and her Privy Council. Yet the queen's caustic quip to her husband that 'he is so much your son that I fear for him hereafter' would never entirely convince the world at large or spare her heir the barbs of ne'er-do-wells in years to come. As a child, indeed, the boy would weep in mortification at the slander, and the occasional taunts of the Scottish mob did nothing to ease his misery. 'Come down, thou son of Seigneur Davy', a baying Perth rabble would jeer in 1600 as he stood at the window of Ruthven House, and much later still, the King of France would chuckle at the boy who had by then become both James VI of Scotland and James I of England, dismissing his fellow ruler as 'Solomon the son of David who played upon the harp'. There were even creeping whispers that Mary's child had died at birth, to be replaced by a child of John Erskine, Earl of Mar – empty legends which were nonetheless given a further lease of life in the eighteenth century when the skeleton of a newborn child, 'wrapped in a rich silken cloth … belonging to Mary Stuart, Queen of Scots', was uncovered in a wall of Edinburgh Castle's banqueting hall next to the castle courtyard.

Yet, aside from hurtful jibes and murky tales, there was never serious doubt about the legitimacy of the child who presently occupied the royal cradle. At the time of his conception, after all, the boy's mother was still wildly infatuated with her lawful husband – so much so, indeed, that she appeared to have sacrificed all judgement on his behalf. 'The queen,' wrote the English diplomat Thomas Randolph, 'is so altered with affection towards Lord Darnley that she has brought her honour in question, her estate in hazard, her country to be torn in pieces.' And the child's resemblance to his father in an early portrait depicting him with a sparrowhawk on his arm remains striking. His flaxen hair, finely contoured features and, above all, his distinctive widely spaced eyes left little doubt about his parentage. Nor, in any case, was Darnley, for all his twisted bitterness, the new heir's greatest liability. Instead, it was the very mother who had borne him, for though she was bold, courageous and gracious, with a charm and allure that still captivates across the centuries, she was also headstrong, careless and ambitious – a passionate, high-spirited and ultimately self-centred creature who yearned for adulation but could neither bridle her emotions nor curb her whims. It was she, above all, who barred the way to long-term peace within her realm and she too, who, in spite of initial successes, menaced the fortunes and security of her longed-for son.

Sent away to France in 1548 at the age of 6, after a planned betrothal to young King Edward VI of England finally proved intolerable to her countrymen, Mary had spent nearly the whole of her life abroad. In her absence, English bullying would increase Scotland's traditional reliance on the French Crown, as the Queen Mother, Mary of Guise, herself a Frenchwoman and staunch Catholic, served as regent, holding Scotland somewhat precariously to the old religion and alliance with her homeland until her death in 1560. As part of this alliance, the absent queen had been betrothed and finally married to the dauphin, and from 1559 to 1560, the absorption of the Scottish Crown, which had eluded the English, consequently became a reality for their enemies across the Channel. By 1560, moreover, Mary Queen of Scots was not only Queen Consort of France, but rightful Queen of England in the eyes of every loyal Catholic in Europe by virtue of her paternal grandmother, Margaret Tudor, elder sister of Henry VIII.

But the early death of her husband, Francis II, and the animosity of her mother-in-law, Catherine de Medici, confirmed the fickleness of Mary's fortunes as a 19-year-old widow and drove her back to Scotland in 1561 after an absence of thirteen years, which had seen the overthrow of the Roman Catholic Church and growing division at the very heart of the Scottish political nation. For reasons of policy and for the sake of a more secure future she now moderated her direct claim to the English throne, in the hope that Elizabeth might recognise her as heiress without duress. And she remained uncommitted, likewise, to any specific party or policy in Scotland when she landed at Leith on 19 August 1561, to reclaim her realm. Indeed, Mary had announced in advance to the Scottish Parliament, the so-called 'Estates', that its members were free to establish whatever religious settlement they chose, though her own faith was to remain non-negotiable. She, personally, would adhere to the Church of Rome come what may – and hope, in doing so, to straddle the coming storm unruffled.

In this, however, Mary had not counted upon the influence and bitter hostility of John Knox, the most formidable of all the Calvinist missionaries from Geneva, whom Queen Elizabeth had just transferred from England under safe conduct with the deliberate intention of undermining the Catholic 'party' among the Scottish nobility. A thundering Scots Elijah, who had served as a French galley slave in payment for his Protestant faith, Knox now called upon his countrymen to forsake the false prophets of Baal and, in doing so, declared a single Catholic Mass more awful than the landing of 10,000 foes. Nor, above all, would he spare the sensibilities of Scotland's newly arrived ruler. On the contrary, he would blast her as an idolatrous Jezebel and bewail her very coming. Upon her return, which was marked by a curiously ill-omened mist lasting some five days, 'the very face of heaven', wrote Knox, '... did manifestly speak what comfort was brought into this country with her: to wit, sorrow, dolour, darkness and all impiety.' And sure enough, on the very first Sunday after Mary's landing, a riotous demonstration broke out at Holyrood when Mass was said within the royal chapel for the queen and her predominantly French household.

Yet, for the first four years of her reign, it seemed that Mary might prevail. Though she was no stateswoman and her intelligence was often at the mercy of her passions, she was nevertheless dogged and determined and could often more than hold her own in the tangled world of shifting alliances and affrays that were such a notable feature of Scottish politics. And though there were restive murmurings among 'the godly', her secretary, William Maitland of Lethington, himself a Protestant, was able to argue convincingly that she might well be brought round to 'sweet reasonableness'. Not least of all, there were early signs of common sense and tolerance. Other members of her council, for instance, were also staunchly Protestant and she was prepared, to her credit, to countenance the funding of the reformed church. Moreover, on the occasion of her arrival in Edinburgh, she not only accepted the gift of a vernacular bible and prayer book, but witnessed the burning of a priestly effigy unmoved. Accordingly, a calmer atmosphere soon descended. As one ardent Protestant declared, 'At first I heard men say, "Let us hang the priest", but after that they had been twice or thrice in the Abbey [of Holyrood, at the Queen's Court], all that fervency was past. I think there be some enchantment whereby men are bewitched.' And that enchantment was undoubtedly the queen herself.

Nor was Mary's early success confined to religious affairs, for, in spite of the undoubted glamour of her court, she avoided taxation and largely paid for the regular cost of her household from the income of her French lands. She was visible, too, covering a distance of some 1,200 miles in various progresses across her realm from August 1562 to September 1563: something which demonstrated not only her vitality but also her determination to unite the nobility, the mainstay of her government. Until the very end of her reign, indeed, the backbone of her noble support, for whom John Knox remained a largely marginal figure, would hold steady. And though she was a female, she gained considerable authority from her status as both dowager Queen of France and prospective heir to the throne of England. Almost as important, she was an adult after a prolonged and troubled period of Scottish history in which the throne had been bedevilled by minority government. If, therefore, she married prudently, gained loyal and competent counsel from the men on whom she now relied, and duly circumvented the intrigues of her wily royal cousin south of the Border, the prospects were far from bleak. But her head was proud, her spirit restless and ambition welled within her. The result was the crowning disaster of her marriage to her first cousin, Henry, Lord Darnley.

The son of Matthew Stewart, 4th Earl of Lennox, whose family was closely related to the Scottish royal line, and Margaret Douglas, daughter of Margaret Tudor by her second marriage, Darnley appeared an ideal candidate for Mary's hand in terms of his lineage, boasting a direct claim to the throne of England in his own right. Furthermore, though a Catholic by upbringing, he had toyed with Protestantism and was not associated initially with any dangerous cause either at home or abroad. But, although he was 'accomplished in all courtly exercises' and a gifted lutenist who penned elegant Scottish verses, he was also stupid and treacherous – 'a man of insolent temper' who swiftly alienated most of his potential allies in Scotland, though not, it seems, the queen, who had soon fallen madly in love with the 'fayre, yollye yonge man' and married him on 24 July 1564. Accordingly, when Mary chose the day before the wedding to declare her husband 'King of Scots' – a title which she could not legitimately bestow without the consent of the Scottish Estates – her proclamation was received in stony silence at Mercat Cross by all save the bridegroom's father who offered up a sturdy cry of 'God save his Grace!'

Thereafter, the elements of the final tragedy, which created such an unfavourable start to the life of the future James I, unfolded with a remorseless momentum. Within three weeks of her marriage, Mary's scheming half-brother, the bastard Earl of Moray, whose considerable influence had been threatened by the queen's marriage, came out in open rebellion in the name of the Protestant Kirk, backed by £3,000 from England's Queen Elizabeth. And though he was eventually defeated and driven into exile south of the Border after a chaotic engagement known as the 'Chaseabout Raid', in which a pistol-toting Mary rode in armour and plated cap, 'ever with the foremost' of her troops until 'the most part waxed weary', the price was heavy. For the Queen of Scotland was now placed in a position of open hostility to both Scottish Protestants and her English cousin. Edinburgh, it is true, had ignored Knox's fervent appeals and remained loyal when Moray entered the city in August, but Mary had survived rather than solved her underlying problems, and both her husband's and her own indiscretions would multiply uncontrollably with the mutual antipathy that now exploded between them.

'No woman of spirit', wrote Sir James Melville, 'would make choice of such a man', and whether it was she who first spurned Darnley or he who rejected her remains unknown. But what began as an overwhelming infatuation on Mary's part degenerated within six months into outright and irremediable repulsion, as Darnley cavorted with loose women and, on occasion, behaved with great brutality towards his wife. Refusing absolutely to grant him the Crown Matrimonial, which would have allowed him to rule co-equally and keep the throne in the event of her death, Mary turned increasingly for counsel and consolation to her 'evil-favoured' Italian minion, 'a man of no beauty or shape', and the altogether more dangerous James Hepburn, Earl of Bothwell. While Riccio – 'that great abuser of the commonwealth, that poltroon and vile knave Davie', as Knox graciously dubbed him – flaunted the queen's good offices more and more injudiciously; he courted, of course, the kind of mortal disaster which duly befell him on the night of 9 March 1566. Dragged from the queen's apartments while at supper with her at Holyroodhouse, the helpless secretary was stabbed some fifty-six times within earshot of his horrified mistress. The men responsible included Darnley himself and a motley crew of disgruntled Protestant lords, which numbered the Earl of Morton and his Douglas cronies, the old and dying Lord Ruthven and the exiled Moray, who had been loitering darkly in Newcastle with his fellow rebels awaiting the first available opportunity to conjure trouble.

Whether the intention was also to kill the queen herself or at least encourage her to miscarry from the trauma involved remains uncertain, though some accounts suggest as much. Certainly, Mary appeared to be in danger of miscarrying soon afterwards and the whole event may well have prompted what appears to have been her mental collapse the following year. But if her child was nearly lost and her judgement was to disintegrate catastrophically not long afterwards, for the time being she would show remarkable resources of inner strength and resourcefulness. With the power of Huntly in the Highlands and of Bothwell on the Borders still intact, she could, after all, fight back with every chance of victory and before the bloody night was done her cringing husband, whose very own blade had been left in Riccio's shredded corpse, became so terrified by the possible consequences of his actions that he swiftly deserted his fellow assassins and agreed to take her the 25 miles to the safety of Bothwell's castle at Dunbar. 'Come on! In God's name,' Darnley urged along the way. 'By God's blood, they will murder both you and me if they can catch us … If this baby dies, we can have more.' And when Mary's double-dealing brother, Moray, rode in prudently late next morning, he too was graciously detached from an ill-conceived plot that had so clearly failed to achieve its purposes. A pardon and a cynical guarantee of reinstatement were all that was required.

So there had occurred, even before he was born, the first mortal threat to the future James I of England. Within a week, however, his mother was back in her capital and apparently secure. Already Riccio's murderers were scattered in hiding or in exile and the queen's outward reconciliation with both her husband and Moray was complete. Moreover, for the six months that elapsed from baby James's birth to his christening at Stirling, the surface calm remained intact. Much, if not all, depended upon the child's security, of course, for if he should fall into the hands of the queen's enemies, the pretence of protecting the child would lend a sheen of respectability to any would-be rebel. With this in mind, therefore, James was duly whisked into the guardianship of the Earl of Mar at Stirling Castle when two months old and would remain there for the next twelve years. Mar's family had, in fact, been frequently trusted with similar charges in the past and could claim with some justification to be the hereditary guardians of Scotland's infant royalty, though in this case the boy was largely entrusted to the less than capable hands of a wet nurse named Helena Little. While his father detested him and his mother fought for her political life, Lady Mar, it is true, exercised a genuine, if superficial, tenderness for the child. But Little would remain both everyday overseer of his welfare and a drinker, too, it seems, who is sometimes alleged to have either dropped the prince or neglected an attack of rickets which left him with weakened legs – his right foot 'permanently turned out' – and a shambling, much-mocked walk for the rest of his life.

Yet, at the time of his birth, James's health and appearance left nothing to be desired. Sir Henry Killigrew, the new English ambassador, saw the infant when he was only five days old, and described him as 'a very goodly child'. First, he watched him 'sucking of his nurse' and afterwards saw him 'as good as naked … his head, feet and hands, all to my judgement well proportioned and like to prove a goodly prince'. The new heir could, moreover, even charm his mother's religious rivals, for on the day following his birth, John Spottiswoode was given the privilege by Mary of holding the child, whereupon he fell to his knees, utterly disarmed, and proceeded to play with him, attempting to teach the infant to utter the word 'Amen'.

And while Spottiswoode's request for a Protestant christening was met with resolute silence, there was nothing coy or even remotely restrained about the ceremony that did eventually follow. The child, after all, was of critical national importance and the lingering slur upon his legitimacy made it doubly necessary that his baptism at Stirling Castle, in December 1566, should be suitably splendid. For the few days involved, therefore, and in spite of the fearful strain upon the Crown's meagre resources, the Scottish court would give free vent to its mistress's extravagance and rival the standards of its French counterpart. Though the child's godparents – the King of France, the Queen of England and the Duke of Savoy

– were unable to appear in person, the embassies and gifts they sent with their proxies were nevertheless suitably impressive. The Comte de Brienne arrived with an entourage of thirty gentlemen and a necklace of pearls and rubies, the Earl of Bedford presented a golden font on Queen Elizabeth's behalf, and the Duke of Savoy, represented by Philibert du Croc, the resident French ambassador, delivered a jewelled fan, trimmed with peacock feathers.

At the service itself, which was to prove the last great Catholic ceremony in sixteenth-century Scotland, the prince was borne from his chamber to the chapel by Brienne, who walked between two rows of barons and gentlemen and was followed by a number of Scottish nobles – all Catholics – proudly bearing the baptismal emblems of their religion: the great 'cierge', or ceremonial candle, the salt, the rood, the basin and the laver. Waiting at the font to officiate was another strident symbol of the old religion, Archbishop Hamilton of St Andrews, attended by the Bishops of Dunkeld, Dunblane and Ross in full episcopal regalia – 'such as had not been seen in Scotland these seven years' – and the entire college of the Chapel Royal. At the font, meanwhile, it was the Countess of Argyll, Elizabeth I's representative as godmother, who held the baby while Hamilton christened him 'Charles James' – 'Charles' after Charles IX, the current King of France, and 'James' in recognition of his five Scottish predecessors of that name. In one respect only did the ceremony vary from ancient Catholic practice, since Hamilton was widely known to be stricken by venereal disease and the queen herself refused to have a 'pocky priest' smear his saliva on her son's mouth, as time-honoured custom normally dictated.

Thereafter, the Lord Lyon King of Arms proclaimed the prince's name and titles – Charles James, Prince and Steward of Scotland, Duke of Rothesay, Earl of Carrick, Lord of the Isles and Baron of Renfrew – and the celebrations ensued. There was triumphant music and dancing, a masque devised by the prince's tutor-to-be, George Buchanan, Latin verses, a torchlight procession, two magnificent banquets and spectacular pyrotechnics of 'fire balls, fire spears and all other things pleasant for the sight of man'. For three whole days, in fact, Mary Queen of Scots allowed herself the illusion that the gentility, carefree excitement – and security – of her Gallic past was still intact, as she danced and charmed with her old familiar energy and aplomb, speaking French at every opportunity, while studiously ignoring the rising tide against her.

And much, indeed, was far from well behind the scenes. The fact that the Earl of Bothwell – dressed in shoes of cloth and silver, and a new suit of 'taffetie' provided at the queen's expense – would not venture beyond the door of the Chapel Royal was proof in its own right of Scotland's religious divisions. But the fact that the Countess of Argyll was subsequently forced to do penance by the Protestant Kirk for her participation in the papist ritual spoke no less eloquently of the simmering discord. Even at the junketing which followed, there was ill feeling. According to Sir James Melville, his fellow ambassadors were affronted, because they believed that the English had been treated 'more friendly and familiarly used than they'. But Melville would claim that the English, too, were no less offended when several men dressed as satyrs, 'running before the meat' at one of the banquets, had 'put their hands behind them to their tails, which they wagged with their hands, in such sort as the Englishmen supposed it had been done and devised in derision of them'. Ultimately, it seems, only the Earl of Bedford's timely intervention prevented an ugly incident.

Much more ominous, however, was Darnley's conspicuous absence from all proceedings, for although he was present at Stirling, 'neither was he required nor permitted to come openly' – or so, at least, it seemed. In fact, he had been furnished by his wife with a splendid suit of cloth of gold and had made his own decision to boycott a ceremony at which none were prepared to accept him as king. Indeed, although his father, the Earl of Lennox, continued to scheme on his behalf for the Crown Matrimonial and a genuine share in government, the queen's husband was now treated with open contempt. The Earl of Bedford, for example, was under strict orders to show Darnley 'no more respect in any way than to the simplest gentlemen present' and when one of the Englishman's assistants happened to encounter Darnley by chance, he was severely reprimanded for referring to him as king. Brienne was under similar instructions and, after three attempts had been made to summon du Croc to Darnley's chamber, the ambassador admitted that he had been told 'to have no conference with him', since he 'was in no good correspondence with queen'.

Mary, moreover, did indeed remain at deepest odds with her husband. Soon after the prince's birth, she had taken herself to the pleasant, airy retreat of Craigmillar Castle and lamented to Maitland, Moray, the Earl of Bothwell and others that she could see no 'outgait' from her marriage, since she dared not consider divorce for fear of affecting her son's legitimacy. And now, perhaps, she was more vulnerable to her husband's bitterness, not to mention the suspicions of her nobles and the venomous denunciations of the Kirk. When summoned to the queen's presence on 22 December, for example, du Croc found her 'laid on the bed weeping sore', complaining of 'a grievous pain in her side' and the effects of a riding accident, in which she had 'hurt one of her breasts'. She was, it is true, as resolved as ever to consolidate her power, particularly against her husband, and du Croc recognised as much. 'The injury she received is exceeding great,' he commented, 'and her majesty will never forget it.' But, as Sir James Melville, one of her few entirely faithful servants, observed, 'there were overfew to comfort her'. And it was in these circumstances that Mary turned to the Earl of Bothwell – 'a man', according to Lord John Herries, 'high in his own conceit, proud, vicious and vainglorious above measure, one who would attempt anything out of ambition'.

A reckless and acquisitive adventurer who was widely thought to be 'of no religion' and who attracted women as effortlessly as he discarded them, Bothwell had at first won Mary's trust and swiftly ascended to become the controlling passion of her life, though he possessed none of the good looks or superficial graces which had first made Darnley so attractive to the queen. On the contrary, he was a short, broad man, whom George Buchanan saw fit to describe as a 'purple ape'. He did, however, exhibit a rugged strength, which had burnished his reputation as a fighting Border magnate and which, to an embattled and infatuated female ruler, might well pass for reliability at a time of flux and crisis. Likewise, though he was no courtier or man of letters, he was nevertheless well educated. Indeed, he had acquired an impressive veneer of French culture during his time on the continent as commander of the King of France's Scottish Guards and was known to be widely read. Had he not had these qualities, it is doubtful whether even the impulsive Queen of Scots might have become such a slave to her own passions and determined to 'go with him to the end of the world in a white petticoat'.

Yet it was undoubtedly as a man of action – the bold, rock-like, canny and decisive manipulator of men and events – that Bothwell made his mark upon Mary. He was, it is true, without scruple, but he was also without fear – 'a rash and hazardous young man' in the words of Sir Nicholas Throckmorton, who scorned both the spiteful effeminacy of her husband and slippery double-dealing of a Maitland or a Moray. And he had proven his mettle already when his quick thinking and notorious private army had plucked the queen from disaster in the aftermath of Riccio's assassination. That Bothwell, who was so dismissive of all convention, should have had so many enemies was something that Mary might certainly have considered before rashly throwing in her lot with him. But that such a character would manage to exercise so overpowering an influence over so vulnerable and notoriously impressionable a ruler is not nearly as surprising as is often assumed.

With or without Bothwell, however, the noose was tightening rapidly for Darnley. On 24 December, only a week after his son's christening, Mary's panic-stricken husband learned that his wife had pardoned the survivors of Riccio's murderers, whom he had blatantly betrayed. At least half the nobility of Scotland were now slavering for his blood, and by this time, too, the queen had certainly been shown at least one of the 'bands' that Darnley had signed with the assassins. Knowing now that his feeble pretence of acting against his wife's Italian favourite on a blind and passionate impulse could no longer be sustained, Darnley made at once for his father's house in Glasgow, then a small village on the River Clyde, and the hoped-for safety of Lennox territory, while the atmosphere all around thickened with plots and ugly whispers.

Back at Craigmillar Castle before Christmas, Maitland of Lethington had already assured Mary that a 'mean', not involving divorce, might be devised to rid her of her husband both neatly and without prejudice to her son's legitimacy and that Moray, who was a 'little less scrupulous for a Protestant than your Grace is for ane Papist', would 'look through his fingers thereto'. No specific decision, it seems, was actually taken at that time, but in that same month Mary managed to restore Archbishop Hamilton's authority to pronounce decrees of divorce by nullity, whether for her own marriage or perhaps Bothwell's. At the same time, there were unsettling reports – originating with William Hiegait, town clerk of Glasgow – of a counterplot by Darnley and his father to kidnap the baby prince from Stirling.

By now, as matters reached a climax, Darnley was convalescing at Glasgow after an attack of either syphilis or smallpox, which had overtaken him during his flight. Blue blisters had broken out upon him and he was said to be 'in very great pain and dolour in every part of his body'. Then, at his father's home, Lord Herries tells us, 'his hair fell off'. But this did not, it seems, impair either his gall or his libido. On 14 January, Mary's request to visit him had been met by a rude verbal answer and when Mary arrived nonetheless a week later, her husband's main concern appears to have been that she should restore his conjugal rights as soon as he was fit once more. Already, more than a year earlier, according to the records of Catherine Maxwell Stuart, 21st Lady of Traquair, Darnley had disgraced himself when Mary excused herself from accompanying him on a hunting expedition for fear that she might again be pregnant. 'What,' retorted Darnley, 'ought we not to work a mare well when she is in foal?' And now, with his disfigured face still covered by a taffeta mask, the husband's prurience remained as undiminished as his bile.

Once more, however, Mary was quick to re-establish her ascendancy and Darnley – notwithstanding the sensible forebodings of his father – swiftly conceded to be brought to Edinburgh in a litter that his wife had brought with her, so that he should not be 'far from her son'. Upon his arrival, he was eventually lodged outside the walls, not only 'in a solitaire place at the outmost part of ye town' but in a squalid, ruined neighbourhood approached by a street known as 'Thieves' Row', where there lay a small and wholly unsuitable four-roomed house known as the Kirk o' Field, in which the queen had nonetheless established a magnificent bedroom for herself on the ground floor. It was there on the night of Sunday, 9 February that Mary visited her husband, by torchlight, for what would prove to be the last time. Leaving shortly before midnight to attend a masque in honour of the wedding of her French servant, Bastien Pagès, she may well have done so in full knowledge of what would shortly ensue. For Bothwell and his most trusted retainers had secretly packed the basement of Kirk o' Field with gunpowder, which was duly ignited two hours later. Though Darnley's naked corpse showed no marks of the explosion when it was later discovered in the garden, along with that of one of his pages, there was no doubting that the 'deid was foully done', and there was no doubt either of the gravity of what followed. If, wrote one contemporary, Darnley 'had not been cruelly vyrriet [strangled], after he fell out of the aire, with his own garters, he had leived'. And in subsequently flaunting all serious pretence at justice, Mary duly incurred outright moral disgrace, not only in Scotland but throughout Catholic and Protestant Europe alike.

Though Bothwell underwent a spurious form of trial before fifteen hand-picked peers and lairds, which conveniently foundered for lack of evidence, since Darnley's father dare not enter Edinburgh while 6,000 of Bothwell's armed Borderers remained in firm control, Mary herself made no effort to clear her name, and then, to crown all, carelessly embarked upon her ultimate folly. With the capital awash with denunciations of her lover and the abuse of Protestant preachers ringing in every kirk, the queen remained impervious. She remained equally unmoved, too, by Moray's final warning in early April that she was courting disaster 'because of the great trouble seeming to come to the realm'. Moray himself had already cannily distanced himself from Darnley's murder by removing to Fife on the actual night of the deed and now he carefully avoided any involvement in what he clearly perceived to be his half-sister's imminent ruin, leaving for England on ostensibly amicable terms with Bothwell, to wait as he had before for events to turn decisively in his favour.

Nor was Moray's reasoning anything other than sound. On 17 April, as the ultimate insult to the whole of Scotland, Bothwell was selected to carry the crown and sceptre before the queen at the opening of Parliament. And upon the very evening that the Estates dissolved, he finally revealed the full scale of his ambitions for the first time. Appropriately, perhaps, it was at Ainslie's Tavern, which had been thoroughly surrounded by his armed retainers, that Bothwell brazenly forced his noble guests to pledge their belief in his innocence and commit their support to a 'marriage betwixt her Highness and the said noble lord'. The pledge, of course, was as empty as the dreams of the man who imposed it, but the die had now been irretrievably cast – both for him and for the queen who had given herself over to him so entirely.

Around this time, Mary would pen for Bothwell a series of sonnets and appallingly indiscreet love letters, but even now, it seems, her passion for the frenzied earl had not entirely blinded her to the interests of her infant son. Accordingly, on 21 April, she rode to Stirling to pay what was to prove her last visit to the prince. Such, however, was the general collapse of her credibility that not even that most loyal of servants, the Earl of Mar, would surrender the child to her. On the contrary, knowing full well that the whole future of the realm was inextricably tied to the boy's well-being, Mar would only allow the queen to bring two of her ladies with her into the castle, and her two-day sojourn brought little consolation to all concerned. It was even rumoured, albeit wholly improbably, that, at one point during her stay, Mary had attempted to coax the child to stop screaming by the offer of an apple, which he brusquely rejected. Whereupon the apple was subsequently eaten by a greyhound bitch that promptly swelled and died.

Certainly, Mary left Stirling without the prince and swiftly succumbed to the final episode in her disgrace and downfall, for at Linlithgow – with almost farcically suspicious ease – she was duly 'kidnapped' by Bothwell and carried off to Dunbar. Deluded by the pledges delivered at Ainslie's Tavern, the earl seems to have fondly imagined that the majority of Scots nobles would actually condone his action, while Mary, if her subsequent behaviour is any guide, was apparently too besotted to care. If, moreover, her claim that she had been raped carried no conviction, her decision to marry the perpetrator on the grounds that she had been irretrievably compromised by her violation, was the ultimate act of political madness. Nevertheless, during the three weeks that Mary remained at Dunbar, Archbishop Hamilton rushed through an annulment of Bothwell's marriage to his current wife, Janet Gordon, and the new union of queen and earl was formally sanctioned. Though even Mary would not dare to grant her husband the title of king, he was nevertheless created Duke of Orkney and Shetland, which caused sufficient scandal it its own right. And as if to seal the scale of Mary's current derangement, the wedding itself was duly performed according to Protestant rites.

Soon, moreover, it was patently clear that Mary had not only abandoned her son but placed his life in dire peril, since Bothwell determined at once to gain possession of the prince, and the mother was ready to comply. 'She intends,' wrote one Scots lord, 'to take the prince out of Mar's hands and put him in Bothwell's keeping, who murdered his father.' In the meantime, for the next three weeks the couple honeymooned unhappily at Borthwick Castle in a state of near siege, while the earl continued to ply Janet Gordon with letters and steadily honed his plans for ultimate mastery. If the prince should fall into his new stepfather's hands, it was generally acknowledged that the earl would 'make him away … as well to advance his own succession, as to cut off the innocent child, who in all probability would one day revenge his father's death' – all of which seems to have escaped Mary herself. Likewise, as the remnants of her supporters, including Lethington, deserted the rapidly sinking ship and the situation in the capital grew steadily more menacing, the queen continued to hope against hope that events might yet turn decisively in her favour. Accordingly, by the time that she and Bothwell slipped away once more to Dunbar, everything already depended upon a final trial of arms, which would not be long in coming.

The Protestant 'Lords of the Congregation', who had all been implicated more or less in Bothwell's assassination of Darnley, had already committed themselves by bond to protect Prince James. And now these 'True Lords' had the brazen effrontery to march against the mastermind of the murder under a banner depicting the victim's naked body. Yet Mary remained undaunted and on 15 June, both she and Bothwell moved out to confront their enemies at Carberry Hill. For most of the day, in fact, the two armies faced each other, very reluctant to fight, and while the uneasy posturing continued, there were last minute attempts at mediation by the French ambassador, du Croc. Mary, however, refused to accept the condition that she leave Bothwell and refused, too, to concede to her husband's outlandish request that matters be resolved by single combat with any one of the opposition lords. In the event, as Bothwell frothed and she in her turn clutched at any straw to hand, the royal army slowly disintegrated.

Surrendering herself, therefore, in return for an agreement that her hated spouse be permitted to escape with his life, Mary now encountered at first hand the full wrath and resentment that her actions had stirred. Kirkaldy of Grange, a brave and chivalrous soldier who negotiated the queen's final surrender, had guaranteed her respectful treatment, but he had not counted upon the venom and violence of the Edinburgh mob. Indeed, such was Mary's reception that Kirkaldy did well to keep her alive and eventually joined her cause in disgust at her enemies' hysteria. Surrounded by the victorious Protestant lords, whom she continually cursed and threatened, she rode into the capital for the last time to cries that she be lynched or drowned as an 'adulteress' and 'murderess', and further howls that 'the whore' be burnt. Utterly distraught at the full, brutal shock of her new condition, she was ultimately detained at the provost's lodging in the High Street, where she made a fleeting appearance at a window to issue a final appeal for aid. With her hair loosened and her clothes indecently torn and disordered, she appeared for the moment to have lost her reason.

Even at this critical pitch of despair, however, Mary nevertheless represented an ongoing threat to her captors. With Huntly in arms in the north and the Hamiltons secure in the west, the future was still uncertain, and the queen, for all her faults and wretchedness, was queen nonetheless. At 26, moreover, she was likely to be a danger for many years to come. Nor was this all, for open rebellion by the Protestant lords against the anointed queen could still be guaranteed to raise a majority of Scots against them. Elizabeth, too – though she wrote privately to her cousin sharply condemning both the murder of Darnley and subsequent marriage to

Bothwell — left no doubt that she would declare war, if there was any attempt to stage a deposition. The lords' ostensible target had therefore always been Bothwell, and when Mary was now closely confined among the Fifeshire bogs surrounding the island castle of Loch Leven, it was firmly emphasised that she was undergoing 'seclusion' rather than imprisonment.

Yet her treatment, predictably, left much to be desired. A month or so after her arrival, she capped her misfortunes by miscarrying Bothwell's twins, though this, it must be said, did nothing to soften her captors. Her gaoler, for instance, was Sir William Douglas — a 'depender' of the Earl of Morton, one of Riccio's assassins, and her disaffected half-brother, Moray — who did little to protect her from the insults and ill-treatment, which were blatantly intended to break her spirit. Lord Ruthven, on the other hand, the son of the corpse-like old murderer whose appearance had so alarmed the queen on the night of Riccio's murder, oppressed her with his lust, while Moray's mother, Lady Margaret Erskine, also took every opportunity to insult her. Worse still, perhaps, Lord Lindsay, a brutal and unscrupulous bully, subjected her to outright threats of physical force. Sick, imperilled and wholly beyond the reach of any 'friends', Mary was therefore hardly equipped to withstand these ultimate assaults upon her fast-waning reserves of resilience and utterly shredded integrity.

Ironically, however, the Queen of Scots was once more undone by her own indiscretions. Only one week after her capture, in fact, the silver casket in which Bothwell had kept her secret sonnets and letters to him, as well as her pledge to marry him, was duly discovered. The earl, it seems, had left them behind for safekeeping in Edinburgh, only to furnish his enemies with the most explosive of weapons. Indeed, irrespective of any tampering that may have occurred, the 'Casket Letters' not only stirred John Knox and his fellow pulpiteers to new heights of invective by their 'coarseness' but also placed Mary in alarmingly real danger for her life, for according to an ancient Scottish statute, which had been recently revived, adultery was not only a capital offence but punishable in the case of females by burning.

All sober, moderate opinion now accepted, in any case, that 'a Queen hath no more liberty or privilege to commit adultery or murder than any other private person, either by God's laws, or the laws of the realm', and even Elizabeth's heartfelt appeals on Mary's behalf had already lost much of their force. In these circumstances, therefore, the only remaining option, as the queen grudgingly accepted, was abdication and a subsequent minority government on behalf of her son. Finally broken by her captors' naked intimidation and threats that she would be brought to trial and execution, Mary duly provided the lords and, above all, her half-brother with the documents they required. Though Moray was not present at the final disgraceful scene, where threats of throat-cutting, drowning in the loch and even marooning on a desert island were aimed at his half-sister, he was duly authorised to assume the regency, with the assistance of a commission of seven noblemen, 'in caisse', it was almost laughably claimed, 'he should refuse to exercise ye same alone'. No such refusal was, of course, forthcoming and on 24 July, Mary duly acquiesced. 'When God shall set me at liberty again,' she declared through bitter damned-up tears, 'I shall not abide these, for it is done against my will.' But her reign as Queen of Scots was over forever.

# 2 ✧ King and Pawn

'I was alone, without father or mother, brother or sister, King of
this realm, and heir apparent of England.'

*James VI*, 1589

On 29 July 1567, in the church of the Holy Rude at Stirling, on a craggy
hillside rising to the castle, Prince James was crowned King of Scots,
the sixth ruler of his kingdom to have governed with that name. With its
commanding view over the River Forth and the Ochils, no fortress in
Scotland boasted a more dramatic setting. A few miles to the north, rising
sharply from the plain, lay the 'Highland Line', one of the great geological
faults to which Scotland owed not only its shape but its history, while
to the north-west spread the expanse of bog land, across which meagre,
sluggish streams ambled off to supply the river Forth. To the south, on the
other hand, spread the humbler ridge of the Campsies, though these, too,
reflected the wall of the Highlands and the more imposing peaks of Ben
Ledi, Ben Vorlich and Ben Lomond. All in all, no spot resonated more
with Scottish prowess, Scottish pride and Scottish royalty. Bannockburn,
the most decisive battle in Scottish history had been fought for the castle
and the vital bridge below it, and both James IV and James V had left
their indelible mark upon the place, improving its buildings and
rendering the six main apartments of its royal palace comparable to any
in Northern Europe. Plainly, then, Stirling was the benchmark for high
Scottish culture and the clearest possible statement of Stuart legitimacy
and permanence – the traditional home of the current dynasty, and the

right and natural starting-point, by any standards, for the reign of the 13-month-old James VI.

Yet the new king's inauguration was a mean and meagre affair, tainted by circumstance and shrouded in fears for the future. Staged only five days after the deposition of the former queen, it was the worst-attended coronation in Scottish history. In the opinion of one of her spokesmen, indeed, only seven lords, no more than a tenth of the Scots nobility, were present, and even the English ambassador – a fervent Protestant – was obliged to boycott the ceremony, since it was the act of an illegal regime that had challenged the sovereignty of the Crown and threatened the established order of things. As such, the Elizabethan government, which was still engaged in establishing its own respectability, could not afford to become entangled with it. This was not the only oddity, since the ceremony also involved a change of name for the monarch – something that had occurred only once before in the whole of Scotland's past. For, as a result of its association with the French king, the king's baptismal name of 'Charles James' was suitably clipped and the new monarch would henceforth be known only as James VI.

Crowned, then, not in the castle's Chapel Royal, where his mother had been enthroned in 1542, but in the altogether humbler setting of the burgh's parish church, the infant king found himself at the centre of a ritual which fully reflected the tensions existing in his realm. Though anointed in the style of previous Scottish monarchs, there were neither candles nor copes nor incense on hand. Nor were there fanfares or heralds to proclaim the new king in what was consciously presented as an aggressively Protestant reaction against the former regime. Latin, too, was carefully avoided; instead, all prayers were 'in the English tongue'. It was not without irony either, of course, that the infant ruler's crown was placed upon his head by Robert Stewart, the former Catholic Bishop of Orkney, who had last appeared in public to marry the child's mother to Bothwell. And it was an equally curious footnote to Queen Mary's reign that the subsequent sermon should have been preached by none other than John Knox. Taking his text from 2 Chronicles 23: 20–21, he declaimed with characteristic candour and at typical length upon the coronation of the child king Joash, whose mother, Queen Athaliah, had rent her clothes and cried 'treason, treason', before being taken out and slain by the sword.

To seal the transformation at the heart of government, however, it was none other than James Douglas, 4th Earl of Morton, who read aloud the new king's coronation oath. Arguably the most crooked and treacherous of the whole shifty crew that had brought Mary Stuart to her ruin, it was red-headed Douglas who had held Holyrood for Riccio's murderers and signed Bothwell's bond against Darnley. And now it was he who pledged the king not only to maintain the 'lovable laws and constitutions received in this realm' and 'to rule in the faith, fear and love of God', but also to 'root out all heretics and enemies to the true worship of God that shall be convicted by the true Kirk of God of the aforesaid crimes'. For his efforts on the new government's behalf, Morton was duly nominated as chancellor in the Regency Council that now assumed power.

It was composed, said George Buchanan, the king's future tutor, 'of nourishers of theft and raisers of rebellion', who were characterised by 'insatiable greediness' and 'intolerable arrogance'. 'For the most part', it seems, its members were men 'without faith in promises, pity to the inferior, or obedience to the superior'. 'In peace', moreover, they were 'desirous of trouble, in war thirsty of blood'. But the power of Buchanan's beloved Kirk depended on these men. Like him, his fellow preachers desired that the whole government of Scotland, civil and ecclesiastical, be subordinated to their charge, while the nobles were determined merely to maintain power and wealth for themselves. To say, therefore, that both parties were uneasy allies is an understatement of some magnitude, though their mutual dependence was unavoidable, and both were also driven to an equally distasteful dependence upon England, which most thinking Scotsmen had long been struggling to avoid. It was English intrigues and often English subsidies, after all, which had assisted the present clique of Protestant lords into power and now, as conservative and moderate opinion alike recoiled from the implications of this dependence, it was English influence that would hold their opponents at bay.

In the meantime, the infant boy who might one day serve to guarantee good order and government was entrusted once more, at Regent Moray's behest, to the Earl and Countess of Mar. Formally appointed on 22 August, Moray had chosen the obvious candidates, since the earl in particular was a nobleman of the highest order, respected by both his own party and its opponents. And until his death in 1572, when his role was assumed by his brother, Sir Alexander Erskine, his conduct was exemplary. Like the earl himself, moreover, Erskine was another genuinely benign influence – 'a nobleman', observed Sir James Melville, 'of a true, gentle nature, well-loved and liked of every man for his good qualities and great discretion, in no wise factious or envious' – though in the countess, King James was not perhaps so fortunate. For while, as we have seen, she played his foster-mother with due conviction initially, referring to him always as 'the Lord's Annointed' and occasionally objecting when he was beaten, she was nevertheless a stern enough governess in her own right – especially after her husband's death when she continued in her post and held the king in 'great awe' of her. Rather more worryingly, she also continued to delegate too much of the child's everyday care to Little, his tippling wet nurse. Though hardly the fount of all objectivity, Knox described Lady Mar as 'a very Jesabell' and a 'sweet titbit for the Devil's mouth', and if his intention on this occasion was to highlight her connection to the former queen, it was true, nevertheless, that the king's own feelings about the countess were always likely to have been mixed.

Certainly, the provision of the royal household, though less than extravagant, was adequate to its purposes. Four young women, for example, were employed to rock the king in his cradle – perhaps the wooden cradle of Traquair which is traditionally supposed to have been his – and there were also three gentlemen of the bedchamber, two women to tend the king's clothes and two musicians, Thomas and Robert Hudson, though James himself exhibited no ear for music in later life. And while the Master of the Household, Cunningham of Drumwhassel, was not only Moray's cousin but in Melville's view an ambitious and greedy man, even he appears to have devoted himself effectively enough to the day-to-day management of the king's domestic arrangements, which changed little throughout his early childhood. Food and drink were ample, with an allowance for the 'King's own mouth daily' of two and a half loaves of bread, three pints of ale and two capons. And though most of the former queen's furniture lay idle at Holyrood, three fine tapestries were nevertheless brought to Stirling for her son's comfort, notwithstanding the fact that his bed in the Prince's Tower was a gloomy contrivance of black damask, with ruff, head-piece and pillows also fringed in the same colour.

In the king's bedchamber too, fittingly enough, hung a portrait of his ill-fated grandfather, James V, who, along with boy's great-grandfather, James IV, had played such a key role in fashioning Stirling Castle. The latter, arguably the most heroic of all Scotland's kings, had been killed in battle by the English at Flodden Field in 1513, and although the boy would show no such prowess whatsoever, his great-grandfather's legacy was all around him – not only in stirring tales of his martial deeds and tragedy, but in the very walls of the castle itself. Known as 'James of the Iron Belt', the fallen hero of Flodden had, for instance, erected the great defensive bulwark across the main approach to the fortress around 1500 and had probably completed the royal courtyard, now dubbed the Upper Square, about the same date. James V, meanwhile, influenced by two marriages to French princesses and his own sojourn in France, imported French masons and built the palace on the south side of the square and the Great Hall. In particular, the sculpted figures of the Devil and King James himself on the south walls displayed the same French influence that had characterised the whole reign, so that while the new monarch would be reared in expectation of the English Crown, he was left in no doubt either that his roots were plainly Scottish and that the realm across the Border was both foreign and fierce.

Before he could unify his God-given kingdoms, however, James would have to grow and learn – two tasks that, in his case, were far from carefree. Before he had reached his fourth birthday, in fact, two scholars were appointed by the government to supervise the king's education: the formidable George Buchanan, poet, humanist, historian and unyielding taskmaster, and the altogether more temperate Peter Young, a young man fresh from his studies under Theodore Beza at Geneva. Learned, gifted, and widely lauded for his accomplishments, Buchanan, on the one hand, had lengthy experience as a tutor to some of the best families in both Scotland and France, and had once been the instructor of none other than the great French essayist, Michel de Montaigne. But he was ill-suited to deal with a child scarcely out of the nursery – and especially one as nervous, excitable and overstrung as James would prove to be. Melville, for his part, described the elderly master aptly enough as a 'stoic philosopher' and fully acknowledged the 'notable qualities' of his learning and knowledge which were 'much made account of in other nations'. There was ample recognition, too, that Buchanan was 'pleasant in company, rehearsing on all occasions moralities, short and forceful', 'of good religion for a poet' and a man of commendable frankness and honesty – someone who, as Melville put it, 'looked not far before the hand'. Yet Buchanan was also, we are told, 'easily abused', 'factious in his old days' and, worse still, 'extremely vengeable against any man that offended him which was his greatest fault'.

Nor, it seems, was Buchanan especially inclined to forgiveness of his royal pupil. More than sixty years older than the king and in declining health, he found his new role, at times, an irritating distraction from his own more serious studies – even though he had lobbied for the post before it became available – and in consequence occasionally vented his frustrations upon both the boy and reputation of his mother. Curiously enough, the old man had once taught Mary herself and at that time commended 'the excellency of her mien, the delicacy of her beauty, the vigour of her blooming years, all joined in her recommendation'. But he had become Protestant in 1563 and thenceforth the schoolmaster's antipathy to that 'bludy woman and poisoning witch' knew no bounds. Indeed, his scornful *Detectio Mariae Regina Scotorum* not only rivalled Knox's vitriol but possibly surpassed it, ranking perhaps as one of the most powerful pieces of rhetorical invective of its day. She, who had formerly been the object of Buchanan's elegiacs, a woman 'of nobility rarer than all her kindred', now suddenly degenerated, into the caricature reviled in every Edinburgh gutter – the personification of 'intemperate authority' whose 'immeasurable but mad love' for Bothwell had brought her to shame. In due course, indeed, it would be Buchanan who confirmed Mary's handwriting in the Casket Letters before both the Scottish Parliament and Elizabeth I's court at Westminster.

And now James would be exposed to the full blast of Buchanan's sulphurous wrath. On one particularly notorious occasion, a boyish tussle had, it seems, broken out between the king and his playmate, John Erskine, son of the Earl of Mar, over possession of a sparrow, in the course of which the unfortunate bird met its end. When Buchanan heard of the incident, he went to work with characteristic gusto, slapping the king's ear and adding, with all his usual venom for the former queen, that the boy himself was 'a true bird of that bludy nest'. On another occasion, when James and Mar were somewhat noisy at their play and the master was at his books, he warned the king that 'if he did not hold his peace, he would whip his breach'. The result was a not altogether unprecedented attempt at cheek from the king and the due delivery – 'in a passion' – of a thorough thrashing by Buchanan. When, moreover, the Countess of Mar came to the boy's rescue, rebuking the old man for laying hands upon 'the Lord's Annointed', Buchanan responded in style. 'Madam,' he replied with the kind of cudgel subtlety that could be guaranteed to thwart his protagonist in mid-flight, 'I have whipped his arse, you may kiss it if you please.'

It is small wonder, then, that James should have recalled his tutor in considerable awe, if not outright fear. Years later, at the age of 53, he would tell one of his officials 'that he trembled at his approach, it minded him so of his pedagogue'. Nor, perhaps, is it altogether surprising that the king should have gone on to challenge so roundly some of the political lessons that his tutor had been at such pains to instil. In 1579, when James was 13, not far from the time of his personal rule in Scotland, Buchanan wrote *De Jure Regni apud Scotis* (*The Rights of the Crown in Scotland*). Dedicated to James, the book emphasised that kings should be lovers and models of piety, bringing dread to the bad and delight to the good. But though, in Buchanan's view, a king was the father of his people, he was also accountable to them, existing for their benefit rather than vice versa. Kings, he asserted, were bound by the power which first made them kings, which was not God, but the people. More radically still, Buchanan also taught that it was desirable not only to resist tyrants, but to punish them. Clearly, James's later claims that kings were God's lieutenants on earth would have appalled his former teacher. And it was not insignificant, of course, that, years later in 1584, when James had achieved his majority, Buchanan's work was duly condemned at the king's behest by Scotland's Parliament.

Yet James would also boast, with full justification, of his training under a teacher of such renown. In 1603, for instance, he told Nicolo Molin, the Venetian ambassador, how his tutor had instructed him in the excellence of Venice's constitution, and when, as King of England, an English scholar praised the elegance of his Latin, James was quick to acknowledge his debt: 'All the world knows that my master, Mr George Buchanan, was a great master in that faculty,' he observed. 'I follow his pronunciation both of the Latin and the Greek, and am sorry that my people of England do not the like; for certainly their pronunciation utterly spoils the grace of these two learned languages.' Nor, surprisingly enough, was James entirely unforgiving of his former tutor's temperament. 'If the man hath burst out here and there into excess or speech of bad temper, that must be imputed to the violence of his humour and heat of his spirit, not in any wise to the rules of true religion rightly by him conceived.' Buchanan's character was, after all, a curious mixture of opposing qualities, as James's backchat prior to his most notorious beating and the master's subsequent encounter with Lady Mar clearly confirm. Like other men of intellect before and after, in fact, George Buchanan was both humane and vindictive, mirthful and morose, cultured and coarse, full of prejudice, but above all fond of truth.

And there was in any case, of course, the mollifying influence of the king's other tutor to compensate for the older man's more pitiless approach. Born in 1544, the gentle, lovable, 'wise and sharp' Peter Young believed in praise and encouragement rather than the rod as the foundation of learning, and was probably James's first real friend, bringing a note of genuine humanity to his childhood that the pupil would not, it seems, forget. Young, indeed, would remain about the king's person to the end of the reign, by which time he had served as chief almoner, performed ambassadorial roles in Denmark, and been endowed with a long string of ecclesiastical preferments which eventually left him the Master of the Hospital of St Cross in Winchester. Whether, of course, Buchanan consciously encouraged the contrast in teaching styles with his younger protégé is unknown but, as Young's superior, he made no effort to discourage it and spoke of his colleague in the highest terms in his *Epistolae*. Given Buchanan's advancing years, moreover, Young may well in any case have undertaken the major share of teaching duties.

There were, it is true, notable gaps in the king's upbringing. Not least of all, James ate and drank carelessly, making slovenliness almost a virtue at a time when courtly graces were widely considered the hallmark of true nobility. Doubtless, Buchanan regarded such refinements as largely unimportant and, from some perspectives, even reprobate. A king's main business, after all, was simply to rule and to rule simply at that. Dour democrat that he was, the old man frequently reminded his pupil how affectation as well as flattery were loathsome vices, in much the same way that titles – 'majesties', 'lordships' and 'excellencies' – served only to erect an artificial barrier between rulers and ruled. More worryingly, however, the very rigour of the king's education propagated at least two long-term side effects, for while the iron self-discipline of scholarship and the equally rigid spiritual constraints of Calvinism were intended to fashion a careful and conscientious ruler, the adult James was only truly diligent under compulsion and never consistently so. Nor, for that matter, could his tutors eliminate the devious secret side that became compounded by the broader circumstances of a childhood beset by political insecurity and physical threats.

Of considerable significance, too, was the fact that Buchanan's own particular brand of pedantry and intellectual intolerance seems to have left another lasting imprint. Certainly, the young king was not without wit – albeit of the more sardonic kind often characteristic of the highly educated. Indeed, under Buchanan's guidance, the exercise of a quicksilver, caustic tongue may well have taken the form of an ancillary education in its own right. But what might have developed into an accomplishment of sorts became in this particular case a clear-cut vice. On one occasion, the tedium and imperfection of the young Earl of Mar's French became too much for James, who detonated with all the vigour worthy of his tutor.

'I have not understood a single word you have said,' declared James, adding, 'what the Lord Regent has said of you seems to be true, that your French is nothing and your Scots little better.' And when Peter Young reminded the king that he should never lose his temper, he was met with more irritation still. 'Then,' came the response, 'I should not wear the lion on my arms but rather a sheep.'

Equally regrettable, in some respects, were the effects of James's comparative isolation from women. Apart from the Countess of Mar, whom he affectionately dubbed his 'Lady Ninny', there was little female company on hand to leaven his childhood – something, it is sometimes said, that may partly explain his later, almost tragic, yearning for affection. On one occasion we hear of him thanking the countess for a gift of fruit, but apart from Lady Mar herself, who was also capable of a distinct hardness in her own right, the influence of female society was little in evidence. Significantly or otherwise, Young mentions a game of *trou-madame* in which the king made a small wager with several young ladies but, upon losing, brusquely displayed his irritation at having to pay the forfeit. It is tempting to speculate, too, that Buchanan's own influence as a hard-bitten bachelor and outright misogynist may well have reinforced the strength of James's opinions in later life about the role and station of women.

Meanwhile, the distance – both physical and emotional – between the king and his mother remained considerable. It was not only Buchanan, for instance, who impressed upon James that his sole surviving parent was a murderer and adulteress, and it was not long either before the boy realised that this alone was why he was king. More disconcertingly still, perhaps, he soon appreciated that from her exile in England, Mary remained a threat to his status, since she had never accepted her enforced abdication and had therefore never recognised her son's assumption of the Crown. And while she hoped for reconciliation with the child she still considered her heir, the gulf between them would be carefully maintained by those controlling James's destiny. After Mary attempted, for instance, 'to remind him of his afflicted mother', she explained to Queen Elizabeth in England how her efforts had been ruthlessly suppressed. She had sent him when he was 2 years old an ABC and a pony complete with saddle and bridle, but because the affectionate letter that went with them was addressed to 'my dear son, James Charles, Prince of Scotland' rather than to the king, they were never given to him. Two years later, moreover, the Scottish Parliament formally decreed at Stirling that there should be no contact between James and his mother, except through the Council. No feeling of tenderness or pity was therefore ever forthcoming from the boy until Mary's head was finally severed from her shoulders, at which point he would denounce the Casket Letters as forgeries and bemoan the fate of 'that poor lady, my mother'.

In the meantime, however, the king would continue to persevere in his studies, which, in spite of any shortcomings and for all their rigours, were not without considerable virtues. 'First in the morning', Young tells us, 'he sought guidance in prayer, since God Almighty bestows favour and success upon all studies', and having been 'cleansed through prayer and having propitiated the Deity', James then devoted himself to Greek, practising the rules of grammar and reading either from the New Testament or Isocrates, or from the apophthegms of Plutarch. But this, in fact, was still only the start of his daily programme, since breakfast was followed by readings in Latin, either from Livy, Justin, Cicero, or from Scottish or foreign history. After which, with dinner over, he devoted himself to composition, before spending the remainder of the afternoon, if time permitted, upon arithmetic or cosmography – which included geography or astronomy – or dialectics or rhetoric. As a result, the young king was soon able not only to compose both competent verse and accurate pithy prose in English, French and Latin, but to hold his own in argument against many much older men – especially when discourse in Latin was involved.

In the process, however, James acquired a love for rigorous logic and incontrovertible argument that would later smack of dogmatism and pedantry – and nowhere more so, perhaps, than in the field of religion where his familiarity with Calvinist theology and methods of reasoning left an indelible mark upon his general outlook. Calvin's whole system, in fact, was based upon the notion of absolute truths derived by remorseless reasoning from infallible premises. This narrow, rigid and dialectical method, with its cast-iron approach to divinity, leaving no room for compromise, appealed to James strongly, for though he was fully versed in the humanism of the Renaissance, he was also immersed in the thinking of the medieval schoolmen and, above all, the power of the syllogism. Having thereby derived his thoughts on any matter, whether theological or secular, James could be guaranteed to hold firm against all comers, especially upon the subject of Catholicism.

At the age of 11, indeed, the king was declaiming against the Catholic controversialist Archibald Hamilton with all the vigour that one would expect from a pupil of both Buchanan and Young. The king 'marvelled', it seems, that Hamilton's most well-known book 'should be put forth by a Scotsman'. 'I love him not so evil because he is a Hamilton,' he concluded, 'as that I do because he is an apostate.' On another occasion, in speaking of the papal claim to the keys of heaven and hell, James readily quoted St Luke (6:52): 'Woe unto you lawyers: for ye have taken away the key of knowledge: ye enter not in yourselves, and them that are entering ye hindered.' He was more than capable, too, of exercising his learning at the expense of the younger of his tutors. When Peter Young punished a small fault by forbidding the king to read the lesson for the day, which was the 119th Psalm, James quoted the text very aptly: 'Wherewith shall a young man cleanse his way?' It was this kind of 'smartness' and yen for having the last word, too, which may well have encouraged some of Buchanan's more brutal responses to his precocious young charge.

Yet James's supreme confidence in his own opinions was by no means entirely vacuous, for rarely has a youthful royal mind been so successfully filled by his tutors, and his 'great towardness in learning' was widely and rightly acknowledged by those well placed to judge. 'At this early age,' Buchanan told the king when he was 16, 'you have pursued the history of almost every nation and have committed many of them to memory.' And a passing comment scribbled in one of James's copy-books – possibly a flash of penetrating protest or at least exasperation – bears ample testimony to the intensity of his education. 'They gar me speik Latin,' he observed, 'ar I could speik Scotis'. Furthermore, the Protestant minister James Melville tells in his autobiography how he encountered the king at Stirling in 1574 when he was only 8 and found him 'the sweetest sight in Europe that day for strange and extraordinary gifts of wit, judgement, memory and language'. 'I heard him discourse,' said Melville, 'walking up and down in the old Lady Mar's hand, of knowledge and ignorance, to my great marvel and astonishment.' Nor did Sir Henry Killigrew, who observed the king regularly, harbour any doubt whatsoever about the boy's considerable ability. Writing to Queen Elizabeth in the same year that Melville recorded his observations, the ambassador informed the queen how the boy 'was able extempore … to read a chapter of the Bible out of Latin into French, and out of French after into English, so well as few men could have added anything to his translation.' The boy was, concluded Killigrew, 'a Prince sure of great hope, if God give him life'. And in 1588 the Jesuit James Gordon would also highlight James's intimate knowledge of biblical texts – something which both Buchanan and Young, wholly predictably, had placed at the top of their educational agenda. A chapter of the Bible was read and discussed at every meal and the effects were notable. James, said Gordon, 'is naturally eloquent, has a keen intelligence, and a very powerful memory, for he knows a great part of the bible by heart'. 'He cites not only chapters,' Gordon added, 'but even the verses in a perfectly marvellous way.'

Buchanan in particular, however, was determined to fashion a king as well as a scholar – an individual endowed with self-knowledge as well as book learning, and a grasp of the ways of the world as well as the narrower realm of the classroom. In an attempt, for instance, to halt James's tendency to grant favour too freely and to neglect the content of requests, the master implemented a test, which involved presenting the boy with two stacks of papers that he subsequently signed without reading. As a result, Buchanan then spent the next few weeks declaring that he rather than James was actually King of Scotland. When questioned by James about his behaviour, the old scholar duly produced one of the documents previously signed by his pupil. 'Well,' declared Buchanan, 'here is the letter signed in your hand in which you have handed the kingdom to me.'

Nor did the intensity of the king's education entirely stifle his broader development. Though his childhood was comparatively solitary, he was not, for example, without companions of his own age, the closest of whom was John Erskine – the young Earl of Mar – whom he nicknamed Jockie o' Sclaittis (pronounced slates) in recognition of his knack for mathematics. There was also John Murray, a nephew of the Countess of Mar, Walter Stewart (a distant relative) and Lord Inverhyle. And though the rough and tumble of normal boyhood games was off-putting to James, it is not insignificant, perhaps, that so many of his early portraits depict him with a hawk on his wrist or that a beautiful hawking glove was gifted to him. There were many presents of bows and arrows, too, as well as two golfing gloves. But in spite of his undoubted physical awkwardness, it was his love of horses and hunting that came to dominate his leisure. Two relations of the Earl of Mar, David and Adam Erskine (lay Abbots of Cambuskenneth) were employed to train him, and in early life he acquired a passion for stag hunting, which would never desert him, notwithstanding a loose seat in the saddle and poor hands that may have caused his near death in the summer of 1580 when his mount fell upon him. Last but not least (and somewhat surprisingly, perhaps, in light of his reputation for ungainliness), he was also a competent dancer in childhood. 'They also made his Highness dance before me,' observed Killigrew, 'which he likewise did with a very good grace.' And the encouraging hand behind this particular aptitude was, it seems, none other than the redoubtable old Buchanan himself.

But if James's everyday circumstances were not, then, quite so pathetic and bleak as is frequently suggested, the political circumstances of his early childhood were altogether a different matter. The upheavals of Queen Mary's reign, which had awoken old feuds and created new ones among the Scots nobility, and finally forced her into flight in 1568, were followed not by peace but by five years of civil war. Both she and her supporters refused to recognise her deposition or the legality of her son's government, and though these wars left him untouched physically, they created nevertheless an insidious atmosphere of mistrust and insecurity, as the Earl of Moray and the three regents who followed him attempted to protect the Protestant settlement and alliance with England over twelve troubled years. One of those regents – James's own grandfather, the Earl of Lennox – would become the victim of a sudden raid on Stirling by Marian lords, and James would recall years later how the earl was borne into the castle and died the same day. It was not for nothing, perhaps, that James was said to have been 'nourished in fear', beset, in his own words, by 'daily tempests of innumerable dangers', or that in 1605, in the aftermath of the Gunpowder Plot, he would explain that his 'fearful nature' had been with him 'not only ever since my birth, but even as I may justly say, before my birth: and while I was in my mother's belly'.

In August 1567 Moray had made a characteristically well-timed return from England to exploit the growing chaos in his homeland. Already invited to become regent by the rebel lords, this cold, calculating and double-dealing individual would not accept the role until he had visited his half-sister at Loch Leven, where he fuelled her fears and posed as her only saviour from trial and the subsequent 'fiery death', which she believed was bound to follow from the publication of her letters to Bothwell. In this way, he was able to assume the regency at nothing less than the urgent request of both sides, while also retaining the support of his patron, Elizabeth, south of the Border. 'That bastard', as King James called him later, 'who unnaturally rebelled and procured the ruin of his owne sovran and sister', then proceeded to sell the English queen the majority of those very pearls that Mary had delivered into his hands for safe-keeping, while giving most of the remainder to his wife. To complete his betrayal, moreover, one of his first acts as regent was to order that the Casket Letters be read aloud to the Scottish Parliament and their handwriting authenticated to preclude any likelihood of the former queen's restoration.

Even so, Moray's problems were far from over. In particular, he was forced to contend with the bitter enmity of the powerful Hamilton and Gordon families who, with their great following of warlike dependants, continued to support the queen. And when Mary finally escaped from Loch Leven in May 1568, it was they who not only welcomed her, but endorsed her revocation of the abdication and prepared to do battle with Moray and the 'True Lords Maintainers of the King's Majesty's action and authority'. But notwithstanding the queen's apparently inexhaustible supply of personal magnetism, Moray moved swiftly to corner her in the south-west before her forces could gather in overwhelming force. The result was victory for the regent at Langside, and a disastrous decision by his sister – taken against the advice of her closest advisers – to abandon the fight and cross the Border at Workington to seek sanctuary within England, where she would remain a constant, overwhelming threat to her royal cousin Elizabeth and, in consequence, face various forms of house arrest, humiliation and harsh imprisonment for the rest of her life.

In Scotland, thereafter, the queen's cause encountered a slow death over five more years, which merely perpetuated the rancour and insecurity of the new reign. In 1573 Huntly and the Hamiltons surrendered at last, and only Edinburgh Castle held out under Kirkcaldy of Grange and Maitland of Lethington who had been forced at last to come down on Mary's side, since she possessed the only absolute proof that he had been 'art and part of Darnley's murder'. That same year the walls of the castle were at last breached with guns borrowed from England and the remnants of the former queen's support finally mopped up. Grange, one of the few truly honourable men around, was hanged, while Lethington, already a very sickly man, may well have killed himself rather than face trial.

Before that time, however, the Earl of Moray was already cold in the grave, along with two more of his successors. For all his mendacity, perhaps in part because of it, the earl remained a more than capable statesman who bolstered the king's authority, consolidated the progress of Protestantism and sought to cement relations with Scotland's southern neighbour by looking ultimately towards the union of the two Crowns. At the time of his coronation, James had been unacknowledged by a sizeable proportion of his subjects, but by the end of Moray's regency, he was appreciably nearer to acceptance – not least of all because it was Moray who guaranteed that the Queen of Scots was firmly imprisoned once and for all in England. And in the meantime, the earl's desire for peace and good government had actually won him the respect and even the love of the majority of Scottish people who dubbed him, it seems, 'the Good Regent'. Even in the lawless Borders, for that matter, 'there was', wrote one contemporary observer, 'such obedience made by the said thieves to the said regent, as the like was never done to no king in no man's day before'.

Yet if harsh times necessitated harsh government, they also spawned the kind of grudges and gangsterism that eventually brought Moray to his doom. And almost inevitably, therefore, he was assassinated at Linlithgow on 23 January 1570, by his enemy James Hamilton of Bothwellhaugh who had already stalked him from Glasgow to Stirling and finally 'pierced him with one ball, under the navel, quitt through'. Though Moray had 'leapt from his horse and walked to his lodging on foot', the initial optimism of the surgeons proved ill founded and he 'gave up the ghost', we are told, 'that same night', mourned as the 'defender of the widow and the fatherless' and revered by John Knox who preached his funeral sermon on the text 'Blessed are the dead which die in the Lord'. And while James himself would never accept such glowing judgements of his uncle, there was still no doubt that the king's interests had been desperately compromised by the death.

There followed, indeed, six months of civil war and large-scale Border raids, which culminated ultimately in a muddy compromise that pleased no one. For no good reason beyond lack of alternatives, therefore, it was the king's grandfather – the elderly, treacherous and nominally Catholic Duke of Lennox – who assumed the role of regent. Widely believed to have ordered a massacre of children many years previously, which had left him unable to endure his own conscience, the former exile and father of Darnley was thoroughly mistrusted by Protestants for his religious views, and vilified, too, by Catholics for his present opposition to Mary, which had finally earned him the assistance of England in securing his new role. Only James himself, in fact, seems to have harboured any real affection for the duke who was now widely dismissed as the 'sillie Regent'. And in 1572 he too paid the ultimate price for Scotland's present divisions – fatally wounded during a wild raid on Stirling, conceived by Kirkcaldy of Grange and led by the Earl of Huntly and Lord Claud Hamilton.

Leaving Edinburgh just after sunset on 3 September, the raiders reached Stirling with the first grey light of dawn, when the town was still in the silence of sleep, 'so quiet as not a dog was heard to open his mouth and bark'. Whether their intention was to kidnap the king remains uncertain, but the violence, clamour and general disorder that followed would certainly leave their own indelible imprint upon both the boy and the man he would later become. In the initial onslaught, a dozen of the king's lords, rudely awoken by the war cries, the clattering of horses' hooves and the clash of weapons quickly submitted to demands that they should 'render themselves'. Among them was Lennox. But the tide was eventually turned, first by the Earl of Morton's resistance, who defended his burning residence until 'two of his men were slain and the lodging filled with smoke', and then by the Earl of Mar who sallied forth from the castle with a handful of harquebusiers and 'set upon the attackers, who then realised that all was lost'.

Lennox, in fact, had first surrendered to David Spence, Laird of Wormeston, on condition that he be spared, but the laird's guarantee, honestly given, proved worthless in the noisy chaos that now prevailed. 'So afraitt that they took the flight, and going out at the port trod upon others for throng', the raiders were attempting to make off with their captives, when a certain Captain Calder discharged his gun into Lennox's back, leaving the valiant Wormeston, who was 'shott through also' with the same bullet, to be dragged from his horse and hacked to pieces, while the dying regent 'cryed continually' that the man 'who had done what he could for his preservation' should not be killed. His only other concern, it seems, was the safety of the king. 'If the bairn be well, all is well,' he is said to have muttered after he had been returned to the castle, slumped in his saddle, for the last time.

The king was, indeed, unharmed, though not unmoved. For the clash of naked steel, the clatter of firearms, the acrid smell of burning timber, the frenzied cries of anguish and fury, and the general buzz of danger that accompanied them had not eluded him. On the contrary, the memories of that night of gunfire and confusion would remain with him – and none more so than the image of his wounded grandfather carried directly past him to the bed where he died later that afternoon. When Lennox 'called for a physician, one for his soul and another for his body', it marked the end not only of his brief regency but of his grandson's innocence, as bars now went up over the windows of his apartments and elaborate measures were devised to surround him whenever he rode out. Henceforth, more than ever, he was brought up under a blanket of suspicion and unease, for fear of what 'the lords of the Queen, his mother' might do at any moment. For a studious, sensitive and imaginative child, who lacked the physical resources to outface the savage, unruly men whom God had called him to govern, the prospect was truly daunting.

Within the year, moreover, the duke's successor John Erskine, Earl of Mar was also dead. An honourable, grave and mainly peaceable man, Mar may well have died of simple exhaustion, outright desperation or a mixture of both. One contemporary suggested that the main cause of his 'vehement sickness' was that 'he loved peace and could not have it'. But there were whispers, too, of poison, administered by the very man who replaced him: James Douglas, Earl of Morton – in some respects the most blackguardly of all the former queen's enemies, but a strong ruler nevertheless who, for all his ruthlessness and lack of scruples, would eventually gain the king's respect. Often intimidated by the fourth and last of his regents, James still acknowledged that 'no nobleman's service in Scotland was to be compared to Morton's'. There may even have been a shred of affection on the king's part for the fearsome earl, for on one occasion when Morton bemoaned his advancing years, James's response was as kindly as it was sincere. 'Would to God you were as young as the Earl of Angus [Morton's nephew] and yet were as wise as you.'

For six years of Morton's regency, moreover, James remained unmolested – at liberty to pursue his studies under Buchanan's guidance while regaining some modicum of inner peace after the murder of his grandfather. Even his enemy, Sir James Melville, who rightly regarded Morton as avaricious and much too fond of the English, acknowledged that 'he held the country under great obedience in an established state'. And if tyranny rather than pity or remorse was his trademark, it was only by tyranny, after all, that Scotland could for the moment be governed. Yet by the spring of 1578 a formidable coalition had formed against him, led by the Highland earls of Atholl and Argyll, and the quiet routine of the king's schoolroom was once more rudely interrupted. Appearing at Stirling on 4 March, the two earls explained to James that they wished him to summon the nobility to judge a dispute between them and the regent, who was presently in Edinburgh. When, however, Morton responded by demanding that the king must either punish the earls or accept his resignation, his boldness backfired, since a formidable body of lords at Stirling, supported by the king's guardian, Alexander Erskine, quickly advised the young monarch to adopt the latter course. 'The king,' it was said, 'liking best the persuasions that were given to him to reign (a thing natural to princes), resolution was taken to discharge the regent of authority and to publish the King's acceptance of government.'

But while Morton temporarily retired to Loch Leven to tend his gardens and on 8 March the 12-year-old monarch was formally installed at the head of his council table, the counter-revolution was not long in coming. Cleverly exploiting family tensions to persuade James's former playmate, the young Earl of Mar, that his uncle, Erskine, should be supplanted, Morton had stirred an ingenious *coup de main* even before April was out. As Mar took control of the castle, the king was once more woken in the small hours by the clash of arms in the castle courtyard and filled with uncontrollable terror when word reached him that Erskine, whom he loved, had been killed. In fact, the news was false, though his son was trampled to death in the confusion and the king remained inconsolable. 'He was in great fear,' wrote Sir Robert Bowes, the English ambassador, 'and teared his hair, saying that the Master [Erskine] was slain.' And his distress was soon compounded by the realisation that Morton had swiftly ridden to the scene to assume control of government once more.

There followed a brief flurry of threatened civil war as Morton began to raise an army and the people of Edinburgh turned out under a banner depicting a boy behind bars, with the motto 'Liberty I crave and cannot have it'. The king's mother, too, was eager to interfere from her confinement in England as she intrigued with her uncle, the Duke of Guise, to remove her son to France, yet English influence remained dominant and foiled not only Mary's madcap schemes but the prospect, too, of civil war in Scotland. Even George Buchanan featured, to the tune of £100, on Elizabeth's list of prominent Scottish nobles and gentlemen to be bribed in her kingdom's interest, though Peter Young alone chose to reject the £30 earmarked for him. When, therefore, the English ambassador intervened with the offer of a patched-up compromise whereby Morton gave up the regency in return for 'first roome and place' in the Privy Council, with Atholl next in dignity, even Argyll complied with the arrangement. Nor was there any appreciable resistance from Erskine whose earlier flash of resistance had been broken by grief at the death of his son.

On 16 April, however, there was one more dastardly footnote when Morton delivered a great banquet at Stirling to which all his current allies and erstwhile enemies were invited, among them Atholl, who left the celebration 'very sick and ill at ease', in much the same way that Regent Mar had left a previous banquet of Morton's at Dalkeith. Just like Mar before him, moreover, Atholl was soon to die amid rampant rumours of poisoning. When subjected to persistent 'rhyming libells' that he was the culprit, the earl's initial response was merely to hang the unlucky authors. But such was the persistence of the Countess of Atholl that he was obliged to sanction a post mortem, which proved a drama in its own right. Conducted by several eminent doctors, only one physician, a certain Dr Preston, saw fit to dismiss the charge of poisoning, though in doing so he opted for a method of proof that would cost him dear. For, as a masterful expression of scorn at his colleagues' conclusions, he decided to lick the contents of the corpse's stomach 'and having tasted a little of it with his tongue, almost had died, and was after, so long as he lived, sicklie'.

Such, then, was the flavour of Scottish politics as James VI neared adolescence. Nevertheless, Morton's rivals remained temporarily hamstrung, though the king himself had already fluttered his fledgling wings and failed to fly. Now, therefore, for one more year at least, he would have to shelter in his classroom refuge and brood upon the violence that had once more broken his slumbers. He was silent and outwardly compliant, but anxious nonetheless for change and troubled by his prospects. For, as Bowes noted in the aftermath of Morton's coup, 'his Grace by night hath been by this means so discouraged as in his sleep he is therefore greatly disquieted'.

# 3 ⚑ Love and Liberation

'His Majesty, having conceived an inward affection to the Lord d'Aubigny, entered in great familiarity and quiet purposes with him.'

*David Moysie*, Memoirs of the Affairs of Scotland, 1577–1603

Though James's first attempt at independence had been roundly frustrated, it preceded nonetheless the onset of a marked alteration in his status – a subtle but decisive sea change in his role and influence, which saw him transformed by turns from a regent's plaything into the outright guide and instigator of his kingdom's affairs. By the time of his thirteenth birthday in June 1579 – within only fourteen months of Morton's 'triumph' – the lords of Scotland, and in particular the former regent himself, were having to accept the inevitable: that the helpless child of yesteryear was growing older and that, for all his previous frailty, he was plainly king by God's decree. To confirm the point officially, on 17 October James made formal entry into the capital, where he was greeted at the West Port by a pageant depicting King Solomon rendering judgement. It was his first visit in more than a decade, and provosts, baillies and councillors turned out in force to greet him, along with 300 prominent citizens clad in silks and velvet. At the port of the Strait Bow, a boy descended from a great globe to present the king with a set of massive silver keys to the city worth 6,000 marks, while Latin orations, stirring sermons and music from viols flowed as freely as the puncheons of wine at the Mercat Cross. From Canongate to

Holyrood, the front of every house was draped with fine tapestries and, to cap all, further pageants celebrated the genealogy of Scotland's kings and the favourable conjunction of the planets at James's nativity. It was, in short, a spectacle fit for any ruler – not only a heartfelt statement of civic pride, but, much more important still, a clear-cut sign of changing times and an equally emphatic rite of passage for a freshly empowered king.

James, it is true, remained separated from most of those who had originally set him against Morton. But by no means all Morton's enemies were gone, for the council was still an uneasy coalition which contained some, like Argyll, who were reconciled only in appearance to the present status quo. These dissenters were keen, moreover, to find a suitable replacement for Atholl – a figure acceptable to both the king and those who still favoured his mother. And across the Channel in Esmé Stuart, Seigneur d'Aubigny, they found not only someone to alter the entire situation in Scotland radically, but a figure who was to exercise a profound influence upon the king's whole life. Any alternative to Morton was by now, after all, infinitely attractive and d'Aubigny's nearness in blood to the king was a distinct recommendation, as was his right to the earldom of Lennox which made his arrival in Scotland inevitable sooner or later. The fact that he was more French than Scottish, Catholic and something of a moral reprobate into the bargain was neatly overlooked. 'The King has written to summon his cousin the Lord d'Aubigny from France,' the Bishop of Ross noted in a letter dated 15 May, before adding in all apparent innocence that he was 'a man of sound judgement and marked prudence, a constant upholder of the Catholic religion, and one whom the king is anxious to have at his side'.

For nearly two centuries, in fact, the Lennox Stuarts had held lands and titles on either side of the Channel, rendering distinguished service as soldiers and diplomats to the kings of both France and Scotland, but especially the latter. As the son of John Stuart, brother of James's grandfather and former regent, the Earl of Lennox, Esmé Stuart was therefore first cousin to Lord Darnley. And although his ostensible aim upon his arrival in September 1579 was the re-establishment of his family's Scottish position as the last male representative of his line, his real intent was altogether more sinister. As the secret agent of the Guises, he would win, he hoped, the confidence of the king, promote the cause of France and Mary Stuart – by whom he had been given 'fourtie thousand pieces of gold, in crowns, pistolets [coins] and angels' for use at the Scottish court – and save the Catholic faith before it was too late. Such was his gift for deception and intrigue that before long he would declare himself a convert to Protestantism and play an astonishing series of double games, not only with Guise, Mary and the Catholic powers, but with Elizabeth of England and the Scottish Kirk as well. Yet one consistent thread ran through the machinations of this fascinating but sinister figure all the while: the pursuit of personal ambition. And once established, though not unmindful of his original mission, he soon discovered that his control over the king's affections had infinitely more to offer than the favour of France or Spain, the interests of the old religion, or, even more obviously, the vanishing hopes of the former queen.

For more than a decade the ministers of the Scottish Kirk, propped up by a body of powerful and unscrupulous nobles, had maintained a deceptive dominance in Scotland, which poorly reflected the wishes of even most Protestants. Catholicism, in fact, had collapsed largely by default under the strain of a series of weaknesses, structural and circumstantial, which had left it easy prey to its enemies. The remoteness of the centres of Catholic power in the north and west, on the one hand, coupled to the calamity of Mary Stuart's example, the collapse of France into religious civil war and, perhaps most of all, English interference, had all taken a mighty toll, and the lands of the old Church had been readily seized, along with four-fifths of her revenues, by the ever-watchful Lords of the Congregation. But the subsequent Presbyterian settlement, pushed through by Knox and his fellow zealots, was far from universally welcome. Nor did the English alliance it entailed or its dependence upon ruffians like Morton enhance the new Kirk's moral authority. Starved of funds and short of educated recruits, it could actually find no more than 289 ministers for Scotland's 1,000 parishes, and in such circumstances, it was far from inconceivable that the Counter-Reformation might yet gain a foothold or that the 'auld alliance' with France might still be revived – especially if the king himself could first be won to d'Aubigny and then to Rome.

The dashing French courtier's arrival could not therefore have been better timed, since James, at the age of 13, was maturing both politically and physically, if not emotionally. At that moment, as Morton's stifling authority waned, any influence exerted over the king might indeed prove decisive, and none more so than the easy-going, affectionate glamour which Esmé Stuart exhibited in such abundance. A man of great personal magnetism around 37 years old – 'of comely proportion', 'civil behaviour' and 'honest' conversation – this elegant, red-bearded visitor, whose piercing black eyes spoke eloquently of his Italian ancestry, brought colour, amusement and gaiety to the dour Scottish court. He delivered civilisation and learning, too, of the kind that contrasted starkly with Buchanan's bleak instruction and was bound to appeal to the scholarly young king. The cultural amenities at Stirling were, after all, limited to say the least. As a concession to gentility, the Scottish Parliament had funded the employment of four fiddlers at the castle and this, in effect, was the limit of James's exposure to the frills of high refinement. D'Aubigny and his train of twenty gentlemen, however, brought with them not only grace and elegance, but respect and deference in the sharpest possible contrast to Morton's gruff and Spartan realism. Above all, however, as events would prove, they brought intimacy and love.

James's weak physique, shambling gait and slovenly manners left much to be desired, of course. But an earlier attack of smallpox had left him unscarred and his appearance was generally considered 'not uncomely'. According to Sir Henry Killigrew's description of 1574, indeed, he was 'well grown, both in body and spirit'. Even kings, however – especially those as perceptive as James – are sometimes capable of grasping realities and reflecting upon their own limitations. And James was no physical paragon. But the newcomer who now doted upon him was all this and more – so much so that even his potential enemies were at first wholly taken with him. Sir James Melville, for instance, thought him 'upright, just and gentle', and though he spoke only French and made little attempt to learn 'Scottis', even the hard-edged Scottish nobility were soon won over, in the main, to what John Spottiswoode called his 'courteous and modest behaviour'. And if d'Aubigny could allure even ministers of the Scottish Kirk, how much more susceptible to his attention would be the awkward, graceless youth upon whom he had set his sights?

Certainly, the Frenchman's rise to influence was instantaneous. Indeed, by the time that James made his grand 'entrie to his kingdome' at Edinburgh in October, he had already insisted that d'Aubigny accompany him, and honours followed thick and fast. Given first the rich Abbey of Arbroath and a sizeable endowment from the Hamilton lands forfeited after Mary's final downfall, he was then created Earl of Lennox in 1580 after his ineffectual uncle Robert was encouraged to renounce the title in exchange for that of March. He was admitted, too, to the council, awarded custody of Dumbarton Castle – the key to western Scotland and gateway to France – and before long had become both Lord Chamberlain, responsible for the king's safety, and First Gentleman of the Bedchamber. Most important of all, however, on 5 August 1581, he was finally made Duke of Lennox – the only duke in Scotland at that time and the first in Scottish history, apart from Bothwell, not to be a 'prince of the blood'. 'Lennox's greatness is greatly increased,' wrote Robert Bowes, 'and the king so much affected to him that he delights only in his company.' 'Thereby,' Bowes added, 'Lennox carries the sway.'

And as the duke's star ascended, so Morton's, of course, continued to wane, though he remained curiously unmoved at first by the mounting threat. Predictably, the preachers of the Kirk were from the outset deeply suspicious of the 'papistes with great ruffs and side bellows' that Lennox carried in tow with him. But, as the king continually hung on his favourite's shoulder and fiddled with his fine clothes and jewels as they walked together, Morton appears to have dismissed the spectacle as little more than a boyish enthusiasm – a passing fad resulting from 'the flexible nature of the king in these tender years'. In forsaking the regency and pushing the king to the political forefront as the figurehead for his own power and policies, he had in any case burned his bridges. In October 1579, just after his triumphal entry into Edinburgh, James had presided over Parliament and formally presented himself as the governor of his kingdom in his own right. And though Morton continued to pull the strings, the prospect of any return to a regency had gone forever. More ominously, however, by May 1580 the earl had retired in frustration to his estates at Dalkeith gathering his friends about him, while the English ambassador whispered in turn of the king's 'great myslyknge' for his former regent.

Nevertheless, throughout 1580 the holiday atmosphere continued, though events were steadily darkening behind the scenes. While James stayed at Holyrood, Lennox was even prepared to undertake some informal instruction in Calvinism at the king's hands, seeing the plain advantage that such a move might yield with the Kirk's preachers, many of whom were apparently 'much overtaken with the conceytt of his reformation'. By May, indeed, Lennox had officially committed himself to the new faith at St Giles in Edinburgh before returning to Stirling and signing the so-called 'Articles of Religion' in the Chapel Royal. Naturally, his action was dismissed by many as a cynical manoeuvre. 'Those who wish to rule,' Mauvissière, the French ambassador in London commented at the time, 'must learn to conceal themselves.' But Lennox would die a Protestant and his conversion captured the king entirely – so much so that the two travelled together on progress around the kingdom that summer.

In the meantime, the impending showdown between Morton and his supplanter edged ever closer. In England, Elizabeth and Burleigh were increasingly alarmed at the prospect of Franco-Scottish reconciliation, and the queen hinted to James for the first time that he might succeed her as king, while advising him 'rather to fear for his ambition than to comfort and delight his affection'. She encouraged Morton, too, to 'lay violent hands' on Lennox and sent Bowes north once again to galvanise the Protestant lords into some kind of effective action. In a secret midnight meeting with the former regent, the ambassador learned how the king was indeed beginning 'to commend and be contented to hear the praises of France'. And as rumours of kidnapping mounted once more, James also revealed a new and curious turn of mind. When Argyll spread word that Morton was bent on abducting him to Dalkeith, the king swiftly abandoned his hunting and returned to Stirling. Lacking the resources to fund a permanent armed guard, he was, after all, sorely exposed – especially as his councillors' attendance was irregular, since they too lacked funds and had to pay their own charges while they remained at court. But now, as Bowes noted, the king was more convinced of his ability to frustrate his enemies, since 'into whose hands soever he should fall, they should note in him such inconstancy, perjury and falsehood' that they would swiftly regret their action. Plainly, the king had learnt an important lesson from past experience, but in doing so he had also drawn a suspect moral conclusion for the future.

The opportunity for kidnap or any other decisive act of self-preservation on Morton's part had, however, already passed, for he was, it seems, 'loved by none and envied and hated by many, so that they all looked through their fingers to see his fall'. When James conducted his leisurely royal progress that summer, with Lennox and a bevy of loyal lords in attendance, Morton was left behind, laid up with a leg injury after a horse had kicked him. And when the embattled earl, 'indifferently well recovered', finally joined the king at St Andrews, where a convention of the Scottish Estates was due to assemble, he received a harrowing warning from the most unlikely of quarters. At a play performed before the king at the New Inns of the Abbey, a mad seaman – 'a known phrenetick man' who, Morton was convinced, could not have been put up to it – warned him that a plot was afoot and his 'doom in dressing'. Always susceptible to superstitious fears, the earl remembered, no doubt, how 'a lady who was his whore' had already shown him 'the answers of the oracles' and told him 'that the king would be his ruin'. And both the sailor and the whore would now be proven right.

Weary of his self-interest, even the Kirk had grown disillusioned with its former champion, and the English, too, who had previously seen the earl as such a worthwhile asset, now hesitated, relying on threats and intrigue when only armed assistance would do. When Bowes warned James of the dangers involved in preferring 'any Earl of Lennox before a Queen of England', however, and demanded that his favourite be removed from the Privy Council, the Scottish king's alarm was nevertheless palpable, whereupon the only remaining question was how the blow to Morton might be delivered most conveniently – and profitably. For Lennox, with typical Gallic finesse, now devised a masterstroke not only to remove his enemy, but to do so in a way that suited both his French patrons and the former queen more admirably than even they might have hoped. The method would involve raising the ghost of Darnley from its sordid resting place, while the instrument for the dirty work in hand was to be not Lennox himself, but a bold, ambitious opportunist who had been hovering watchfully about the court for some years past.

The murder of the king's father at Kirk o' Field more than a decade earlier was, of course, fertile ground for any ill-intentioned intriguer. Still mired with mystery and wrapped in rumour, any number of candidates might plausibly be connected with the deed and countless 'witnesses' found to attest to order. In the event, it was Sir James Balfour, brother of the owner of Kirk o' Field, and himself a suspect, who now came forward to furnish the evidence that finally did for Morton. Claiming to possess the bond signed by the conspirators, and assuring Lennox that Morton's signature was on it, Balfour had provided the duke with a tool that would finally exonerate the former queen from any involvement and, in doing so, both heartily relieve her son and place the English in serious embarrassment at her current treatment. With James Stewart, Captain of the King's Guard – a newly formed body of sixty men at arms, specially commissioned by Lennox – more than willing to undertake the task, all was set for Morton's final erasure.

Gloriously self-assured, splendidly handsome and exuding what might best be described as a coarse variety of magnificence, Stewart, the son of Lord Ochiltree, revelled in his courage and resolution and 'thought no man his equal'. He was, moreover, as capable as he was confident – intelligent, educated and, no less importantly, politically astute. And his connections with John Knox, who at the age of almost 60 had married his sister, also gave Stewart a cachet of sorts with the Kirk. But his swaggering conceit and single-minded brutality had already earned him the suspicion of many and hatred of some. To Sir James Melville, who detested him, he was 'a scorner of all religion, presumptuous, ambitious, covetous, careless of the commonwealth, a despiser of the nobility, and of all honest men'. His confederates, too, were aware of his baser motives. Even Lennox, for that matter, was conscious that his henchman in what now followed was 'eager to win credit by what means soever'.

Nor would the king pass up his own duplicitous part in the tawdry circumstances of Morton's downfall. On 30 December, the former regent was taken hunting by James and treated cordially throughout the day. Upon their return that evening, indeed, the lily was perfectly gilded by the canny youth, who delighted increasingly, it seems, in the finer points of double-dealing. 'Father,' he told his quarry, 'only you have reared me, and I will therefore defend you from your enemies.' Whether Morton had somehow elicited the comment by expressing his concerns is unknown, but soon enough the hollowness of James's assurance was starkly exposed. For only the next day, during a meeting of the Privy Council at which the king was personally present, Captain Stewart burst into the room, fell on his knees before his sovereign, and pointing histrionically at Morton, accused him of being 'art and part for knowledge and concealing' of Darnley's murder. When, moreover, Morton dismissed him as the 'perjured tool' of his enemies, uproar ensued with both men grasping their swords whilst being held apart by Lords Cathcart and Lindsay.

In the meantime, Lennox had feigned incomprehension at the furious exchange conducted in a foreign tongue. But his purposes had been served to perfection. The king, now thoroughly implicated in the whole affair, made no move whatsoever on behalf of Morton, who was duly arrested and eventually removed to Dumbarton, where he became not the first to declare that 'if he had been as upright to his God as he was faithful to his prince, he had not been brought to this pinch'. Thereafter, on 1 June 1581, he was tried and condemned in Edinburgh and executed the next day. In the process, he admitted foreknowledge and concealment, but declared Bothwell to be Darnley's principal murderer. And though the Queen of England had sent 2,000 men to the Border to save him, her young Scottish counterpart, revelling in his newfound confidence, brazenly outfaced her. 'Though he be young,' wrote Thomas Randolph the English emissary, before being sent packing by a pistol shot through his window, 'he wants neither words nor answers to anything said to him.'

Morton's Douglas kinsmen under the Earl of Angus, meanwhile, had plotted only half-heartedly to save him, and though the condemned earl conducted himself bravely throughout his ordeal – claiming as he mounted the scaffold that he was 'entering into the felicity of Almighty God' – only a few members of the Kirk would ever mourn him. Indeed, the night before he was beheaded by the so-called 'Maiden' (an early prototype of the guillotine) that he had earlier brought from Halifax for his own nefarious purposes, he had written letters to James defending his conduct, 'but the king would not look upon them, nor take heed what they said; but ranged up and down the floor of his chamber, clanking with his finger and thumb'.

Not surprisingly, when Elizabeth heard of James's betrayal of his erstwhile regent her fury was undisguised, deriding him as 'that false Scotch urchin' from whom only 'double dealing' could be expected. And while the Scottish king might rightly bridle at the 'urchin' epithet, the second claim in the queen's outburst was hardly deniable. The circumstances of his childhood had, of course, already taught him cunning but, in procuring Morton's ruin, Lennox had also taught him to regard the practice of duplicity as something much more: a necessary skill and intellectual art, if not outright virtue. 'The king's fair speeches and premises,' wrote an English noble, 'will fall out to be plain dissimulation, wherein he is in his tender years better practised than others forty years older than he is.' He 'is holden among the Scots for the greatest dissembler that ever was heard of for his years'. And Lennox was attempting to teach him too, perhaps, that condign justice gave little cause for regret. For though James had commuted the verdict that Morton should be hanged, drawn and quartered, and was absent from the execution, Lennox's supporters showed few such qualms. Indeed, Lord Seton 'stood in a stair' close by, while Ker of Fernihurst, gained the best view of all 'in a shott over against the scaffold, with his large ruffs, delighting in this spectacle ...'

To James's credit, this last lesson was never lasting, if ever learned at all, though there were others delivered up by Lennox that would indeed prove permanent. It was under Lennox's influence, for instance, that the king came to scorn the more radical elements of the Scottish Reformation as anti-monarchical rebels against properly constituted authority, and, more generally, to question the broadly democratic principles that George

Buchanan had been so keen to imbue in him. From this time forth, the Scottish clergy became to James what the Huguenots were to his French counterparts – seditious disturbers of the peace. In that time of confusion, the king would write, 'some fiery-spirited men in the ministry got such a guiding of the people as finding the gust of government sweet they began to fancy a democratic form of government'. He was made to think evil, wrote the minister James Melville, against those who served him best and to regard the Reformation as 'done by a privy faction turbulently'. And in this regard Lennox would also, it seems, frequently discuss the absolutism of the King of France and emphasise its virtues, working along with Captain James Stewart, another driven by his own self-interested purposes, to encourage the king to assert his God-given authority more stridently. Suppressed and disregarded for so long, and treated hitherto like a chattel of the high and mighty – a political talisman to be controlled and brandished at convenience by contending rivals – it is easy to see how an impressionable 15 year old might well have reacted to such advice and adulation.

The Kirk, after all, had been flexing its muscles more and more stridently, and in James's current state of growing confidence, it was hardly surprising that he should listen so readily to more gratifying alternatives. John Knox had set the mould in the first instance by affirming in his own inimitable way that the laws of God, needing no confirmation from any king or parliament, were to guide the state, and that kings who resisted the Kirk's injunctions should be swept from office. Under the pressures of practical politics, however, Andrew Melville, who led the Kirk in James's reign adapted the strongly theocratic emphasis in Knox's thinking into what would become known as the 'doctrine of the two kingdoms', whereby secular authority should be exercised by the reigning monarch, while the clergy assumed sole responsibility for religious affairs. On this view, all spiritual authority flowed from God the Father through Jesus Christ the Mediator directly to his Kirk, by-passing entirely both king and state, since the Kirk had 'no temporal head on earth, but only Christ, the only spiritual king and governor of his Kirk'. The king, therefore, had no higher place within the Kirk itself than any private person and must obey the clergy in all matters of the spirit.

By 1581, moreover, with the publication of the *Second Book of Discipline*, the authority of the secular ruler was being further undermined. For now the king was encouraged to follow the clergy's advice even in matters that had hitherto been deemed his own. 'The ministers,' it was still accepted, 'exercise not the civil jurisdiction.' But the new departure was apparent in the proposal that they should nevertheless 'teach the magistrate' how that jurisdiction should be exercised, and in the further claim that 'all godly princes and magistrates ought to hear and obey'. To compound the king's frustration, the same radical ministers were conducting a steady assault on the authority of his bishops. Calvinism, after all, plainly asserted the equality of all pastors and called for church government through an ascending structure of presbyteries, synods and General Assemblies. Bishops in any form were, from this perspective, symbols of Roman error and instruments of royal tyranny – the king's agents in controlling and perverting the exercise of God's true design for his people. No monarch, however, could lightly accept such an assault upon his prerogative and, under the influence of his new friends, James was therefore quick to seize upon what he eventually termed in 1604 the 'No bishop, no king' principle – a notion that would have such significant long-term consequences.

It was not coincidental either that James's relationship with his mother appeared to warm significantly after Lennox's arrival. As a boy he had always resented Buchanan's spiteful attacks upon her reputation and continued to exhibit a sentimental interest in her story. Now, however, he equated criticism of his mother with the more general assault upon royal authority in progress at that time, and in 1584 he prevailed upon Parliament to condemn his elderly tutor's writings formally. Years later, too, he would advise his son to read history but not 'such infamous invectives as Buchanan's and Knox's chronicles'. Nor, he urged, should his son countenance malicious words against his predecessors, since those who speak ill of a king 'seek craftily to stain the race and to steal the affection of the people from their posterity'. He had found his most loyal servants, he added, among those who had been faithful to his mother – once again combining filial devotion with notions of loyalty to the Crown and hatred of all challenges to the political and religious hierarchy which, in his view, guaranteed order and stability throughout the body politic.

For her part, of course, Mary was still unable to acknowledge the legality of her abdication or the resulting transfer of sovereignty to her son, and Morton's execution had only served to heighten her hopes of rehabilitation. In 1581, therefore, she proposed the so-called 'Association' whereby James, who had not hitherto been recognised by the Catholic powers of Europe, should 'demit' the Crown to her, after which she would immediately bestow it upon him again with her full blessing. As a result, he would be universally acknowledged as King of Scotland and become joint sovereign with her while ruling the country in their joint names. And this was not all, since the former queen also suggested that her abdication be formally annulled, that James be crowned anew and that her supporters be pardoned. Catholics were to be granted liberty of worship and it was clear, too, that no important decision was to be delivered without her approval. Above all, Mary asked that her son be reconciled to Rome. Only by means of the Association, she suggested, could he ever hope to secure the English throne.

It was a scheme shot through with wild improbabilities and, as such, wholly worthy of the woman who fashioned it, but the temptations it offered were not without appeal to the young king. And though the former queen was actually coldly suspicious of Lennox and the plan itself a palpable threat to his own primacy if ever implemented, the favourite had little actual option but to mask his concerns and indulge the queen's machinations. His links to the House of Guise, after all, made it virtually impossible to reject the Association outright, and at his instigation, therefore, James now began to correspond with Mary, writing her brief but affectionate letters, which assured her that he continued to hold her in high honour and would act at all times as her obedient son. Always a lover of animals, he referred in one instance to 'the fidelity of my little monkey, who only moves near me', though the strength of the renewed link between son and mother should not be measured by such disarming comments. On the contrary, the king offered no concrete concessions and though he addressed his mother as Queen of Scots, he never neglected to sign himself as 'James R'. In fact, he seems to have been playing, at Lennox's behest, what was becoming an increasingly familiar game of gracious deception, and in doing so he now slithered into an elaborate ruse, which would allow him (in theory) to retain his options while paying lip service at one and the same time to his obligations as son.

In the meantime, however, the king's everyday behaviour as well as his character and relationships had gradually begun to evolve upon new and questionable lines. Though swearing was something that he severely condemned in his writings, deeming it all the more reprehensible since it was a sin 'clothed with no delight or gain', he nevertheless developed a habit for it and sought to justify the vice in part, so long as it sprang from sudden, unpremeditated anger. Certainly, his conscience does not seem to have troubled him unduly in this respect, for when admonished by the clergy 'to forbear his often swearing and taking the name of God in vain', he merely replied, 'I thank you' – with a little laughter. Nor was this the only vice to trouble the ministers, for now, it seems, he was sometimes remiss in attending the Kirk, no longer called for preaching at dinner and supper, and disliked to hear his shortcomings rehearsed from the pulpit. He indulged too, we are told, in pastimes on the Sabbath. And though James's sports and entertainments were mainly innocent enough, bawdy jests became another facet of his behaviour, which accorded aptly with the declining tone of the company he now kept. Lennox's colourful French associates were as free in their language as they were in their habits, and Captain James Stewart, soon to be rewarded with the stolen title of Earl of Arran, did nothing to moderate their influence. On the contrary, his wife, a daughter of the Catholic Atholl and now the chief lady at the Scottish court, had gained a pungent reputation for licence and immorality. Formerly married to the king's uncle, the Earl of March, she had subsequently divorced him on grounds of impotency – or, as Moysie put it, 'because his instrument was not guid' – though she was pregnant at the time with Stewart's child.

Much more significantly still, however, the king's love for Lennox was widely thought to have contained a sexual element. Indeed, it was Lennox, according to many, who first awakened James's lifelong interest in beautiful young men. Courtiers and ambassadors were fully aware, of course, that the king was 'in such love' with his favourite 'as in the open sight of the people often he will clasp him about the neck with his arms and kiss him'. But the ministers of the Kirk were also ready to suggest that the relationship had passed beyond affection and that Lennox had 'provoked' the king 'to the pleasure of the flesh' and drawn him to 'carnal lust'. And if the indignation of earnest Calvinist clergymen may be treated with due suspicion, there were other voices of the same opinion. 'His Majesty,' wrote the chronicler David Moysie, 'having conceived an inward affection to the Lord d'Aubigny, entered in quiet purposes with him,' a phrase bearing special connotations in the Scots idiom of the time. And the English clergyman John Hacket, writing many years later, reinforced the like conclusion for posterity by observing how James had clasped Lennox 'Gratioso in the embraces of his great love above all others'.

The precise nature of the sexual relationship can never, of course, be known. But in later life, James condemnation of sodomy was certainly unwavering. Whether he accepted the biblical prohibition of this specific act only and regarded other homosexual activity as pardonable remains debatable. However, his book on kingship, *Basilikon Doron*, published in 1599 at a time when he was without a serious attachment to any specific male favourite, would categorise sodomy itself among those 'horrible crimes which ye are bound in conscience never to forgive', and thereafter he followed the same line with total and unembarrassed consistency, using the full force of English law to reinforce his opinions. Writing to Lord Burleigh, for instance, he would give a forthright directive that judges were to interpret the law rigidly and were not to issue any pardons, stating unequivocally that 'no more colour may be left to judges to work upon their wits in that point'. And while Lennox confided to the king that he had given up his wife and children 'to dedicate myself to you entirely', such statements were entirely consistent with the usual conventions of the day – as, indeed, was much of the emotional excess exhibited on James's part. Prematurely old in some respects and painfully naive in others, the king's ability to baffle and mislead modern observers remains palpable, and even allowing for the objectivity of certain contemporary commentators, which is by no means self-evident, the precise extent of the king's sexual involvement with Lennox remains as doubtful as ever.

In any event, the year after Morton's death was probably the happiest of James's life as he basked in the freedom and gaiety of the new status quo. Six pairs of fine horses, a gift from the Duke of Guise, filled him with joy, and he revelled, too, in the company of a leading light in Lennox's entourage, a certain M. Momberneau – 'a merry fellow, very able in body, most meet in all respects for bewitching the youth of a prince'. Not least among Momberneau's winning accomplishments was his skill as a horseman. 'Tuesday last,' wrote Thomas Randolph, capturing the regular routine of James's newly liberated existence, 'the king ran at the ring, and, for a child, did very well. Momberneau challenged all comers. The whole afternoon and great part of the night were passed with many pleasures and great delights. The next day the king came to Edinburgh to the preaching. That afternoon he spent in like pastimes as he had done the day before.' Meanwhile, at Leith, where he dined a few days later, a castle, derisively dubbed the 'pope's palace', had been built on boats to be burnt before him for his entertainment. Horse racing on the sands followed, along with a ludicrous joust between courtiers. All, in fact, was a continual round of jollity and high spirits as James left the business of government, or what he called 'auld men's cummer' to Lennox and Arran.

But such a state of affairs could not, of course, last. Predictably, the first rumours of the proposed Association between James and Mary had caused a fresh wave of anti-Catholic hysteria, in which Lennox was the principal object of suspicion. His signature of the so-called 'Negative Confession', which denounced 'the usurped authority' of the pope, 'that Roman Anti-Christ', had left his enemies unconvinced of the 'frutes of his conversion', while the composition of the pro-Lennox party in government, which contained a number of the former queen's supporters, such as Seton, Maxwell, Fernihurst and Maitland of Lethington, only confirmed existing doubts. Worse still, the arrival of Spanish and Guisard agents during 1581 and 1582 now brought matters to a pitch, placing Lennox in the most intractable of dilemmas. For, despite the implications of the Association for his own personal influence, the mounting tide of opposition at home made the prospect of foreign assistance in a far-ranging Catholic scheme to convert the king and subdue England increasingly irresistible.

From the time of his earliest memories, in fact, James had been encouraged to set his sights upon the English succession, but for good reason Elizabeth had never acknowledged his right to the throne. When his mother, in the days of her good fortune, had pressed her royal cousin to recognise her as successor, Elizabeth had made it clear that to do so would provide a ready focus for the plots of her enemies. 'Think you,' she is said to have responded, 'that I should love my own winding sheet?' And the same reservation applied with equal force, of course, to the claims of Mary's son. During 1581, therefore, as his own position in Scotland became increasingly fragile, Lennox suggested to James that the key to the English throne might lay not so much in Elizabeth's approval as in her removal. If the king was prepared to compromise with his mother and accept the Catholic faith into the bargain, the rewards for both he and Lennox would be considerable. A successful Catholic invasion of England which resulted in the release by force of Mary Stuart and the deposition of Elizabeth would bring Mary and James jointly to the thrones of both Scotland and England, leaving Lennox rescued at last from his Scottish Protestant enemies.

As the plots thickened on all fronts, James meanwhile continued to exhibit both craft and coolness, committing himself to no one but entertaining overtures from, in some cases, the most unlikely of sources. In the summer of 1581, for instance, Bernardino de Mendoza, the Spanish ambassador in London, dispatched two Jesuits to Scotland for a secret interview with the king, who received them cordially and gave his assurances that, though he deemed it advisable to appear pro-French in public, his heart was inclined to Spain. At the same time, however, he showed no inclination to change his religion, for early the next year the same Jesuits were mentioning a plot by Catholic noblemen to secure his conversion by force. Nor it seems, did the arrival of another Jesuit, William Crichton, yield any more progress, as Lennox's position continued to decline steadily. Clearly, the king was prepared to probe a range of possibilities upon his favourite's advice, but only on his own terms. And if the favourite was reduced to fear and frustration as a result, then that would have to be. For, as Mendoza made plain, Lennox's continual troubles and terror at feeling himself in the daily presence of death were reducing him to 'a deplorable condition'.

Even former allies now became threatening, in fact. On 22 April 1581, Captain James Stewart had been created Earl of Arran and was now firmly fixed in the king's affections in his own right. Like Lennox, he too had good looks to recommend him, though in this case the king's attraction was primarily psychological rather than physical, since Arran exhibited a gift for leadership and imperious mastery of events to which James's subservient personality paid natural homage. Where Esmé Stuart offered reverence, James Stewart, by contrast, exuded raw, untempered masculinity. In some respects, the contrast between the two personalities tapped into the contrasting needs of James's own character: on the one hand, the hankering for love and deference; on the other the wish for security and control of events. But it was still to the duke, his 'dearest cousin' and 'nearest heir male', that the king looked first and when Arran quarrelled with his rival in the winter of 1581 he found himself forbidden to attend the Chistmas celebrations at Dalkeith, which James enjoyed with Lennox instead.

Predictably, the reconciliation that occurred soon afterwards was nothing more than a matter of mutual convenience, as Lennox continued to connive with Spain and Mary, and, in doing so, courted disaster ever more freely. His schemes took no account of Franco-Spanish rivalry, and ignored both James's deep-seated Protestantism and innate reluctance to share his throne. They underestimated, too, the very forces that had broken the king's mother and would break him far more easily still: the volatility of the Scottish nobility and, just as important, the vengeful wrath of the Scottish Kirk. His power, after all, was based upon nothing more dependable than the doting favour of a 16-year-old king. And while James had grown in confidence and learnt some kingcraft in rapid time, he remained no match for concerted dissent and, above all, that tried and trusted trump card of Scotland's nobility: abduction.

In attempting to encourage his Catholic allies while feathering his nest financially, therefore, Lennox now made a crucial blunder which would actually pave the way to his downfall. When Robert Montgomery, a minister at Stirling who had publicly denounced episcopal government of the Kirk, was appointed Archbishop of Glasgow under terms that left the revenues of the see in Lennox's hands, the result was general outrage, which demonstrated all too conclusively the practical limits of royal authority. Angry deputations of ministers, threatening to excommunicate Montgomery, descended upon James, whose attempts at resistance proved painfully futile. 'We will not suffer you,' declared James in a forlorn attempt to outface his clergy, but the response was predictable. 'We must obey God rather than men,' retorted the small but fiery John Durie whom Lennox had dubbed 'a little devil'. 'And we pray God,' the minister continued, 'to remove evil company from about you. The welfare of the Kirk is your welfare; the more sharply vice is rebuked the better for you.' Smarting and angry, the king was close to tears. But his only familiar weapons – deceit and subterfuge – were now, of course, useless. The dam had been breached and, in the process, James once again stood exposed as a raw and vulnerable youth. Nor was this the end of his chastening. For, not long afterwards, Andrew Melville preached a famous sermon in which 'he inveighed against the bloody gully [knife] of absolute authority, whereby men intended to pull the crown off Christ's head and to wring the sceptre out of his hand'.

Coming, as it did, at a time when James had been experiencing the first stirrings of genuine pride in his regal status, the whole experience could not have been more significant, and matters reached a harrowing crescendo when a certain 'Seigneur Paul', an emissary of the Duke of Guise arrived on Scottish soil. When Durie encountered the Frenchman, he responded with an act of histrionics that, for all its excess, nevertheless captured the intensity of disgust and sense of betrayal experienced by his clerical colleagues in general. Pulling his bonnet over his eyes, he declared that his eyes should not be polluted by the sight of the Devil's messenger, and then berated James to his face – with no trace of deference or hint of restraint. James should adhere to his religion, refuse a Catholic marriage and, added Durie for good measure, keep his body unpolluted. He had, in short, misbehaved as errant children are wont to do, and now he must amend. It was a put-down of withering proportions. But what made the pill more bitter still – and the memory so indelible – was his own response. Overawed by the white heat of the minister's invective, the king merely conceded on all points – in hushed voice and plainly broken spirits.

By now, too, the anti-Lennox faction among the nobility had grown to critical proportions. Staunch Protestants, such as Lord Lindsay and the Earl of Glencairn, were joined in conspiracy by two of Morton's kinsmen, the Earl of Angus and Douglas of Loch Leven, as well as the new Earl of Gowrie, who had supported Lennox initially, only to balk before long at his selfishness. And when the young Earl of Mar and the 5th Earl of Bothwell lent their support, the 'band' of discontented nobles was complete. Emboldened by English bribes and bolstered by the Kirk's delivery to James in July of a set of articles, denouncing his traffic with France and reliance on 'bloody murtherers and persecutors', the plotters shaped for action.

Nor would they have to await their opportunity long, for in the summer of 1582 James and Lennox were for once apart. The king, in fact, was hunting in Atholl, while Lennox remained in Edinburgh to preside over a court of justice in his judicial capacity as Lord Chamberlain. And on 22 August, as James was riding south once more towards Perth, he was met by Gowrie, who cordially offered him hospitality at his nearby castle of Ruthven. Suspecting that something was afoot, James 'yet dissembled the matter', we are told, 'thinking to free himself the next day when he went abroad with his sports'. Next morning, however, he discovered the full extent of his predicament. For, as his departure neared, Mar and Gowrie accosted him with a list of grievances which he at first attempted to ignore. But when making for the door of the parlour in which the exchange occurred, the Master of Glamis barred his way with a leg. Storms of rage and floods of tears were unavailing. 'Better that bairns should weep than bearded men,' observed Glamis.

Arran, meanwhile, had foolishly chosen to ride on his own to the king's rescue on the false assumption that he would be released on demand, whereupon he too was promptly arrested; and though Lennox 'lurkit' at Dumbarton throughout the autumn, 'waiting upon opportunity' and conceiving hopeless schemes for the king's rescue, his cause was thoroughly lost. James, in fact, was so closely guarded that when a secret message arrived from Lennox, the bedchamber attendant who delivered it, a certain Henry Gibbe, could do so only in the privacy of the king's 'close stool' or toilet. And, even then, all that James dared answer to his favourite was that he should send no more dangerous messages of this kind. In any case, only a few days after his abduction, James had been forced to sign a proclamation declaring himself a free king and wrote to Lennox ordering him to leave the country. He had, it is true, protested bitterly and continued to speak out against his oppressors. 'His Majesty,' wrote Sir James Melville, 'took the matter further to heart than any man would have believed, lamenting his hard estate and mishandling by his own subjects, and how he was thought but a beast by other princes for suffering so many indignities.' It was only with great difficulty, too, that the Ruthven lords obtained the king's consent to a proclamation acknowledging the freedom of the Kirk. 'He spared not to say,' wrote David Calderwood, 'that the ministers were but a pack of knaves, that he had rather lose his kingdom than not be avenged upon them, that the professors of France [the Huguenots] were but seditious traitors, rebels and perturbers of commonwealths.' Yet the intensity of James's bitterness only reflected the hopelessness of his position – a fact which Lennox, too, reluctantly accepted when he finally returned to France on 21 December. 'And sa the King and the Duc was dissivered,' wrote one contemporary laconically, 'and never saw [each] uther againe.'

Before his departure, however, James had been forced to accuse Lennox of 'disloyalty and inconstancy' in not leaving Scotland in accordance with previous orders, and in answer the duke had penned a last message which must have made the most painful reading for the shamed and heartbroken young king. James was, Lennox declared, his 'true master, and he alone in this world whom my heart is resolved to serve'. 'And would to God,' he continued, 'my body should be cut open, so that there should be seen what is written upon my heart; for I am sure there would not be seen there those words "disloyalty and inconstancy" – but rather these, "fidelity and obedience".' The message, moreover, was as gratefully received as one might expect, for in the winter of 1583 James would write a long and poignant lament for his exiled loved one, entitled *Ane Metaphorical Invention of a Tragedie called Phoenix*. As Lennox came from France, James wrote, so the phoenix flew to Scotland from Arabia; as Lennox was converted to Protestantism, so the phoenix was tamed; as Lennox experienced the enmity of the ministers of the Kirk and of Gowrie and his associates, so the phoenix; and as Lennox was exiled to France, so the phoenix flew off to Arabia and sacrificed itself on its own pyre.

By the time the verse was complete, however, Lennox was already dead. According to one report he had been suffering from an 'affection [i.e.

infection] hepatick and dissenterick', and according to Calderwood from 'a dysentery, or excoriation of the inner parts, engendered of melancholy, wherewith was joined gonnorhea'. As he lay on his deathbed on 26 May 1583, he had refused the last rites from the Catholic priest who came to attend him. And though his wife, Catherine, would bury his remains at Aubigny according to full Roman ritual, he had also – perhaps as the supreme compliment to his former royal master – declared his commitment to the Protestant faith. Already, of course, he had provided James with the inviolable memory of his first love and liberation, in return for which the king would exhibit a lifelong concern for his children. And now, last of all, he would bequeath his former royal master his embalmed heart.

# 4 ❧ Lessons in Life and Kingcraft

'A king will have need to use secrecy in many things.'

*James VI*, Basilikon Doron, *1599*

Though the so-called 'Ruthven Raid' had been conducted with minimal violence, its effect upon the king could not have been more profound. On the one hand, it had dealt a withering blow to his youthful self-esteem and newfound confidence in his own potential as a ruler. But it had also represented the grossest insult to his sovereignty – that God-given authority, as he saw it, marking him out so absolutely from each and every one of his subjects. In October, moreover, the forty-sixth General Assembly of the Kirk had described the raid as 'a good and godly cause', and James's continuing captivity in the months that followed merely compounded both the insult and his fury. On one winter evening at supper, we are told, he had toasted Lennox in his captors' presence amid stony silence, 'wherewith the king moved after he had drunk and hurled the rest over his shoulder'. Like his mother before him, he was clearly not without the stomach for a bold gesture when desperation left him no option. And, like her too, two objectives now consumed him. One, of course, was escape, and he confided to Sir James Melville, the only associate of Lennox allowed to approach him, that he had 'taken up a princely courage either to liberate himself fully, or die in the attempt'. The other was revenge for, as he told Sir Robert Bowes later, he would never forget how 'greatly wounded' his honour had been.

But if James found himself in what Bowes described, with remarkable understatement, as a 'ticklish situation', the quandary facing his enemies was, if anything, even more daunting. On the one hand, the Earl of Gowrie and the other so-called 'Lords Enterprisers', though temporarily in control, were finding, even more than Morton before them, that the king was now too old to be held under compulsion convincingly. The unwritten conventions governing royal abduction required that the act should be conducted for the king's own good and that there should be some degree of co-operation between kidnappers and kidnapped. He would also have to be paid for and even this was not the end of their predicament, for unless the king endorsed their authority, they could make little use of it, and to gain his support in the first place, they would have to grant him the kind of additional freedom that might well lead to his escape. In short, the mouse had cornered the cat and was now faced with the consequences.

The pro-English sentiments of James's captors were also proving of limited value. Indeed, Elizabeth's support for the Ruthven Raid was characteristically equivocal, which was amply demonstrated by her lukewarm response to the embassy that she received in London in April 1583. With typical artfulness, the king had already written to tell her how he wished to follow her counsel in all matters of importance, since ingratitude was the vilest of vices. But without her financial support, the Lords Enterprisers could not hope to remain long in power, and they were subsequently offered only a quarter of the £10,000 they claimed to require for guarding and maintaining the king. Elizabeth was even prepared to guarantee their subservience by threatening to consider Mary's earlier suggestion of establishing joint sovereignty with her son. In reality, of course, the Association that now briefly reappeared on the agenda remained the non-starter it always had been, not least because James was less prepared than ever in Lennox's absence to entertain the notion, but it nevertheless demonstrated England's coolness to the Ruthven conspirators aptly enough, and fuelled the unease which was soon steadily infecting their ranks.

The future, then, was on James's side, and in preparation for his bid for freedom, he now adapted his behaviour accordingly. Dissimulating with all the skill that was now his trademark, he duly probed and prodded for opportunity. In early 1583, he had given his secret assurance to two French ambassadors that 'although he had two eyes, two ears, two hands, he had but one heart, and that was French', and in response the Marquis de Mainville busied himself in harnessing an anti-Gowrie party among the Scottish nobility. Huntly, Atholl, Montrose, Rothes, Eglinton, Seton and Maxwell all readied themselves to support the king's escape, while James himself duly swallowed his pride and set out to exploit his enemies' wishful thinking. Now, therefore, he ceased his ill-tempered outbursts and acts of defiance and progressively convinced the Lords Enterprisers that he was willing to remain in their hands, even conceding on one occasion that Lennox was 'not wise'. Indeed, his stubborn sullenness gave way by degrees to feigned good humour as he exhibited a surprising graciousness even to the Earl of Gowrie himself, whom he considered to be as directly responsible for Lennox's death as if he had murdered him outright. The duke's privations at Dumbarton in the autumn of 1582, coupled to the rigours of the subsequent winter journey back to France, had served, or so James believed, to break his favourite's health irreparably. But even this would not impair the king's performance. And in playing his role to perfection, he duly seized his chance.

It was not without some irony that in allowing James the liberty of a hunting trip in June 1583, his captors made the same mistake as Lennox before them. And James's escape plan was in essence no less straightforward than the one that had led to his capture in the first place. He was to be issued a sudden and apparently innocent invitation by his great-uncle, the Earl of March, to come to St Andrews – where other members of the anti-Gowrie faction would be on hand – and 'make good cheer with him'. In the meantime, the need for hasty attendance was emphasised on the far from compelling grounds that the elderly earl had prepared a feast of 'wild meat and other fleshes that would spoil' in the event of delay. No greater artifice than this was required, it seems, and even the trip from his designated hunting spot at Falkland to St Andrews was conducted in apparently holiday spirits as the exultant king rode towards safety 'passing his time in hawking by the way'. 'His Majesty,' wrote Sir James Melville, already 'thought himself at liberty, with great joy and exultation, like a bird flown out of a cage … Albeit I thought his estate far surer when he was in Falkland.' And though Gowrie rode hard to catch him, he was indeed too late to stop the king.

There followed a day and night of tension, as St Andrews Castle eventually teemed with armed supporters of both factions – each ready to unleash murderous riot, though neither anxious to take responsibility for doing so – until the Earl of Gowrie's nerve finally broke. Accepting the inevitable and falling to his knees in James's presence, the earl 'in all humility asked pardon of the King's Majesty … and showed himself penitent in particular in the offences he had made and uttered against the late Duke of Lennox … and above all, against his Majesty's own person'. Reproaching him first before pardoning him, James duly savoured the occasion, though the sweetest moment of all did not arrive until the morning of 28 June when the king duly announced to his assembled nobles that he was now free from all faction and intended to rule henceforth as a 'universall king' in his own right, drawing them 'to unity and concord' and 'impartial to them all'. Faithful and ever willing to listen to good counsel, he would forgive past offences and surround himself with wise and virtuous advisers, though decisions would be delivered ultimately according to his princely judgement alone.

There was now, wrote Sir Robert Bowes, 'a great alteration both in his mind and also in his face and countenance', and there was indeed no denying that Scotland had, at long last, an independent ruler with a mind of his own, for James was not only free from those who had held him captive physically. Gone, long since, was the first man in whom he had hoped to find a mentor, the Earl of Atholl. Gone too, for the time being, was his childhood companion, the factious Earl of Mar. Gone finally, though long since repudiated by his pupil, was George Buchanan who had died in October 1582 and now, like Morton and Lennox, was forever out of influence. Henceforth, as Bowes also reported resentfully, James would keep the key of the box containing his private papers himself, so that the English ambassador could not 'get any certainty of the contents'. For the time being, too, freedom from faction at home would also free the king from subservience to England. From now on, it seemed, neither pungent letters from England's queen or embassies from her ailing Secretary of State, Sir Francis Walsingham, could alter the fact that Scotland had, in effect, a new king – albeit, as Walsingham observed angrily, 'a dissembling king, both with God and man,' and one who 'with a kind of jollity said that he was an absolute king'.

Yet James still had call for a man of strength about the throne, and that man now would prove to be the Earl of Arran, whose temporary imprisonment in the aftermath of the Ruthven Raid served as no more than a minor setback. Certainly, it had done nothing to temper either his ambition or his ruthlessness. 'Quick, penetrating, subtle, desirous of goods and greatness, arrogant, confident and capable of many things', Arran was ideally suited to become the king's chosen instrument for government at this critical juncture – not only willing to wield the cudgel, but also to lift the broader burden of rule from his royal master's shoulders. For though James was wilful and headstrong – so much so, indeed, that 'he could hardly be withdrawn from the thing that he desired' – and was resolute in his determination to direct policy, the daily routine of government remained repugnant to him.

The Frenchman Fontenay, who came to Scotland in 1584, considered the young king 'too lazy and thoughtless about business, too devoted to his pleasures, especially to hunting, leaving all his affairs to be managed by the Earl of Arran'. But James, it seems, was keen to explain away any hint of indolence. Excessive work was inclined to make him ill, he claimed, but nothing of importance occurred without his knowledge, since he had spies at the chamber doors of his councillors and was told everything they said. And though he lacked endurance, he could, or so he believed, do more work in an hour than others might do in a day when he applied himself. Indeed, it was his boast that he often accomplished more than six men together. He watched, listened and spoke simultaneously and sometimes did five things at once, he said. Besides which, he had advanced only simple soldiers and gentlemen rather than high-ranking nobles, since he could easily ruin them if they proved either inefficient or disloyal. In consequence, the king remained as confident of his application as he was of his ability – and it was this naive faith in his superiority over ordinary mortals that would, of course, explain many of his subsequent miscalculations.

For the while, however, James could indulge his passions while Arran steered the ship of state. Now, for instance, his 'vacant hours' were spent increasingly in the company of the poet Alexander Montgomerie – a hard-drinking, witty man, whose raffish exterior hid a nature of considerable sensitivity and sweetness – and a coterie of other literary 'brethren', consisting of figures like Sir Patrick Hume of Polwarth, Alexander Scott, Thomas Hudson, John Stewart of Baldyneiss, and the poetess Christian Lindsay. To the king, Montgomerie was 'Belovit Sanders, master of our art', and while Lennox may well have been the first to ignite James's love for the muse, it was Montgomerie who seems to have been most influential in guiding his pen. Yet, far from living up to the flattery heaped upon him by his 'bretheren' as the 'royal Apollo', the king's was in fact a 'dull Muse', as he himself admitted, and he was actually at his most inspired only rarely – usually when writing on political themes. His famous sonnet 'God gives not kings the style of gods in vain' was certainly creditable enough but he remained, in general, no more than a competent spinner of verses, to the extent that his first published work, *The Essayes of a Prentice in the Divine Art of Poesie*, was published anonymously in Edinburgh in 1584.

Nevertheless, the king's reliance upon his dutiful earl remained of considerable overall benefit in personal terms. Freed at last from fear and insecurity, he found himself liberated, too, from tiresome obligations and able, in the process, to fly in whatever direction – artistic, social, recreational – his fancy took him. But more generally, his relationship with Arran was at the very best a mixed blessing, for while the latter's drive and efficiency made up for the king's frailties in these areas, he was generally unpopular and widely mistrusted both at home and abroad. Certainly, there could be no charges laid against Arran of the kind made previously against Lennox by the minister Andrew Melville that the king had been kept 'in a misty night of captivity and black darkness of shameful servitude'. On the contrary, James was no puppet and Arran no 'pseudo-regent'. Indeed, the king remained in a position of commanding partnership, employing the earl, who became chancellor in 1584, as both executor and enforcer where his own youth might otherwise have compromised his wishes. But Sir Francis Walsingham and the English as a whole were deeply suspicious of Arran's former link with Lennox's pro-French inclinations, and the execution of the Earl of Gowrie on 3 May 1584, ostensibly for a new act of treason, confirmed the worst fears of many Scottish noblemen. For it was believed in most quarters that Gowrie's death was an act of revenge – a final token of retribution for the Ruthven Raid and Lennox's untimely demise. There were ominous rumours, too, that Arran had tricked his victim into making a full confession of his second plot by a promise to obtain for him the king's clemency, which, if true, was a shabby return for the mercy that Gowrie had earlier extended to him in the aftermath of the king's abduction.

Subsequent to the execution, moreover, the earls of Angus and Mar and the Master of Glamis were all forced to fly to England while others were ruined, it seems, for no better reason than that they were worth ruining; some, it was rumoured, at the prompting of Arran's capable but greedy wife who was roundly hated by the preachers and reviled as a witch. 'These cruel and rigorous proceedings,' wrote Sir James Melville, 'caused such a general fear, as all familiar society and intercourse of humanity was in a manner lost, no man knowing to whom he might safely speak or open his mind.'

And it was not long either, of course, before Arran was delivering the king's revenge upon the Kirk and its ministers, the more outspoken of whom had already fled to Berwick in anticipation.

James's first interview with the Kirk's leaders after his escape from the Ruthven lords had already augured ill. When they entered his presence at Falkland, he glared at them in silence for fully fifteen minutes before rising and leaving the room and eventually calling them to his cabinet. Furthermore, the exchange which followed was hardly less strained. 'No king in Europe,' James informed his audience, 'would have suffered the things that I have suffered,' and in response to complaints about developments at his court, he went on to declare that he alone might choose 'any that I like best to be in company with me'. Equally provocatively, the king justified his stand on the grounds that he was 'Catholic King of Scotland' – a somewhat injudicious choice of terms, to say the least, which finally prompted the intervention of one of the more amenable ministers. 'No brethren,' said David Ferguson, 'he is universal king and may make choice of his company, as David did in the 110th Psalm,' though few of his colleagues were satisfied. 'We will look no more to your words,' James was informed, 'but to your deeds and behaviour; and if they agree not, which God forbid, we must damn sin in whatsoever person.'

Even worse was to follow, however, when John Durie and Andrew Melville refused to recognise the council's jurisdiction after being called before it to answer for their sermons. Appointed moderator of the General Assembly in 1578, Melville had led the Kirk's onslaught against Robert Montgomery and emerged by turns as the leading and most vociferous critic of secular interference in religious affairs. Famed for his unashamed irreverence towards the 'anointed monarch', it was he, who in defending the principle of the 'two kingdoms' dismissively referred to James as 'God's sillie vassal', and now he would prove no less defiant. The king, Melville pointed out, perverted the laws of God and man, and councillors possessed no authority over the messengers of a king and council far more powerful than they. Taking a Hebrew Bible from his belt, Melville then, it was said, 'clanked it down on the board before the king and chancellor'. 'There,' he declared, 'are my instructions and my warrant,' after which he was ordered into confinement at Blackness Castle before fleeing to England.

Under such circumstances, it was hardly surprising, perhaps, that James and Arran would move to subdue the more radical elements within the Kirk, and in May 1584 the Presbyterian system in Scotland was temporarily ended by what would become known as the 'Black Acts'. In a sweeping attempt to bring the Kirk to heel, the king was henceforward declared head of religious affairs and given jurisdiction over ecclesiastical cases. The courts and assemblies of the Kirk were now to convene only with royal permission and, more provocatively still perhaps, the authority of bishops was confirmed in preference to that of 'pretended presbyteries'. From this point forward, moreover, all affairs of state were to be reserved purely for the judgement of the secular ruler without interference or comment from the pulpit. And to drive the message home, Arran added his own inimitable brand of subtlety. To those ministers who persisted in their objections, the chancellor's message was clear. Their heads would be shaven as an example to all who held their sovereign ruler in contempt.

There was, however, altogether less scope for assertiveness abroad, where James and Arran found themselves forced to renew the foreign intrigues begun by Lennox. Estranged from England, menaced by exiled lords and ministers, and threatened by growing rumbles of discontent at home, the appeal of French or Spanish assistance grew increasingly irresistible, and James slithered accordingly into a further bout of diplomatic intrigue. Sir Francis Walsingham, in a fit of pique, had already told the Scottish king that his power was insignificant and that rulers as young as he were apt to lose their thrones, so when James was contacted by the Duke of Guise after his escape from captivity, his response was predictable: thanking the duke warmly for his friendship and offers of protection, James praised him as the first soldier of the age and expressed his willingness to join an enterprise to release his mother and bring vengeance upon her captors. Further appeals for help were also addressed to the Kings of France and Spain, and most surprisingly of all to the pope himself. 'I trust,' wrote James, clearly implying the possibility of his conversion at some later date, 'to be able to satisfy your Holiness on all other points, especially if I am aided in my great need by your Holiness.'

But this, as the king well knew, was playing with fire, and the pitfalls of his dealings were soon painfully exposed with the appearance of a new favourite – the eminently plausible and equally treacherous Patrick, Master of Gray. As with Arran, no portrait of Gray exists. But he was a handsome, urbane and polished nobleman, who had been appointed a gentleman of the privy chamber in October 1584 and made Master of the King's Wardrobe and Menagerie, in charge of James's jewels, clothing and tapestries, and the employment of tailors and shoemakers. He was also fully conversant with the affairs of Queen Mary, especially in France, where he had been her agent. Indeed, he was still in her service – or so she believed at least, for he had lost faith in a Catholic assault upon England and now considered her situation hopeless. Spain, it seemed, was hesitant, the pope niggardly and Guise weighed down by domestic concerns, and if the Spanish king should actually succeed in unseating his English counterpart, he was hardly likely to bestow her throne upon the Protestant King of Scots. Thus, Gray reasoned, James should abandon his mother's interests and ally instead with Elizabeth.

In the meantime, of course, James would continue to court his mother's representatives, though he had already confirmed to Sir Robert Bowes that her entanglements with Catholic powers rendered her unfit to rule either Scotland or England. The Association, he told Bowes, had been offered by Mary only out of self-interest and was 'tickle to his crown'. Yet when M. de Fontenay, the former queen's representative, arrived in the summer of 1584, James protested loudly that he would never abandon her. On one occasion, in fact, he went so far as to summon Arran, before whom he delivered an oath in Fontenay's presence that Scottish policy would support Mary's cause unwaveringly – an astonishing example of James's tendency to overact in his attempts to deceive. In fact, he fooled no one – least of all the canny Frenchman who noted nonetheless that 'he has a remarkable intelligence, as well as lofty and virtuous ideals'. And this was not all that Fontenay said of the king, for in accepting that his mission was a failure, he also went on to present us with one of the most illuminating brief portraits of the king on record.

The envoy's letter to the Queen of Scots was, in effect, a long and rambling report of his mission in which he attempted to offer some assurance of her son's goodwill towards her. But the letter he addressed to his brother-in-law, Claude Nau, Mary's secretary, was altogether more informative and bore the superscription 'My brother, the letter which follows will remain secret between you and me'. From one other observation, too, it is clear that Fontenay was planning to furnish a full and frank account of his judgements that should not be seen by Mary herself. The king, he noted, 'has never asked anything about the Queen, neither of her health, nor of the way she is treated, nor of her servants, nor of what she eats or drinks, nor of her recreation, nor any similar matter, and yet, notwithstanding this, I know that he loves and honours her much in his heart'. Clearly, the intention was to provide the kind of intimate pen portrait not necessarily normal or appropriate to a visit of the kind that Fontenay had just conducted, though the account that resulted was both fair and perceptive.

On the one hand, the young king's merits were listed in full. 'Three qualities of the mind,' Fontenay observed, 'he possesses in perfection: he understands clearly, judges wisely, and has a retentive memory.' His questions were 'keen and penetrating', his replies 'sound' and 'in any argument, be it about religion or any other thing, he maintains the view that appears to him most true and just' – to the extent that 'I have heard him support Catholic against Protestant opinions'. The king was also 'well instructed in languages, science and affairs of state; better, I dare say, than anyone else in his kingdom' … 'In brief,' Fontenay concluded, 'he has a remarkable mind, filled with virtuous grandeur,' though it was also noted that he possessed 'a good opinion of himself' and had a great desire to hide his deficiencies, attempting anything in pursuit of virtue. Though timid, for example, he wished greatly to be considered courageous. Nor could he bear to be surpassed by other men. Upon hearing that a Scottish laird had passed two days without sleep, the king passed three, and if ever a man succeeded in outstripping him, he abhorred them, it seems, forever.

But Fontenay furnishes us, too, with a range of comments on the king's habits and everyday demeanour that not only picture him minutely, but came to shape the standard image of him down the centuries. We hear, for instance, how he disliked 'dancing and music in general', along with 'all the little fopperies of court life', whether they involve 'amorous talk or curiosities of dress'. In particular, he had 'a special aversion to ear-rings'. There is also the earliest reference on record to James's lack of physical grace. 'He never stays still in one place,' commented the Frenchman, 'taking a singular pleasure in walking up and down, though his carriage is ungainly, his steps erratic and wandering, even in his own chamber.' And there is mention, too, of other less savoury characteristics. His manners, it seems, were 'aggressive and very uncivil, both in speaking and eating and in his clothes and sports', as well as in 'conversation in the company of women'. His body, meanwhile, was 'feeble' – though he was not himself 'delicate' – and his voice 'loud'; his words 'grave and sententious'. 'In a word,' concluded Fontenay, 'he is an old young man,' though he loved hunting, of course, above all other pleasures, 'galloping over hill and dale with loosened bridle' for six hours at a time.

Most intriguingly of all, however, Fontenay explored what he considered James's chief deficiencies as a ruler. In all, he noted 'only three defects which may possibly be harmful to the conservation of his estate and government'. The first was his 'failure to appreciate his poverty and lack of strength, overrating himself and despising other princes' – common enough faults in many contemporary rulers, of course, and especially understandable, perhaps, in so young a king, though James's reputation for extravagance would be heavily reinforced by his behaviour in later life. Then Fontenay observed how James 'loves indiscreetly and obstinately despite the disapprobation of his subjects', and criticised him further for being 'too idle and too little concerned about business'. In this latter connection, the king was apparently 'too addicted to pleasure, principally that of the chase, leaving the conduct of business to the Earl of Arran, Montrose [the treasurer] and the Secretary [John Maitland of Thirlestane]'. And here, too, the criticisms levelled against James would be raised repeatedly in years to come. In all, he emerges as a ruler of considerable intellect but one of limited political intuition: cunning but also prone to naivety; thoughtful but of limited self-awareness; proud and determined to lead but lacking in self-control and charisma – in short, a king of many commendable qualities without the hallmarks of genuine majesty.

Above all, perhaps, Fontenay had been shocked by James's carelessness concerning money. 'The king is extremely penurious,' it was noted. 'To his domestic servants – of whom he has but a fraction of the number that served his mother – he owes more than 20,000 marks for wages and for the goods they have provided.' 'He lives,' said Fontenay, 'only by borrowing.' Yet this, we are told, in no way restricted his spending. The confiscated lands and revenues of the Ruthven lords were frittered upon courtiers and when James was quizzed about money his answers were unsatisfactory. Having claimed at first that his finances were sound, he then confessed they were not, only to add that his youth and nature made it difficult for him to fulfil his promises to be less liberal. It was hardly surprising, then, that in his official letter to Mary, Fontenay warned her against sending any money to her son.

Nor was it without irony that one of the major beneficiaries of the king's generosity at this time was none other than the Master of Gray, who had recently received the tidy sum of 6,000 marks that had been gifted to James by the Duke of Guise. More and more influential with each passing day, Gray was now sent to England in August 1584 with the specific intention of forging an Anglo-Scottish alliance that would not only ignore but directly undermine the interests of the former queen, upon whose behalf he had supposedly been working. With the expulsion of the Spanish ambassador from London in January 1584, the English were increasingly anxious to secure their northern border, and the very choice of Gray as emissary to Elizabeth gave the clearest possible proof that James was prepared at last to abandon his mother openly, since his mission was to reveal her secrets and reach an agreement with her English rival in which she had no part. In the process, he would also reveal the existence of his own enmity with Arran, which could be exploited as Elizabeth saw fit.

When, however, Mary discovered the true nature of Gray's dealings, she was both appalled and terrified. Fondly believing that he had been negotiating for her release, it was not until March 1585 that she discovered his treachery. Worse still, the duplicity of her son was soon equally manifest, for, upon receiving an English pension of some £4,000 a year and a gift of fallow deer, James would finally disavow the Association upon which Mary had staked so much. Politically, of course, his decision was a wise one, since the way now stood open not only for a league with England but even, perhaps, the prospect of his nomination as Elizabeth's successor. The English Parliament's Act of Association in 1584 had, after all, explicitly disqualified from the succession any candidate who supported armed action in favour of Mary Stuart. And there was no denying either that Mary's dealings had been equally tainted with self-interest from the outset. When the time came, then, the choice was a clinical one. Though, as John Colville pointed out, James seemed 'not to have lost all affection to his mother … yet (as those about him will speak) hee had rather have hir as shee is, then him self to give hir place'. Coldly informing her in 1586 that he could not 'associate her with himself in the sovereignty of Scotland' and that he could not in future 'treat with her otherwise than as Queen-Mother', James duly delivered his parting thrust. Nor, when Mary condemned her 'ungrateful son' and threatened to disinherit him in favour of Philip of Spain, did he offer any further justification. Instead, he offered only silence and never wrote to her again.

The Queen of England, meanwhile, was fully aware of James's slippery tendencies. 'Who seeketh two strings to one bow,' she told him, 'they may shoot strong, but never straight.' And she knew no less certainly that any alliance with Scotland could never be confidently secured without the departure of Arran who was now increasingly exposed on all fronts. To the long list of his former enemies he had added the Catholic nobles, alienated by the negotiations with Elizabeth, and the insolence of both him and his wife had by now even upset the king. William Davison, the English ambassador 'observed the strangeness of their behaviour towards the poor young prince, who is so distracted and worried by their importunities as it pitied me to see, and, if I be not abused, groweth full of their fashions and behaviours, which he will sometimes discourse of in broad language, showing he is not ignorant of how they use him'. Men of influence like Maitland of Thirlestane, who had initially supported Arran, were also no longer his friends, and with the Master of Gray now bolstered by English approval as well as the king's high favour, the time of crisis was not far off.

With Gray's return from London, his enmity with Chancellor Arran was soon spawning mutual fears of assassination, as the court sank into confusion and uproar and English agents awaited their opportunity. For all his panache and earlier bravado, Arran's heart, in particular, began to fail him as he awaited the inevitable onslaught. Armed at all times as he went about his daily affairs distractedly, he brought armour and supplies into Edinburgh Castle, and strengthened his apartments at Holyrood. On one occasion, too, he showed that even for a soldier of such dashing reputation as he, discretion could yet be the better part of valour. Returning one night through the ill-lit alleys of the capital, he let his wife pass on up the High Street, it seems, while he, with torches extinguished and a cloak about his shoulders, stole into the fortress by a secret route accompanied by a lone servant.

When the time came, however, neither the dark, nor disguises, nor secret passageways could save Arran. Nor, for that matter, could innocence on the one occasion, perhaps, when he had done no real wrong. For when Lord Francis Russell (the son of the Earl of Bedford) was shot and killed by Sir Thomas Kerr of Ferniherst (the Scottish warden of the Middle March) during a 'day of truce on which Scottish and English representatives were meeting to settle Border disputes', the blame was at once ascribed to the chancellor. Since Ferniherst owed his appointment to Arran, it was easy for the latter's enemies to accuse him of instigating the killing as a means of preventing the conclusion of the proposed Anglo-Scottish treaty. And even though Arran had actually been responsible for initiating the discussions with England's representatives, this was conveniently ignored. Indeed, Sir Edward Wotton, the leading English negotiator, later admitted without embarrassment how he had 'aggravated the matter not more than it deserved but as much as he could'.

The Scottish king, moreover, offered comparatively little resistance in abandoning his leading enforcer. Though greatly distressed by the incident, his concerns were at least as much for his pension and prospects for the English succession as they were for the fate of his chancellor. Wotton found him, in fact, alone and melancholy in his chamber, his eyes swollen by weeping. In discussing Russell's murder, the king, wrote Wotton, 'shed tears over it like a newly beaten child, protesting by his honour and crown that he was ignorant of this practice, desiring her Majesty not to condemn him for other men's faults'. The Queen of England, he lamented, would consider him a dissembler, though he would rather lose all the kingdoms of the world than be found untrue to his word. Indeed, for twenty-four hours he neither ate nor drank nor slept before proving his good faith to Elizabeth by ordering Arran's imprisonment in St Andrews Castle. And though the earl was released a week later on James's orders, it was no difficult task for Elizabeth to finish the matter by agreeing to 'let slip' the exiled 'Ruthven Raiders' across the Border, where they could be guaranteed to tip the balance. On 2 November, therefore, the earls of Mar and Angus, the two Hamiltons and the Master of Glamis duly appeared before Stirling Castle with sufficient forces to compel the king to abandon his favourite once and for all.

Inside the castle there followed the usual chaotic scenes of violent recrimination in which Gray and Arran and their rival gangs squared off in the king's presence without actually striking, since both 'suspected falsehood in friendship' and feared to trust his own side. Arran denounced Gray as the instigator of the exiles' return, while Gray protested his innocence – apparently with some success, for in the morning Arran's nerve broke and he opted to save himself by flight, closely followed by the king who, in nervous terror, attempted unsuccessfully to exit through a postern gate previously locked by Gray. In the event, the decisive factor in Arran's submission was probably the powerful threat posed by the new Earl of Bothwell, nephew of Mary's lover and already, with his considerable Border power, one of the key political players in Scotland. With his political career at an end, Arran had no other choice, in fact, than to head west and to settle for the obscurity that must have been so intolerable for such a one as he. In conceding that he should forfeit his earldom and the chancellorship, James did, however, protect him as best he could from the further vengeance of his enemies – until, that is, a nephew of the former regent, Morton, with a long memory and lingering sense of duty finally butchered him in 1595.

# 5 ❧ The Headsman and the King of Spain

'[We are] in a despair to do any good in the errand we came for, all things disheartening us on every side, and every hour giving us new advertisement that we deal for a dead lady.'

*George Young, member of the Scottish embassy to plead for the life of Mary Stuart, 10 January 1587*

On 3 November 1585, the King of Scotland was once again obliged to 'make vertue of a need' and receive the returned Ruthven raiders in the Great Hall of Stirling Castle. Though they sank to their knees and professed their loyalty, it was surely a scene to excite mixed emotions in the young man occupying the throne. In Sir James Melville's view, the king spoke pertly and boastfully, as if victorious over the rebels. Yet Calderwood suggests that he addressed them 'with cheerfulness, it seemed', thanking God that they had returned with so little bloodshed. In any event, he pardoned them and proceeded to demonstrate not only his rapidly developing adroitness by the speed with which he turned the new situation to his advantage, but the extent to which he was now – as he had been for some time – the master of policy in his kingdom. By finally accepting the overthrow of Arran and reinstating the exiled lords in Scotland, James had, indeed, freed himself from further compulsion. And the continuance of his recent measures was the firmest possible proof that he rather than his favourite had been the instigator of policy all along. The 'Black Acts', for example, though modified in application, were not rescinded, and when the exiled ministers returned in the wake of the lords, who now distanced themselves from religious radicalism, the king maintained the initiative. Andrew Melville, for his part, found himself ordered north to spend his time looking for Jesuits and to 'travail, so far as in him lies, to reduce them to the true and Christian religion', while others even more vociferous, such as John Gibson, were given short shrift. After denouncing James for maintaining 'the tyranny of bishops and absolute power', Gibson was swiftly dispatched to prison, but not before the king enjoyed the final word. 'I give not a turd for thy preaching,' he had howled in derision at the hapless minister.

Nor, when James pressed forward with the Anglo-Scottish treaty was he succumbing to external pressure. Though the arrival in January 1586 of a resident French ambassador, M. de Courcelles, was consciously designed to assert his independence, alliance with England remained his first best option, and was actually the most assured means of nullifying once and for all any potential threat from the old Gowrie faction. For, by becoming England's ally, James would curb at once the baleful effects of that country's interference in Scottish politics, which had led to his abduction at Ruthven in the first place. He had, moreover, already shown his teeth over Gowrie's execution, so that the earl's former associates were more than grateful to receive their lands anew, in spite of their past misdeeds. And the loss of Arran was never, in any case, the blow to royal prestige – let alone the emotional trauma - that Lennox's departure had represented. Indeed, James had become increasingly concerned in his own right about the chancellor's over-mighty posturing and may even have tacitly encouraged Gray in his machinations. Much more importantly still, Arran's removal left the way clear at last for the promotion of John Maitland of Thirlestane, the gifted and comparatively selfless statesman, who eventually succeeded to the vacant chancellorship in July 1587: a man who was more ready than most to shun the limelight and more able than any to harness the king's talents while masking his weaknesses.

All was set fair, then, for the effective restoration of royal authority with considerable smoothness and in exceptionally quick time. James's former enemies were, after all, not so much a compact party as a coalition drawn from discordant elements, united for a brief space of time only by a common hatred of Arran. While Angus, Mar and Glamis had headed the coup which restored them to power, the returning exiles also included

Lord John Hamilton, an old adherent of Mary though a Protestant, his younger brother, Lord Claud, a Catholic, and the Catholic Border chieftain, Maxwell. Other Catholics joined the council, too, alongside men like Maitland, Gray and Sir Lewis Bellenden, who were all retained from the previous government. In consequence, there was a healthy balance which gave the king, for the time being at least, an unexpected degree of independence – especially when it is remembered that most of his former enemies were heavily preoccupied with restoring their private fortunes. Indeed, to humour the king they even yielded precedence to his new young favourite, Ludovic Stuart, the 10-year-old son of Lennox, whom James had recently brought over from France. No guard was placed about the king as he hunted to his heart's content, and in matters of state, too, he would be crossed more rarely than ever before.

By the summer of 1586, moreover, James even had reason to believe that Scotland's age-old enmity with England, which had hitherto proved so damaging to his kingdom might well be ending. His power to negotiate the most advantageous terms had been limited, in fact, by his keenness for official recognition as heir to the English throne, though the most that Elizabeth would offer was a 'firm promise in the word of a queen' that she would never bar him from 'any right or title that might be due to him in any time present or future', unless 'by manifest ingratitude she should be justly moved and provoked to the contrary'. Plainly, the Queen of England was still intent upon exerting the maximum control over James's actions – and for good reason, too, since any assurances given to James while his mother lived involved tacit acceptance of her own place in the succession. But on 5 July, notwithstanding such snares and limitations, the treaty of alliance was indeed formally signed, and James could congratulate himself, it seemed, at another sign that his earlier difficulties were at long last reaching resolution.

The unforeseen complication which now presented itself therefore left James both stricken and dumbstruck, for at the very moment the alliance was being sealed, the final act of his mother's sad drama was about to begin. Her complicity in the Babington Plot had been painstakingly monitored by Elizabeth's spymaster, Walsingham, and on 3 August her papers were seized and she was duly arrested along with her secretaries. Such was his surprise that James had no idea at first of Mary's full predicament. As late as October he was assuring the French ambassador that she was 'in no danger' and should henceforth meddle with nothing beyond prayer and service to God, though he acknowledged that 'as for the conspiracy, she must be content to drink the ale she has brewed'. From his perspective, this was likely to entail little more than closer confinement. Mary's servants, on the other hand, should be hanged, new ones appointed, while she herself should be 'put in the Tower or some other manse and kept from intelligence'. Even by Christmas, moreover, he seemed unable to comprehend that Mary's execution was a distinct possibility, though she had already been brought to trial at Fotheringay on 11 October and sentenced to death before the month was out. Certainly, the thought that he had signed away his mother's life for a pension of £4,000 only months earlier would never occur to him.

Nor was James much inclined to disturb his personal relations with Elizabeth on his mother's behalf. 'The only thing he craves is her life,' wrote Roger Aston, an Englishman in his service. But reports from Archibald Douglas, the Scottish representative in London, whom James had appointed in spite of his implication in the murder of Darnley, gradually seem to have taken effect. Douglas, along with Sir William Keith, a young member of the royal household who acted as the king's messenger, had been instructed by James to pursue two objectives: 'the one to deal very earnestly both with the Queen and her councillors for our sovereign mother's life', the other to ensure that his title to the English Crown 'be not pre-judged'. On 22 November, however, Douglas informed the Master of Gray that Mary was 'in extreme danger of her life' and Gray's response reveals his master's growing concern – as much for himself as for his mother. 'The king nor no man ever believed that the matter would have gone so far,' wrote Gray, before pointing out James's concern not only for his mother but 'the opinion of all his people' and the implications for his 'honour'. And James was indeed largely hamstrung, for, as Douglas had reminded him, the English Parliament's Bond of Association rendered any interference on his part a most risky undertaking, since it might easily be employed against him if any suspicion arose that he was party to Mary's plottings. In effect, therefore, he was faced with an excruciating choice between his mother's life and his hopes for the English throne.

But the pressure was raining in upon James from other quarters, too. That Elizabeth should threaten violence against a Scottish queen was wholly abhorrent to his nobles who assumed automatically that Mary's death would at once bring war with England and alliance with her enemies.

The Earl of Bothwell, who shared, it seems, his ill-fated uncle's knack for plain speaking, told James that if he countenanced his mother's execution, he would deserve to be hanged next day – at which the king is said to have laughed and suggested that he would prepare for that. Lord Claud Hamilton, on the other hand, swore that he would burn Elizabeth's kingdom as far as Newcastle if Mary was injured, and even an enemy of the former queen, like Angus, declared that Mary would be fully justified in slitting Elizabeth's throat. Clearly, when national pride and the old enemy were involved, even the Scottish elite, famous for their blood feuds and divisions, could unite. And this meant only further trouble for their king, since his other subjects, too, were deeply embittered by the presumption of Mary's captors. 'I never saw all the people so willing to concur in anything as this,' wrote Gray. 'All men drive at him.'

Under the growing pressure, James's temper seems, in fact, to have compromised his judgement and he adopted a curiously provocative tone in the first letter of formal protest that he dispatched for Elizabeth's attention on 27 November. Sovereign princes like his mother, 'descended of all hands of the best blood in Europe' could not, he affirmed, be judged 'by subjects' mouths' and on this basis it was to the Queen of England's discredit that 'the nobility and counsellors of England should take upon them to sentence a Queen of Scotland'. Much more strongly still, he reminded Elizabeth that, while her father had stained his reputation 'by the beheading of his bedfellow', that particular 'tragedy' was 'far inferior' to what she was now contemplating. There were further references, too, to James's personal predicament and an ambivalent suggestion that he wished only to protect the English monarch's name and interests north of the Border. 'I desire you to consider,' he appealed, 'how my honour stands engaged, that is her son and a king, to suffer my mother an absolute princess to be put to an infamous death.' 'Guess ye,' he added, 'in what state my honour will be in, this "unhappe" being perfected; since, before God, I already scarce dare go abroad for crying out of the whole people; and what is spoken by them of the Queen of England it grieves me to hear, and yet dare not find fault with it except I would dethrone myself, so is the whole of Scotland incensed with this matter.'

Far from helping matters, however, the letter succeeded only in infuriating Elizabeth. Above all, the reference to Henry VIII's 'bedfellow', her mother Anne Boleyn, raised an altogether unmentionable subject and, in consequence, she was said to have taken 'such a chafe as ye would wonder'.

Indeed, according to Douglas who delivered the letter on 6 December, she 'conceived such a passion as it was a great deal of work to us all ... to appease her'. Nor were the after-effects of her rage in any way to James's advantage. Keith, his messenger, was now informed that if he 'had not delivered unto Her Majesty so strange and unseasonable message as did directly touch her noble father and herself', she would have delayed proceedings against the Queen of Scots, but now would not do so if any emergency should arise. Equally significantly, Elizabeth refused to receive a new delegation of two Scottish noblemen and decided instead that the Queen of Scots' case could only be presented to her by two commoners. In doing so, she ensured that James's final embassy on behalf of his mother would lack the prestige of nobility to lend it weight.

James, then, had clearly miscalculated and, in the meantime, the fact that his baiting of Elizabeth had been framed in a private letter did nothing to convince his subjects of his resolve. Worse still, his determination to defend his mother was, in any case, effectively non-existent. Indeed, on 3 December, the day that James's letter arrived in London, Douglas had conferred informally with the Earl of Leicester in his carriage and left little doubt of the king's real intentions. After offering to support his claim to the English Crown, Leicester had at first made clear the advantages accruing to James in the event of his mother's death. Then, without further ceremony, he had asked Douglas directly whether Mary's execution would put paid to the alliance between the two countries. The response was negative. The treaty, said Douglas, was the king's policy and he would not break it unless the English forced his hand – an implicit but unmistakable reference to his right to succeed Elizabeth.

Meanwhile, the official delegation that James had dispatched in accordance with Elizabeth's conditions was making its way to London, though Douglas's private discussion with Leicester had already largely undermined its mission. The two non-noble ambassadors stipulated by Elizabeth were Gray and Sir Robert Melville of Murdocairney, a steadfast partisan of Queen Mary, assisted by George Young, a member of the king's household and Sir Alexander Stuart, a dubious and fickle character who, along with Douglas, would eventually help to hammer the final nails into the queen's coffin. The plan, in fact, was to present the most cogent plea for Mary's life to date, assuring the Queen of Scots' abstinence from further political intrigue and guaranteeing Elizabeth's deliverance from further conspiracies. This time, moreover, there was to be no risk of offence. 'If neither of the overtures aforesaid be thought sufficient,' wrote James, 'ye shall with all instance press our dearest sister to set down by advice of her wisest and best affected counsellors such form of security as she and they shall think sufficient', making clear that 'we will not only yield for ourselves but also do our best endeavour to obtain the performance thereof'. For good measure, the ambassadors were to 'protest before God' that 'the life of our dearest sister is no less dear unto us in all respects than the life of our dearest mother or our own'.

On 15 December, James had also written to the Earl of Leicester, having learned only the day before of Douglas's dialogue with him. He knew therefore that unless he directly contradicted the impression that Leicester had been left with, his mother would die. But he knew, too, that to break the alliance with England would not only remove all hope of the English succession but throw him into a league with Spain that was potentially more damaging still. The prospect of Spanish victory was, after all, far from guaranteed and if it did indeed occur, dependence on his new Catholic ally was inevitable – with all that this would entail for the religious harmony of his realm. Both alternatives were heavy indeed, but one, as James realised, nevertheless remained preferable. Given that his ambassadors were already, in all likelihood, 'dealing for a dead lady', since Elizabeth's chief ministers had in effect staked their own lives by condemning Mary, he would salvage what he could from a lost cause and look through his fingers at his mother's fate. 'I am honest, no changer of course, altogether in all things as I profess to be', he informed Leicester, before declaring how 'fond and inconstant' he would be 'if I should prefer my mother to the title'.

Whether this last statement was meant to imply that he had no intention of 'preferring', i.e. supporting, his mother's claim to the title or whether he was referring to the choice between his mother's life and his own claim to the succession remains unclear. But such ambivalence is unlikely to have been coincidental and the impression remains that James was content for Leicester to retain the impression given earlier by Douglas. In any case, the letter was not genuinely indicative of the callous indifference of a son to his mother's plight. Instead, it was little more than a resigned footnote to a drawn-out saga that had already all but run its course. The end, if not the ending, had long been sealed in fact. And James was left with little more than empty gestures and hollow professions, concluding his communication with Leicester by pointing out once more how 'my honour constrains me to insist for her life', though 'my religion ever moved me to hate her course'.

There followed another letter to Elizabeth, delivered by Gray and Melville, which was another exercise in dutiful posturing. Employing precisely the kind of bold language that the decorum of abject submission demanded from a monarch in such circumstances, James spoke this time of the divinity of kings, whose sacred diadems were not to be profaned, and suggested once more that rulers were beyond the condemnation of their subjects. Elizabeth should, he said, beware the revenge of his mother's supporters and send her abroad under guarantees of future good behaviour. Alternatively, Mary might be encouraged to sign a bond committing her to a traitor's death if implicated in any further conspiracies. But James, as well he knew, was whistling in the wind and his choice of Gray as ambassador, when the latter had already betrayed the Queen of Scots on a previous mission, did little to hide the fact.

Gray and his partner, Melville, were in any case fighting a losing battle against the very men who were intended to assist them. Sir Alexander Stuart, for instance, lost no time in claiming to possess instructions superior to theirs and, together with Douglas, was soon informing Elizabeth that his royal master would accept Mary's death and 'with tyme digest the worst' – an impertinence which later so infuriated James that, according to Courcelles, he fell into a 'marvellous choler' and vowed to hang Stuart upon his return 'before he put off his boots'. But for all the displays of royal anger and fits of wounded conscience, Stuart's only fault had been to say what the king himself could not, as Elizabeth remained impervious to James's 'earnest suite' and 'friendly advice'. Indeed, when Gray suggested that Mary, in return for her life, should transfer to her son the right to the English Crown, Elizabeth did not hold back. 'By God's passion,' she exclaimed, 'that were to cut my own throat, and, for a duchy or an earldom to yourself, you or such as you would cause some of your desperate knaves to kill me.' When, likewise, the Scottish ambassadors appealed for a reprieve of firstly fifteen days and then eight, the response was unyielding. 'Not for an hour,' declared the queen.

One further letter followed in which James recited the familiar mantra: if Elizabeth knew of his grief and his problems in Scotland, she would spare him this ordeal. Princes, he repeated, were not subject to earthly censure, while Sir Alexander Stuart was to be condemned for exceeding his authority. Elizabeth should consider, too, he added, the 'almost universal hatred' that his mother's execution would evoke among other rulers. Yet such implicit threats were wholly unavailing, and when the Queen of

England hesitated at the last to sign her cousin's death warrant, it was her own private misgivings rather than fear of James that made her do so. When, moreover, the warrant was finally sealed in February she contrived, in any case, to confer the blame for 'that miserable accident' upon her hapless secretary William Davison – irrespective of the fact that she had spent a week attempting to persuade Mary's gaoler Sir Amyas Paulet to arrange the death in secret. 'God forbid that I should make so foul a shipwreck of my conscience or leave so great a blot to my poor posterity to shed blood without law or warrant,' Paulet had responded. But the axe duly fell on 8 February – the very day that Gray and Melville received the official thanks of the king's council for their efforts in England.

Accounts of James's reaction to the news of his mother's death vary markedly, in fact. One contemporary eyewitness, Ogilvie of Powrie, suggested in a letter to Archibald Douglas that the king was apparently indifferent to the report that reached him. 'The king,' we are told, 'moved never his countenance at the rehearsal of his mother's execution, nor leaves not his pastime and hunting more than of before.' David Calderwood, meanwhile, who was a child at the time and therefore not on hand to witness events, suggested that the king could not suppress his delight that his rule in Scotland was now at last unchallenged. 'When the king heard of the execution,' wrote Calderwood, 'he could not conceal his inward joy, howbeit outwardly he seemed sorrowful.' That night, it seems, he expressed his satisfaction at being 'sole king' – a comment which left Maitland so ashamed that he ordered a gaggle of onlooking courtiers from the room. Yet not all descriptions reflect quite so poorly upon James's lack of sensitivity. Moysie, for instance, suggests that he 'was in great displeasure and went to bed without supper' and rode to Dalkeith next morning 'desiring to be solitaire'. And when he was later given a moving description of his mother's final moments by one of her ladies-in-waiting, he was said to have been 'very sad and pensive all that day and would not sup that night'.

An English spy, on the other hand, reported that James reacted 'very grievously and offensively and gave out in secret speeches that he would not digest the same or leave it unavenged'. And the sense of outraged dignity that the king often exhibited certainly seems to have manifested itself in some angry talk about the Queen of England in the period that followed. He suggested in private, for instance, that he would not allow himself to be intimidated by an old woman, who was so unloved by her subjects and in such perpetual fear of assassination that she fled at the approach of strangers. 'He protesteth,' wrote one observer, 'though he be a mean king with small ability, he would not change fortune with her, choosing rather to live securely among his subjects than to seek after the blood of his people of contrary religion as she does.'

As always, however, James's frothy talk gave way to clear-cut deference when dealing with Elizabeth in person, and his response to her letter excusing herself of Mary's death was a further example of the one-sidedness of their actual relationship. In acknowledging her 'long professed goodwill' to his mother, for instance, he also accepted Elizabeth's 'solemn attestations of innocency' and 'unspotted part' in the death of the one whom he merely referred to as 'the defunct'. Not even the slightest veiled hint of disapproval, let alone reprisal, was delivered, and there was no suggestion either that relations between the two rulers would remain anything other than cordial in time to follow. Instead, James merely concluded with the familiar reference to his hopes for the succession: 'And, as for my part, I look that ye will give me at this time such full satisfaction in all respects, as shall be a mean to strengthen and unite this isle, establish and maintain the true religion, and oblige me to be, as before I was, your most loving.'

Overly extravagant professions of grief might well, of course, have opened James to the charge of hypocrisy, but if the king's correspondence with his English counterpart suggested a rational and pragmatic acceptance of political realities, his subjects' response to the death of their former queen was altogether less thin-blooded. The ministers of the Kirk did not, of course, regret her passing. Indeed, when James had exhorted them to pray for her safety, a certain John Cowper needed to be hauled from his pulpit as he railed against him to his face. Inspired, or so he claimed, by the spirit of God, Cowper prophesied trial and tribulation for all who lived in Edinburgh and foretold that the day of calamity when it came would be 'a witness against the king'. When, moreover, the ministers had finally agreed to offer prayers for the former queen, their objective had been not her delivery from death, but only that she 'should become a profitable member of Christ's Kirk'.

Yet there were those, among the nobility especially, who were more inclined to express outrage at Mary's treatment. When James chose, for instance, to follow the custom of wearing a 'dule weid' (mourning garment) of dark purple, he was told by Earl Francis of Bothwell that his only 'dule weid' should be a suit of armour until the Queen of Scots was avenged against a kingdom that had harried Scotland for the last three centuries.

# The Headsman and the King of Spain

Likewise, at a Parliament in July 1587, Maitland made an impassioned plea for retribution – after which each and every one of the attending nobles swore on bended knee to assist James in military action. And while James the poet remained studiously silent on the subject of his mother's death, other quills were busily at work, one of which – wielded by an outraged Scots versifier, who was wise enough to remain anonymous – promised nothing less than the gift of a noose for the Queen of England:

> To Jesabel that English heure [whore]
> receive this Scottish cheyne,
> As presages of her gret malhoeur
> for murthering of our Quene.

Such, indeed, was the clamour that James had little option other than to countenance some token gestures against his kingdom's time-honoured oppressors. Even when Sir Robert Carey first brought news of the execution, he had been refused passage beyond Berwick, where he was met by Peter Young, and now James was prepared to take the cynical step of consenting to the temporary imprisonment and banishment of Gray who became the scapegoat for the government's failure to save Mary. He also wrote lamely to Henry III of France, to Catherine de Medici and to the Duke of Guise to appeal for aid in avenging his mother, and retained the Archbishop of Glasgow as an ambassador to the French court - while refraining, predictably, from signing or dating his appointment. Nor did his correspondence with Henry III prevent him from cultivating the friendship of the French king's arch-enemy, Henry of Navarre. Though he had promised the king that no Scot would serve in the armies of Navarre, the promise was at once broken when he allowed Sir James Colville of Easter Wemyss to raise a company of soldiers for precisely that purpose. And it was at this time, too, for good measure, that James chose to entertain the Huguenot poet, Guillaume Sallust du Bartas, at the Scottish court with no expense spared, irrespective of the fact that the poet came as an unofficial envoy for Navarre's cause.

Plainly, then, James's gestures were no less hollow in the wake of his mother's death than they had been all along, albeit justifiably, perhaps, in light of his limited resources and equally limited options, though his subsequent quest for the English Crown – not to mention his hungry pursuit of his English pension and the English estates of the Lennox family which Elizabeth had annexed – often seemed less than becoming. To remove all suspicion of evil after the infernal proceedings against his dearest mother, he proposed, for instance, that Elizabeth should give him the requisite lands in northern England, along with the title of duke. But the queen remained coldly dismissive of his desire for material advantage. Though it eventually averaged the stipulated £4,000 per year, James's pension was paid only irregularly and when, on one occasion, he managed to screw the sum of £2,000 from her special envoy, William Ashby, Elizabeth later repudiated the payment on the grounds that Ashby had exceeded his authority in agreeing to it. As such, a quip recorded by a Spanish diplomat was much to the point. For when the Spaniard related to an English counterpart how the wound to James's honour could only be healed by a declaration that he would succeed Elizabeth, the Englishman was said to have replied that 'this was rather a point of profit than of honour'.

But it was not so much repartee as the growing Spanish threat that was soon preoccupying James and diverting his subjects' attention away from the treatment of their erstwhile queen. Although he had initially intended to make Mary Queen of Scots the lawful Catholic ruler of England, King Philip's position had altered fundamentally during the last months of her life as he decided to press Spanish claims to the English throne on the strength of Mary's alleged will, the existence of which was never proven, and his own somewhat tenuous descent from Edward III's son, John of Gaunt. In consequence, the queen's execution had been met with no small relief in Madrid, as Philip now pushed forward with plans to place the crown of England upon the head of his own daughter. In fact, on 11 February 1587, four days before Mary was actually beheaded, he wrote to the Count of Olivares, his ambassador at the papal court, revealing his intentions plainly. 'You will impress upon His Holiness,' he informed Olivares, 'that I cannot undertake a war in England for the purpose merely of placing upon the throne a young heretic like the King of Scotland, who, indeed, is by his heresy incapacitated to succeed. His Holiness must, however, be assured that I have no intention of adding England to my own domains, but to settle the crown upon my daughter, the Infanta.'

Nor was James's prospective loss of the English throne his only fear, for if England were conquered by Spain, Scotland would be bound, as an English envoy had once reminded him, to follow close behind. And there were, of course, powerful interests within the Scottish kingdom that might be ready to assist such a turn of events. The Catholic nobility in the north, led by the Earl of Huntly, head of the great clan of Gordon, and his supporters, Crawford and Montrose, had grown to considerable strength, and by 1588 were hatching plans, along with Lord John Maxwell and Lord Claud Hamilton in the south, to welcome a Spanish army and force James's conversion. Yet the king, in spite of the obvious dangers, would remain loath to tackle Huntly and his associates, for though the earl had been educated in France as a Roman Catholic, he was nevertheless James's kinsman through the marriage of two of his forebears and had signed the Presbyterian confession of faith in 1588. More importantly still, he had helped free James from the Ruthven raiders and become an object of considerable affection to the king, who, with characteristic lack of inhibition, was sometimes seen to kiss him in public – 'to the amazement of many'.

There were considerations of policy, too, firmly underpinning James's indulgence towards the Catholic lords in general, since they enhanced his bargaining power with Elizabeth, formed a counterpoise against the Kirk, and offered hope of survival in the perfectly likely event of Spanish victory. He was even, it seems, prepared to meet with their spies in an attempt to hedge his bets. Robert Bruce, for example, met the king on three occasions and suggested that he was willing to negotiate with Philip. The same was said, for that matter, by a Spanish agent from the Netherlands named Colonel Semple, who was subsequently permitted to move freely around Scotland, plotting at leisure. Nor, of course, would this be the last occasion that James was prepared to flirt with Rome, for, long after the current crisis of imminent invasion had passed, he would pursue a similar approach in the hope that English Catholics would support his quest for their kingdom's Crown.

Early in the same year, however, he published a short meditation on selected verses of the Book of Revelation which confirmed his unwavering commitment to the Protestant faith by 'commenting of the Apocalypse' and setting out 'sermons thereupon against the Papists and Spaniards'. After all, 1588 had been ushered in by an ominous tide of baleful prophecies, and the final book of the New Testament, with its lurid descriptions of the end of times and compelling reflections on the fragility of all earthly power, had, in any case, always been a particular source of fascination to him. 'Excellent astronomers,' observed David Calderwood, foretold that the year would be 'fatal to all estates'. 'And if the world did not perish,' he added, 'yet there should be great alterations in kingdoms and empires, so that thereafter it should be called the year of wonders.' In Spain, too, as preparations for the Armada proceeded apace, 1588 was deemed to be 'pregnant with misfortune'. In such circumstances, nothing, it seems, could have been a more fitting subject for the royal pen than the loosing of Satan and the rise and fall of empires. And the villain throughout is the papal Antichrist – the king of locusts, the beast rising from the sea and the woman in scarlet sitting upon the waters.

If, however, any further proof were required of James's inherent hostility to Rome, he was not long in providing it, for, around the same time, he saw fit to challenge James Gordon, the Jesuit uncle of the Earl of Huntly, to a public disputation at court. Over five long hours, in fact, James held his own with both rigour and the utmost courtesy, leaving his rival to reflect afterwards that no one could 'use his arguments better nor quote the Scriptures and other authorities more effectively' than the king. And while one of his courtiers, flushed with pride at his sovereign's knowledge, suggested that even the most learned papist in Europe would never trip him, Mendoza, too, the Spanish ambassador, acknowledged James's performance. 'I hear,' wrote the Spaniard, 'that after the disputation the king said in his chamber that Gordon did not understand the Scripture, which is a fairly bold thing to say, except that the king has the assurance to translate Revelation and to write upon the subject as if he were Amadis of Gaul.'

When the day of reckoning arrived, moreover, and the Spanish Armada conducted its vain enterprise upon England, James remained faithful to his alliance with Elizabeth throughout. Though he had offered her 'his forces, his person and all that he commanded against yon strangers', his assistance was in fact never more than moral, but he had stirred no coals upon either the Border or in Ireland and could fairly record that Spanish forces 'never entered within any road or haven within his dominion, nor never came within a kenning near to any of his coasts'. Nor, typically, was he stinting in his estimation of the significance of the English victory. Indeed, in a meditation upon the fifteenth chapter of the Book of Kings that he penned soon afterwards, James suggested that the victory over Spain was far greater even the David's victory over the Philistines – though his personal gains were predictably limited, as Elizabeth remained silent on the succession issue and, in August, refused him the payment of further money previously promised by William Ashby. 'I am sorry to know from Scotland,' wrote the exiled Master of Gray, 'that the king our master has, of all the golden mountains offered, received a fiddler's wages.'

Meanwhile, the news for James at home in 1589 was no more pleasing either. In February, the English intercepted letters signed by the Earl of Huntly and his associates, which expressed their regret at the failure of the Armada and promised their assistance to King Philip in the event of any future invasion attempt. The plan, in fact, was to bring the Duke of Parma over from the Netherlands with a force of 6,000 men as a prelude to a joint invasion of England, and Elizabeth's response was predictable. 'Good Lord! Methinks I do but dream,' she protested, in a letter that left her embarrassed counterpart no choice other than to act, though in what would prove to be, under the circumstances, an astonishingly half-hearted manner. For, after removing Huntly from the captaincy of the royal guard and imprisoning him in Edinburgh Castle, the king went on to dine with him only the next day. Rather than serving up condign punishment, James also wrote pathetically to the traitor to the Protestant faith whom he believed he had converted only the year before. 'Are these,' he inquired lamely, 'the fruits of your new conversion?' And within days, Huntly was not only released but restored to his former position – an action which provoked Maitland to threaten resignation until James reversed his decision yet again, dismissing the earl once more and ordering him to his estates in the north.

It was Maitland, indeed, whom Huntly had blamed for his fall in the first place and it was Maitland, too, who, since his appointment as chancellor, had become the butt for the animus of a noble class sensing that the days of its free play with the Crown might well be numbered. He was, wrote Spottiswoode, 'a man of rare parts … learned, full of courage, and most faithful to his king and master.' 'No man,' Spottiswoode added, 'ever carried himself in his place more wisely, nor sustained it more courageously than he did.' The brother of Sir William Maitland of Lethington, Sir John Maitland of Thirlestane had at first been an adherent of Mary's and had entered the king's service during Lennox's time of influence when the former queen's supporters had been welcome. Thereafter, his rise had been steady, becoming secretary to the king in 1584, vice-chancellor in 1586 and chancellor a year later – an office that he would hold until his death in 1595. And in demonstrating his efficiency and good sense, there was much about him in personal terms that appealed to the king. Not least of all, he wrote English and Latin verse, possessed a sharp Scots tongue, loved raillery and sarcasm, and mingled grave affairs of state with jests and facetiousness. But it was policy and politics that rendered him so invaluable, and his judicious mix of firmness, pragmatism, devotion to duty and loyalty that distinguished him so markedly from his predecessors. Despite his indifference to the religious disputes of the age he urged upon the king the necessity of better relations with the Kirk, and though he bitterly resented Mary's death he saw clearly that moderate friendship with England was essential to Scotland's interests.

More importantly still, perhaps, the chancellor stood for rigorous administration and the firm reassertion of monarchical authority through a series of reforms which, if successful, would modernise and strengthen the whole fabric of government. Tough new measures were intended to purge the kingdom of lawlessness and crime, while James was also encouraged to assume a much more prominent role. At home, too, reforms to the royal chamber and household not only curtailed expense but enhanced the king's dignity – something that was also to be encouraged, albeit with only limited success, by tightening the notoriously lax rules governing access to the royal presence. The Borders, Highlands and Western Isles were to be pacified, and the 'ordinary and daily' council of the king, as well as the Court of Session, were to be shorn of noble influence, even if this meant packing them with his friends and relatives, most of whom were lairds like himself.

And it was for precisely this reason, of course, that Maitland found himself surrounded by such a formidable ring of noble foes. The office of chancellor had long been one that the nobility regarded as their perquisite, so that his comparatively low-born origins only added to the anger caused by his policies. In Bothwell's view, indeed, he was nothing more than a 'puddock stool [toadstool]' in contrast to the ancient cedars of the traditional ruling caste, and James was therefore pressed to govern 'with his nobility in wonted manner, not by private persons hated'. Both Catholic and Protestant lords despised the chancellor, in fact, and in such circumstances it was imperative that the king should have done everything possible to protect him from the individual he had recently offended so grievously. For, though a man of no great ability, Huntly had developed nevertheless into a competent military leader. And quite apart from his ambition and disloyalty, beneath the earl's superficial charm, there lurked an even darker side that the king had so far neatly ignored. It was Huntly, after all, who butchered the Earl of Moray in a foul murder and made a pastime of summary justice. Not content with roasting alive two cooks from an enemy clan as an example to all, the turrets of his castle at Strathbogie were proudly adorned with the severed limbs of those who crossed him.

Yet even after James had deprived Huntly of the captaincy of his guard and dismissed him to his lands in the north in accordance with Maitland's ultimatum, he still found it impossible to break the knot of friendship cleanly. On the contrary, his intention, it seems, was to maintain both Huntly and Maitland in some inexplicable state of balance, and merely by the exercise of his own influence bring them somehow to accept each other. 'The King,' wrote one of the English government's informants, 'hath a strange, extraordinary affection to Huntly, such as is yet unremoveable … The Chancellor is beloved of the king in another sort, for he manages the whole affairs of this country … The king had a special care to make and keep these, his two well-beloved servants, friends, but it never lasted forty days without some suspicion or jar.' And Huntly had surrounded the king with his friends at court, not the least of whom was the king's current 'best-loved minion' and 'only conceit', Alexander Lindsay, younger brother of the Earl of Crawford. Eventually ennobled as Lord Spynie in 1590, 'Sandie', as the king called him, was actually a mediocre figure of few political ambitions in his own right, but he had been made vice-chamberlain and, though the emotional connection between the two was comparatively limited, was known to be James's 'nightly bedfellow'.

With influences such as these at work, it was not entirely surprising therefore that James should cling so doggedly to Huntly and his circle. 'There is not one in the chamber or of the stable,' the English intelligencer Thomas Fowler told Lord Burghley on 28 March 1589, 'which two sorts of persons are nearest attending on the king's person, but are Huntly's […] and the Chancellor cannot mend it, for the king will not change his mind, he loves them so well.' Even before the earl left Edinburgh, therefore, James foolishly insisted upon hunting in his company and only narrowly escaped capture when he awoke in panic the morning after a banquet and left the scene of danger in haste. A few weeks later, moreover, while spending the night in the countryside outside the capital, he was suddenly warned of Huntly's approach in the company of Errol and Crawford. All three, it seems, were marching down from the north with a view to apprehending him, while Bothwell, who had joined the enterprise in the hope of ruining Maitland, was advancing from the Border. Ultimately, in fact, it was only the chancellor's timely intervention that saved the day. For at midnight the king took flight and by 3 a.m. was safe in Maitland's house in Edinburgh, after which Bothwell retired to Dalkeith and the Catholic earls to Perth.

It would take an outright act of rebellion, therefore, before James could be stirred to some semblance of decisive action. Yet now, faced with absolutely no alternative, he reacted with surprising vigour in what would prove to be the only military action in which he directly participated. Assembling his Protestant nobles and supporters in the south, he mustered his forces in Edinburgh and marched north so rapidly that within two weeks he was approaching Aberdeen where, at the Bridge of Dee, Huntly and his 3,000 or so supporters prepared for battle. Indeed, throughout the brief campaign James's personal involvement was exemplary. When it was rumoured, for example, that the royal army was about to be attacked by night the king responded with commendable resolve. 'That night we watched in arms,' wrote a member of the expedition, 'and his Majesty would not so much lie down on his bed, but went about like a good captain encouraging us.' And in other respects, too, James warmed to the task with surprising energy and commitment, as his officers pressed in upon him with information, advice and requests. 'These people,' wrote Fowler, 'must have free access to the king's presence. If there were no more but the continual disquiet of such a throng from morning to night and their entertainment, it were too much toil for any prince; but he must visit their watches nightly, he must comfort them, be pleased with them passing from place to place, that day or night the good king has little quiet or rest. He hath watched two nights and never put off his clothes.'

James's efforts would ultimately prove needless, however, as Huntly's followers lost heart and melted away into the hills. Nor would James sustain the resolve that had largely carried the day. Instead, the opportunity to follow up a worthy bloodless victory with a long-overdue assertion of his regal authority was sacrificed yet again for sentiment's sake and the fond hope that clemency might heal rather than chafe his kingdom's wounds. It was true, of course, that his position was still less than entirely secure. When James finally entered Aberdeen, for example, his men remained weary from forced marches and the ravages of Highland weather, and supplies were low. Huntly, moreover, was still at large, and without a captive ringleader, the king could hardly return to his capital in triumph. Overreaction on his part would also surely lead to calls for vengeance later.

But the line between moderation and weakness is a thin one and if discretion is always laudable, hesitancy and appeasement are rarely so – especially when the seeds of further disruption remain intact. The king's decision to offer Huntly a secret deal was therefore as unwise as it was typical. So long as he surrendered himself, James told the earl, he would be treated mercifully, and this he did, along with Crawford and Bothwell. In consequence, what might have represented a defining triumph for the king became another lesson that treason was a low-risk enterprise, especially when the culprits were objects of affection for the man they sought to control. Tried for treason and found guilty, the rebels were merely placed in leisurely confinement until, a few months later, they were released – much to the exasperation of the Kirk, the chancellor and the Queen of England, though much to the comfort and pleasure of one particular malcontent who still awaited his moment.

# 6 ✤ 'Cupide Blinde' and Wyches' Waies

'The king's impatience for his love and lady hath so transported him in mind and body that he is about to commit himself, Leander-like, to the waves of the ocean ...'

*William Ashby to Elizabeth I, 23 October 1589*

Though 1589 had begun with threats, intrigue and calls to battle, by spring the tide had turned and happier thoughts had scope at last to blossom. With varying degrees of seriousness, the king's marriage had been discussed since 1582, but now, after the merits of various Catholic and Protestant princesses had been duly weighed and pondered, two alone stood out as fitting candidates: the Calvinist Catherine of Navarre and the Lutheran Anne of Denmark. Those who regarded tight-knit community of religion with the Kirk of Scotland as essential to the maintenance of a united front against the Counter-Reformation supported the case for Catherine, while those who were broadly anti-Catholic, yet more pragmatically concerned with preserving commercial links between Scotland and Denmark, naturally backed her Danish rival. Nor was the king prepared to dally further in making his choice, even though it had been generally acknowledged until now that he 'never regards the company of any woman, not so much as in any dalliance'. At first, in fact, it was to be a question of form and duty rather than a matter of a young man's yearning. 'God is my witness,' the king admitted at the time, 'I could have abstained longer than the weal of my country could have permitted, [had not] my long delay bred in the breasts of many a great jealousy of my inability, as if I were a barren stock.' But before long, as fresh horizons opened and new adventure beckoned, all would be headlong love and reckless courtship.

Even so, the choice of bride had not been straightforward. Marriage to a daughter of the Danish king, Frederick II, had been one of the earliest options to be discussed and during the spring of 1587 an embassy headed by the king's former tutor, Peter Young, and the wealthy merchant, Sir Patrick Vaus of Barnbarroch, had been empowered to negotiate the terms, though talks finally faltered over the perennial bugbear of the princess's dowry. A French marriage, on the other hand, was supported by both Chancellor Maitland and a majority of Scottish nobles, and had influential enthusiasts across the Channel too, including the poet du Bartas who had urged the match unofficially during his earlier visit to Scotland. Yet once again there were pitfalls, since Henry of Navarre, Catherine's brother, was initially fighting to become King of France and hoped for Scottish assistance. In the meantime, moreover, he remained relatively penurious and, like Frederick in Denmark, proved reluctant to supply the kind of hefty financial settlement for which James thirsted.

Ultimately, however, a decisive combination of policy, pressure and personal attraction had swung the balance in Princess Anne's favour. Denmark was, after all, an orderly, well-established and Protestant kingdom which had largely escaped the religious tensions engulfing Europe and served as a convenient bridge for Scottish alliance with the Protestant princes of north Germany. Furthermore, Anne seemed to have the edge over her rival in terms of both age and beauty. In September 1587, William Melville of Tongland had been sent to France to inspect Catherine and returned with 'a good report of her rare qualities'. But while Anne was eight years younger than James, Catherine was eight years older, and though prudent, virtuous and capable of offering good counsel, might therefore prove less than perfectly biddable in her dealings with a considerably less experienced husband. Worse still, it was said that 'the sister of Navarre was old and crooked and something worse if all were known'. And when the worthy burgesses of Edinburgh, 'having their most necessary trade with the Easterlings', decided to press the case for a Danish match by rioting in May 1589 and actually breaking in upon Maitland's chambers with threats of death amid fears that he was about to prevail once more with the king and seal a marriage to Catherine, the die was finally cast. James, it seems, had already 'conceived a liking in imagination' for Anne, and now his imagination could be given full vent.

The passion for the 15-year-old princess which presently consumed him had not, however, been apparent from the outset. According to Sir James Melville, the king had agonised at length over his decision, arming himself with portraits of the two princesses and making up his mind only after 'fifteen days' advertisement and devout prayer'. When, moreover, Young and Vaus had opened negotiations in 1585, the king was hoping mainly for the hand of King Frederick's elder daughter, Elizabeth, but when asked what should be done if the princess were betrothed already, James's sound business sense swiftly prevailed. 'Forfend the omen,' he exclaimed, 'but if it happen, ask for the other.' And the fears of James's ambassadors were indeed well founded, for, as they soon found upon arrival, the Duke of Brunswick had already been promised Princess Elizabeth's hand. Only when the Danes had finally convinced the Scots that, though Elizabeth 'was the more beautiful', Anne was nevertheless far from unlovely and for her age taller and more fully developed than her sister, were negotiations allowed to go forward.

Even at this point, however, the deal was far from done. When James's representative, George Keith, the Earl Marischal, sailed for Denmark in mid-June, he went with a series of Scottish demands that remained excessive. In addition to a dowry of £1,000,000 (Scots), he was to ask that the Danes offer naval and military assistance should Scotland be invaded or James had cause to fight for a foreign title due to him by just inheritance. Furthermore, Denmark was to abandon its ancient claim to the Orkneys and join an anti-Catholic league, along with Scotland and other Protestant states. Such, indeed, was the scale of Scottish demands that only the increased influence of the Danish Queen Mother, Sophia, after King Frederick's death the year before, saved the match at all. Five hundred tailors and embroiderers had, it seems, been at work on Princess Anne's trousseau for the best part of three months, and her bridal coach – fashioned throughout from silver rather than iron parts – was already underway. The proud mother refused, therefore, to see her efforts sacrificed at the last, and, as she pushed and the would-be bridegroom pined, Scottish demands not only abated but virtually disappeared.

In succumbing to Cupid's dart, therefore, James finally conceded that he 'would not be a merchant for his bride', and settled in August for a dowry of 75,000 thalers. But his generosity, gracious though it may have been, did not befit his straitened circumstances. The king, remarked Thomas Fowler, 'has neither plate nor stuff to furnish one of his little half-built houses, which are in great decay and ruin. His plate is not worth £100, he has only two or three rich jewels, his saddles are of plain cloth. He is served with six or seven dishes of meat but eats but of two; no bread but of oats, and cares not what apparel.' And Fowler's gloomy observations were echoed by another of his counterparts. 'Surely Scotland was never in worse state to receive a Queen than at present,' commented William Ashby, 'for there is not a house in repair.' A tax of £10,000, raised by the Scottish Parliament for the marriage, had already been spent and fears grew that the Danish princess might arrive before James's wedding garments had been prepared – leaving him no alternative, ultimately, but to scrounge the necessary funds from England. 'It is now time to give proof of affection,' he wrote to Burghley in London. 'No time must be lost, for tempus deals most straitly with me.' The result was a grudging gift of £2,000 worth of plate and £1,000 from the Queen of England, who, for her own good reasons, was already infuriated by his decision to choose the Danish princess over the French.

In the event, James's concerns about his loved one's premature arrival proved unfounded. For although the marriage was celebrated by proxy in Copenhagen on 9 August 1589, and she set sail for Scotland on 1 September, her passage was swiftly halted by raging west winds. Three times, in fact, her ship the *Gideon* was forced to turn back, and at the tempest's height the young princess herself had narrowly escaped death when three cannon broke loose from their housings and careered back and forth across the deck. Two clergymen on board, Drs Knibbe and Kragge, had been urged to pray for calm and safety, and when their prayers went unanswered, Peter Munk, the Danish admiral, assumed that the weather was witches' work. For prior to the journey he had quarrelled with the husband of a woman reputed to be skilled in the black arts, and now felt impelled to seek safe haven in a Norwegian fjord.

James, meanwhile, waited with increasing anxiety as the same storms prevented news of his bride from reaching Scotland. Indeed, as the days lengthened into weeks he became deeply alarmed, ordering public prayers for Anne's safekeeping and retiring to Craigmillar near Edinburgh to ponder her fate in isolation. 'The king, as a true lover, wholly passionate and half out of patience with the wind and weather, is troubled that he hath been so long without intelligence of the fleet and thinketh every day a year till he see his joy and love approach.' By 8 October, moreover, the same lovelorn king had dispatched a messenger of his own to determine her whereabouts, and convey to her a tender note in French – the language that Anne had been learning rather than Scottish to communicate with him. He wrote to his 'only love' of 'the fear which ceaselessly pierces my heart' and ended by 'praying the Creator' to grant her a 'safe, swift and happy arrival upon these shores'. There were verses, too, in which he compared his plight to Leander and lamented the 'full manie causes suire' that now increased his 'woe and caire'.

Ultimately, it was only on 10 October that James finally received word of Anne's safety in Oslo, and it was now, beside himself with a mounting surge of chivalry, romance and raw impatience, that he decided to opt for bold solutions. His Lord High Admiral the Earl of Bothwell would achieve by Scottish seamanship what the Danes had failed to do and bring Anne home with minimum delay. James, moreover, would accompany him, just as his grandfather had braved the Channel to bring home Madeleine de Valois, his first queen. He would 'commit himself and his hopes Leander-like to the waves of the ocean, all for his beloved Hero's sake', and prove his manly credentials at last. The 'nakedness' caused by his lack of parents and siblings, he admitted somewhat extraordinarily in the proclamation announcing his decision, had always left him weak before his enemies. Likewise, 'the want of the hope of the succession breeds disdain'. So he would seize the opportunity to prove once and for all that he was 'a true prince' and no 'irresolute ass who could do nothing of himself'. Nor, he claimed, should Maitland continue to be blamed 'for leading me by the nose, as it were, to all his appetites, as if I were an unreasonable creature or a bairn that could do nothing of myself'.

What the king had not counted upon, however, was the cost of such an expedition, and it was left to Maitland to volunteer to foot the bill as a means, no doubt, of compensating for his earlier opposition to the marriage. Neither was James unduly concerned about the potential political consequences of his adventure. In the aftermath of Huntly's abortive rising that spring and two unsuccessful 'bands' to remove Maitland from office in 1588, Scotland was still hardly at rest, and it was for fear of concerted opposition to his decision to absent himself from his troubled kingdom so needlessly that James kept his planned departure secret from his councillors. By now, after all, his bride's safety had been confirmed, and her arrival was hardly likely to be long delayed with or without his intervention. South of the Border, meanwhile, Elizabeth was immediately anticipating the worst when she heard of James's impending absence – warning her wardens of the Marches to be on the alert for trouble and sending Sir Robert Bowes north again to hold the Protestant interest together against a possible Catholic *coup de main*.

In the event, the queen's worst fears did not materialise, since the arrangements for Scotland's governance while James was away proved surprisingly effective under the circumstances – albeit, perhaps, with a healthy slice of good fortune. To James's credit, administration was left in the hands of a Privy Council in which all interests were represented and in which authority was so delicately shared out among the most dangerous nobles on both sides that they would act as automatic checks upon each other. A number of key roles were also filled by men of proven ability. Melville of Murdocairney was deputed to carry out the routine duties of the chancellor, in the absence of Maitland who was to accompany the king and be thereby spared the hostile attention of both Huntly and Bothwell. Lord John Hamilton, in his turn, was given overall responsibility for military affairs and policing the Borders, while Robert Bruce, a minister of the Kirk whom James actually liked and trusted, was accorded the role of independent watchman and whistle-blower, ensuring where possible that correct behaviour prevailed and guaranteeing that loud protest would follow in the event that it did not.

Assuming, then, that no divisive issue arose during the king's absence and that comparatively little positive action was required, such an arrangement would prove effective. Yet Bruce's ability to calm troubled waters if discord arose remained questionable, and there remained other aspects of the plan that could hardly be justified even in the event of James's incapacity through a genuine crisis, as opposed to the current self-imposed quandary. The president of the council was, after all, to be the 15-year-old Ludovic, son of Esmé Stuart and now Duke of Lennox in his own right. Though he was actually Scotland's 'Second Person' or heir presumptive at this time and his new role was largely nominal, he was far from ideal as a figurehead at a time when the whole tenor of government, particularly in the prevailing Scottish context, was so intensely personal in nature. More problematic still, however, was the decision to install the Earl of Bothwell as his advisory associate. As wild and fractious as ever and, as events would later demonstrate, no less disgruntled, Bothwell's inclusion as de facto conciliar head was purely the result of his arrant unwillingness to settle for anything less.

Yet James, as he now joyfully confessed, had been 'inflam'd and 'pearc'd' by 'Cupide Blinde', so that even the tempestuous autumn seas, let alone considerations of state, would not deter him. Six ships, the largest of which was the king's own at 130 tons, had been well provisioned with food, wine and livestock, and he therefore set sail on 22 October, reaching Flekkefjord on the southern coast of Norway six days later. Fortunately, the prevailing winds had eventually favoured his passage as much as they had hindered his bride's, and he arrived safe and unruffled. 'All in good health,' reported one of the party, 'the king's majesty was never sick.' But he too had at first been driven back upon the coast of Pittenweem by tempests and the last day was so stormy that his flotilla was thought to be in great hazard. Clearly, though James's outright terror of any naked blade was already well established, his willingness to face danger on the high seas would have matched many a hardened sailor's. The cold, moreover, was bitter and the final stage of his journey required a further voyage along the Norwegian coast to Tönsberg, which was not completed until 11 November, followed by a hard cross-country journey of more than 50 miles before he arrived in Oslo.

But in spite of his long and arduous travels, it seems James still managed to cut an impressive figure upon his arrival in Norway's capital. Travelling in considerable pomp, accompanied by Danish, Swedish and Norwegian noblemen and preceded by heralds, he was described by eyewitnesses as tall and thin, with deep-set eyes. He was also splendidly dressed in a red velvet coat decorated with gold and a black velvet cloak lined with sable fur. Nor did he tarry in meeting – and somewhat embarrassing – his bashful bride. According to David Moysie, he made his way at once to the old Bishop's Palace where Anne was lodged and 'passed quietly with boots and all to her Highness', whereupon he 'minded to give the queen a kiss after the Scottish fashion', only to be told by his loved one that such intimacy upon first meeting was not the custom of her own country. The rebuff was not sustained, however, and 'after a few words privately spoken' there did indeed pass 'familiarity and kisses' between them. Next morning, moreover, dressed in blue velvet bespangled with gold and preceded by local nobles and six heralds clad in suits of red velvet, he visited Anne again and spent the day in her company.

Still almost a month short of her fifteenth birthday, the princess was a tall, blue-eyed, slender, graceful girl whose white skin and radiant, golden hair were the subject of much admiration as she grew steadily towards the statuesque beauty of her adult years. Like James, she enjoyed hunting and was also amused by dressing up and acting. And though her intelligence has often been treated dismissively, her ability to speak French and her rapid acquisition of Scottish go some way to countering this view, as indeed does her eventual patronage of Inigo Jones and other artists. If, moreover, she was not politically minded, this was to prove more of a virtue than a vice from her husband's perspective and her subsequent intrigues at the Scottish court were driven more often than not by mainly worthy motives, such as a wish to exercise greater control over the upbringing of her children. Her wish to manipulate behind the scenes declined in any case over time, and she became over the years a shrewd judge of character and a generally discreet influence, who never lapsed from faithfulness herself and was prepared, though not always patiently, to tolerate James's relationships with good-looking young men. Certainly she could not sustain a great passion in her husband, mainly perhaps, because she never became what he most wanted throughout his life – an intellectual peer and genuine confidante with whom he could share his innermost secrets. But there was much early optimism at the match. 'I trust,' wrote David Lindsay, the minister of Leith, who had been brought over from Scotland to officiate at the wedding, 'she shall bring a blessing to the country, like as she giveth great contentment to his Majesty.'

After their earlier proxy marriage, the two were finally wed in person at St Halvard's church in Oslo on 23 November. The ceremony, lasting just over an hour, was conducted in French and included a sermon upon the significance of Christian marriage delivered by the local bishop, after which Anne left and James remained to receive the congratulations of all attending, to whom he replied in Latin. Thereafter, the wedding banquet was held in the imposing hall at Akerbus, a fortress standing in a rocky promontory in Oslo fjord. Almost one month later, moreover, on 21 January 1590, the marriage was solemnised once again in the presence of the Danish royal family in the castle of Kronborg, by which time the winter was so far advanced that James rejected the opportunity to return to his own kingdom and opted instead to accept an invitation to stay with his new relatives. Leaving Scotland to fend for itself, he travelled overland by sledge and arrived at Bohus Castle near the Swedish frontier on New Year's Day, before being finally welcomed in state at Kronborg Castle once again on 21 January amid volleys of cannon fire lasting a whole half hour.

What had begun as an absence from his homeland intended to last some twenty days had therefore already extended to some three months, but this was still by no means the end of James's stay. Indeed, he would remain in Denmark until late April 1590 where his new brother-in-law, the 12-year-old Christian IV, was already impressing by the tremendous pace of his hospitality. Predictably, there was much hunting and hawking, and there was also much drinking. 'Our king made good cheer and drank stoutly till the spring time,' wrote David Lindsay. And James made no secret of his activities, informing his current favourite, Alexander Lindsay, how he was writing to him 'from the castle of Kronborg where we are drinking and drinking over in the old manner'. But there was also entertaining company and more intelligent conversation on offer from his Danish hosts than he had been used to in Scotland and this, too, encouraged James to prolong his stay. He was worried, indeed, that the inadequacy of preparations in his own kingdom for his return might well be an embarrassment. 'For God's sake,' he urged Robert Bruce, 'take all the pains ye can to tune our folks well now against our homecoming lest we be all shamed before strangers. Thus recommending me and my new rib to your daily prayers, I commit you to the only All-sufficient.'

In the meantime, however, James would not only revel to his heart's content, but enjoy every opportunity both to extend the scope of his learning and display his own scholarship to best possible advantage. At Roskilde, for example, he discussed the Calvinist doctrine of predestination with Dr Nils Hemmingsen, and later treated the theological faculty at Copenhagen University to a three-hour oration in Latin, after which he received a silver goblet in honour of the occasion. At the Danish royal academy, meanwhile, he attended lectures on both theology and medicine, and was complimented on his learning by Dr Paulus, Superintendent of Sealand. Nor was James's response anything other than honest. 'From my most tender years,' he replied, 'I have been given to books and letters, which even today I willingly acknowledge.' And finally, most famously of all, there was also an encounter between James and the intensely difficult, but brilliant, astronomer Tycho Brahe, who had lost the bridge of his nose in a duel over the legitimacy of a mathematical formula and now sported an eye-catching brass prosthetic in its place. At his castle and observatory on the island of Hven, Brahe had spent a day with James in learned discourse upon Copernican theory, which so impressed the young king that he subsequently composed a sonnet in his host's honour.

Even the King of Scotland could not, however, remain away from his homeland indefinitely, and once a new round of festivities was concluded after the marriage of Anne's elder sister Elizabeth to the Duke of Brunswick on Easter Day, the time for fond farewells inevitably arrived. On 21 April, therefore, James and his bride sailed from Denmark, and on 1 May they landed at Leith where they lodged in 'the King's Wark', a building which doubled as both a customs house and a royal or ambassadorial lodging, while preparations were laid for the state entry into Edinburgh of a brand new Queen of Scots. In the event, James's worries about the reception proved largely unfounded – not least because Anne had brought with her the coach of silver, drawn by eight white horses, prepared by her mother. It was in this fairytale carriage that she entered the capital on 6 May accompanied by her husband, with the Duke of Lennox, the Earl of Bothwell and Lord John Hamilton riding beside her. All had co-existed in apparent harmony throughout the winter and all now played their parts with due grace and loyal flourishes.

Eleven days later, Anne was duly crowned in the Abbey Kirk of Holyrood, though the ceremony had, it seems, been the cause of not inconsiderable dispute beforehand. There had already been complaints from the Kirk that the queen had made her entry upon the Sabbath, but the main bone of contention now concerned the practice of anointment, which the more radical clergy dismissed as a pagan and Jewish custom and one more objectionable remnant of 'popish' superstition in need of eradication. James, however, had himself been anointed by the Bishop of Orkney and insisted that it confirmed the sacred character of kingship. Nor would the king be thwarted in his plans for the forthcoming coronation of his wife, even if it did involve the kind of impish provocation to which he now resorted. With characteristic adroitness, he selected none other than Robert Bruce to perform the ceremony and when the ministers objected that one of their own brethren should be chosen to conduct such an offensive ritual, the king's response bore all the hallmarks of his familiar wile and wit. Firstly, he pointed out that anointing had been instituted in Old Testament times, and then – no doubt relishing his opportunity to enjoy such an elegant checkmate – he duly added that, if Bruce's involvement was so wholly unacceptable, he would call upon a bishop to act as a replacement.

On 17 May, therefore, after sermons and orations in Latin, French and Scots, the Countess of Mar duly opened the queen's gown and Robert

Bruce poured on her breast and shoulder 'a bonny quantity of oil'. Whereupon the king took the crown in his own hands – one of which was said to be painfully swollen by the pressure of correspondence which had given him scarcely three hours' sleep at night since his return and left him 'much disquieted in mind and spirit' – and honoured his chancellor, Maitland of Thirlestane, by passing it to him to place upon Anne's head. Having received the sceptre from Lord John Hamilton and the sword of state from the Earl of Angus, the young Lutheran queen finally swore an oath 'to procure peace to the Kirk of God within this kingdom,' and, notwithstanding the holy oil on her breast, to 'withstand and despise all papistical superstitions and ceremonies and rites contrary to the word of God'. The oath, however, was little more than a formality from Anne's perspective and carried little real conviction, it seems, for she soon grew to resent the freedom with which ministers of the Kirk deigned to instruct royalty and eventually converted to Catholicism in the 1590s, though never, it must be said, devoutly – rather, perhaps, as a diversion from some of the drearier aspects of her new existence.

Certainly, the Scottish court was not ideally suited to the new queen's high spirits and gaiety. Her childish love for games, masques and pageants earned her an unwelcome reputation for empty-headedness and the ministers were soon condemning her 'want of godly and virtuous exercise among her maids', her 'night-walking and balling' and, most disconcerting of all, her absence from the services of the Kirk. Furthermore, her pastimes were expensive – so much so that the king was appreciating before long that her household cost more than his. And when frustration overwhelmed her, childish tantrums, spiteful words and acts of indiscretion were not always sufficiently resisted. Above all, she nurtured a smouldering resentment towards Maitland of Thirlestane for first opposing her marriage. Yet the queen was also, of course, a vivacious, naive and sometimes innocent figure in a dour and largely colourless setting, where ill-willed courtiers, embroiled in faction and intrigue, often proved more than willing to involve her in the rivalries of the moment and set her, if necessary, against her husband or his councillors.

Even when the heat of James's ardour for his 'Chaste Diana' and 'Cytherea faire' had gradually abated and they had finally ceased to live together, he would continue to treat her with tenderness and affection. And the marriage was undoubtedly a procreative success, for Anne bore her husband seven children. Henry Frederick, a somewhat long-awaited heir was born in 1594, and two daughters followed – Elizabeth in 1596 and Margaret in 1598 – before the future Charles I arrived in 1600. Robert, Mary and Sophia were also born over the next six years, though only Henry, Elizabeth and Charles actually survived childhood, and the promising and greatly loved Prince Henry would eventually die on the cusp of manhood at the age of 18. Nevertheless, when the scale of Anne's task in satisfying a husband of James's complexity and outright oddity is considered, it was no small credit to her that the marriage fared as well as it did. Nor was it any discredit to James either, for his inherent sentimentality and sense of fair play continued to the end of his wife's life. Ultimately, in fact, it was only after the birth and death of his last child that the homosexual preference of his youth seems to have reasserted itself. Until that time, there is every reason to accept Bishop Goodman's later observation that the royal couple managed to 'live together as well as man and wife could' without 'conversing'.

Yet one of the bishop's other contentions – namely, that James was 'never taxed nor tainted with the love of any other lady' – is thrown into doubt by some sparse though tantalising evidence. In a letter written by Sir John Carey to Lord Burghley on 10 May 1595, for instance, there is a reference to 'fair mistress Anne Murray, the king's mistress', and on 3 June Carey also reported a marriage to be solemnised in the near future at Linlithgow 'between young Lord Glamis and the king's mistress'. The lady concerned was once again Anne Murray of Tullibardine who did indeed go on to marry Patrick Lyon, Lord Glamis. Nor are these fleeting comments the only foundation for suspicion, since James composed a narrative poem entitled 'A Dream on his Mistress my Ladie Glammes', in which he dreams that the God Morpheus transports his loved one to him as he lies sleeping and finds it impossible to resist her charms, for 'onlie mot I conquered be'. Beyond this, in fact, the records are silent, but it remains difficult to avoid the impression that the first flush of nuptial bliss may well have carried James forward to at least one extra-marital excursion with a member of the opposite sex.

Soon after his return from Denmark, however, and long before his ardour for his wife had begun to pall, the king was deeply engrossed with women of an altogether different kind. The notorious Scottish witch-hunt throughout the winter of 1590-91 may well have originated in the belief of the Danish admiral Peter Munk that the storms which forced his fleet to abandon the passage of Princess Anne to England had been raised by witchcraft. Doubtless James will have heard this opinion when he arrived to collect her in person, and the unusually stormy weather accompanying the return journey to Scotland the following spring will only have served to reinforce the conviction that malevolent powers were being invoked to harm both him and his wife. For, like other contemporaries across Europe, James displayed an implicit faith in the powers of evil and the efficacy of black magic – along with a cold terror of witches in particular – that would continue to hold even a well-trained, rational mind like his in thrall. Supplied by the devil with enchanted stones, magical powders, poisons and other even more mind-boggling paraphernalia, witches could raise storms, induce insanity, impotence or exaggerated sexual desire, raise spirits to plague mankind and even cause death. They could also (as James recorded in his own book on the subject) fly – so long as they held their breath. Their gender, moreover, made them especially susceptible to diabolical influence, since, in James's view, Satan had seduced Eve in the beginning and been 'the homelier with that sex' ever since.

Towards the end of the king's life, it is true, he was to admit some doubts, though chiefly on the question of the actual prevalence of witchcraft and the nature of confessions wrung under torture from plainly deluded persons. At some time, it seems, he had gone on to read Reginald Scot's *The Discoverie of Witchcraft*, which was brave enough to express scepticism at the efficacy of the black arts. But when his own *Daemonologie* was published in 1597, James left no doubt that his intention was to refute the 'damnable opinion' of Scot, who was 'not ashamed in public print to deny that there was such a thing as witchcraft'. And in his preface to the reader he spoke too of the 'fearful abounding at this time, in this country, of these detestable slaves of the devil, the witches and enchanters'. He wrote that he claimed, 'not in any way (as I protest) to serve for a show of my learning and ingenuity, but only (moved by conscience) to press thereby, so far as I can, to resolve the doubting hearts of many both that such assaults of Satan are most certainly practised, and that the instruments thereof merit most severely to be punished'.

Presenting his book in the form of a dialogue between the uninformed and questioning Philomanthes and the didactic Epistemon, James's discourse encompasses the entire range of supernatural phenomena known to the ancients – ghosts, wizards, spirits, demons, possession, fairies, even werewolves – but it is most concerned with witches and their abilities. Nor had he moderated his opinions significantly by the time he became King of England in 1603. Indeed, although he eventually mitigated the ferocity of some aspects of his new kingdom's witch trials, one of his first priorities upon his succession was to have his book republished and to follow this with a significant stiffening of the old Witchcraft Act in employment under Elizabeth I. Still seeing himself as the white knight of the biblical Book of Revelation, wielding a flaming sword of justice, his personal crusade as the terror of witches and scourge of the devil continued to serve in his view as proud testament to his love of virtue and thirst for justice, so that crimes that had previously been punishable by a prison term now became punishable by death. Furthermore, the explanation James offers in his *Daemonologie* for the contemporary upsurge in the practice of witchcraft is extremely significant. 'The consummation of the world, and our deliverance drawing near,' he wrote, 'make Satan to rage the more in his instruments, knowing his kingdom to be so near an end.' For James, then, as with other contemporaries, the occult was not so much a belief as an ideology in its own right, encompassing the direction of history itself, as well as his own place within it – and in the winter of 1590–91 he would find more than ample scope for indulging his fixations to the full.

The end of the sixteenth century had seen the onset of a tenacious 'witch craze' across Northern Europe and nowhere more so than in Scotland where the Kirk was more or less continually engaged in detecting and combatting dark forces. Deriving in part from ancient Scottish fertility rituals in which devotees worshipped an incarnate deity that appeared before them, the so-called 'Dianic Cult' of James's time was said to be celebrated in midnight orgies by covens of men and women who claimed to have sold their souls to the Devil. It was perceived, moreover, as an organised phenomenon, permeating every class. In Scotland, for instance, the Ruthven family had dabbled in it for generations and their activity had certainly added to James's aversion to them. But it was the exposure of a far-reaching occult conspiracy in East Lothian that now fired the king's interest so keenly that for almost a year he would work at little else than 'the sifting out of them that were guilty' – attending some of the more important examinations in person and, at one point, even making one of the key participants, a maidservant from Tranent called Geilie Duncan, play to him the tunes she had piped for the Kirkyard dances on All Hallows' Night at North Berwick.

It appears, then, to have been the king's trip to Denmark, and conceivably his discussions with the theologian Hemmingsen, author of the well-known treatise on witchcraft *Admonitio de Superstitionibus Magicis Vitandis* that turned James's interest in the occult into something of a more consuming passion culminating in the publication of his own *Daemonologie* in 1597. Though references to Satan are plentiful in his early published works, for instance, he made no direct statements at all on the subject of witchcraft before 1590. But throughout the North Berwick witch trials – and especially from April 1591 when the evidence mounted that James had been the object of a concerted satanic conspiracy involving some 300 individuals of various social ranks – his involvement reached a peak. In all, three or more organised covens were said to have been working against the king and his marriage with all the usual tools and tricks of the craft: toads hanged and roasted to extract their venom 'for his Higness' destruction'; cats bound to the severed joints of dead bodies cast into the sea; magical practices upon the king's bed linen that he and it might waste away together; and finally the fashioning of a waxen image to be melted slowly in a fire to ensure an agonising death.

James listened, it seems, with all the credulity that might have been expected from any contemporary, but also exhibited a particularly acute fascination with some of the more outlandish and curious features of the spectacle. A bewitched man, for example, was said to have capered and shrieked before the king 'to the great admiration of his Majesty', while one of the witches, Agnes Sampson, 'a renowned midwife', whispered things he had said to the queen on their wedding night, at which he 'swore by the living God that he believed all the devils in hell could not have discovered the same'. There were tales, too, of the devil incarnate for James to pore over. According to Sir James Melville, the witches met by night in the kirk of North Berwick, 'where the devil, clad in a black gown, with a black hat upon his head, preached unto a great number of them out of the pulpit, having light like candles round about him'. 'His face,' Melville recorded, 'was terrible' – his nose 'like the beak of an eagle', his eyes 'great' and 'burning' – and his limbs were hairy, 'with claws upon his hands and feet like a griffin'. 'He spoke,' we are told, 'with a hollow voice.'

Certainly, James was prepared on occasion to countenance some of the more notorious excesses associated with witchcraft investigations of the period. When, for example, Agnes Simpson was examined by the king in person at Holyroodhouse, she was fastened to the wall of her cell by a 'witch's bridle', an iron instrument with four sharp prongs forced into the mouth, so that two prongs pressed against the tongue and two others against the cheeks. She was also kept without sleep, and thrown to and fro with a rope around her head before confessing to fifty-three indictments against her and suffering subsequent death at the stake. Likewise, a local schoolmaster at Saltpans by the name of Dr John Fian, who was accused of being a coven leader, was subsequently tortured with both the rack and 'the boot', and by having needles inserted under his fingernails before the fingernails themselves were eventually torn out with pincers.

Above all, though, it was the case of Barbara Napier, widow of Earl Archibald of Angus, that best demonstrates the king's tenacity in rooting out and punishing what he himself considered 'the highest point of idolatry, wherein no exception is admitted by the law of God'. When Napier had the good luck to be acquitted with the help of influential friends like her brother-in-law, the Laird of Cairschoggil, on the pretence that she was pregnant, James was not prepared to let the matter rest. On the contrary, when he learned of her escape from being 'put into a great fire', he insisted that her jurors be arrested and tried for delivering a false verdict. At issue in part, of course, was the credibility of the king's justice, which James correctly emphasised in making his decision. For the practice of corrupting juries and witnesses, and flouting due process on grounds of kinship and friendship was a long-established evil in Scotland – one, said James, that 'here bairns suck at the pap' – and he made clear that it must be dealt with 'not because I am James Stuart and can command so many thousands of men, but because God hath made me a king and judge, to judge righteous judgements'.

Yet the same conviction that witchcraft was a living, potent force also firmly underlay the king's determination that Barbara Napier must pay the ultimate price for her sacrilege. 'We are taught,' James told her jurors, 'by the laws both of God and men that this sin is most odious; called *Maleficium* or *Veneficium*, an ill deed or a poisonable deed, and punishable likewise by death.' 'As for them who think these witchcrafts to be but fantasies,' he continued, 'I remit them to be catechised and instructed in these most evident points.' And although the number of executions resulting from the North Berwick witch trials remains uncertain, Napier was only one among some seventy in all who found themselves prosecuted. She, along with Geillie Duncan, Dr Fian, Agnes Sampson and Euphemia MacCalzean, daughter of Lord Cliftonhall, the well-known judge, were each executed for conspiring to bring about the death of the king, who had only, it seems, survived against all odds. When, according to one confession, the devil inquired of the witches whether their magic was proving effective, he was disappointed 'and because a poor old silly ploughman named Grey Meill chanced to say that nothing ailed the king yet, God be thanked, the Devil gave him a great blow'. When, moreover, the witches sought the devil's explanation for the king's unexpected survival, the Devil could offer only one explanation, which he chose to deliver, curiously enough, in French. '*Il est*,' retorted the Prince of Darkness, '*un homme de Dieu*.'

In the meantime, a sea captain called Robert Grierson, who was also implicated, had died as a result of torture before he could be formally tried. But it was the alleged involvement of none other than Francis Stewart Hepburn, 5th Earl of Bothwell, which gave the entire case a new and altogether more sinister significance. For on the evidence of various suspects it was suggested that Agnes Sampson had proposed the king's destruction at the earl's request, and that a waxen image melted before the devil at Acheson's Haven near Prestonpans had been the result. Indeed, on 5 May 1591, in what seems to be an increasingly manipulated statement to be used for political ends, Bothwell was specifically incriminated by Geillie Duncan and others as the one who commissioned and provided funds for the making of the wax image to be used to destroy the king. As the image was passed from hand to hand, it was said, all present uttered the following words: 'This is King James the Sixth, ordained to be consumed at the instance of a nobleman – Francis, Earl of Bothwell.'

Nor was this the only claim of its kind. The king was to be destroyed, said one defendant, 'that another might rule in his Majesty's place and the government might go to the Devil'. And there was talk, too, of the English queen's involvement with Bothwell's treachery after the unfortunate Grierson had let slip that 'it would be long before the gold came out of England'. When, moreover, the warlock Richard Graham, who had already produced a garbled Latin conjuration invoking syphilis upon the king, agreed to tell all he knew about the earl, so long as he was granted the protection of Edinburgh Castle's walls, the scene was well and truly set for what would prove to be the prelude to a final and decisive reckoning between the King of Scotland and his turbulent nobles.

# 7 ⚹ The Wrath of Earls and Kirk

'He hath oft told me the wickedness of his nobility and their evil natures, declaring himself weary of his life among them.'

*Thomas Fowler, chief agent of Sir Francis Walsingham's intelligence network in Scotland, writing of King James in 1589*

In May 1587, more than two years before his marriage, James had made a characteristically eccentric attempt to heal his kingdom's endemic divisions by staging a curious feast of reconciliation at Holyrood. Gathering his nobles together for what was ostensibly a celebration of his twenty-first birthday, he urged them with all the earnestness and naivety that was so typical of his nature to foreswear their feuds and hatreds by drinking three solemn pledges to eternal friendship. Thereafter, in a gesture that smacked more of dark humour than statecraft, the king delighted his capital with a procession of the entire nobility of Scotland walking two by two, each man coupled with his most notorious enemy, to take their place at banquet tables laid out by the Mercat Cross amid fireworks, salvoes of cannon fire and a lavish distribution of free wine for the onlookers who gaped with a mixture of astonishment and cynical amusement at the spectacle before them. It was, in its way, a not unlovable gesture from a king whose good intentions were rarely in overall doubt. But it was symbolic, too, of the same king's impotence in the face of powerful forces that he lacked both the material and personal resources to quell. Within a year, there had been two unsuccessful attempts to lever his chancellor out of office and in the spring of 1589 a major rising was narrowly averted. Now, as a final straw, one of those self-same would-be rebels, who had been treated with such remarkable leniency, was firmly implicated in a sacrilegious murder plot against his sovereign master.

In many respects, the character and career of Francis Stewart Hepburn, 5th Earl of Bothwell and godson to Mary Queen of Scots, mirrored the influence of his two more famous uncles. Born in 1563, the earl was the son of John Stewart, one of the numerous bastard sons of James V – a distinction that he shared with his dead brother, the Earl of Moray. And from his uncle Moray, who had served as regent before his assassination, Bothwell derived a somewhat anomalous alliance with the Kirk, which could not afford to be overly fastidious in its choice of political champions and whose interests he strove to maintain at court, in spite of his own waywardness and his family's stained reputation. For the mother of the current Earl of Bothwell was Jean Hepburn, sister of the Bothwell who had been Queen Mary's notorious lover, and it was this latter uncle, rather than the erstwhile regent, whom the fifth earl most resembled in conduct and character. Fierce, dissolute, profligate and lawless, wedded to feuds and loose living, he lived, in fact, in a world of fantasy, which encompassed a vain hope that the Crown might one day be his, irrespective of the superior claims to the succession of the Lennox-Stuart families.

Not surprisingly, the king had already been greatly incensed by Bothwell's participation in the recent revolt. But largely as a result of blind hopefulness born of weakness and superstitious fear, he had opted for mercy and reconciliation, telling the volatile earl that just as he 'had resolved to be a reformed king, so he would have him to be a reformed lord'. As so often in such instances, however, James was merely postponing the inevitable. Bothwell's companions in his Border district of Liddesdale were, almost to a man, thieves and murderers, and the general lawlessness and disorder that pertained there was a constant source of tension not only for James but for the English, too. In January 1591, for instance, during the trial of one of his cronies, Bothwell abducted a witness from the Tolbooth at Edinburgh, regardless of the fact that the king himself was residing in an adjoining chamber. That same night, moreover, James found himself forced to ride to Kelso in the hope of preventing a brawl in which he knew Bothwell was bound to take part. But when the earl was summoned before him the following day, James issued only threats. He had loved and favoured Bothwell, he said, only to be treated to insults in return. Unless there was a change in behaviour, the law would be enforced with full vigour.

Even so, like his uncle before him, Francis Stewart Hepburn was a man of contradictions. 'There is more wickedness, more valour, and more good parts in him,' wrote Thomas Fowler, 'than in any three of the other noblemen,' and there was no doubting either his brains or his charm, for he was handsome, dashing and eloquent. On one occasion, for example, he found himself the guest for a few nights of the Bishop of Durham, the shrewd and perceptive Tobie Mathew who made no secret of the earl's more admirable qualities. 'This nobleman hath a wonderful wit,' wrote Mathew, 'and as wonderful a volubility of tongue as agility of body on horse and foot; competently learned in the Latin language, well versed in the French and Italian; much delighted in poetry, and of a very disposition, both to do and to suffer; nothing dainty to discover his humour or any good quality he hath.' In the early days, indeed, Bothwell's considerable charm had made him a favourite of the king, who liked to embrace him tenderly and hang about his neck in the familiar fashion. Had he never become a symbol of aristocratic hatred for Chancellor Maitland and of Protestant loathing for the Catholic earls, it is doubtful whether he could ever have created nearly so much trouble. But his support within the Kirk increased his threat and behind the veneer of French culture, which attracted James, there remained a lingering undercurrent of terrifying unpredictability that made his involvement in the black arts wholly plausible and now rendered him an object of the king's deepest fear and hatred.

It had not been until 15 April 1591, however, that Bothwell was finally accused of plotting with Agnes Sampson and Richard Graham against the king's life. And he would prove a difficult man to bring to heal, for on 24 June at two in the morning, according to David Moysie, he succeeded in escaping from Edinburgh Castle, where he was being held pending his trial. Outlawed, he remained in his fortresses along the Border, sometimes spending the nights in the woods. But although the council proclaimed him a traitor, many secretly sympathised with the man who promised to be Maitland's nemesis and who had, in any case, been charged on the evidence of witches. Certainly, the contrast between Bothwell's treatment and the leniency accorded earlier to Huntly was widely acknowledged, while the Kirk, on the other hand, remained willing to apply a remarkable degree of tolerance to its influential ally. Emboldened, therefore, by widespread support, he flouted the king's threats by appearing at Dalkeith, Crichton and Leith, and most blatantly of all at the Canongate in Edinburgh, where he issued a taunting challenge to the chancellor. In response, the king made a token effort to raise troops and capture Bothwell at Kelso, though his own superstitious dread of the nobleman and nagging fear of capture continued to militate against more decisive action. Without a standing army and with few dependable friends of his own, James's position remained, to say the least, precarious. Indeed, such was the earl's power that no appointments were made to the offices he had lost and no one accepted his forfeited estates.

On the dark night of 27 December he made a daring raid upon Holyroodhouse, along with Archibald Douglas, son of William, Earl of Morton, and some forty or fifty 'murderers and broken men'. Forcing their way into a stable, they seized the keys of the porter and succeeded in pursuing Maitland to his chamber, while the king, whose poverty had deprived him of an adequate guard, was left to seek shelter in a remote tower as the intruders careered through the buildings amid cries of justice for their leader. As doors were broken with hammers and set on fire, it seemed yet again, then, that the king's authority was about to be rudely shattered by a show of brute force and daring. And only when the common bell of the City of Edinburgh was rung and the local citizenry, armed with pikes, rallied to the royal cause did the attempted coup finally collapse. In the confusion, Bothwell and four of his associates fled downstairs and after killing John Shaw, master of the stables, and a number of others, made good their escape on horseback. Even then, however, the king's embarrassment was not over, for, although Bothwell had been publicly proclaimed a traitor at the Mercat Cross, the king was reprimanded by the ministers officiating at a thanksgiving service held at St Giles only one day afterwards.

In the process, the self-same ministers had all but justified the raid, notwithstanding the fact that eight of Bothwell's Borderers were now hanging from the palace's walls. And the earl himself, of course, remained at large, while James and his household were forced to move for safety into cramped lodgings at the top of the city, under the shelter of the castle guns. The king's words, it is true, were brave enough, and in a sonnet composed to commemorate the slain John Shaw he wrote of his intention 'with deeds, and not with words to pay'. But it was easier to posture than deliver, and though he pursued Bothwell courageously enough, his bleak

January chase through the Border country, where the canny, hardy earl had so many friends, was unavailing. Indeed, far from apprehending the outlaw, James nearly perished when his own horse fell and plunged him into the icy waters of the Tyne.

While Bothwell was flaunting his liberty quite openly at race meetings and football games and accepting invitations to card parties with the English Border gentry, the king's other tormentors were also making hay. The earls of Huntly, for their part, had long been at odds with the earls of Moray, since Mary Queen of Scots had first bestowed that title upon her half-brother Lord James Stewart, the later regent, thereby depriving the 4th Earl of Moray of an earldom that he considered to be rightfully his. And to compound matters, James Stewart Doune, who had married the elder daughter of the regent and become Earl of Moray in right of his wife, had been contriving to extend his influence in the north-east of Scotland at the expense of Huntly's family, the Gordons, thus continuing the policy followed earlier by the queen's half-brother. Handsome and high-spirited, Doune was the 'Bonny Earl' of the famous ballad, 'comely, gentle, brave and of a great stature and strength of body', beloved by both Kirk and people as the heir and son-in-law of the former 'good' regent.

He too, however, would now fall victim to the kind of bloody retribution in which Huntly specialised so avidly. For when the Queen of England's representative, Lord Worcester, came north in 1590 with a belated wedding present, to confer on James the Order of the Garter, he also brought with him some of Huntly's treasonable correspondence that had been intercepted by English intelligence. It was widely known, too, that Huntly and his confederate, the Earl of Errol, had remained in continuous touch with Spain, even in 1590 when they had not yet been formally pardoned for their last act of defiance which had ended at the Bridge of Dee. In spite of all, though, James had nevertheless pressed ahead with a formal pardon for Huntly in December of that year, and by early 1592 was resolved, or so he claimed, to settle the earl's outstanding feud with Moray. By calling the latter to his mother's castle at Donibristle on the northern bank of the Firth of Forth, the king intended, ostensibly, to reconcile the two contending nobles by little more than the magic of his own personality. The result, however – both sadly and predictably – was one of Scotland's most spectacular and notorious aristocratic murders.

Certainly, there was no shortage of speculation at the time that James himself had duped Moray into attendance at Donibristle and thereby connived in his death. The king's dislike for all connected with 'that bastard', the former regent, remained as intense as ever, of course, and Moray's friendship with Bothwell will have done nothing to soften relations. There were rumours, too, reflected in the ballad that celebrated him, that the earl was 'the Queen's love' and had been privy to the Christmas raid at Holyrood. Add to this the evidence that Maitland was eyeing some of Moray's lands and had procured a warrant for Huntly to arrest him as a Bothwell partisan, and the case suggesting an orchestrated assassination is far from implausible.

If, moreover, James himself was not in fact directly involved, the timing of his invitation could not have been more conducive to Huntly's purposes. For on 7 February 1591, while accompanying the king on a hunting trip with an escort of forty horsemen in the vicinity of Donibristle, Huntly absented himself on the pretext of pursuing a group of Bothwell's accomplices who were known to be close by. In the encounter with Moray that actually followed, however, the earl and his clansmen held the castle all day until it was finally set ablaze and he was left with no option but to flee from the inferno towards the shore – his hair and helmet plume, it was said, in flames. Thereafter, he was finally dispatched among the rocks by Huntly himself who delivered a killing dagger blow to the face. 'You have spoilt a better face than your own', the handsome victim is said to have declared before he expired.

Yet the deed in itself was still not one that appears to have troubled James unduly, for he was capable of a striking indifference to injustice or suffering when his own affections were not directly involved. At the same time, of course, the assassination was wholly in keeping with a long-standing culture of political violence in Scotland, with which James was already all too familiar, and in addition to liking Huntly and despising Moray, he also needed the former's strength in the north of the kingdom to counterbalance both Bothwell and the Kirk. But any wishful thinking on James's part that the martyring of a popular Protestant hero by Scotland's premier papist might somehow pass unmarked was soon disproven, as Moray and his dead friends lay in state in Leith church and the earl's bloodstained shirt was paraded on a spear around the Highlands amid the growing clamour for clan war. In the meantime, for further dramatic effect, his mother made her own demand for vengeance, presenting James with a picture of her murdered son and a musket ball plucked from his body.

The subsequent treatment of Huntly, moreover, only added fuel to the flames that had already consumed Donibristle. Having placed himself voluntarily in the king's custody at Blackness, he was yet again indulged and released almost immediately amid howls of disapproval from both Presbyterian ministers and common people alike. 'Always,' James told Huntly, 'I shall remain constant,' while Maitland rather than the king was left to incur the main blame of the rioting Edinburgh mob. 'Since your passing from here,' James informed Huntly, 'I have been in such danger and peril of my life as since I was born I was never in the like, partly by the grudging and tumults of the people, and partly by the exclamation of the Ministry whereby I was moved to dissemble.' But while such dissembling came naturally to James, on this occasion it counted for little as rumours of his involvement in the murder mounted. Above all, there was talk that when Moray's kinsman, Andrew Stewart, 2nd Lord Ochiltree, attempted assistance, he had found the ferry to Donibristle barred by the king's express order from all but Huntly.

If so, then James had clearly miscalculated grievously. On the one hand, England's Queen Elizabeth was not slow to repeat her familiar warnings that no good could come from clemency to Catholics, and least of all a wholly wild card like Huntly. Balancing rival forces was, of course, a key art of political management, but when Huntly and Bothwell were the counterweights involved, no happy outcome was ever conceivable. And now relations with the kingdom to the south worsened significantly, as James complained to Elizabeth of his English pension and how she was not 'content as freely to pay it as freely ye promised it'. Equally gallingly, the queen showed no inclination to stem the flow of unofficial assistance tendered to Bothwell, who had now been gifted a popular cause to fight for and the wholehearted backing of the Kirk to wage it. Even with his connection to the black arts, he was nevertheless viewed as what one minister, John Davidson, termed a 'sanctified plague' whose divinely ordained mission was to cause the king to 'turn to God' and deliver Scotland from the papist scourge. Far from damaging him, then, the loss of Moray would actually render Bothwell the bane of James's life for at least two more years to come.

Certainly, in June 1592 the turbulent earl was once more snapping at the king's heels – this time with a midnight attack upon Falkland Palace involving 300 men and a battering ram, during which James had to withstand a seven-hour siege before help arrived. Furthermore, when two of Bothwell's accomplices, the Lairds of Logie and Burley, were subsequently arrested in early August, even this would prove a source of vexation to the king. For Logie's escape on the night of his interrogation was apparently achieved with the help of his sweetheart Margaret Vinster, one of the queen's Danish maids. Leading her lover through the royal sleeping-apartments at Holyrood in the dead of night, Vinster, we hear, 'conveyed him out at a window in a pair of sheets'. And such was James's dejection at his betrayal by those so near to him that he not only had his first recorded quarrel with his wife, but, more significantly, entered another of those phases of dejection to which he was always prone, 'lamenting his estate and accounting his fortune to be worse than any prince living'. Indeed, he would spend the rest of that summer and autumn fleeing from place to place in dread of Bothwell, and was further weakened in August when Maitland, finally bowing to the raft of opposition existing at court, retired temporarily to his estates at Lethington.

In the summer of 1593, moreover, Bothwell came again, and this time successfully. By now, he had enlisted almost all the leading Stewarts to his cause – Lennox, Atholl and Ochiltree – and when Lady Atholl allowed a raiding party to slip into Holyroodhouse between eight and nine in the morning of 24 July there followed the familiar royal surrender. Half-dressed and trapped in his bedchamber, James was finally confronted by Bothwell in person, emerging from behind the hangings in the king's ante-room and 'craving mercy and pardon most humbly' but carrying nevertheless a naked sword which he laid at the king's feet. Nor could James's apparent boldness in telling the kneeling earl that he could not, like Satan dealing with a witch, obtain his immortal soul, conceal his actual impotence. For, in spite of his brave words and cries of 'treason', he found himself forced all the same into what was arguably the most humiliating capitulation of his entire life. Compelled on the one hand to dismiss his friends at court, the king was also left to agree a compromise whereby Bothwell himself would withdraw from court in return for an acquittal at his forthcoming witchcraft trial and an additional pardon for all other offences.

When, furthermore, a crowd of armed citizens soon gathered around the palace, they were duly informed by the king, leaning from his window in thoroughly unkingly fashion, that all was well. Pretending that the band of Bothwell's ruffians now escorting him everywhere were being employed at his own request, James then duly acquitted Bothwell of witchcraft on 10 August notwithstanding the fact that the earl's chief accuser, Richard Graham, had already insisted upon his involvement up to the very time that he himself was executed in February. Worse still, the verdict of outlawry previously proclaimed three times against the earl at the Mercat Cross was soon formally revoked. With greater bravado than ever Bothwell then wrote in triumph to Queen Elizabeth, whom he addressed as 'Most Renowned Empress', expressing his hope that he might become Lord Lieutenant of Scotland and assuring her how, with continued help, he could 'manage the estate about the king'.

Yet, like other subverters of royal will before him, Bothwell's sway and swagger concealed weaknesses of his own that would grow increasingly apparent in the longer term. Not least of all, the uneasy confederation of malcontents that he had conjured around him would eventually demonstrate tensions of its own, as hidden rivalries resurfaced, mutual suspicion crept in and the inherent need for stable monarchical authority became apparent once more. Furthermore, Bothwell's agreement to abandon the court in return for his acquittal and pardon proved a costly error, since it allowed James to improvise a middle party to protect him from both the earl himself as well as Huntly's supporters. Lennox and Mar, for example, were easily detached from their temporary ally, and Maitland was now reconciled with a group of his former adversaries, including Glamis, the treasurer, Lord John Hamilton, one of the greatest of all the nobles, and the Catholic Homes and Maxwells, with whom Bothwell was at feud along the Border. Such an alliance could not last for long itself, of course, but it enabled James to slip away from a parliament at Stirling to the security of Loch Leven and subsequently raise an army of sufficient strength to banish Bothwell once again.

The delinquent earl's campaign of force and fury had, in any case, no real substance of policy behind it beyond his own self-aggrandisement and the general contempt for Maitland that would ultimately evaporate with the chancellor's death in 1595. Bothwell's final rebellious fling in 1594 was therefore little more than a swansong – the act of an increasingly frustrated man who had no more imaginative cards in his pack. By now, his usefulness to the Queen of England was largely spent and the paltry subsidy of £400 that she offered to assist his latest adventure was wholly inadequate to the task, though in April he would drive a force of royal cavalry back upon Edinburgh, leaving the king, as David Calderwood observed, to retire into the city 'at full gallop with little honour'. This initial setback was only the prelude, however, to an altogether more decisive rally on James's part as 1,000 Edinburgh citizens, aided by three great cannon from the castle, turned out to resist Bothwell's assault. After a short engagement which cost the king's forces no more than twelve men killed, Bothwell promptly disbanded his army and withdrew to England, where he was this time promptly disavowed by Elizabeth.

The ultimate folly was yet to come, however, as Bothwell now threw in his lot with none other than Huntly, which guaranteed at a single stroke the loss of any residing sympathy from the Kirk. For at this time Huntly was further mired by his involvement in one more Catholic intrigue, dating back to the end of 1592. The conspiracy of the so-called 'Spanish Blanks' had been hatched, it seems, by two Jesuits, James Gordon and William Crichton, and involved a heady project to land a force of 30,000 Spaniards from the Netherlands, of whom 5,000 were to establish Catholic control in Scotland, while the rest marched south to England. Since the plan contained no provision whatsoever for securing the necessary sea lanes, the immediate danger posed by the plan was actually negligible. But the Scottish messenger, George Ker, who was arrested while embarking for Spain, was in possession of mysterious blank papers signed by Huntly, Angus, Errol and Sir Patrick Gordon of Auchindoon committing them to some unspecified, but no doubt dastardly, compact with the enemy. The upshot was another largely futile punitive campaign in the north, in which James got as far as Aberdeen and hanged a few peripheral players, while the main culprits went into hiding in Caithness.

Throughout 1593, in fact, the king continued to treat the conspirators with extraordinary leniency. Plainly despairing of his ability to bridle them, he had admitted to Sir Robert Bowes, the English ambassador, that Huntly, Errol and Angus were three of the most powerful nobles in the kingdom and that 'if he should again pursue them and toot them with the horn he should little prevail'. In summer, the parliament which was expected to pass an act of forfeiture against the conspirators' lands did not do so, and, in November, James obtained from a sparsely attended convention of nobles held at Edinburgh an Act of Oblivion by which the guilty earls were forgiven their involvement in the conspiracy of the Spanish Blanks on condition that they made a formal submission to the Kirk – which was never in fact forthcoming. To obtain the Act, James was, it seems, quite willing to manipulate the convention blatantly and, ultimately, even to tamper with the text of the submission itself. In consequence, the whole clergy raved and the Synod of Fife excommunicated the earls without consulting the king, while Elizabeth added to the chorus of disapproval by dispatching a stinging letter in which she rued the sight of a 'seduced' king and a 'wry-guided' kingdom.

So it was, then, in the summer of 1594, that Bothwell and Huntly finally found themselves consorting in the most consummately unholy alliance of all. On Midsummer Eve, indeed, the Catholic earls, buoyed by common cause with Bothwell and assurances of Spanish help, held a great feast with dancing and drinking to celebrate their impending triumph, notwithstanding the fact that at the time of Bothwell's raid at Leith, James had already issued a solemn pledge to his subjects that could not be easily repudiated. 'If ye will assist me against Bothwell at this time,' he had pleaded, 'I promise to prosecute the excommunicated lords so that they shall not be suffered to remain in any part of Scotland.' And now, faced with Bothwell's escalating defiance, even James could not avoid a more decisive response. Drawing his strength not from the nobility, but from the lairds, the burghs and the Kirk, the king therefore prepared for a punitive expedition, and by sheer good fortune resulting largely from the Crown's inherent advantage as the only potential source of long-term order, it at last proved possible to cut the cancer. For unlike his enemies, James had always been able to rely ultimately upon a feudal army to contain any threat for a month or two at any given moment, and it was this advantage which now rendered him well placed to employ such a contingent for offensive purposes.

In September 1594, therefore, a royal force led by the king himself and accompanied by Andrew Melville, whose presence confirmed the Kirk's determination to deal with Huntly once and for all, duly marched into Aberdeenshire. And though the royal army received an early reverse on 3 October when the young Earl of Argyll's advance detachment was defeated in a minor skirmish, known rather more grandly than it deserved as the 'Battle of Glenlivet', Huntly's men had no stomach for a further encounter with the main force following close behind. Instead, they returned once more to the wilds of Caithness and, at the Kirk's insistence, agreed to the destruction of their dwellings. In the process, Huntly's fortified stronghold was blown up, along with Errol's in Buchan and another half-dozen Catholic fortresses, after which, within only a few months, James duly obtained an agreement whereby Huntly and Errol accepted exile abroad, leaving only Angus to lurk, albeit impotently, among his Highland cronies.

Bothwell, meanwhile, had also escaped to Caithness after Huntly purposely delayed in surrendering him to the king, and a certain Edinburgh merchant named Francis Tennant refused to betray him. Yet it was not long before Bothwell, too, settled for exile in Europe and a later career mirroring, in many respects, that of his more famous uncle. While the last husband of Mary Queen of Scots fell into the hands of Scandinavian enemies and died imprisoned and insane, his deranged young successor would, it is true, manage to retain his freedom. But by April 1595, he was in France without friends or means, since his Scottish estates had been forfeited to powerful neighbours, and before long, like many an unsuccessful rebel unlucky enough to evade the more merciful retribution of the executioner's blade, he faced lingering loneliness, ignominy and penury. Journeying forlornly through Spain and Italy, he would eventually die in poverty in Naples in 1612.

By contrast, Huntly once more returned to Scotland and yet again enjoyed the king's boundless clemency. He remained, of course, a tool of sorts for James to juggle against the Kirk and a grubby diplomatic counter to hold in reserve in his dealings with Elizabeth, but the absence of more decisive punishment remains profoundly puzzling. His return to Scotland in 1596 was, in fact, accompanied the following year by his reception into the Kirk at the king's insistence. Yet even James's further decision to reward him with a marquisate would not ensure his religious compliance in the longer term. For, while he conformed in James's lifetime, he would nevertheless choose to die a Catholic in 1636. And it was only fitting, perhaps, that a man who had both milked and abused royal favour so consistently should have staged this final gesture of defiance upon his deathbed. Insofar as he had ever been defeated in any really meaningful sense of the term, Huntly had, after all, been largely vanquished by default – a victim not so much of the king's strength or political wisdom as of his own latent weakness and eventual lack of options. The king he could shake with ease, the Crown ultimately he could not.

The same, moreover, was true of the nobility in general and James's apparent 'victory' over them, for although his authority would never again be flouted quite so flagrantly, James triumphed over his aristocratic enemies as much by luck and circumstance as by judgement. Bothwell, for his part, had ruined himself by his own outrageous behaviour, while Huntly and his henchmen were chasing a vain Catholic cause that had long since had its day in Scotland. As a result, their peers were largely tired of a stream of disorder, subsidised too often from abroad, which could only harm both Scotland's and their own interests in the longer term. Ultimately, indeed, Bothwell and Huntly had not only demonstrated the futility of politics by kidnapping, but served to underline the central importance of monarchy in a Scottish context. More, in fact, was to be gained from partnership with the Crown than opposition at a time when James's succession to the throne of England appeared to be drawing closer and as James, in any case, employed the lands once held by Rome to coax compliance. Above all, however, it was the growing political leverage of the kingdom's emerging middle classes – the very lairds, burgesses and moderate clergymen who had rescued James on more than one occasion – that persuaded even the most die-hard advocates of noble rule that their interests were best conserved by co-operation rather than conflict.

In the event, it was more radical sections of the clergy who still potentially posed the most fundamental challenge to the king's authority, and at Maitland's instigation every effort had been made to achieve an accommodation between Church and state. Shortly after his return from Denmark, for example, James was encouraged not only to visit the General Assembly to thank God that he had been born into the Scottish Kirk, but to deride its Anglican equivalent, which had already been bitterly attacked by the likes of James Melville for its 'bell-god bishops' who were anxious to advance the cause of episcopacy in Scotland. 'As for our neighbour Kirk in England,' the king declared, 'it is an evil said Mass in English, wanting nothing but the liftings.' Nor did he entirely buckle under Elizabeth's pressure to prevent the Kirk from extending support to English Puritans, three of whom – John Udall, Robert Waldegrave and John Penry – had sought asylum in Scotland. 'There is risen both in your realm and mine,' Elizabeth protested:

> a sect of perilous consequence, such as would have no kings but a presbytery and take our place while they enjoy our privilege. Yea, look well unto them. I pray you stop the mouths or make shorter the tongues of such ministers as dare presume to make orisons in their pulpits for the persecuted in England for the Gospel.

But while prayers for English Puritans were indeed forbidden and Penry was expelled in accordance with Elizabeth's wishes, James moved slowly and retained Waldegrave as his printer. As a further gesture towards the Kirk, he would also continue to intercede on behalf of Thomas Cartwright and other English Puritans during 1591.

In the wake of Huntly's assassination of Moray, Maitland and James seemed especially keen to placate the Kirk, with the result that in May 1592 Parliament had been encouraged to agree a string of concessionary measures. The so-called 'Golden Act', for instance, had fully and clearly established ecclesiastical government by presbyteries, synods and General Assembly, while a law of 1584, which confirmed the status of bishops, was now rescinded.

Nor were James's actions altogether as surprising as they may at first seem. Though he feared and hated any threat to his own primacy in secular affairs and equated incursions on his ecclesiastical authority as the first step on a broader slippery slope, he had been raised and educated nevertheless within the structure of the Kirk, and respected both the institution itself and the Calvinist theology upon which it was based. He was also, by nature, averse to confrontation where compromise might be attainable and, with Maitland's guiding hand, it was still conceivable that the Kirk could itself be finessed into compliance. The process, of course, might well be long and fraught, as was demonstrated later in 1592 when an Act asserting the royal supremacy led to a string of attacks from the pulpit, while the king himself 'chafed and railed' against the ministers after a law to silence them was rejected. But if the Kirk could be bought off ultimately by substituting the term 'presbyter' for 'bishop', might not the Crown's control of secular affairs continue unchallenged?

All depended, in fact, upon a firm stand against Catholicism, and the goodwill of key personnel at the summit of the Kirk's hierarchy: neither of which would be forthcoming. Indeed, James's lame response to the affair of the Spanish Blanks appeared to be one more demonstration of his indecision and ambivalence regarding his treasonous Catholic subjects, and left his Presbyterian ministers demanding condign punishment not only for the Catholic earls involved, but for the entire body of Scottish Catholics who took it for granted that the king himself had been a party to the plot. At the time of George Ker's arrest, after all, a private memorandum drawn up by James had been found among his papers in which the king discussed the scheme's merits as a means of assisting his succession to the English throne. And although the memorandum concluded that any invasion of England was impractical at that time, its inclusion in a portfolio bound for Spain nevertheless offered fertile ground for speculation. Nor did James altogether rule out dealings with the Spanish in his efforts to achieve his dynastic ambitions. 'In the meantime,' he noted:

I will deal with the Queen of England fair and pleasantly for my title to the Crown of England after her decease, which thing, if she grant to (as it is not impossible, howbeit unlikely), we have attained our design without stroke of sword. If by the contrary, then delay makes me to settle my country in the meantime and, when I like hereafter, I may in a month or two (forewarning of the King of Spain) attain to our purpose, she not suspecting such a thing as she does now, which, if it were so done, would be a far greater honour to him and me both.

At this very time, meanwhile, the Kirk was once more firmly under the influence of the man whose fire-breathing capabilities exceeded, by reputation, even those of John Knox. For since March 1586, Andrew Melville had been back at his post at St Andrews where he would continue for the next twenty years. By 1590, indeed, he had become the university's rector. A veteran of Calvin's Geneva, and an expert not only in theology but in Hebrew, Chaldee, Syriac and Rabbinical languages, Melville had already fled to England to escape a treason charge in November 1584, but returned within twenty months to champion the liberties of the Scottish Kirk against all encroachments of the government. And from that time forth his opportunities multiplied. Not least of all, James's Act of Annexation of 1587, which had appropriated episcopal temporalities to the Crown, had contributed significantly to the decline of the very episcopacy upon whose continued existence he staked so much. In principle, the measure had represented a sensible attempt to win the sympathy of the General Assembly, but the subsequent action of the Catholic earls and James's weak-kneed response meant that no such sympathy could ever be sustained.

More worryingly still, James's Act of Oblivion in 1593, which forgave Huntly for his involvement in the affair of the Spanish Blanks, merely reinforced Melville's primacy. Indeed, while James continued to insist upon employing the Catholic earls as a counterweight to his Presbyterian clergy, so Melville's leadership of the Kirk became not only increasingly critical but, if anything, even more radical. The king, it is true, was caught in a vicious circle. Yet it was one at least partly of his own making, since it was dictated not only by politics but by his ongoing personal favouritism for Huntly. Leniency towards Scottish Catholicism resulted too, no doubt, from James's wish to enlist the support of Catholics south of the Border in his quest for the English succession. However, his eventual decision to allow

Huntly and Errol to return from exile in the summer of 1596 would stretch far beyond the bounds of subtle signal-sending to potential sympathisers south of the Border. On the contrary, it would needlessly flout all political common sense, and flagrantly inflame those very forces within the Kirk that he most feared – the self same forces indeed that he would help to entrench in the absence of Maitland's moderating influence after the latter's death in October 1595.

# 8 ⚶ 'King and Sovereign Lord' of Scotland

'Here is a strange country. I would say a most vile people.'

*Thomas Fowler, chief agent of Sir Francis Walsingham's intelligence network in Scotland, May, 1589*

The birth of James's first child at Stirling on 19 February 1594 was a welcome glint of sunlight amid the storm clouds surrounding him on all sides. According to David Moysie, the news was greeted with joy by 'the whole people', and 'moved them to great triumph, wantonness and play … as if the people had been daft for mirth'. And though the child's arrival opened up fresh possibilities to every plotter in Scotland seeking another target for kidnap and an alternative power source of exactly the kind that James himself had become during his mother's reign, the king could not be anything but happy overall. For the birth not only strengthened his dynastic position in Scotland, but also improved his eligibility for the Crown of England as the father of a healthy heir. Notwithstanding the continued presence of Bothwell and ongoing preparations for the campaign against Huntly, the baptism, therefore, was to prove a particularly grand and symbolic event. In memory of his two grandfathers and with the founder of the Anglo-Scottish marital alliance, Henry VII, no doubt also in mind, the prince was to be named Henry Frederick, and by the time festivities were staged in August, enough money had been scraped from the meagre Scottish treasury to do full justice to the spectacle – the highlight of which was a masque in which James appeared as a Christian Knight of Malta, alongside the Border lord, Buccleugh, and other nobles in female garb, representing Amazons.

The christening itself, moreover, was no less colourful – and ritualistic – than James's own twenty-eight years earlier, including to the fury of Andrew Melville and his like-minded ministers an anointing with holy oil by the Bishop of Aberdeen. But while the full-bearded clergy fumed, the king could take due satisfaction from other aspects of the event. Certainly, the exotic arrangements for the serving of dessert at the evening banquet left no doubt of James's determination to emulate the opulence of more prestigious courts in Europe. A great chariot, 12ft long and 7ft broad, was drawn, it seems, by a single Moor, while Ceres, Fecundity, Faith, Concord, Liberality and Perseverance, dressed in silver and crimson satin, dispensed fruit from it. Then, to symbolise James's voyage to Denmark to claim 'like a new Jason, his new queen', there followed a great ship 18ft long and with 40ft masts, taffeta sails and silken rigging, which discharged a volley of thirty-six cannon into Stirling Castle's great hall before distributing all kinds of fish and shell fish 'made of sugar and most lively represented in their own shape'. Whereupon, the choir sang a fourteen-part harmonised version of Psalm 128: 'For thou shalt eat the labours of thine hands: O well is thee, and happy shalt thou be'.

In other respects, however, not all was quite so satisfying for the king's ego. It had been wise of James to invite the English queen to be godmother and in spite of her notorious parsimony in present-giving, she had shown her appreciation generously enough with 'a cupboard of silver-over-gilt, cunningly wrought' and some massive gold cups. Much more importantly still, she would soon sharply forbid Bothwell to 'show banner, blow trumpet, or in any way live or breathe in England'. At the baptism itself, however, Elizabeth was represented only by the young and surprisingly unsophisticated Earl of Sussex, while the French ambassador failed to turn up at all. Carried away, moreover, by an overwhelming desire to impress, the king had taken too much responsibility for the subsequent junketing at a time when he was already overstretched, and increased his burden further by attempting grandiose and ultimately fruitless negotiations with a number of his foreign guests – too many of whom had actually been invited in the first place. In consequence, many details of the event went awry, and in the midst of proceedings he was further perplexed by a malicious rumour that the Duke of Lennox was the child's real father. While James was busy commissioning spectacular entertainments involving oversize chariots, mock ships and muscular Moors, another seed of dissatisfaction with his marriage was being vindictively sown. And within a year, in the summer of 1595, this same seed had blossomed into a full-blown quarrel between king and queen over the new prince's custody.

It was not long after she had settled in to her new home and accustomed herself to the astonishing vicissitudes of her husband's kingdom that Queen Anne first developed an occasional taste for political intrigue. Above all, and much to her spouse's irritation, she had become enmeshed in all the recent combinations against Maitland. Now, however, when James decided that his son should be brought up, just as he himself had been, in the comparative seclusion of Stirling Castle, and under the guardianship of his old playfellow, the Earl of Mar, Queen Anne was given a deep grievance with which to underpin her broader misgivings. The king's decision was in fact firmly upheld by long custom, but there was more involved in the decision than the emotional ties between mother and son or the maintenance of tradition, since Anne, as James well knew, perceived the custody of Prince Henry as a source of political leverage and an opportunity to achieve the importance that had so far been denied her. Notwithstanding Mar's own history of wavering loyalty, the king was therefore determined to abide by his decision, irrespective of his wife's rages and entreaties.

When husband and wife were together at Linlithgow in May 1595, she reminded him 'how she had left all her dear friends in Denmark to follow him, and that King Christian her brother, for love of her, had ever been his sure friend'. But when Anne tried to obtain possession of her son while James was on a hunting expedition at Falkland, it was clear that she would not necessarily confine herself to moral blackmail. Returning in a furious temper, James did indeed take her to Stirling and grant her access to her son for several hours, but beyond this the king would not compromise, as a letter he dispatched to the Earl of Mar makes quite clear. 'Because in the surety of my son consisteth my surety,' wrote James, 'and I have concredited unto you the charge of his keeping upon the trust I have of your honesty, this I command you out of my own mouth, being in company of those I like, otherwise for any change or necessity that can come from me, you shall not deliver him. And in case God call me at any time, see that neither for the queen nor estates, their pleasure, you deliver him till he be eighteen years of age, and that he command you himself.'

Finding her protests unavailing, therefore, Anne found her only solace in a largely irrational antipathy for Mar, which lasted for the next eight years and led her, ironically, to heal her old quarrel with Maitland and intrigue at the same time with a string of other nobles who happened to be Mar's enemies. Maitland, meanwhile, was so devoid of friends at court that he was prepared to make common cause with the queen, albeit without benefit ultimately to either himself or her. When she claimed to be ill, moreover, her pleas that James should visit her at Holyrood fell on deaf ears, since the king had become so suspicious, even of his chancellor, that he feared being made captive, and only after her illness was proven to be genuine did James relent. Ultimately, his greeting for her was both tender and admonitory. 'My heart,' he told her, 'I am sorry you should be persuaded to move me to that which will be the destruction of me and my blood.' For Maitland, however, he had only anger and reproof, declaring heatedly that 'if any think I am further subject to my wife than I ought to be, they are but traitors and such as seek to dishonour me'. In the event, the whole affair gradually subsided as the queen's faction departed for their homes and James and Anne headed for Falkland, apparently reconciled.

Yet 1595 marked a watershed of sorts in the royal marriage, as Anne and her husband continued to drift apart. Four years later, in his *Basilikon Doron*, James would make a number of revealing observations on marriage in general and his own in particular. 'For your behaviour to your wife,' he wrote, 'treat her as your own flesh, command her as her lord, cherish her as your helper, rule her as your pupil, and please her in all things reasonable.' But there were hints of the king's experience of his own domestic quarrels: '... Be never angry both at once, but when ye see her in a passion ye should with reason danton [subdue] yours: for when both ye are settled, ye are meeter to judge of her errors; and when she is come to herself, she may be best made to apprehend her offence ...'

True, a good deal of tenderness continued to exist between the two and James urged his heir to have the greatest respect for the woman who would eventually learn to accept even the increasingly overt homosexual dalliances of her husband. 'If it fall out that my wife shall outlive me,' he wrote, 'as ever ye think to purchase my blessing, honour your mother.' Yet, for all James's genuine regard and sympathy, there remained a cultural, temperamental and intellectual gulf, which was bound to grow, on both sides, with long familiarity.

Maitland's days as the king's right-hand man were also numbered in more senses than one, for on 3 October he died – comparatively unlamented by the man whom he had striven so hard to guide and nurture. James did, it is true, compose a graceful sonnet to his former chancellor, which was carved on a marble memorial tablet above his tomb in Haddington church. There was reference to the 'vicious men' who rejoiced at his fall and praise, too, for Maitland's 'wisdom and uprightness of heart' as well as his piety, intelligence and 'practice of our state'. But the king had always found Maitland more inclined to lead than listen and his sorrow at the loss of a gifted servant was tempered by a distinct sense of liberation. 'His Majesty,' one courtier noted, 'took little care for the loss of the Chancellor'. Shortly after Maitland's death, moreover, James went so far as to declare that he would 'no more use chancellor or other great men in those his courses, but such as he might convict and were hangable'. And though the last comment was partly made in jest, it was indeed more than three years before the chancellorship was finally filled.

Now James was determined and able, it seemed, to rule in his own right. Though he refused to acknowledge it, he would build upon the foundations that Maitland had laid over all of eleven years, but he was also to apply an energy and discipline to the task that the chancellor himself would doubtless have admired. Fully aware of his altered status in the wake of his victories of 1594 and 1595, James firstly wasted little time in issuing a proclamation warning all men to obey the law. 'As he is their king and sovereign lord', the proclamation affirmed, so the king's subjects should know 'that he will be obeyed and reverenced as a king, and will execute his power and authority against whatsoever persons' as shall 'contemn his Highness, his authority or laws'. Nor did James settle merely for words and noble sentiments. On the contrary, the registers of the council were filled from this time with rules and orders against common criminals and outlaws and those within the law who nevertheless carried 'pistols and dags'. Feuding lairds were hauled before king and council, while, 'at his own pain and travail', James sought by a mixture of force and persuasion to heal long-standing enmities. It was to be, in fact, nothing less than a political coming of age for the king. And it was a time, too, when James generally succeeded in imposing his stamp upon a country that was at long last finally ready to accept it.

There were even attempts to tackle the wilder regions of the kingdom, though success here was predictably more limited. Three areas above all –

the Highlands, the Isles and the Borders — had always proved resistant to the consistent imposition of law and order, and, as James would find, the geographical isolation and cultural idiosyncrasies of these areas meant that only long-term strategies were likely to be of any effect. Taken together, the Highlands and the Northern and Western Isles lie beyond the meandering 'Highland Line' which bisects Scotland from north-east to south-west. North of the 'Line', the clans pursued a pattern of life that had changed little with the centuries: pastoral, sparse, heroic and warlike. Wringing a meagre existence from primitive agriculture and fishing, the Highlanders' private wars and feuds and raids upon their neighbours' cattle and sheep had made them, at every social level, among the fiercest fighting men in Europe. And though many of their chieftains had acquired feudal titles from the king, the intense closeness of the clan meant that for generation upon generation the power and independence of a Huntly or an Argyll had been effectively unbreakable.

Earlier endeavours to assert some degree of centralised control had even proved counter-productive. By the time of his death in 1286, Alexander III had, for example, come close to uniting his subjects, who then included English, Norman-French and Gaelic speakers, in a common loyalty which might over time have created some degree of homogeneity. But the Wars of Independence against England, which followed Alexander's death, actually entrenched and magnified existing differences, so that when Scotland eventually confirmed its status as a free nation, it had become firmly divided not only on broadly cultural lines, but even linguistically with the Gaelic-speaking Highlands and Isles, and Scots or Anglo-Scots-speaking Lowlands. Even the decision to move the royal court to Perth in the fifteenth century and to make the town a capital for both regions proved unavailing as the political centre of Scotland shifted inexorably towards Edinburgh. It was the narrow belt of the Lowlands, after all, that contained some of the country's richest agricultural land and witnessed the first shoots of commercial and industrial development, largely concentrated in Edinburgh itself.

To compound matters, the later Stewart kings had looked to France, the Low Countries and Scandinavia for their political alliances, and indulged in frequent enmity with England. On Scotland's southern border, therefore, they were forced not only to tolerate but actively encourage the fierce local magnates whose task was to engage in ceaseless conflict with the 'auld enemy'. In the process, the priorities of central government were increasingly directed eastward and southward, while the north-west of the kingdom was largely abandoned to its own devices. As such, it was hardly surprising perhaps that the last Gaelic-speaking Scottish king should have been James IV, who died fighting the English at Flodden Field in 1513, or that George Buchanan had never felt the need to teach the king's great grandson the language. James VI, indeed, was in no doubt regarding the inherent superiority of the agrarian and trading communities of the Lowlands over their Highland cousins. 'As for the Highlands,' he wrote in *Basilikon Doron*, 'I shortly comprehend them all in two sorts of people: the one that dwelleth on our main land, that are barbarous for the most part, and yet mixed with some show of civility: the other, that dwelleth in the Isles, that are utterly barbarous without any sort of show of civility.'

Yet James, flushed with his newfound confidence as Scotland's outright 'King and Sovereign Lord', was determined, as best he could, to make his own efforts to impose the law-abiding culture of the Lowlands across his kingdom. Twice he announced his intention of visiting the Isles and Western Highlands, though lack of provisions aborted his first expedition and on the other occasion he reached no further than Glasgow and Dumbarton. Similarly, when he summoned the western chieftains to Edinburgh to render proof for their titles to land, few attended. And his efforts to tame the Isles 'within short time' by implanting Lowland culture through colonisation was equally unsuccessful. For, far from influencing the Islesmen, whom James scornfully compared to wolves and wild boars, the few settlers who managed to stay put were mostly absorbed into Gaelic ways and habits. Most, in fact, like those planted in Kintyre and Lochaber and on the island of Lewis as a result of an Act of 1597, became victims of botched planning and a smattering of bad luck. On the one hand, the gentleman-adventurers, who established the Lewis settlement on the present site of Stornoway, soon abandoned their project, while a further attempt to revive the project in 1605 resulted in equally dismal failure. Indeed, James's most significant success in pacifying this part of his realm only resulted eventually from the application of 'general bonds', by which chieftains accepted responsibility for the conduct of their clansmen – a policy that had in fact been established even before he became king.

On the Borders, however, there was ultimately better progress, though not before the king's wavering hand had been forced by fierce English protests about the antics of a celebrated Scottish freebooter named William Armstrong of Kinmont – more commonly known as 'Kinmont Willie'.

When Armstrong was finally apprehended by the deputy of the English Warden of the West March on a 'Day of Truce in 1596' and held captive at Carlisle Castle, the blue touchpaper had been lit for a major test of wills between not only the governments of England and Scotland, but between James and his notoriously feisty Border magnates. Swearing that he would avenge English treachery, the Scottish Warden, Sir Walter Scott of Buccleugh, had scaled the castle's walls on a dark and stormy night and with a few others bore 'Kinmont Willie' off in his irons, whereupon the Queen of England herself 'stormed not a little', insisting that Buccleugh should be delivered into her hands. And though his subjects urged him to resist, James nevertheless felt compelled to detain the valiant rescuer at St Andrews before delivering him to Elizabeth as requested. Indeed, so alarmed was James at Elizabeth's anger that in a curious display of anxiety before his council, he produced and formally entered in the register a letter she had written him some years earlier promising not to oppose his lawful right of succession. And when his second child, a daughter, was born at Dunfermline in August, he deferred to England's queen once more by naming the girl Elizabeth.

It was no small irony, therefore, that the queen's eventual meeting with Buccleugh suggested that James's panic was largely unnecessary. When asked how he had dared to storm one of her castles, the Borderer responded with characteristic boldness. 'What is there, Madam,' he inquired 'that a brave man dare not do?' And far from being infuriated by Buccleugh's quip, the queen seems to have been impressed by his courage. 'With a thousand such leaders,' she told her onlooking courtiers, 'I could shake any throne of Christendom.' It was hardly indicative of uncontrollable outrage – any more than the entire affair of 'Kinmont Willie' was the kind of episode that might have seriously compromised James's succession to the English throne. In truth, the King of Scotland's tunnel vision concerning the English Crown had on this occasion merely undermined his already limited credibility as a leader of genuine resolution, and in the immediate aftermath of his climb down, there was a significant increase in lawlessness along the Border.

Yet James was rescued once again by circumstance and good fortune, for in 1597 a joint commission drew up an Anglo-Scottish agreement, which laid the foundation for forty years of comparative tranquillity in the Border territories. With the prospect of a Scottish King of England drawing closer and with firebrands like the Bothwells well and truly spent, the era of raids and culture of military bravado and brooding bitterness seemed increasingly irrelevant, and it was agreed that a treaty should be drawn up to provide means of bringing notorious offenders to justice. With the conditions for disorder thus eradicated, James was subsequently well placed to play a vigorous role in steadily exerting the control of central government, visiting the Borders frequently and hanging large numbers of the dying breed of ruffians who had thrived for so long. And it was in these circumstances, as the turmoil receded, that men came to speak – albeit somewhat simplistically – of 'King James's Peace'.

In the meantime, Maitland's place in government had been taken by a group of eight ministers, known as the 'Octavians': James Elphinstone, Lord Balmerino; Walter Stewart, Lord Blantyre, who had been educated as a boy at Stirling along with the king, and would eventually become Lord Treasurer; Sir David Carnegie of Colluthie; Sir John Skene of Curriehill, the Lord Justice Clerk; Thomas Hamilton, Lord Drumcairn, a shrewd and versatile, though ultimately corruptible individual, who was known by the king as 'Tam o' the Cowgate' as a result of his residence in that street; John Lindsay, Lord Menmuir, who received worthy praise from John Spottiswoode as a man of 'exquisite learning and sound judgement'; Sir Alexander Seton, Lord Urquhart, who was President of the Court of Session, and went on to become Chancellor and Earl of Dunfermline; and Sir Peter Young, the king's former tutor. Though sometimes assumed to be of broadly middle-class origins, the Octavians were all substantial landowners and drawn in certain cases from highly respected noble families – 'Tam o' the Cowgate', for instance, being a kinsman of the royally connected Hamiltons. All, moreover, were capable and experienced servants of the Crown. Some, indeed, had succeeded in helping the queen to become more solvent, and it was in their capacity as financial troubleshooters that the group were initially employed by the king.

The state of the royal treasury had already been brought home to James particularly starkly on New Year's Day 1596, when Queen Anne's advisers had been able to provide her with a purse containing £1,000 in gold. Keen to exploit such an opportunity to tease her husband, Anne therefore approached him, shook the purse in his face, and condescendingly delivered half its contents to him, inquiring in the process when his council would give him as much. For some time, in fact, James had been reduced to increasingly miserable shifts by his financial predicament, and the queen's gesture was, in effect, the final embarrassment. In November 1588, indeed, the English agent, William Ashby had reported to Sir Francis Walsingham that the King of Scots was 'so poor [that] he can neither reward nor punish', and things had not changed since. His household had continued to be maintained from the private means of his officials, and his debts to moneylenders were common knowledge. He had long suspected, too, that that the officers of the Exchequer were making considerable profits at the expense of the Crown, though his only recourse had been to plunder the mint and debase the coinage. And without exception, his lack of personal magnetism had failed to galvanise – and in particular intimidate – those conducting his business. 'I have been Friday, Saturday and this day waiting upon the direction of my affairs, and never man comes,' he told Sir John Skene in an undated letter. 'Them of the Exchequer that were ordained to take accounts, never one. The affairs of the household should have been ended this day, no man comes down … In short no tryst or meeting is kept. What is spoken this night is forgot the morn …'

But while James bemoaned the self-interest and inefficiency of those around him, he did little to curb his own extravagance and remained the root cause of his problems. 'He gives to everyone that asks, what they desire,' wrote Thomas Fowler to Walsingham in December 1588, 'even to vain youths and proud fools the very lands of his crown or whatever falls, leaving himself nought to maintain his small, unkingly household. Yea what he gets from England, if it were a million, they would get it from him, so careless is he of any wealth if he may enjoy his pleasure in hunting, the weather serving.' James, moreover, freely acknowledged what amounted to a pathological profligacy on his part. 'I have offended the whole country, I grant, for prodigal giving from me,' he noted to Maitland in 1591. And like many victims of addiction, he vainly professed his determination to amend. In the same letter to Maitland, he noted how 'the two aids of the kitchen ran out yesterday and would not make the supper ready, saying condition was not kept'. The chancellor therefore was to remedy the situation. 'Suppose us be not wealthy, let us be proud poor bodies,' James instructed, though a true measure of his pride had already been furnished by his insistence in 1590 that the laird of Caldwell make the customary gift of a hackney, so that Queen Anne's ladies could be transported in some kind of style, and by his earlier alleged plea to the Earl of Mar for the loan of a 'pair of silken hose' before he received a foreign ambassador.

As a token of the king's new resolve, however, Glamis, the treasurer and other members of the Exchequer were now to be summarily dismissed and replaced by the Octavians in the hope that the king's finances might receive the same much-needed attention as his wife's. The preamble to their commission made clear that the king's income was declining as a result of 'unprofitable dispositions out of the property and collectory', increased pensions, falling customs revenue in spite of increased shipping, decay of the coinage, and general neglect and improper management, 'so that all things are come to such confusion ... that there is not wheat nor barley, silver nor other rent, to serve his Highness sufficiently in bread and drink'. Furthermore, the new body was to be granted an unprecedented degree of control to achieve its stated objective of augmenting the king's income by £100,000 – so much, indeed, that the king pledged himself to do nothing in financial matters without the consent of at least five of its members. He agreed, too, to abide by any directives they might draw up, though his zeal to place so much power in his ministers' hands was governed, predictably, at least as much by a wish for self-preservation as by any wholehearted commitment to the kind of financial discipline that they might impose. Knowing full well the kind of general odium that his new watchdogs were bound to incur, it made sound political sense – especially to a ruler of James's disposition – to distance the Crown from the impact of their decisions.

And the Octavians neither shirked their task nor retreated from the hostility that came their way. On 18 January 1596, the English agent Roger Aston was not exaggerating when he observed that 'these new Checker men begin very sharp'. By the end of the month they had discharged seventy people from the king's household and required the Earl of Mar to present a list of Prince Henry's retainers so that his household might also be pruned. Pensions were reduced, appropriate rents were set for Crown lands, and the first general customs duty on imports in Scottish history was successfully imposed. They also tried to obtain financial assistance from the General Assembly of the Kirk, though this would prove a step too far even for their considerable drive and ingenuity. Nonetheless, for the short period that they held the reins, their impact was considerable. During 1596 and 1597, while their reforms were in effect, the expenses of the royal household averaged about £3,650 a month. However, for the half-year period from November 1598 to May 1599, by which time their restraining grip had already been removed, the average was £4,580. Likewise, the comptroller's excess expenditure for the fiscal year 1596–97 had been merely £258 – only to rise to £3,141 during 1597–98. Indeed, during 1598–99, the year of Duke

Ulrick of Denmark's expensive and alcohol-fuelled visit, it would rise to over £26,000. Still needing to pawn his jewels occasionally, by 1598 James was in debt to two Edinburgh merchants to the tune of £160,000. Nor would he be spared the final indignity in 1599 of receiving the resignation of one of his ministers on the grounds that service to the Crown entailed almost certain financial ruin.

In fact, the Octavians would be overcome ultimately by a mixture of religious intolerance and the machinations of a somewhat amorphous opposition group known as the 'Cubiculars', not to mention the king's own short-lived enthusiasm for the rigours of financial austerity. Hostility from holders of Crown land whose tenures were of dubious validity and courtiers whose pensions had shrunk or vanished altogether was of course inevitable, and chief among the disaffected was Sir George Home of Sport, Gentleman of the Bedchamber – a fat and ruthless *faux bonhomme* who wished to be rich and retained the king's ear by mending his quarrels with the queen. 'Where Sir George declares himself either a friend or an enemy,' wrote Samuel Cockburn to Archibald Douglas in June 1596, 'there is none to stand to the contrary.' And though the days of deadly feuds were largely over, Home had little difficulty in exploiting Alexander Seton's sympathy for Catholics for his own ends. Faced with outcries from the Kirk at a time when the return of Huntly and his crew was already causing consternation, James therefore decided to neutralise Seton's moderating influence and place financial control in the hands of Lord Treasurer Blantyre – the most impeccably Protestant and mediocre of all the Octavians – with the result that the body, which had promised so much, swiftly disintegrated.

The battle with the Kirk on the other hand, far from abating, was fast approaching its climax. Even the birth of the king's daughter on 19 August 1596, had served to highlight the growing tension, since the Presbyterian ministers did not on this occasion offer even token congratulations. Nor would the queen herself escape the wrath of indignant clergymen. Once more she lost the custody of her child when Princess Elizabeth was handed over to the excellent guardianship of Lord and Lady Livingstone – both suspected Catholics – who gave the child seven happy years of care and affection in Linlithgow Castle. But the queen could not have anticipated the remorseless assault upon her own religious beliefs, which duly rained down upon her from the pulpit of the Reverend David Black of St Andrews, who had already acquired a dubious celebrity in

1594 by damning the king's councillors as 'atheists of no religion' and declaring the nobility to be 'degenerate, godless dissemblers, enemies to the Kirk'.

At some point, in fact, the queen, who had never taken kindly to Presbyterianism nor to the freedom of the Scottish clergy in instructing her husband, had become a secret, though half-trifling, Roman Catholic under the influence of her intimate friend the Countess of Huntly, the former Lady Henrietta Stuart, sister of Esmé Duke of Lennox. And as a result, on 19 October 1596, she was openly insulted by Black who proclaimed in a sermon that 'the devil was in the court' and professed that though required to pray for the queen, 'we have no cause', since 'she will never do us good'. To add salt to the wound, Black did not hesitate to deny that the king had any right to judge him when called before the council a month later to account for his behaviour. In the pulpit, he declared, he was subject only to Christ's word, and answerable for that solely to 'the prophets', i.e. ministers or 'the ecclesiastick senat'.

The fuel for this high-octane clash had been provided, of course, by the king's decision to seek the return of Huntly and his confederates that summer. 'Papists,' James told his clergy at that time, 'might be honest folks and good friends' to the Crown, since his mother was a Catholic and yet had been 'an honest woman.' And with precisely this kind of specious reasoning, which could be guaranteed to outrage the likes of Andrew Melville, James duly obtained the assent he desired from a convention of the Estates held at Falkland in August – though the victory was not achieved without a price. Melville himself, for instance, had appeared at the convention and, in spite of the king's command that he withdraw, proceeded to accuse the entire assembly of treason against Christ. A month later, moreover, Melville was back at Falkland for a conference, during which he 'broke out upon the king in so zealous, powerful and irresistible manner, that howbeit the king used his authority in most crabbed and choleric manner, yet Mr Andrew bore him down', referring to the ruler as 'but God's sillie [i.e. simple] vassal' and taking him by the sleeve to tell him, 'Sir, you are brought in extreme danger both of your life and crown'. 'There are,' Melville continued, 'two kings and two kingdoms in Scotland. There is Christ Jesus the King and his kingdom the kirk, whose subject King James the Sixth is, and of whose kingdom not a king, nor a lord, nor a head, but a member.'

With a mixture of self-control, incredulity and cold apprehension at the prospect of a face-to-face confrontation with a zealot of Melville's cast-iron will, James nevertheless managed to dismiss the ministers pleasantly, promising that the Catholic earls would receive no favour from him until they had satisfied the Kirk of their good intentions. But the preachers of Edinburgh, far from responding with similar latitude, merely, we are told, 'pressed forward and sounded mightily' – with the result that the capital was soon rocked by mayhem and disorder. On 17 December during the course of violent sermons at St Giles the cry went up for the congregation to defend itself, with the result that rioting broke out, mainly around the Tolbooth, where the king was meeting with his lords of session. In the event, the crowd was beaten back and the rioters swiftly calmed by the provost, who then hastened the king down the Canongate to the security of Holyroodhouse. But it had been another chastening experience for the king, and it was small wonder, perhaps, that he would warn his heir so forcefully about the threat posed by 'Puritans' within the Kirk.

'Take heed therefore, my son,' he wrote in *Basilikon Doron*, 'to such Puritans, very pests in the church and commonweal, whom no deserts can oblige … breathing nothing but sedition and calumnies, aspiring without measure, railing without reason and making their own imaginations the square of their conscience. I protest before the great God that ye shall never find with any Highland or Border thieves greater ingratitude and more lies and vile perjuries than with these fantastic spirits.' Yet by the time that James wrote, the threat of which he spoke was already in full retreat, for the radicals within the Kirk, like the errant nobles before them, had overstepped the mark and finally detached themselves from their more moderate counterparts – a fact which the king, to his not inconsiderable credit, appears to have at least partially appreciated at the time. For the very next day after a rioting mob had massed outside the Tolbooth, he, his queen and the entire court, along with judges, lawyers and other government officials, removed from the capital to Linlithgow, leaving a herald to proclaim at the Mercat Cross that their departure was to be permanent, since the city was no longer fit to be a royal residence. Nobles, likewise, were ordered to withdraw to their country estates.

Far from staging an ignominious retreat, however, James had literally turned the tables overnight, for if Melville, Black and their counterparts were more than capable of outfacing their monarch in eyeball-to-eyeball confrontation, they lacked his political skill and, in particular, his sense of timing. Faced with certain ruin as a result of the king's departure, the capital abandoned its preachers and within a fortnight James was back on his own terms, accompanied by a daunting troop of Border ruffians who now were more than willing to do his bidding. Not only did Edinburgh's citizens make their peace with the king, moreover, they did so gladly, even accepting a fine of 20,000 marks without demur. Those guilty ministers who were not imprisoned, meanwhile, had no recourse but flight, leaving the Kirk largely purged of its most troublesome elements. Thereafter, a General Assembly of the Kirk called to Perth later in the year formally withdrew the extreme claims made by Melville and, by accepting representation in Parliament, acknowledged their status under the authority of the secular ruler in precisely the way that James desired. Henceforth, a clerical commission set up by the General Assembly would merely advise the king on ecclesiastical affairs rather than dictate policy in the way that had been suggested previously. Kings, it seems, were no longer 'Satan's bairns' and all talk of theocracy was conveniently shelved.

Yet even this was not the limit of James's victory, for he used this heaven-sent conjunction of circumstances to install moderate ministers of the Kirk in Parliament as 'bishops' – a term hitherto anathematised because of its papist connotations. He had long opposed the so-called 'parity' of ministers 'whereby the ignorants are emboldened to challenge their betters', and proposed instead to 'advance the godly, learned and modest men of the ministry to bishoprics and benefices [and thus] not only banish their conceited parity but also re-establish the old institution of three estates, which can no otherwise be done'. 'I mind not,' he said, 'to bring in papistical or Anglican bishoping; but only to have the best and the wisest of ministers to have place in Parliament.' And in October 1600, notwithstanding Andrew Melville's opposition to the king's 'Anglo-piscal, papistical conclusions', James duly appointed three diocesan bishops to the sees of Caithness, Ross and Aberdeen. Though their influence lay only in Parliament and they had no defined function within the government of the Kirk, which remained thoroughly Presbyterian in nature, it was a triumph no less significant in its way than the taming of the earls. Now at last, perhaps, James could genuinely rule as 'king and sovereign lord' in the way that he had always intended. 'Alas,' lamented James Melville, 'where Christ guided before, the court began then to govern all'.

# 9 ❦ Towards the 'Land of Promise'

'St George surely rides upon a towardly riding horse, while I am bursten in daunting a wild, unruly colt.'

*James VI of Scotland, comparing the kingdom of England with his own*

Against all expectation, including his own, the King of Scotland had taken only nine months from the time of his son's christening to be rid of all the major torments that had plagued him for so long. And for the next six years, aside from one rogue tremor of noble insolence in 1600, there was comparative peace. In the wake of Maitland's death, moreover, there remained no single, dominating figure at the centre of politics beyond the king himself. Such, indeed, was the king's confidence and sense of liberation that he found himself able to indulge his intellect in a spate of writing that fully reflected his determination to be master of his kingdom and of all estates within it, be they noble or clerical. In September 1598, his *Daemonologie* was quickly followed by a 1,000-word pamphlet entitled *The Trew Law of Free Monarchies: Or The Reciprock and Mutual Dutie Betwixt a Free King and his Natural Subjects*, in which he forcefully expounded what would become known to history as the principle of 'divine right'. Then, one year later, there occurred the publication of *Basilikon Doron* or 'Kingly Gift' – the book dedicated to his infant son, into which he attempted to pack all the wisdom and understanding of kingship that he had accumulated over the previous sixteen years or so. Written exclusively for the prince and falling only accidentally into the hands of a wider readership, it was the least self-conscious, most transparently honest and pleasantest of all James's works to read. Thus diverted, he was able at last to recover and take stock before his last, all-absorbing and defining project as King of Scotland: the acquisition of the Crown of England.

Though *The Trew Law of Free Monarchies* was published anonymously, its regal tone and direct and informal style left little doubt about its authorship. Aside from his *Daemonologie*, James had already published four books: *The Essays of a Prentise in the Divine Art of Poesie* (1584); *Ane Fruitfull Meditatioun on the Seventh, Eighth, Ninth and Tenth Versies of Chapter XX of Revelations* (1588); *Ane Meditatioun upon the First Buke of the Chronicles of Kings* (1589); and *His Majesties Poeticall Exercises* (1591). But now his aim was more directly didactic: to teach his subjects in no uncertain terms the nature of their duty to their king, and in particular the necessity of obedience and the wickedness of revolt. Nor was the simplicity of James's language coincidental, since *The Trew Law* was never intended as an abstract academic treatise. On the contrary, it was meant to be understood in the most lucid, vivid and cogent way, and read in entirety without labour or ambiguity. James's purpose, he said, was 'to instruct and not irritate'.

Yet the tract certainly contained statements that would rightly or wrongly redound to his discredit down the centuries. Though there was nothing novel in the claim, for instance, that it was the duty of subjects to obey even tyrannical kings and to accept that, as God's appointees, anointed rulers were open only to divine judgement, his words, like those of many a well-intentioned pedant, frequently have an unfortunate ring to them. 'Kings are called Gods by the prophetical King David,' he reminded his readers, 'because they sit upon God His throne in the earth, and have the [ac]count of their administration to give unto him'. In Scotland, he argued, kings existed before any Parliaments were held or laws were made and so 'it follows from necessity, that the kings were the authors and makers of the laws, and not the laws of the kings'. Likewise, the coronation oath involves no compact between ruler and ruled. Instead, the good king is a loving father to his people, cherishing their welfare, tempering punishment with pity, and safeguarding order and harmony for the mutual benefit of all concerned by the very inviolability of his authority. As for evil kings, their chastisement will follow in the afterlife. Indeed, they will be punished far above other men, 'for the highest bench is sliddriest to sit upon'.

All such claims were, in fact, standard political fare for any Early Modern ruler and accepted, at least implicitly, by the vast majority of contemporary men and women. But whether James had on this occasion chosen the most judicious time to air such views is more open to question. And although the whole concept of the tract may well have sprung in part from the decidedly quixotic streak in his nature, it continues to carry with it certain uneasy resonances. In practice, James's kingship would often prove eminently pragmatic and flexible – much more so than *The Trew Law* implied was actually necessary. From his own perspective, moreover, James was genuinely committed to the principles of good kingship – establishing good laws, ministering justice, advancing good men and punishing wrongdoers, ensuring peace and security, and guaranteeing sound religion. Yet 'divine right' was a double-edged sword for a king like James. Though it offered order, it sat ill with impatience, timidity and, above all, any lack of genuine 'majesty' on the part of the ruler who attempted to exemplify its virtues. And as a principle founded, or so James claimed, upon the bedrock of ancient biblical precedent, its application would need to be especially subtle at a time of such pronounced social, political and religious change.

*Basilikon Doron*, meanwhile, further emphasised the patriarchal nature of kingship and the virtues required by the godly ruler. Originating, or so James claimed, from a dream which left him fearing that his life would be short, the book is a moral and didactic work outlining a series of precepts for his son's guidance. And once again the divinely ordained authority of kings features prominently. 'God gives not kings the stile of Gods in vain', runs the well-known sonnet opening the work. But James's primary concern remained the 4-year-old Prince Henry's education, or as he himself put it, 'timeously to provide for his training up in all the points of a king's office'. In this regard, Henry was firstly to attain a knowledge and fear of God by study of the scriptures, by prayer, by preservation of a sensitive conscience, and by learning to distinguish between essentials and non-essentials in matters of faith. It was also essential that he avoid not only pride per se but 'the preposterous humility' of those who demand parity in religious affairs. 'Surely,' James tells his son, 'there is more pride under such a one's black bonnet than under Alexander the Great his diadem.'

There are observations, too, on the Scottish Reformation and predictable sideswipes at the troublesome Scottish nobility. We hear, for instance, how 'some fiery-spirited men of the ministry got such a guiding of the people at that time of confusion, as finding the gust of government sweet, they

begouth [began] to fantasy to themselves a democratic form of government
… and after usurping the liberty of their time in my long minority, settled
themselves so fast upon that imagined democracy as they fed themselves
with the hope to become Tribuni plebis: and so in a popular government by
leading the people by the nose, to bear the sway of all the rule'. The prince,
therefore, was not to tolerate the pretensions of these 'fantastic spirits …
except ye would keep them from trying your patience as Socrates did an
evil wife'. Nor should he forget the problems associated with over-mighty
subjects. For while James accepted that 'virtue followeth oftest noble blood'
and urged his son to employ those that are 'obedient to the law among them
… in all your greatest affairs', he could not forget the damage that had been
done to the kingdom by the arrogance of some and the feuds they had
waged. 'The natural sickness that I have perceived this estate subject to in
my time,' he observed, 'hath been a feckless arrogant conceit of their own
greatness and power.' All too often, it seems, they were prepared to 'bang it
out bravely, he and all his kin, against him and all his …'.

Slightly less expected, however, are the passages concerning warfare.
The image of James as a man of peace is, of course, firmly established in
popular perceptions, and rightly so. Yet when he wrote *Basilikon Doron*, his
own expeditions against Huntly and Bothwell had recently demonstrated
that, when extremity demanded, he appreciated well enough the need for
any prince to wield the sword. And though his advice on warfare was
conventional, it did not merely rehearse the advice of others or limit itself
to generalities. 'Choose old experimented captains and young able soldiers,'
he told his heir. 'Be extremely strait and severe in martial discipline, as well
for the keeping of order, which is as requisite as hardiness in the wars, and
punishing of sloth, which at a time put the whole army in hazard.' There
was praise, too, for the renowned discipline of the Spaniards and especially
the Duke of Parma's infantry. Caesar's *Commentaries* were also cited as
recommended reading, 'for I have ever been of that opinion that of all …
great captains that ever were, he hath farthest excelled, both in his practice
and in his precepts in martial affairs'.

The third and final section of the book, meanwhile, dealt with the ruler's
public image and everyday behaviour – matters which James considered
of no small significance, since 'a king is as one set on a stage whose smallest
actions and gestures, all the people gazingly do behold'. Though James
appreciated the elegant manners of those who had resided at the French
court, he nevertheless considered it preferable for his son to adopt the

simpler ways of Scottish practice. Indeed, he spoke of 'the vice of delicacy, which is a degree of gluttony' and recommended that eating, for instance, be conducted 'in a manly, round and honest fashion'. A king should likewise 'keep a proportion' in his dress appropriate to the occasion, and 'look gravely and with a majesty' when sitting in judgement, while remaining 'homely' in the private company of his servants and 'merry' during his pastimes. Perfume, long hair and unclipped nails were to be avoided, along with the idle company of females 'which are no other things else but irritamenta libidinis'. The language of a king should be plain, honest, natural and brief, avoiding crudeness, and he should engage in all types of athletic exercise, especially riding, 'since it becometh a prince, best of any man, to be a fair and good horseman'. Chess, on the other hand, was to be discouraged on the grounds that it was 'over fond and philosophic a folly'.

Taken as a whole, *Basilikon Doron* is rightly considered to contain the best prose ever written by James, for in spite of certain artificialities of style it is generally fresh, natural and spontaneous, abounding in racy phrases and picturesque passages that are tinged with James's characteristic dash of wry humour. Nor does the king's frequent inability to follow his own advice detract from its significance. If anything, indeed, it adds to the book's interest. Certainly, it was an immense success, and reappeared constantly in the publications on courtesy produced for the education of upper-class young men during the seventeenth century – notwithstanding the fact that it had first been printed in a secret edition of seven copies which the king distributed among his most trusted servants. Above all, he had wished to keep it from the knowledge of the clergy and later feared that passages might reach England in garbled form and cause suspicion. Yet within days of Elizabeth's death it was on sale in London, eagerly snapped up by Englishmen and foreigners alike. Ultimately, it would be translated into most of the languages of Western Europe and remains one of the most intriguing windows into the king's attitudes and personality.

Only two years later, moreover, a brief sketch of James and his court, penned by the English diplomat and poet Sir Henry Wotton, would offer us a further series of perspectives on the man and his ways, which are among the most well known of all contemporary descriptions. According to Wotton, he was 'of medium stature, and of robust constitution', 'fond of literary discourse, especially of theology', and 'a great lover of witty conceits'. His speech, we are told, was 'learned and even eloquent' and rather more surprisingly, perhaps, he was also described as 'patient in the work of government' – a claim belied, of course, by his incorrigible tendency to neglect the routine tasks of administration in favour of pastimes, especially hunting. Another of his admirable qualities, it seems, was his chastity which, Wotton suggested, 'he has preserved without blemish, unlike to his predecessors who disturbed the kingdom by leaving many bastards'. Above all, however, though James enjoyed 'listening to banter and to merry jests, in which he takes great delight' and was 'extremely familiar' with the gentlemen of his bedchamber, there was another, less personable side to him. Indeed, he was 'said to be one of the most secret princes of the world', Wotton informs us, and capable of 'bitter hatred, especially against the Earl of Gowrie' – though the king's antipathy to this particular individual was hardly surprising in light of what had transpired only the year before.

The Gowrie conspiracy of August 1600 was, in fact, the only violent episode to disturb the peace that had descended so unexpectedly on the closing years of James's direct rule in Scotland. But it was also, arguably, one of the most impenetrable of all the baffling mysteries associated with James's reign, not least of all because the king alone, of all the principal persons involved, lived to tell the tale, and because his skill at concealment and subterfuge had been honed by this time to little less than an art form. As Wotton correctly realised, of course, James had already consciously drawn a veil of uncertainty over many episodes in his life, and in *Basilikon Doron* he had freely admitted how 'a king will have need to use secrecy in many things'. Even some of his portraits – and none more so than that produced by Adrian Vanson in 1595 – suggest a ruler who, in spite of frequent flashes of bonhomie, possessed a deeper, warier, cannier and more circumspect side. Vanson had, in fact, been patronised by James for all of fourteen years by the time the portrait was produced, and in 1584 had succeeded Arnold von Bronckhorst as official painter to the Scottish court. One year later, he had also produced a portrait of the king for the Danish court as part of the ongoing marriage negotiations. He knew the king intimately over time and he knew the king's circumstances: his insecurities, the indignities to which he had been subjected and the methods by which he survived them. As such, the brooding immobility of the face he painted in 1595 was surely no more of a coincidence than the suspicious watchfulness of the heavy-lidded eyes, conveying a mind full of concealed and private thoughts – a mind more than capable, too, of confounding posterity about the precise nature of what passed at Gowrie House during a summer hunting trip that had begun so routinely.

According to James's account, he was met outside the palace of Falkland in the early morning of 5 August by Alexander, Master of Ruthven, the younger brother of the Earl of Gowrie, who related a strange tale to him. The day before, it seems, a man had been apprehended while attempting to bury a pot of gold coins in a field outside Perth. Since buried gold was forfeit to the Crown as treasure trove and since the man had been about to bury it, the implication was that the king could legitimately claim it for himself. But James, in his version of events, was too preoccupied for the time being with his day's sport and left the matter to the local magistrates. Only later in the day, after the master's story had been running in his mind, did he ride back to Perth, accompanied by ten to fifteen lightly-armed attendants – at which point he was met by Gowrie, who had been informed by his brother of the king's approach, and invited to dinner at Gowrie House. There was, James suggested, a certain uneasiness and lack of cordiality in the earl's invitation, and the poor quality of the meal itself, which consisted of moorfowl, mutton and chicken, suggested that no trouble had been taken to ensure adequate preparation for the royal visit. Even so, the king was apparently ready afterwards to accompany Ruthven to a remote turret chamber, in order to interview the mysterious captive who had been found with the treasure.

What followed from this point becomes more curious still, for as the king made his way to the turret through a series of chambers, Ruthven allegedly locked each door behind him before bringing him to a small apartment where, to the king's horror, there was no pot of gold, but a retainer of Ruthven's, named Henderson, clad in armour and bearing a dagger in his belt. While James's own men were eating cherries in the garden below, awaiting his return, Ruthven, we are told, seized Henderson's dagger, accused the king of murdering his father, and declared that he must die. Whereupon, with Ruthven's dagger at his breast, James entered upon a long discourse on the wickedness of shedding innocent blood and thereby persuaded his assailant to consult his brother before proceeding. While he was gone, James also, it seems, managed to prevail upon Henderson who denied any foreknowledge of an assassination plot and obligingly opened a window – which would prove mightily convenient when Ruthven returned and announced to his captive once more that he would have to die. 'By God, sir,' Ruthven is said to have declared, 'there is no remedy'.

In the struggle that followed, however, James claimed to have got the better of his would-be assassin, dragging him to the open window and crying for help to his attendants below, one of whom, young John Ramsey, made his way to the turret and found Ruthven on his knees before the king, his head under James's arm, and his hand raised over the king's face as though to stifle his cries for help. Striking him from behind, Ramsey wounded the would-be assassin severely and then called for help to Thomas Erskine and Dr Hugh Herries who subsequently finished him with their swords. Gowrie, meanwhile, was also killed after rushing upstairs in wild excitement, sword in hand, to confront his brother's killers. Thus, it seems, were both Ruthvens done to death.

It was, however, a highly improbable story, riddled with inconsistencies and outright falsehoods, which has fed the opinion down the centuries that James himself somehow engineered the episode. Certainly, there was no love lost between the king and the handsome, 22-year-old Gowrie, who had just returned from six years of travel and study on the Continent. The earl's grandfather, Patrick, had of course been Queen Mary's enemy and Riccio's assassin, while his father had been beheaded for treason following the Ruthven Raid of 1582, which had robbed him of his beloved Esmé Stuart. Before his departure abroad, moreover, Gowrie had been not only a supporter of the Kirk but a suspected sympathiser with Bothwell. And to cap the king's resentment, he was also popular, both in Scotland and England. He had been warmly received in London upon his return from his travels, and there was even talk in certain quarters that he might come to rival James as a potential successor to the English throne. His subsequent entry into Edinburgh, meanwhile, attracted such enthusiasm that James was unable at the time to resist a stinging remark. There had been even more people present to mark the earl's return, he quipped, than there had been at the scaffold for the execution of his father. So undisguised, indeed, was the king's hostility that Gowrie was forced to retire briefly to his estates before returning to anger his royal master once more by opposing him at the June parliament. When it is considered, too, that James owed the earl some £80,000 and knew that he had dabbled in magic and astrology during his travels, it is not hard to appreciate why suspicions of subterfuge and skulduggery have been so prevalent down the years.

But there is another side to the story, which continues to render events largely inexplicable. The Master of Ruthven, for instance, was a handsome young man of only 19, who was a known favourite of the king, and whose sister Beatrix was one of Queen Anne's leading ladies. Much more curious still, however, is the fact that James could have brought about Gowrie's downfall at far less personal risk to himself. A decision to put paid to Gowrie in his own house is no more inherently plausible than the unlikely image of James overcoming him in a hand-to-hand brawl while a third party, Henderson, was seemingly close at hand. The king, indeed, explained away any involvement in plots of his own by employing precisely such an argument. 'I see, Mr Robert,' he said to the Edinburgh minister who refused to believe his account, 'that ye would make me a murderer. It is known very well that I was never bloodthirsty. If I would have taken their lives, I had causes enough; I needed not to hazard myself so.' Besides which, though James was actually wholly capable of removing his enemies by violence, he would always do so through due legal process – from the death of Morton in 1581 to the execution of Sir Walter Raleigh in 1618.

Was Gowrie therefore guilty, after all, notwithstanding the question marks beside James's version of events? If so, his blunders and miscalculations remain nothing short of remarkable. Above all, he appears to have made no attempt to enlist the support of other influential nobles, whose consent would have remained so crucial in the unlikely event of his success. The only incriminating evidence to this effect was in fact provided by the discredited lawyer, George Sprot, who confessed on the verge of his own execution in 1608 that he possessed letters confirming Gowrie's plan to spirit James away to Fast Castle, Sir Robert Logan of Restalrick's impregnable cliff-top fortress on the Berwickshire coast. But while the government made every effort to prove the complicity of others in the plot and Logan, a particularly notorious and dissolute conspirator, was eventually convicted posthumously of complicity in the Gowrie conspiracy on Sprot's evidence, the incriminating letters concerned were clear-cut forgeries made in imitation of Logan's handwriting. Whether, as has been claimed subsequently, they were copies of original letters that actually did exist, remains unknown.

With such a dearth of certainties, however, other more imaginative theories have occasionally emerged to exploit the vacuum. One particularly flimsy hypothesis has suggested, for instance, that James may have been responsible for the incident by committing an indecent assault on Alexander Ruthven, which he then attempted to conceal by improvising his strange account of what happened. There has also been play upon the whispers of scandal linking Ruthven's name to Anne of Denmark, though here, too, the king could certainly have dealt with any problems more rationally and conveniently. And then, of course, rather less implausibly, there remains the possibility of a sudden and unpremeditated quarrel which spiralled out of control when the king panicked and called upon his attendants for protection. Significantly, this was the explanation accepted by Sir William Bowes, the English ambassador, who believed that James had referred to Ruthven's father as a traitor when they were alone together 'whereat the youth showing a grieved and expostulatory countenance' caused the king, 'seeing himself alone and without a weapon', to cry 'Treason!' 'The Master,' Bowes continued, 'abashed to see the king to apprehend it so … put his hand to stay the king showing his countenance in that mood, immediately falling upon his knees to entreat the king …', at which point Ramsey entered and 'ran the poor gentleman through'. James, moreover, certainly seems to have called upon Ramsey to deliver the wounding blow. 'Strike him high,' he is said to have cried, 'because he has a chain doublet upon him.'

Whatever the true story, equally interesting in its own way is the speed and astonishing effectiveness with which James turned the incident to his advantage. On 7 August, two days after the deaths, the Privy Council ordered that the corpses of Gowrie and his brother should remain unburied until further investigations had been made, and also that no person of the name of Ruthven should approach within ten miles of the royal court. In the meantime, the bodies of the dead brothers were disembowelled and preserved, and on 30 October sent to Edinburgh to be produced before the bar of Parliament. Thereafter, on 20 November the Ruthven estates were declared forfeit and the family name and honours extinct. For good measure, the corpses of the earl and his brother were then hanged and quartered at the Mercat Cross – their heads being placed on spikes at the Old Tolbooth, their arms and legs likewise placed on spikes at various locations round and about Perth. Ultimately, another Act would be passed which abolished the name of Ruthven forever and laid down that the barony of Ruthven should henceforth be known as the barony of Huntingtower. Gowrie House in its turn was levelled to the ground, while 5 August was henceforth designated a day of solemn thanksgiving. Ramsey, too, was not forgotten, for he was not only knighted upon James's eventual accession to the English throne but became Earl of Holderness.

Nor was propaganda and the skilful application of political leverage neglected. James's account of the episode was published within the year under the title *Gowrie's Conspiracie: A Discourse of the Unnaturall and Vyle Conspiracie, Attempted against the Kings Majesties Person at Sanct-Iohnstoun [Perth], upon Twysday the Fifth of August, 1600* (Edinburgh, printed 1600, *cum privilegio regis*). And James followed this publication by commanding his clergy to offer public thanksgiving for his deliverance from assassination – a gesture which resulted initially in the refusal of five Edinburgh ministers to comply. Their refusal, moreover, was hardly a surprise – and least of all to a shrewd political manipulator like James – since the Ruthvens had, of course, been consistent supporters of the Kirk and the official explanation was, after all, wholly worthy of scepticism. But James was determined to make his account of events both a test of clerical loyalty and a new weapon with which to cow the Kirk. Not only were compliant clergy found to fill the places of the five recalcitrant ministers, but four soon relented, whereupon they were dispatched to various parts of the country to repeat their submissions publicly. The only figure to stand firm, in fact, was Robert Bruce, a man of great dignity and authority, and former confidant of the king, who found himself banished from Scotland on pain of death.

But even after James had obtained approval for his actions from a convention of the clergy, there remained, not altogether surprisingly, a good deal of scepticism, though the conspiracy itself, in its broader historical context, is best perceived as little more than a final, largely meaningless episode, typical of an unfortunate phase of Scottish history that was already effectively at an end. Courtiers continued to whisper of foul play, while the queen, angry at the banishment of Beatrix Ruthven, sulked in her rooms and refused to be dressed, insisting that she required the assistance of her former lady-in-waiting. And though James finally placated his wife by spending considerable sums upon a tightrope walker for her entertainment, the courts of both England and France continued to doubt and sneer. Elizabeth, for instance, upon congratulating James at his escape, remarked nevertheless that since Gowrie had so many familiar spirits she supposed that there were no longer any left in hell. In France, meanwhile, as the diary of James Melville makes clear, the king's account of the Ruthvens' death was met with such ridicule that the Scottish ambassador was forced to suppress it in his reports.

If, then, there really were any doubt before the events of 1600 that James was indeed 'one of the most secret princes of the world', the Gowrie conspiracy had clearly banished it once and for all. Yet even the closeness with which James played his hand at that time pales by comparison with his sustained efforts to ensure that he would succeed to the English throne upon the death of Elizabeth I. Much of his anxiety and subterfuge was, in fact, unnecessary, for although he was a foreigner and therefore technically excluded by the common law, he remained throughout the only really plausible candidate – notwithstanding even the additional problem that Henry VIII's will had excluded his sister Margaret Stuart's descendants. For some time before 1592, it is true, Arbella Stuart, great-great-granddaughter of Henry VII via his daughter Margaret, had been considered one of the natural candidates to succeed Queen Elizabeth, her first cousin twice removed. But, in spite of her inherent advantage of having been born on English soil she, like the other potential candidates on offer, was effectively a non-starter, since her hereditary claim was much inferior to James's and she displayed in any case no appreciable desire for the Crown. Already, then, between the end of 1592 and the spring of 1593, the influential Cecils – Elizabeth's Lord Treasurer, Lord Burghley, and his son and future successor as Secretary of State, Sir Robert Cecil – had turned their attention away from Arbella towards her cousin, the King of Scotland.

The other potential claimants were, moreover, even less worthy of consideration. Certainly, none of the Englishmen whose Plantagenet or Tudor blood gave them some right to consideration – Lord Beauchamp, Lord Derby and the Earl of Huntingdon – had either the prestige, character or ambition to make them acceptable, and continental options were no more inviting. For English Catholics living abroad and nurturing hopeless dreams of an imminent resurrection of the old religion in their homeland, the pretensions of Philip III of Spain, either by descent or as Mary Stuart's nominated heir, seemed to offer some sustenance, though not even Father Robert Persons, the Jesuit responsible for English affairs at Rome and a strong Spanish partisan, seriously believed that another Spanish king could occupy the English throne. Instead, attention focused mainly upon Philip's sister, Infanta Isabella Clara Eugenia, who was married to Archduke Albert and ruled the Netherlands with him. As a descendant of Edward III through John of Gaunt, her claim to the English throne had already served as a pretext for the sailing of the Armada. But even in her case the chances of success were effectively non-existent, for the 150,000 or so Catholics remaining in England were vastly outnumbered and in any case preferred patriotism to the prospect of foreign rule. Nor, for that matter, had the infanta and archduke any real desire to exchange the comfort and security of Brussels for a reckless English gamble at the longest possible odds. And Persons' fond belief that Sir Robert Cecil was more sympathetic to a Spanish claimant than the father whom he had succeeded as Elizabeth's leading minister in 1598 would prove to be wholly mistaken.

# James I: Scotland's King of England

Unless King James foolishly alienated both the Queen of England and her leading ministers by some act of major indiscretion, his succession was therefore virtually assured. Even to Elizabeth herself, indeed, James remained the best available option, irrespective of his Scottish background and links with Mary Queen of Scots, and the same was true not only for Elizabeth's ministers but for those of her subjects that especially cared. He was, after all, an experienced ruler who had always appeared amenable to English interests and whose accession would unite the two kingdoms, and thereby achieve, in effect, the subjugation of Scotland that had always been a long-term goal of her southern neighbour. Most important of all, however, James's bloodline underpinned his claim more convincingly than any other alternative, and at a time when the hereditary principle was still paramount, this alone was enough. With or without Elizabeth's direct say-so, then, the King of Scotland would soon be Scotland's King of England, and if he were to behave as bid and promptly discard the Scottish baggage accompanying his succession, he would doubtless suffice.

In the meantime, however, such was James's fever of anxiety and impatience that he buzzed around Elizabeth with continual requests for confirmation of her intentions and far-fetched schemes to force her hand. In 1596, for instance, he had taken offence at certain passages in Spenser's *The Faerie Queen* which reflected upon his mother, and demanded at once from Elizabeth that the poet be tried and punished. He nagged the queen continually, too, to grant him the English estates of his grandparents, the Earl and Countess of Lennox, on the assumption that these would circumvent the English common law's bar to the succession of aliens. And then there was his near obsession with the act passed by Parliament in 1584 debarring plotters against the queen from inheriting her Crown, notwithstanding the fact that he had not been involved in any of his mother's intrigues. When, for instance, Valentine Thomas, a villainous Catholic ruffian to whom James had once unwisely granted an audience, was later arrested in England and claimed that the King of Scotland had encouraged him to assassinate Elizabeth, James's panic far outstripped the bounds of common sense.

Though Elizabeth assured him that she did not believe Thomas's tale and kept him quietly in prison without trial, so that his slander against the King of Scotland would not be broadcast in public, James nevertheless demanded that she erase his name from all records connected with the accusation and issue a declaration of his innocence. When she refused, moreover, James then declared not only that Thomas had been bribed into making a false confession but that he would issue a public challenge to do battle with anyone doubting his own innocence. Rather less eccentrically, he also saw fit to print the letter in which Elizabeth had expressed her disbelief in Thomas's story and to circulate it on the Continent. Once again, however, James's protests fell on deaf ears, since he lacked the cold conviction and depth of respect to make good his wishes. For James, as so often, did not so much threaten as bleat, and, in consequence, Thomas would remain in prison until Elizabeth's death when he was promptly tried and executed at James's command. Furthermore, the queen's imperious and frequently caustic dismissal of the Scottish ruler's protests demonstrates once more that the fundamental flaw in James's kingship was not so much a deficiency of political acumen or even of material resources but a more intangible, though nevertheless crippling, deficiency of vigour and resolve. As long as sheer pluck and backbone remained a crucial ingredient of majesty, therefore, James would always be ultimately lacking.

Nor was pestering the only method employed by James to achieve his longed-for goal. Much more provocatively, embassies were dispatched to various Protestant courts in northern Europe to conjure up armed support for his wish to 'be declared and acknowledged the certain and undoubted successor to the Crown'. Confident of his links with Denmark, Holland and the Protestant princes of Germany, James had been considering a league of Protestant states as early as 1585, although Elizabeth disliked his pretensions and in consequence purposely snubbed him by excluding Scotland from her own alliance with France and the Netherlands in 1596. Even so, in 1598 James dispatched his ambassadors to Denmark and Germany in what would prove to be the vain hope that a new league might be formed both to oppose the Turks and, much more importantly, prevail upon the Queen of England to name him her successor. Worse still, armaments were purchased and the Scottish Parliament of 1600 found itself faced with an apparently earnest request for the funding of an army to enforce his claim – all of which only succeeded in provoking Elizabeth's anger and prompting the English ambassador to wonder whether James was meaning 'not to tarry upon her Majesty's death'. 'He hasteth well,' Elizabeth warned her would-be successor, 'that wisely can abide', before assuring him that she would favour his claim 'as long as he shall give no just cause of exception'. But in spite of his squirming and posturing, she would not nominate him officially and thereby endure the indignity of seeing her courtiers and ministers turning away from her towards the 'sun rising' in the way that she herself had witnessed during the last months of her sister's life.

Such reasoning was, of course, eminently sensible and palpably transparent – to all, it seems, but James. Indeed, it is a measure of the soundness of James's claim to Elizabeth's throne and the needlessness of his behaviour that she chose to countenance his harassment at all. Certainly, his embassies abroad, his dalliances with papal and Spanish agents, and his attempts to fashion a party for himself at the English court all provided the Queen of England with precisely the 'just cause for exception' of which she warned him, had there been any conceivable alternative. James had even, it was believed in some quarters, secretly dispatched John Ogilvy of Powrie to Rome to angle for loans and promise toleration to English Catholics if he became their king. And even if Ogilvy's claims that he was James's accredited mouthpiece cannot be substantiated, there is no such question mark beside Lord Robert Semple's mission to Madrid in 1598, which sought recognition for James's claim. Nor, for that matter, was Elizabeth likely to have been any more reassured by a mysterious letter allegedly sent to Pope Clement VIII in 1599 in which her would-be successor was said to have addressed Clement as 'Most Blessed Father', before signing himself the pontiff's 'Most Obedient Son'.

James, in all fairness, consistently denied responsibility for the letter and eventually obtained a not altogether convincing confession from his secretary, Lord Balmerino, that it had been written without the king's consent and that his signature had been appended only by placing the document among others which had been signed hurriedly as he departed upon a hunting expedition. But it was enough to prompt a reply from Pope Clement in April 1600 in which he implored James to convert, and further fuelled concerns that the king's penchant for double games might well be carrying him into ever deeper waters. 'He practises in Rome, in Spain, and everywhere else, as he does with me,' wrote Henry IV of France, 'without attaching himself to any one, and is easily carried away by the hopes of those about him without regard for truth or merit.' And in the meantime he was equally prepared to offer restricted toleration on his own terms to English Catholics, corresponding with one of their most influential figures, the Earl of Northampton, and in the process nullifying any residual support for the Spanish infanta's candidacy. 'It were a pity,' wrote Northumberland, 'to lose so good a kingdom for not tolerating a Mass in a corner.' And James's reply confirmed his correspondent's high hopes: 'As for the Catholics,' he told Northumberland, 'I will neither persecute any that will be quiet and give but an outward obedience to the law, neither will I spare to advance any of them that will by good service worthily deserve it.'

That such reassurances should be offered at a time when intolerance was everywhere the norm might well seem to confirm the view that James's wish for compromise and moderation placed him ahead of the age in which he lived. But his contradictory signals to English Puritans suggest a ruler who was prepared to offer whatever potential supporters might wish to hear in a way that could only lead to trouble in the longer term when vague hints and outright guarantees would have to be made good. His agent in London, James Hamilton, was instructed in 1600, for instance, to assure all honest men that the king would 'not only maintain and continue the profession of the gospel there, but withal not suffer or permit any other religion to be professed and avowed within the bounds of that kingdom'. This, it should be remembered, was the king who had already set out his stall so firmly against all 'Puritans', dangerously conflating in the process the considerable range of opinion and outlook encompassed by the very term. Now, however, Puritans too, it seems, were to be courted at the very time that Elizabeth was resisting their pressure at every opportunity.

Equally provocative, in any case, was James's failure at this time to assist Elizabeth in her desperate struggle with the Earl of Tyrone's Irish rebels, which had flared up in 1595 and would rage for the next nine years, resulting in the disaster of the Battle of Yellow Ford and the overall loss of some 30,000 of Elizabeth's troops during the course of the whole campaign. As her ally, James sought outwardly to meet his obligation by issuing proclamations which forbade the clansmen of the Western Isles to assist the Irish rebels or trade with them. When Spanish troops landed at Kinsale, moreover, he offered England military aid. But in perceiving Tyrone as a potentially influential figure after Elizabeth's death, James gladly entered into secret correspondence with him and consistently turned a blind eye when his own royal proclamations were ignored. In fact, the towns of south-west Scotland continued to trade with the rebels as Tyrone recruited Scottish soldiers with apparent impunity. When Elizabeth protested, moreover, James merely issued new proclamations, which again went unenforced. In the meantime, he gladly accepted Tyrone's promises of future service when the time of reckoning finally arrived – though even with this, James's taxing of Elizabeth's patience was not over. Indeed, it was during her own crisis with the Earl of Essex that he came closest to breaking her forbearance once and for all.

Tenuously descended from Edward III, Robert Devereux (2nd Earl of Essex) had steadily emerged as the glittering star of the English court since the time of his arrival in 1584 and within three years had become Elizabeth I's unrivalled favourite. Tall, handsome, brave, ardent and flamboyant, possessing a remarkable capacity for self-dramatisation, which he demonstrated as both a flamboyant courtier and impetuous soldier, he was also the stepson of the Earl of Leicester, who perhaps had been the only man that the queen had ever loved until his death in 1588. But the same peacock brilliance, which made Essex so irresistible to the ageing and increasingly careworn and disillusioned queen, also carried with it an unruly spirit. 'The man's soul,' wrote the queen's godson Sir John Harington, 'seemeth tossed to and fro like the waves of a troubled sea.' And surely enough, while he could dazzle as romantic hero, blaze as dashing adventurer and sparkle as generous patron to writers and artists alike, he was nevertheless spoilt, vain and headstrong: a man incapable of moderation in either his behaviour – or his ambitions. Encouraged by his mother Lettice Knollys and his sister Penelope Rich, who had inspired Sir Philip Sidney's sonnet sequence *Astrophel and Stella*, he believed that by charming or sulking as occasion demanded, he could mould the queen to his will and prevail against those who opposed him.

Chief among these was Sir Robert Cecil who had by now succeeded his father, Lord Burghley, as Secretary of State. No man, in fact, could have presented a greater contrast to his adversary than Cecil to Essex. Immensely hardworking like his father, Sir Robert was, however, an altogether more complex and ambiguous figure than the man who both sired and reared him for high office. About 5ft 3in and suffering from a crooked back that may have resulted from being dropped by a nurse in infancy, the younger Cecil's infirmity seems to have made him all the more sensitive to the grace and panache of Elizabeth's courtiers, and left him curiously detached in spirit from the posturing of those around him. But while he was secretive and reserved by nature, he was also brilliantly clever – a perfect foil for Essex in all respects and someone bound, of course, for collision with him sooner rather than later. For not only were the two men so temperamentally at odds, they were also divided over the crucial political issue of the day: Essex, predictably enough, leading the war party at Elizabeth's court, and Cecil their opponents. In consequence, the late 1590s were dominated by their rivalry.

And it was into this political maelstrom that James now chose to slither. Already, from 1592 onwards, he had made a habit of dealing with Essex rather than Burghley when he had a cause to further at the English court, for no good reason other than his conviction that the latter had been responsible for delaying the pension promised to him by Elizabeth. In due course, indeed, the association had hardened into a dubious alliance of sorts, as Essex convinced him that only he and his friends would guarantee the English throne for him upon Elizabeth's death. It was, of course, an irresistible offer for one of James's impatience and anxieties. But as the relationship unfolded and Essex's own flaws became increasingly manifest, it was an offer that became more and more of a liability: so much so that before 1599 was out, Essex was hoping to oust his rivals at the English court by force, and indulge, with James's help – 'at a convenient time' – in a version of the old Scottish kidnapping game, involving the Queen of England herself.

As it transpired, the wayward earl had already quarrelled seriously with Elizabeth in 1598 when she unwisely decided to give him command of her army in Ireland. The result had been a mismanaged campaign, a disorderly and unauthorised truce with the Earl of Tyrone, and Essex's uninvited return to England. Early one morning in September 1599, moreover, he had forced an entry into the queen's bedchamber at Nonsuch Palace before she was properly wigged or gowned, ostensibly to justify his conduct, but possibly to force her to retain him in favour – in much the same way, curiously, that Bothwell had once coerced James. Arrested and condemned by the Privy Council for his truce with Tyrone and return to England which amounted, it was said, to a desertion of duty, Essex was nevertheless merely committed at first to the custody of Sir Richard Berkeley in his own York House before being convicted and deprived of public office – largely as a result of relentless pressure from Sir Walter Raleigh and Cecil – by an eighteen-man commission in June 1600.

In the interim, however, the earl was able not only to contact James but to ply him with schemes, misinformation and treacherous promises. On Christmas Day 1600, for example, he wrote concerning Cecil and his supporters. 'Now,' the letter runs, 'doth not only their corrupting of my servants, stealing of my papers, suborning of false witnesses, procuring of many forged letters in my name, and other such like practices against me appear; but their … juggling with our enemies, their practice for the Infanta of Spain, and their devilish plots with your Majesty's own subjects against your person and life ...' The clear implication, then, was that Cecil – to whom James himself referred as 'Mr Secretary, who is king there in effect' – was not only favouring another candidate for the succession but had secretly encouraged the Gowrie Plot. Around the same time, furthermore, James received assurances from an agent named Henry Leigh that Essex would tolerate no successor to the English throne other than the King of Scotland who should demand a public recognition of his right. Refusing to rebuff such dangerous baits outright, James replied cautiously that he 'would think of it and put himself in a readiness to take any good occasion'.

And as James temporised, Essex's allies grew bolder. In February 1600, indeed, Leigh came to Scotland once more – this time with a specific plan. The new commander of England's forces in Ireland, Lord Mountjoy, who was himself an intimate friend of Essex, would return with troops and join him in staging a coup at court, while James would gather an army on the Border and dispatch an ambassador to London to demand confirmation of his rights as heir. But whether James seriously entertained the proposal is, in fact, unclear. He had, it is true, been redoubling his efforts to increase his military strength throughout 1599 and on 1 May had ordered a grand muster or 'wapanschowing' and commanded his subjects to supply themselves with arms. In December, moreover, as well as in June of the following year, he had appealed for funds from the Scottish Parliament, only to be met with open derision at any suggestion that Scotland could pose any significant military threat to her southern neighbour, 'at which the king raged'. But while two confessions at Essex's eventual treason trial did imply that James's tardy response to Leigh's offer amounted to a rejection, even this must be balanced by the claim of Henry Wriothesley, Earl of Southampton, and someone on close terms with Essex, that the King of Scotland 'liked the course well and prepared himself for it'.

In all likelihood, James may well have been characteristically ambivalent – hedging his bets and accumulating as many cards in his hands as possible until they inevitably tumbled from his grasp for all to see. Certainly, the scheme was never realised, for when Essex called upon Mountjoy to deliver in the spring of 1600, the latter was already reversing English fortunes in Ireland with a brilliant campaign against Tyrone and duly refused. But even at this point, James was still not free from Essex's overtures as the earl's irrepressible ambition and declining fortunes led him inexorably to the ultimate and fatal gamble. In August, his freedom had been granted, but the main source of his income – the sweet wines monopoly – was not renewed, and he found himself moved 'from sorrow and repentance to rage and rebellion'. Accordingly, he summoned his followers to London in December, with a view to obtaining access to Elizabeth by force, driving Cecil and his other enemies from office, and summoning a parliament that would both endorse his action and recognise James as heir apparent.

This time, however, the King of Scots was not required to call his subjects to arms, but simply to send an ambassador to London by 1 February. 'You shall,' Essex promised, 'be declared and acknowledged the certain and undoubted successor to this Crown and shall command the services and lives of as many of us as undertake this great work.' For good measure, James was even sent a cypher, to be returned as proof of his acceptance, and so it was that the king's dabbling had finally brought him to the brink. Lacking the self-assurance to reject them outright, he had flirted with Essex's schemes and relied upon his skill at the double game to survive the brush with potential disaster. In consequence, the cypher was indeed returned, though no ambassadors made their way south at the appointed time, leaving Essex to stage his insane raid on 8 February and suffer the consequences. Blocked at Ludgate Hill by Sir John Leveson's barricade and forced to surrender soon afterwards, he was tried for treason eleven days later and executed on 25 February – the last man to be executed in the Tower of London. Thomas Derrick, the headsman, would require three strokes to complete the job, as Sir Walter Raleigh, it was said, watched the spectacle from a window on Tower Green, puffing out tobacco smoke in sight of the condemned man.

James, meanwhile, was said to have been 'in the dumps' when news reached him that Essex's rebellion had indeed gone ahead as planned, and by the time that his ambassadors – the Earl of Mar and Edward Bruce, Abbot of Kinloss – finally left for London in mid-March, their instructions had changed significantly. Now, for instance, while ascertaining whether a general rising against Elizabeth was still a possibility, they were to 'dally with the present guiders of the court' and tread the middle ground 'betwixt these two precipices of the queen and the people who now appear to be in so contrary terms'. They were to request, too, that Elizabeth issue a statement that the King of Scotland had taken no part in any rising – though the ambassadors, unlike James, fully realised that such a statement would not only never be granted, but amounted in any case to an effective acknowledgement of James's guilt. Repeatedly in March, on the other hand, James asked George Nicolson, the English ambassador in Scotland, whether he was being mentioned in the trials taking place in the aftermath of Essex's death, while making it clear to his ambassadors that if Cecil would not assist him, he should expect no favours in the future 'but all the queen's hard usage of me to be hereafter craven at his hands'.

Once again, though, James was tearing his hair needlessly. For in spite of his indiscretions, Elizabeth had little choice but to suppress all mention of the King of Scotland in what transpired. Unable to accuse him of conspiracy without excluding him from the succession and thereby opening wide once more the whole knotty issue of who should follow her, she actually accepted the inevitable in precisely the way that even Sir Robert Cecil now did, too. Summoning the Scottish ambassadors to the most secret of interviews the Secretary of State offered, under the most stringent of conditions, to correspond with their ruler and to promote his interests in England. With Essex cold in the ground and James coolly in his pocket, Cecil had ultimately been more than happy to oblige, and what James may or may not have promised Essex was now, in any case, an irrelevance. Before his defeat, the rebellious earl was said to have worn a letter from James in a black bag around his neck, and to have destroyed it along with his other papers before surrendering for the last time. Ashes were therefore all that remained, and from those ashes James's hopes were now revived and renewed.

There followed a truly remarkable secret dialogue between the King of Scotland and the very man who had appeared for so long to be the frustration of his hopes. At his trial, Essex had retracted his claims that Cecil favoured the succession of the Spanish infanta, and the final obstacle to James's trust was effectively cleared. Elizabeth's Secretary of State was soon offering, moreover, not only to support the King of Scotland's claim to the throne, but to advise and instruct him in preparation for it. Carefully constructing his letters to guarantee that no treasonable content could be found in them, Cecil referred to James's impending succession as 'that natural day … wherein your feast may be lawfully proclaimed (which I do wish may be long deferred) …' and went on 'to profess before God that if I could accuse myself to have once imagined a thought which could amount to a grain of error towards my dear and precious sovereign … I should wish with all my heart, that all I have done, or shall do, might be converted to my own perdition.' But he also spoke of Elizabeth's tacit goodwill and assured him that he could rest secure 'as long as we see our way clear from lively apparitions of anticipation' – or, in other words, precisely the kind of intrigues that James had already been much too inclined to countenance.

Cecil showed his considerable shrewdness too by employing Lord Henry Howard to share with him the task of mentoring the future King of England. Mary Queen of Scots had, after all, been closely and tragically connected with Howard's elder brother, Thomas, Duke of Norfolk, who had been executed in 1572 for plotting to marry her. And although Lord Henry was now an elderly, bombastic and penurious Catholic sycophant who had little influence with Queen Elizabeth, he was nevertheless the most able member of a premier noble family, which, as Cecil rightly judged, made him altogether more appealing to James than the coterie of bellicose malcontents who had congregated around Essex. Henceforward Cecil, James and now Howard, too, would all correspond as numerical ciphers – 10, 30 and 3 respectively – easing the way for James's eventual takeover, while continuing to hide their dealings from the existing queen. 'The subject itself,' wrote Cecil, 'is so perilous to touch amongst us as it setteth a mark upon his head forever that hatcheth such a bird.' And now, of course, the dwarfish Secretary of State, with his long, delicate hands, high white forehead and darkly piercing eyes could do no wrong in James's mind. 'My dearest and trusty Cecil,' wrote James, 'my pen is not able to express how happy I think myself for having chanced upon so worthy, so wise and so provident a friend.'

All was now sweetness and light too in James's relationship with Elizabeth. He had been warned by Cecil 'to secure the heart of the highest' by 'clear and temperate courses' and to avoid for this reason 'either needless expostulations or over much curiosity in her own actions'. Instead of the petulant and jibing tone of his former letters, therefore, he now consulted her as an oracle and commended her in his letters as that 'richt excellent, richt heich and michtie princess, our dearest sister and cousin'. More importantly still, he committed himself to securing the English Crown by patience alone. 'It were very small wisdom,' he told the Earl of Northumberland, 'by climbing of ditches and hedges for pulling of unripe fruit to hazard the breaking of my neck, when by a little patience and by abiding the season I may with far more ease and safety enter at the gate of the garden and enjoy the fruits at my pleasure.' And when a doughty Scottish laird drank in his presence to the speedy union of the Crowns, declaring that he had forty muskets ready for the king's use, he was promptly and roundly reproved.

In the meantime, James's mind was steadily moulded by both Cecil and Howard. But it was the latter, above all, whose darker side now surfaced most sordidly. Howard possessed, in fact, a deeply flawed but brilliant intellect which had already displayed its twists and shortcomings more than conclusively. Quite apart from periods of poverty he had, in fairness, experienced other reverses too that might well have taken their toll on men of altogether worthier fibre – at one point even 'suffering the utmost misery' in the Fleet Prison after publishing in 1583 his *Preservative Against the Poison of Supposed Prophecies*, which in addition to attacking astrology also contained allegedly treasonous passages. But he was nevertheless a paid pensioner of Spain, receiving 1,000 crowns annually from the Spanish ambassador, and now wheedled his way into James's trust by both lies and the grossest and most odious forms of flattery that Elizabeth had never fallen for. Henceforth, for instance, James would readily believe Howard's claim that Sir Walter Raleigh, along with Lord Cobham and the Earl of Northumberland, was one of 'a diabolical triplicity' of wicked plotters, hatching treasons from cockatrice eggs that were 'daily and nightly sitten on'.

By such slanders, then, Raleigh was already hopelessly compromised before James's reign in England began, and by such murky counsels, creeping schemes and fulsome flattery did James's 'long approved and trusty Howard' secure a place of prominence at the new king's table when Elizabeth's reign finally ended. Nor would his wait be a long one. For the queen who James had once complained seemed likely to outlive the sun and moon, was showing every visible sign of her own mortality. 'The tallest of ruffs,' it was said, 'could not conceal it, the most glittering of diamonds could not overpower it; voice, action, attitude all disclosed it …'

# 10 ❦ Scotland's King of England

'Forasmuch as it has pleased Almighty God to call to his mercy, out of this transitory life, our Sovereign Lady the high and mighty Princess Elizabeth, late Queen of England, France and Ireland, by whose death and dissolution the Imperial Realms aforesaid are come absolutely, wholly and solely to the high and mighty Prince James the Sixth, King of Scotland …'

*Opening lines of a proclamation read by Sir Robert Cecil at the High Cross in Cheapside on the morning of 24 March 1603*

It was in January 1603 that Queen Elizabeth had first developed a bad cold and been advised by Dr Dee, her astrologer, to move from Whitehall to Richmond – the warmest of her palaces – on what would prove to be 'a filthy rainy and windy day'. Once there, it seems, she refused all medicine and, as the Earl of Northumberland informed King James in Scotland, her physicians were soon concluding 'that if this continue she must needs fall into a distemper, not a frenzy but rather into dullness and lethargy'. The death on 25 February of her cousin and close confidante the Countess of Nottingham had only served to compound her illness with grief, and while all Scotland stirred in happy anticipation of her demise, the queen merely reclined on floor cushions, refusing Robert Cecil's instructions that she take to her bed. 'Little man,' she had told him, 'the word must is not to be used to princes.' She was 69, plagued with fever, worn by worldly cares and frustrations, and dying – so that even she was forced at last to accede to her secretary's pleas. Then, in the bedraggled early hours of 24 March, as the queen's laboured breathing slackened further, Father Weston – a Catholic priest imprisoned at that time in the Tower – noted how 'a strange silence descended on the whole City of London ... not a bell rang out, not a bugle sounded'. Her council was in attendance and, at Cecil's frantic request that she provide a sign of acceptance of James as her successor, she was said to have complied at last.

At Richmond Palace, on the eve of Lady Day, Elizabeth I had therefore finally put paid to her successor's interminable agonising and on that same morning of her death Sir Robert Carey, who had once conveyed her pallid excuses for the demise of Mary Queen of Scots to King James, was now dispatched north with altogether more welcome tidings. Leaving at mid-morning and bearing at his breast a sapphire ring that was the prearranged proof of the queen's demise, Carey had covered 162 miles before he slept that night at Doncaster. Next day, further relays of horses, all carefully prepared in advance, guaranteed that he covered another 136 miles along the ill-kept track known as the Great North Road linking the capitals of the two kingdoms. After a further night at Widdrington in Northumberland, which was his own home, the saddle-weary rider set out on the last leg of his journey, hoping to be with James by supper time, but receiving 'a great fall by the way' which resulted in both his delay and 'a great blow on the head' from one of his horse's hoofs 'that made me shed much blood'. Nevertheless, 'be-blooded and bruised', he was in Edinburgh that evening and though the 'king was newly gone to bed', the messenger was hurriedly conveyed to the royal bedchamber. There, said Carey, 'I kneeled by him, and saluted him by his title of England, Scotland, France and Ireland', in response to which 'he gave me his hand to kiss and bade me welcome'.

James had dwelt upon the potential difficulties of the succession for so long, however, that he could scarcely credit the ease with which it appeared to be taking place and wasted no time in consolidating his position. To the very last, of course, Elizabeth had made no official acknowledgement of the King of Scotland as her heir, and until he had taken physical possession of his new realm, his fear of invasion or insurrection remained tangible. The day after Carey's arrival, therefore, the Abbot of Holyrood was urgently dispatched to take possession of Berwick – the gateway to the south – and within a week, as his English councillors pressed him to make haste, plans for James's transfer to London were complete. Summoning those nobles who could be contacted in the time available, he placed the government in the hands of his Scottish council and confirmed the custody of his children to those already entrusted with them. Likewise, his heir, Prince Henry, was offered words of wisdom upon his new status as successor to the throne of England. 'Let not this news make you proud or insolent,' James informed the boy, 'for a king's son and heir was ye before, and no more are ye yet. The augmentation that is hereby like to fall unto you is but in cares and heavy burdens; be therefore merry but not insolent.' Queen Anne, meanwhile, being pregnant, was to follow the king when convenient, though this would not be long, for she miscarried soon afterwards in the wake of a violent quarrel with the Earl of Mar's mother, once again involving the custody of her eldest son – whereupon James finally relented and allowed the boy to be handed over to her at Holyrood House prior to their joining him in London.

Before his own departure, however, James had certain other snippets of business to attend to. On Sunday 3 April, for instance, he attended the High Kirk of St Giles in Edinburgh to deliver 'a most learned, but more loving oration', in which he exhorted his subjects to continue in 'obedience to him, and agreement amongst themselves'. There was a public promise, too, that he would return to Scotland every three years – though he would ultimately do so only once, in 1617 – and a further suggestion that his subjects should take heart upon his departure, since he had already settled 'both kirk and kingdom'. All that remained thereafter was a plea to the council for money, since he had barely sufficient funds to get him past the Border, and a series of meetings with both English officials on the one hand and a mounting flood of suitors already seeking lavish rewards and promises. In the first category, came Sir Thomas Lake, Cecil's secretary, who was sent north to report the king's first thoughts as he became acquainted with English affairs, and the Dean of Canterbury, who was hastily dispatched to ascertain James's plans for the Church of England. To the second belonged a teeming, self-seeking throng. 'There is much posting that way,' wrote John Chamberlain, an eagle-eyed contemporary reporter of public and private gossip, 'and many run thither of their own errand, as if it were nothing else but come first served, or that preferment were a goal to be got by footmanship'.

In the event, James's progress south might well have dazzled many a more phlegmatic mind than his, since it was one unbroken tale of rejoicing, praise and adulation. Entering Berwick on 6 April in the company of a throng of Border chieftains, he was greeted by the loudest salute of cannon fire in any soldier's memory and presented with a purse of gold by the town's Recorder. His arrival, after all, represented nothing less than the end of an era on the Anglo-Scottish border. In effect, a frontier which had been the source of bitter and continual dispute over five centuries had been finally transformed by nothing more than an accident of birth, and no outcome of James's kingship before or after would be of such long-term significance. That a King of Scotland, attended by the wardens of the Marches from both sides of the Border, should enter Berwick peacefully amid cries of approval was almost inconceivable – and yet it was now a reality for the onlookers whose forebears' lives had been so disrupted and dominated by reprisal raids and outright warfare.

Nor did a sudden rainstorm the following day dampen the king's spirits. The sun before the rain, he declared, represented his happy departure, the rain the grief of Scotland, and the subsequent fair weather the joy of England at his approach. Such, in fact, was his keenness to press forward into his new kingdom that his stop in Northumberland at Sir Robert Carey's Widdrington Castle was deliberately cut short. For he departed, we are told, 'upon the spur, scarce any of his train being able to keep him company', and rode nearly 40 miles in less than four hours. Pausing to slay two fat deer along the way – 'the game being so fair before him, he could not forbear' – he rested over Sunday at Newcastle, and heard a sermon by Tobie Mathew, Bishop of Durham, with whom he joked and jested in high humour. Indeed, the urbane, serene world of the Anglican episcopacy, which so happily combined theological soundness with a proper deference for royal authority could not have been more agreeable to James. Received at the bishop's palace by 100 gentlemen in tawny liveries, he was treated at dinner to a fine diet of delicious food and Mathew's own unique brand of learning, humanity and worldly wisdom, which would bring the bishop considerable rewards three years later when he found himself Archbishop of York and Lord President of the Council of the North. Even before the king left next morning, moreover, Mathew's bishopric had already recovered much alienated property, including Durham House in the Strand, which had been granted previously to Sir Walter Raleigh.

By the time that James entered York on 14 April, however, he had already found much else about his new kingdom to impress him. Above all, he was struck by the apparent richness of a land he was visiting for the first time and knew only by reputation. The abundance of the countryside, the splendour of the great mansions, the extensive parklands through which he travelled, even the quaintness of the villages scattered along his route all proclaimed the contrast with Scotland. Everything, indeed, seemed to lift James into a heady state of expectation after the rigours of his rule in Scotland. According to the eminent lawyer and Master of Requests Sir Roger Wilbraham, the king travelled onwards 'all his way to London entertained with great solemnity and state, all men rejoicing that his lot and their lot had fallen in so good a ground. He was met with great troops of horse and waited on by the sheriff and gentlemen of each shire, in their limits; joyfully received in every city and town; presented with orations and gifts; entertained royally all the way by noblemen and gentlemen at their houses ...'. But the same observer's concerns about what might be awaiting England's new king in the longer term were more revealing still. 'I pray unfeignedly,' wrote Wilbraham, 'that his most gracious disposition and heroic mind be not depraved with ill-counsel, and that neither the wealth and peace of England make him forget God, nor the painted flattery of the court cause him to forget himself.'

And the scale of 'painted flattery' on offer to James, both now and later, was greater by far than anything he had experienced before. Elizabeth I, of course, had skilfully nourished the cult of her own personality. Symbolically represented as a virgin goddess – variously named Gloriana, Belphoebe, Astraea, Cynthia, Diana – she had been the object of much poetic worship. But sober statesmen, no less than poets like Edmund Spenser and Ben Jonson, had also observed the convention of addressing James's predecessor as though she were indeed a goddess. In 1592, for instance, Cecil had referred to the 'sacred lines' of a letter written by the queen, before going on to eulogise her 'more than human perfection'. Now, moreover, as the king continued his journey through England, he would hear in a continual series of panegyrics, similar words which seemed to conform so closely to the theories of kingship that he himself had expressed with such conviction in *The Trew Law of Free Monarchies* and *Basilikon Doron*. 'Hail, mortal God, England's true joy!' ran John Savile's poem, written to salute James upon his acquisition of his new realm.

At York, in particular, he experienced the full gust of exultation at his new status. Initially, at least, his English privy councillors had not planned to supply him with the full trappings of royalty until he had passed through the north of the country and reached Burghley in Nottinghamshire where they were due to meet him for the first time. It was true, too, that up to this point when warm and spontaneous welcomes had been the norm, James's lack of natural dignity had actually been an asset of sorts. His talkativeness and familiarity, and above all his easy, impulsive generosity had been enough to create a favourable impression upon his new subjects. But York was a different matter and James had insisted quite correctly that he enter 'our second city' with appropriate solemnity and magnificence, so that by the time of his arrival on 16 April he was suitably equipped with jewels, regalia, heralds, trumpeters and men-at-arms – though at the King's Manor where he was lodging prior to his entry, he had refused a coach. 'I will have no coach,' he declared, 'for the people are desirous to see a king, and so they shall, for they shall see his body and his face.' Nor, it seems, was he prepared to swap his shabby doublet – specially padded for protection against an assassin's dagger – for a more elegant garment, as he appears to have been wearing it at his first meeting with Sir Robert Cecil, who finally greeted him in York on 18 April.

Cecil had made the journey north, full of his usual cares and perplexity, and troubled in particular by financial affairs and rumours that complaints from Ireland had reached the king's ears. Yet James, it seems, was more concerned with the details of his own journey to London than with weightier matters, and after delivering a jovial greeting and confirming Cecil's ongoing role as Secretary of State, proceeded to his main concerns. He was worried, for instance, that his arrival in the capital might coincide with Elizabeth's funeral – an event he otherwise declined to discuss, since he had a horror of all things relating to death and dissolution. And he was equally anxious to ensure that his coronation did not occur before the arrival of his wife. Indeed, like the lucky lottery winner he was, James became wholly absorbed in the here and now, and in the process allowed his excitement, as so often, to spill over into a characteristic display of rough-hewn humour and familiarity that is unlikely to have been wholly to Cecil's taste. Already dubbed by Elizabeth her 'Pygmy', the secretary became known at once to James as his 'Little Beagle'. 'Though you be but a little man,' James told him, 'we shall surely load your shoulders with business.'

Yet by the time he parted from the king to attend Queen Elizabeth's funeral on 28 April, Cecil had been suitably impressed, recording that James's 'virtues were so eminent as by my six days' kneeling at his feet I have made so sufficient a discovery of royal perfections as I contemplate greater felicity to this isle than ever it enjoyed'. The fact that his new master had already made his first requests for money to the council may have escaped Cecil's attention at this stage, as indeed the additional appeal for jewels and ladies-in-waiting for Queen Anne may have done. Certainly, the king had already laid down that new coins were to be minted, one side of which would join the arms of Scotland to those of his other kingdoms and declare the Latin legend, *Exsurgat Deus Dissipentur Inimici* – 'Let God arise and His enemies be scattered'. And this was only to be expected. But whether Cecil had already begun to guess at the extent to which such coins would soon be slipping through his new master's fingers is a matter of conjecture, for even by the time he reached York, the early signs of extravagance were plain for all to see.

James's journey had already been punctuated, as might be expected, by prolonged and princely civic entertainments, as well as hunting and feasting in the great country houses through which he passed. It had been conducted, too, amidst a growing shower of royal gifts and knighthoods. But by the time the royal progress reached York, matters had become almost unmanageable. For while James had set out with a representative selection of Scottish nobles and an appropriate train of English and Scottish courtiers and officials, numbers were soon swelling as north-bound English place-hunters and impoverished Scots hurrying south for rich pickings, converged from all points of the compass to create a disorderly rabble of more than 1,000 greedy souls. Newcastle had shouldered the whole charge of the royal household for three days and York for two more. But now increasingly the burden fell on private estates like that of Sir Edward Stanhope at Grimstone Hall who extended 'most bountiful entertainment' to 'all comers' – 'every man' eating 'without check' and 'drinking at leisure'.

Nor were desultory attempts to stem the hordes effective. Proclamations ordering home all Scots not in immediate attendance upon the king, and restraining 'the concourse of idle and unnecessary posters' were, for instance, largely ineffectual – not least because James's carefree distribution of gifts, grants and favours continued unabated. The bestowal of knighthoods, for example, which had been so carefully restricted by Elizabeth I was now conducted more and more casually, so that by the time he reached London, James had delivered the title to less than 300 individuals. During the entire forty-five years of the former queen's reign, in fact, only 878 men had been knighted, while in the first months alone of James's reign, there were 906 such promotions. The landlord of the Bear Inn at Doncaster, meanwhile, received the lease of a valuable royal manor as reward for one good night's entertainment. And thus the locust horde continued to swell, consuming all that came in its path and placing an intolerable strain even on such great households as the Earl of Shrewsbury's at Worksop and Lord Rutland's at Belvoir, not to mention the equally impressive resources of a certain Sir Oliver Cromwell at Hinchingbrooke in Huntingdonshire.

Cromwell was uncle, ironically enough, to the future Lord Protector that he eventually came to loathe, and had married the widow of the immensely rich Italian-born financier Sir Horatio Bavarino, whose wealth he had subsequently chosen to lavish in an apparently ceaseless quest for popularity. Now, however, he seems to have exceeded all expectation in terms of both the quality and range of entertainment he provided for the king. Certainly, the dinner provided for James's entourage may well have been the best of the whole journey – 'such plenty and variety of meats, such diversity of wines, and those not riffe ruffe, but ever the best of their kind; and the cellars open at any man's pleasure'. But there were other treats, too, to whet the new king's broader appetites. The Vice-Chancellor of Cambridge University, for instance, along with the heads of the colleges, all attended James, bringing a present of books and proffering speeches and poems of welcome. And just beyond Hinchingbrooke, at Godmanchester, seventy ploughing teams were carefully drawn up, not merely to see the king upon his way, but to emphasise once more the prosperity of his new kingdom.

Nor, it seems, were such displays of goodwill anything less that heartfelt. Just before Worksop, for example, 'there appeared a number of huntsmen all in green; the chief of which, with a woodman's speech, did welcome him', leaving the king, we are told, 'very much delighted'. All the way from Sir John Harington's house to Stamford, moreover, 'Sir John's best hounds with good mouths followed the game, the king taking great leisure and pleasure in the same'. And though James blundered at Newark-on-Trent by ordering that a cut-purse be hanged without trial, he generally charmed and responded in kind as only he probably could. He spoke lovingly, for instance, to Sir Henry Leigh, an honourable old knight, who presented himself 'with sixty gallant men' wearing yellow scarves embroidered with the words *Constantia et fede*. Even a fall from his horse near Burghley, which led Mayerne, his physician, to suggest that he had broken his collarbone, could not stem his enthusiasm or stifle his will to please. Perhaps, indeed,

James would never again feel so entirely and satisfactorily a king as during the three weeks of his journey. With no financial restraints in place and the pressure of state affairs still in front of him, he was able to pose as what he had always wished to be in Scotland – the beneficent, affable, patriarchal dispenser of largesse and justice, responsible only to God for the welfare of a grateful and pliable realm.

And, in the meantime, the fact that James's winning words and lofty promises would one day have to be made good continued, it seems, to elude him. 'Nor shall it ever be blotted out of my mind,' the king would tell his first Parliament, 'how at my first entry into this kingdom the people of all sorts rid and ran, nay rather flew to meet me.' But the resulting effusiveness and spontaneity on his part would come at a price. It was all too tempting, of course, to assure the mayor and aldermen of Hull that they should be 'relieved and succoured against the daily spoils done to them' by Dunkirk pirates, regardless of the naval expenditure involved in properly securing England's lengthy coastline, or to earn popularity cheaply by promising 'their hearts' desire' to Huntingdonshire folk, complaining of the enclosure of their land by a certain Sir John Spencer. Likewise, when 1,000 Puritan brethren presented their so-called 'Millenary Petition', the king could graciously offer to consider and redress their grievances without any real appreciation that English Puritans might prove just as intractable in their own way as their Presbyterian counterparts within the Scottish Kirk.

But when James arrived on the outskirts of London at Theobalds, the home of Sir Robert Cecil, his new kingdom was already dowsed in heady expectation and primed for later disappointment, even if, for the time being, the general euphoria continued. At the border of Middlesex, he had been greeted by 'three score men in fair livery cloaks', before meeting the Lord Mayor and aldermen, accompanied by 500 prominent citizens, all on horseback and clad in velvet cloaks and chains of gold. And by this time the crowds of common people were becoming so dense that when one observer tried to count them, he could not do so, declaring that each blade of grass had become a man. At Stamford Hill, indeed, a humble cart owner was able to charge eight groats for the use of his vehicle as a grandstand for no more than a quarter of an hour. 'The multitude of people in highways, fields, meadows, closes and on trees were such that they covered the beauty of the fields,' wrote John Nichols, 'and so greedy were they to behold the countenance of the king that with much unruliness some even hazarded to the danger of death.' Ultimately, James would have no choice but to avoid the roads on his way to the Charterhouse, where he was entertained for three days by Lord Thomas Howard and saw fit to create a further 130 knights.

But what should have been the climax of the entire trip – James's state entry into the City of London – proved something of an anti-climax, since the death rate from the plague had risen to twenty a day and the king could only skirt the city in a closed coach to Whitehall and inspect the capital from the river on his way to the Tower. There followed, moreover, a further two months of maddening delay while the daily toll of plague victims rose to 700 to 800, and James was left to shuttle uneasily from Greenwich and back to Whitehall, prior to a trip to Windsor, where he presided over a chapter of the Order of the Garter, and a tour of some of the better-stocked deer parks of the home counties. The plague, observed Cecil, 'drives us up and down so round, as I think we shall come to York.' And the king, in the meantime, gave little time to official business. 'Sometimes he comes to council,' wrote Thomas Wilson, an author whom Cecil had been employing as a foreign intelligencer, 'but most time he spends in fields and parks and chases, chasing away idleness by violent exercise and early rising.'

In the end, James's state entry into the City would have to be delayed until the following spring. But the coronation could not wait so long, and the queen was duly summoned from Scotland, causing another northward flood of interested court folk. The 13-year-old Lady Anne Clifford, who was only one of those many noble wayfarers frantically traversing the Great North Road around this time, had already met the king at Theobalds where she, her mother and her aunt had been 'used very graciously' by him, notwithstanding the fact that she had noticed 'a great change between the fashion of the court as it is now and of that in the queen's time, for we were all lousy by sitting in the chamber of Sir Thomas Erskine'. Presently, however, she was in headlong motion once again, recording in her celebrated diary how she and her mother killed three horses in their haste to intercept the queen, and noting that near Windsor, where James eventually met his wife, 'there was such an infinite number of lords and ladies and so great a court as I think I shall never see the like again'.

Even so, the coronation that occurred on 25 July, the feast of St James the Great, was shorn of its usual splendour. By this time, as many as 30,000 Londoners had succumbed to the plague and Prince Henry had been sent to Oatlands near Weybridge in Surrey to avoid infection. 'By reason of God's visitation for our sins,' noted one commentator, 'the plague and pestilence there reigning in the City of London … the king rode not from the Tower through the city in the royal manner …' The pageants in their turn were postponed until the New Year, which was probably for the best as the day was further blighted by pouring rain. And to add the general gloom, the ceremony itself would take place in a sadly empty Abbey, since any concourse of people was forbidden, and nobles and dignitaries were severely limited in the numbers of attendants they brought – an earl, whose rank normally entailed a following of at least 150 for a London visit, being allowed only sixteen, and those of humbler station proportionately less. Among the other more mournful details of the occasion was the queen's refusal to accept Holy Communion according to the Anglican rite from Archbishop Whitgift of Canterbury and James's less than majestic response to Philip Herbert, Earl of Montgomery, who had already established himself as the king's first English favourite. For when the earl flouted tradition by kissing his sovereign as he paid homage, James showed no sign of displeasure but merely tapped him on the cheek indulgently.

By the time that James and his queen set out on a tour of the southern counties shortly afterwards, there were already signs that the brief honeymoon period of the succession was rapidly waning. Royal progresses were, of course, an essential means of popularising the monarchy and maintaining links between king and kingdom. But they were also exceedingly expensive, as James's initial journey from Edinburgh to London had demonstrated, and only the monarchy's ancient right to requisition horses and vehicles and fix low prices for the purchase of local produce made these lavish journeyings feasible at all. Even as he reached the capital in May there were murmurs against the 'general, extreme, unjust and crying oppression' of 'cart takers and purveyors', who, as Parliament was soon to point out, 'have rummaged and ransacked since your Majesty's coming in far more than under your royal progenitors'. The fact was that James had already spent £10,000 on his initial journey south and literally given away another £14,000 in gifts of various kinds at a time when Queen Elizabeth's funeral had already cost £17,000 and the outstanding debt resulting from the Irish campaign stood at £400,000. Before long, in fact, Cecil would be writing anxiously to the Earl of Shrewbury about his royal master's spending in general. 'Our sovereign,' he observed, 'spends £100,000 yearly on his house, which was wont to be but £50,000. Now think what the country feels, and so much for that.'

Another spate of hunting, banqueting, masques and pageants, however necessary in terms of conventions and expectations, was therefore bound to carry with it political as well as purely financial costs, particularly if the king's extravagance were to continue in the longer term – which, indeed, it did. The habit of idleness formed in those early months when plague had kept him out of London was quick to take root, moreover, for after one month of bickering with his first parliament, he had gone off to hunt and left the management of the rest of the session to Cecil. And when official business followed him, he reacted testily. When, for instance, a swarm of local petitioners troubled him with pleas on behalf of nonconforming clergy and what Chamberlain termed 'foolish prophecies of dangers to ensue', James did not hesitate to summon the council to Northampton, with the result that the shell-shocked petitioners were summarily hauled up and rebuked for opposing the king in a manner that was deemed to be 'little less than treason'.

Nor, it seems, was James so inclined to play to the crowds in public or revel in the limelight of official ceremonies once his coronation was finally over. Upon his entry into London, we are told, he had 'sucked in their gilded oratory, though never so nauseous, but afterwards in his public appearances, especially in his sports, the access of the people made him so impatient that he often dispersed them with frowns, that we may not say with curses'. Henceforth, if Thomas Wilson is to be believed, the people missed the affability of their dead queen, since the new ruler 'naturally did not love to be looked on, and the formalities of State but so many burdens to him'. And Wilson was not the only one to suggest as much, for Sir John Oglander observed that the king would now swear with passion when told by his attendants that the people had come of love to see his royal person. 'Then,' said Oglander, 'he would cry out in Scottish, "God's wounds! I will put down my breeches and they shall also see my arse!"'

Yet if James's improvidence and disaffection with some of the more irksome niceties of kingship were already emerging, other facets of his personality continued to create favourable first impressions with many who served him. 'He is very facile,' wrote Sir Thomas Lake, 'using no great majesty nor solemnity in his accesses, but witty to conceive and very ready of speech.' Sir Roger Wilbraham, meanwhile, observed that 'the king is of sharpest wit and invention, ready and pithy speech and exceeding good memory; of the sweetest, pleasantest and best nature that ever I knew; desiring nor affecting anything but honour'. And even the critical eye of Sir Francis Bacon, while quietly hinting at certain vices and indiscretions, remained generally positive about the new King of England during his first interview with him at Broxbourne. In a letter to the Earl of Northumberland, Bacon found James to be 'a prince farthest from the appearance of vainglory that may be'. 'His speech,' Bacon continued, 'is swift and cursory, and in the full dialect of his country; and in the point of business short; in point of discourse large.' Moreover, while he was considered 'somewhat general in his favours', it was also observed that 'his virtue of access is rather because he is much abroad and in press than that he giveth easy audience about serious things'. On the whole, then, there was much praise for James's nimble mind, his loquacity, his affable and homely manner, his good nature and his apparent virtue.

And equally favourable impressions were recorded in the observations of many foreigners. 'The King of England,' wrote the Venetian ambassador, 'is very prudent, able in negotiation, capable of dissimulating his feelings. He is said to be personally timid and averse from war. I hear on all sides that he is a man of letters and business, fond of the chase and of riding, sometimes indulging in play. These qualities attract men to him and render him acceptable to the aristocracy. Besides English, he speaks Latin and French perfectly and understands Italian quite well. He is capable of governing, being a prince of culture above the common.' Even James's physical appearance had its share of admirers, it seems. 'The king's countenance is handsome, noble and jovial,' wrote another Italian, 'his colour blond, his hair somewhat the same, his beard square and lengthy, his mouth small his eyes blue, his nose curved and clear-cut, a man happily formed, neither fat nor thin, of full vitality, neither too large nor small.'

James's carriage and bearing, meanwhile, were far from universally berated. At his first audience with the king in May 1603, the Venetian ambassador 'found all the council about his chair, and an infinity of other lords almost in an attitude of adoration'. Dressed 'in grey silver satin, quite plain, with a cloak of black tabbinet reaching below his knees and lined with crimson', the ambassador also noted that 'he had his arm in a sling, the result of a fall from his horse'. But no mention was made of any lack of grace or manners, and though 'from his dress he would have been taken for the meanest among his courtiers', we hear, too, that this was only the result of 'a modesty he affects'. Nor, for that matter, was such modesty merely a matter of public posturing. On the contrary, James's ability to relate informally and humanely to lesser mortals would remain in evidence long after his early courting of the crowds had begun to wane. Indeed, one of his most sentimental attachments, it seems, was eventually formed with Robin, an old keeper of Theobalds Park, who would make him presents of thrushes and blackbirds' nests and endeared himself to the king by his 'plain, honest and bold speech'.

There were times, it is true, when James's penchant for the common touch crossed the borders of seemliness and decorum, and detracted from the distance and respect properly associated with genuine majesty. Among many trivial, but nonetheless awkward instances of the king's excessive exuberance and over-familiarity was his tendency to visit newly-wed couples after their first night together, even lying on the bridal bed and quizzing the couple on what had passed. When, for instance, his favourite Sir Philip Herbert, whom he had recently created Earl of Montgomery, married the Lady Susan de Vere in Whitehall during 1604, the king was overcome by boyish high spirits as all fashionable London celebrated the event. Giving the bride away 'in her tresses and trinkets', the king, wrote Sir Dudley Carleton, proudly declared that 'if he were unmarried, he would not give her, but keep her himself'. It was a fatherly and endearing comment wholly in keeping with the joy and light-heartedness of the occasion. But whether James's behaviour was quite so appropriate next morning is rather more debatable, for 'the king', John Chamberlain tells us, 'in his shirt and nightgown gave Philip Herbert and his bride a reveille-matin before they were up and spent a good time in or upon the bed, choose which you will'. The fact that Chamberlain commented upon the event and that, more importantly still, the incident had come to his attention in the first place suggests, at the very least, that James's behaviour had once again been less than apt.

Nevertheless, the classic caricature of James, traditionally attributed to Sir Anthony Weldon, stands in stark – and largely unconvincing – contrast to most contemporary descriptions of the man. Weldon came, in fact, from a family that had long associations with the royal household and had served in his own capacity as Clerk of the Green Cloth, auditing accounts and organising the household's travel arrangements. He had, moreover, clearly prosecuted his role effectively, since he was knighted for his efforts, before losing his job, it has always been suggested, in deeply embarrassing circumstances. According to the usual account, he had written an unpleasantly satirical account of the Scots, which subsequently fell into the king's hands when it became carelessly mixed up with official papers. In consequence, even though the king granted him a small pension after his dismissal, Weldon, we are told, apparently went on to pen his famous portrait, *The Court and Character of James I*, although his responsibility for the work is not as certain as is often assumed, since it was not actually credited to him until after his death in 1648 when anti-Stuart feeling was at its height. Curiously, too, the work which allegedly provoked James to sack Weldon – *A Perfect Description of the People and Country of Scotland* – was actually published six years beforehand.

In any event, the description supplied by *The Court and Character of James I* certainly gained currency across the centuries. The king, it suggests, was 'of middle stature, more corpulent through his clothes than in his body, yet fat enough, his clothes ever being made large and easy, the doublets quilted for stiletto proof, his breeches in pleats and full stuffed'. Thus derives the classic picture of a timorous and suspicious monarch, padding his clothing against assassin's knives and pistol shots, 'his eye large, ever rolling after any stranger come into his presence, in so much as many for shame have left the room, as being out of countenance'. We hear, too, of those physically unprepossessing images of the king that have always been so closely associated with him. His beard, in Weldon's account, was 'very thin' and 'his tongue too large for his mouth, which ever made him drink very uncomely', while his skin was 'as soft as taffeta sarsnet, which fell so, because he never washed his hands, only rubbed his finger ends slightly with the wet-end of a napkin'. The king's legs, in their turn, were very weak, 'having as was thought some foul play in his youth, or rather before he was born', with the result that he was 'ever leaning on other men's shoulders'. And his walk was 'ever circular' – 'his fingers ever in that walk fiddling about his cod-piece'.

According to this account, James was also somewhat intemperate in his use of alcohol – with a particular liking, it seems, for 'Frontiniack, Canary, high country wine, tent wine and Scottish ale' – though he was seldom drunk, since he had a 'very strong brain'. 'He naturally loved not the sight of a soldier, nor of any valiant man', the account continues, 'and was crafty and cunning in petty things', and though he was 'infinitely inclined to peace, it was more out of fear than conscience'. 'Wise in small things, but a fool in weighty affairs', he was, we hear, 'very liberal of what he had not in his own grip, and would rather part with £100 he never had in his keeping than one twenty shilling piece within his own custody'. Yet he was 'constant in all things (his favourites excepted)' and 'had as many ready witty jests as any man living, at which he would not smile himself, but deliver them in a grave and serious manner'. 'In a word,' the account concludes, 'take him altogether and not in pieces, such a king I wish this kingdom have never any worse, on the condition, not any better, for he lived in peace, died in peace, and left all his kingdoms in a peaceable condition, with his own motto: *Beati Pacifici* [Blessed are the Peacemakers].'

Whether all this amounted to a species of character assassination in the way that many later authorities came to accept, is perhaps debatable – not merely in terms of the accuracy of the author's claims, but more importantly still, perhaps, in terms of his work's tone. This, after all, was said to be a king who, apart from being a peacemaker, 'loved good laws and had many made in his time' – hardly the kind of declaration that might be expected from a slighted hatchet-wielder. There is no denying, of course, that Sir William Sanderson in his *Aulicus Coquinariae: or a Vindication in Answer to a Pamphlet entitled the Court and Character of King James* (1650) suggested that the author, whom he considered to be Weldon, had later disowned the work 'which with some regret of what he had maliciously writ', he had 'intended to the fire' before 'it was since stolen to the press out of a ladies closet'. But the eleven-page pamphlet remains, perhaps, more sinned against than sinning – more maligned for its supposed subjectivity than any real malignity it actually directed at James himself. That it is a caricature is beyond dispute, but like all successful caricatures it may well have captured more of the essence of the man than it is often credited for. And even if Weldon was indeed the author and carried the grudge for which he is charged, grounds for grievance need not necessarily entail outright ill will. Nor, in the grand scheme of things, can he be said to have exacted his vengeance especially cruelly.

Certainly, some of James's most telling flaws are left unexplored – one of which was his susceptibility to the very 'painted flattery' of which Sir Roger Wilbraham had spoken. Upon his succession, he had passed at once from the brusque and often insolent frankness of his Scottish homeland to the altogether more obsequious conventions of a highly sophisticated court. And unlike the chisel-tongued nobles and ministers of his other kingdom, Cecil was only one among many who now addressed him in terms of the deepest deference, while the likes of Henry Howard grovelled unashamedly. For Howard, in particular, the king was 'your resplendent Majesty', the sun itself – sacred, peerless, wise and learned – and with such a flow of adulation to confirm the king's opinion of his exalted status, it was not long before ceremonies discarded by the former queen now began to reappear once more. An early medal struck in his honour had already borne the title *Caesar Caesorum* ['Caesar of Caesars'], but before long the Venetian ambassador was observing how 'they are introducing the ancient splendours of the English court and almost adoring his majesty, who day by day adopts the practices suitable to his greatness'. State dinners at which the nobility served the king on bended knee – 'a splendid and unwonted sight' – now became increasingly common, while a flood of eulogistic verses flowed with regularity from the pens of Henry Petowe, Samuel Rowlands and Anthony Nixon. 'The very poets with their idle pamphlets,' wrote John Chamberlain, 'promise themselves great favour.'

Even more unfamiliar to James – and equally intoxicating – was the lavish praise heaped upon him by the English clergy. Bullied and derided by the most vocal elements of the Kirk, he now found a halo of holiness placed around his head by bishops who were more than prepared to accede to his claims as God's direct representative on Earth. 'God hath given us a Solomon,' declared Bishop James Montagu, 'and God above all things gave Solomon wisdom; wisdom brought him peace; peace brought him riches; riches gave him glory.' Indeed, Montagu continued, the king surpassed Solomon, since he had been steadfast in the true religion longer than Solomon had reigned, nor had he been immoderate in his dealings with women. And to gild the lily further, there were frequent comparisons with the Roman Emperors Constantine and Theodosius, not to mention Israel's very own sweet singer, David. In the same way that David had united the laws of Judah and Israel, it was proclaimed, so James had bonded the rival kingdoms of England and Scotland.

All in all, then, it was small wonder that James basked in the veneration that now came his way and gave full vent to the inherent sense of his own celebrity that had hitherto been so roughly and so frequently challenged north of the Border. When Roger Aston, a Scottish envoy who had been sent ahead to England upon Elizabeth's death was asked how his master felt about his succession, the answer could not have been more clear-cut or apt. 'Even, my lords,' came the reply, 'like a poor man wandering about forty years in a wilderness and barren soil and now arrived at the land of promise'. Not long afterwards, moreover, James himself would be remarking how little he could find to alter in the state of England, 'which state, as it seemed, so affected his royal heart that it pleased him to enter into a gratulation to Almighty God for bringing him into the promised land where religion was purely professed, where he sat among grave, learned and reverend men, not as before, elsewhere, a king without a state, without honour, without order, where beardless boys would brave him to his face'.

James, of course, could smile at some of the excessive adulation heaped upon him. After losing heavily at cards on one occasion, he commented wryly upon both his predicament and those who would render him near-infallible. 'Am I not as good a king as King David?' he declared, 'As holy a king as King David? As just a king as King David? And why should I then be crossed?' Likewise, he wasted no time in denying the time-honoured claim of English monarchs that they could treat the disease of scrofula merely by touching the affected parts. The age of miracles, he said, was past and he shrank from the ritual whereby the king would place his hands upon the ulcerous sores of those brought before him – a practice, James felt, that savoured far too much of Roman superstition. On one occasion, indeed – 'finding the strength of the imagination a more powerful agent in the cure' – he was actually prepared to belittle the ceremony publicly. When entreated by a foreign ambassador to perform the cure upon his son, the king, it seems, 'laughed heartily, and as the young fellow came near him he stroked him with his hand, first on one side and then on the other, marry without Pistle or Gospel'.

Yet when it came to his new kingdom, James experienced no such doubts about either his powers or his potential or, for that matter, his need to tread carefully, 'inasmuch', wrote one Englishman, 'as he imagined heaven and earth must give way to his will'. As one of his first measures, he did not hesitate, for instance, to install Lord Edward Bruce of Kinloss as Master of the Rolls for life, though the outcry was so great that no other Scot was ever again accorded such high judicial office. More embarrassingly still, even before he had reached London, James attempted unsuccessfully to appoint his old tutor, Peter Young, Dean of Lichfield, notwithstanding the fact that the post was not legally his to give. And these were not the only examples of how his glorious entry into England created what amounted to a curious euphoria that compromised his judgement and heralded the onset of unsettled times ahead. Though the Gowrie Plot was an event of minimal significance south of the Border, the anniversary of its failure, Tuesday 5 August, was nevertheless now set aside in England as well as

Scotland as a holy day of feasting and thanksgiving, while special 'Tuesday sermons', in which, among other things, the Almighty was urged to 'smite the king's enemies upon the cheekbone, break their teeth, frustrate their counsels and bring to naught all their devices' became a regular feature of the court's routine. There was, indeed, even an attempt – albeit unsuccessful – to foist these crude examples of royal propaganda upon both of the kingdom's universities.

Plainly, then, James's limited understanding of his new realm was exacerbated by the false impressions with which he was bombarded upon his arrival. Now, of course, he was free forever from the perpetual fear of kidnapping and assassination, from contending with clan feuds and the individual jealousies of Scottish noblemen, from the maddening pretensions of the Kirk, and the galling sense of unimportance in the great world of European politics that had been his lot hitherto. But far from entering the 'land of promise' – a land described by a Venetian nobleman in 1596 as 'the most lovely to be seen in the world, so opulent, fat and rich in all things, that you may say with truth poverty is banished' – James did not perceive the broader picture. From at least one perspective, of course, he cannot be wholly blamed for this, since England was not only in a state of latent tension, but also in the midst of a series of rapid and painful transitions. Unbeknown to any Italian visitor, it seems, 1596 was also a year of spiralling inflation, widespread agricultural distress, a crippling slump in trade and a heated dispute between Queen Elizabeth and the impoverished south coast ports over the levy of ship money. For even if England was from some perspectives a comparatively rich and splendid inheritance, the Crown itself was poor. Equally importantly, Elizabethan government was founded upon an intricate system of balances and compromises between ruler and ruled that could only be 'caught rather than taught'.

Yet James had neither the time nor, more importantly still, sufficient inclination to learn anew the subtleties of kingship. 'As the king is by nature of a mild disposition and has never really been happy in Scotland,' wrote the Venetian ambassador, 'he wishes now to enjoy the papacy, as we say, and so desires to have no bother with other people's affairs and little with his own. He would like to dedicate himself to his books and to the chase and to encourage the opinion that he is the real arbiter of peace.' But if James desired the Olympian status of a pope in his new kingdom, it would come at a cost, for, though his judgement was often sound, it was sometimes dangerously rash at critical moments. More importantly still, the personal style of government he unwaveringly pursued was only likely to succeed, insofar as the reigning monarch possessed the necessary charisma and majesty, and commanded the love of his or her subjects. Intelligence, high principles and good intentions were actually no more a guarantee of sound leadership at the start of the seventeenth century than they are today. And it was now – after the welcoming, the junketing and the initial friendly posturing were finally done – that James would have to set his stamp upon his restive realm.

# 11 ✣ The King, His Beagles, His Countrymen and His Court

'The English are for the most part little edified with the person or with the conduct of the king and declare openly enough that they were deceived in the opinion they were led to entertain of him.'

*Christophe de Harlay, Comte de Beaumont, French ambassador to England from April 1603 to November 1605*

Though James knew little of England's laws and parliament, he was well equipped to grasp the elements of the struggle for power at Whitehall and the subtleties by which his predecessor had managed to maintain a fragile balance of forces around her council table. The enmity between Cecil and Raleigh, for example, was in any case certainly less noisy than the kind of knuckleduster fuming he had been forced to contend with in Scotland, and he had been kept closely informed of events by the letters of both Cecil and Henry Howard. Moreover, his opening moves on the broader front were wisely non-committal. On the one hand, he at once provisionally confirmed the existing council in office, while choosing to release Lord Southampton and Sir Henry Neville from the Tower, where they had been languishing in the aftermath of the Essex rebellion. As a further gesture towards healing old wounds and offering new beginnings, he also announced his intention of bringing up Essex's heir in his own household – restored in blood and title, and reared in the companionship of Prince Henry. And those who had in any way supported his mother's cause were to be brought back to favour and thus bound to the new dynasty alongside their former enemies.

In the meantime, the immediate shape of the new king's government had been decided on 3 May at Theobalds when he stopped at the home of Sir Robert Cecil on the final leg of his journey from Edinburgh to London. It was there that he had withdrawn with Cecil to a 'labyrinth-like garden, compact of bays, rosemary and the like' for an hour's intimate conversation to confirm the latter's primacy and seal, in the process, the rather more disconcerting triumph of Henry Howard – soon to become Earl of Northampton – and his sailor nephew, Thomas, who was swiftly promoted to the earldom of Suffolk. Charles Howard, too, who had commanded the English fleet against the Spanish Armada as Lord Howard of Effingham, duly retained the office of Lord Steward of the Household under his new title of Earl of Nottingham. In James's view, it would have been the ultimate folly to discard those very men who had so strikingly demonstrated their level-headed competence in securing his succession, and who appeared to embody so strikingly all that typified Elizabethan wisdom and prestige. It was only natural, too, that five of his loyal Scottish lieutenants – Lennox, Mar, Home, Elphinstone and Edward Bruce, Lord Kinloss – should join the reconstituted council, since the court at Edinburgh had effectively ceased to exist, though for Sir Walter Raleigh and his allies, against whom Cecil and Howard had been so successfully poisoning the king's mind, there was to be no crumb of comfort. Indeed, on 15 April James dismissed Raleigh as captain of his Guard, without financial compensation, and ordered him to leave Durham House in the Strand, which had been provided by the former queen for his private use over twenty years.

Dark, saturnine and colossally proud, the 50-year-old Raleigh possessed a swagger that, in spite of his undoubted brilliance, could never endear him to one such as James, who regarded him purely as a reckless old pirate, opposed at any price to peace with Spain. To others he was a 'Macchiavellian' and an 'atheist', but if such terms bore no relation to his actual views, he was certainly no judge of character – and nowhere more so than in the case of the new king. To present James so early on with *A Discourse Touching a War with Spain and of the Protecting of the Netherlands* merely confirmed Elizabeth I's conviction that her favourite was no statesman, and the king lost little time in attempting to put the upstart in his place. When Raleigh presented himself before James at Burghley House, for example, he was merely treated to the kind of clumsy putdown that was the king's stock in trade. 'Rawly, Rawly,' James declared upon their meeting, 'and rawly ha'e heard of thee, mon'. Before long, moreover, the former royal favourite had lost not only the captaincy of the royal guard but the governorship of Jersey, the lord wardenship of the Stanaries and his monopoly on the sale of sweet wines. All in all, it may well have been no more than Raleigh's presumption merited, but it was far more than one such as he could be expected to settle for passively. And, surely enough, this particularly glittering star of a bygone Elizabethan age would neither forgive nor forget.

Yet the flipside of Raleigh's eclipse was the triumph of an altogether more accomplished politician. 'The evidence of a king,' James himself observed, 'is chiefly seen in the election of his officers', and in Sir Robert Cecil, at least, he had acquitted himself most favourably, notwithstanding the fact that the two men had precious little in common. Wholly unlike his new master, the principal Secretary of State stood, in fact, for calculated dignity and restrained decorum. And though he would be able occasionally to share a recondite joke with the king, the rest of their relationship would be largely artificial as his grave and careful judgement was brought to bear upon even the lightest or most minute details. If James wished to tease him clumsily on his puny figure and address him as his 'little beagle', this was a small price to pay for maintaining the reality of power in his own hands, and he was usually more than capable of enduring the king's badinage under an umbrella of urbanity and stoical self-assurance. For there was a gravity and air of civilisation about Cecil that placed even a long-serving king in awe of him – especially a King of Scotland whose provinciality was inclined to surface so frequently. Perhaps, indeed, the very banter that James directed Cecil's way was itself a product of his own innate unease in the minister's presence. Over time, however, the king's awe and sense of debt would turn to frustration against the man who strove so doggedly to curb his impulsiveness and extravagance. Certainly, Cecil's death in 1612 would be a source of considerable relief and emancipation for James – though for the time being that prospect seemed as distant as the notion that the king could somehow cope without him.

As Principal Secretary, Master of the Wards and later Lord Treasurer, the diminutive minister remained until his death the pivot around which the entire machinery of government revolved, controlling all foreign affairs, directing Parliament, and managing every aspect of finance. Within weeks of James's accession, moreover, he had been created Lord Cecil of Essendon before becoming Viscount Cranborne in 1604 and Earl of Salisbury a year later. No living king, James openly acknowledged, 'shall more confidently and constantly rely upon the advice of a councillor and trusty servant than I shall ever do upon yours'. 'Before God,' the king declared, 'I count you the best servant that ever I had – albeit you be but a little beagle.' Such high praise could not, it seems, be delivered by James without a final tweak at Cecil's dignity. But the other key members of the council, which included the courtier earls of Shrewsbury and Worcester, the buccaneer-courtier Cumberland, Mountjoy, who returned from Ireland to become Earl of Devonshire, and the Earl of Northumberland, to whom the king gave credit for the good behaviour of the English Catholics, all accepted Cecil's primacy without question. The same was true for the new Scottish additions and, more importantly still, the ubiquitous Howards, Henry and Thomas, who, along with Worcester, a devout but utterly loyal Catholic, soon formed an inner circle of four on the council with Cecil himself.

But in dominating all areas of government and attempting to bridle the king's impulsiveness as best he could, the secretary would also stretch his physical resources to the limit by almost ceaseless work. As month after month of confusion and uncertainty followed the change in dynasty, Cecil would bemoan to his friend, Sir John Harington, the passing of the former queen. 'I wish I waited now in her Presence Chamber,' he wrote, 'with ease at my food and rest in my bed. I am pushed from the shore of my comfort and know not where the winds of the court will bear me.' 'I know it bringeth little comfort on earth …' he added, before appending his signature 'in trouble, hurrying, feigning, suing and such-like matters'. It was small wonder, then, that Shrewsbury was soon warning Cecil that he would 'blear out his eyes' and 'quite overthrow' his body without some form of respite. For although the king's letters to his principal secretary demonstrate his shrewdness and ability, as well as his capacity to grasp a situation readily and reach a swift decision, he was also disinclined to master details and prone to lapses of concentration – especially when his devotion to hunting and recreation took priority over all else. 'He seems to have forgotten that he is a king,' wrote the Venetian ambassador, 'except in his kingly pursuit of stags, to which he is quite foolishly devoted.'

The room where Mary Queen of Scots gave birth to the future James I of England in 1566. During his only return visit to his homeland in 1617, James ordered that the room be preserved and much of what is now seen is decoration dating from his reign. *(Dave and Margie Hill)*

At the age of 7, James's features bore a close resemblance to those of his father, as this portrait by Arnold van Bronckhorst clearly shows. In particular, the young king's wide-set eyes leave little doubt of his parentage, while his passion for hunting (which lasted throughout his life) is also emphasised by the bird of prey upon his left hand.

Portrayed here by Arnold van Bronckhorst shortly before his death at the age of 76, the tutor to the future James I enjoyed an outstanding reputation, both as a humanist scholar and historian of his native Scotland. Buchanan's rigorous methods would leave lasting imprints upon his most famous pupil's intellect and personality.

The last of the four regents of Scotland during the minority of King James VI, the Earl of Morton was arguably more successful than any of his predecessors, since he concluded the civil war that had been dragging on with the supporters of the exiled Mary, Queen of Scots. Yet his authoritarian rule earned him many enemies and an untimely end when he was beheaded by means of the 'Maiden', a primitive guillotine, which he himself is said to have introduced to Scotland.

IACOBVS · 6 · D · G · R ·
SCOTORVM
ÆTA · 29 ·
1595 ·

In 1584 Adrian Vanson succeeded Arnold van Bronckhorst as official painter to the Scottish Court. Aside from overseeing every aspect of royal painting, including banners used for the coronation of Anne as Queen of Scotland in 1590, Vanson and his studio would have been responsible for the production of the king's official likeness, such as the present portrait, either for prominent supporters, or for foreign courts.

The king's ongoing interest in witchcraft assumed a new intensity in the early 1590s after evidence had appeared to connect Francis Stewart, 5th Earl of Bothwell, with a diabolical attempt upon his life. Here the North Berwick witches, who had allegedly been enlisted by Stewart, meet the Devil in their local kirkyard.

This portrait from 1612 by Marcus Gheeraerts the Younger depicts James I's wife in mourning for her eldest son Henry Frederick, Prince of Wales, who had died from typhoid fever earlier that year.

Potentially the most able of the Stuart line, Prince Henry would not live to succeed to the throne. 'His body was so faire and strong,' declared his chaplain after the young prince's death at the age of 18, 'that a soule might have been pleased to live an age in it.'

**Above**: Between 20 May and 16 July 1604, eighteen conference sessions were held at Somerset House, leading to the Treaty of London, which was signed on 16 August and concluded almost twenty years of warfare between England and Spain. On the left are the members of the Hispano-Flemish delegation, on the right, the English commissioners, which, reading from the far end, included the Earl of Dorset, the Earl of Nottingham, Lord Mountjoy, the Earl of Northampton and Robert Cecil.

Robert Carr was a Scottish noble who met James in 1607 and very quickly rose to a position of considerable authority at court. As the premier royal favourite, Carr received a steady flow of gifts in the form of cash, land and titles, until petulance, jealousy and a scandalous marriage to Frances Howard eventually soured his relationship with the king.

King James's personal relationships are much debated, with George Villiers the last in a succession of handsome young favourites who were lavished with affection and patronage. In 1617, John Oglander wrote that he 'never yet saw any fond husband make so much or so great dalliance over his beautiful spouse as I have seen King James over his favourites, especially the Duke of Buckingham'. Edward Peyton, meanwhile, would note how 'the king sold his affections to Sir George Villiers, whom he would tumble and kiss as a mistress'.

**Left**: On 31 January 1606 Robert Keyes, Ambrose Rookwood, Thomas Wintour and Guy Fawkes were taken to the Old Palace Yard in Westminster to be hanged, drawn and quartered. Rookwood and Wintour were the first to ascend to the gallows. Grim-faced, Keyes went 'stoutly' up the ladder, but with the halter around his neck he threw himself off, presumably hoping for a quick death. The halter broke, however, and he was taken to the block to suffer the remainder of his sentence.

By his mid-50s, James had become increasingly sickly and careworn. A 'cradle king' from the age of 1, he had at various times encountered danger and disrespect, and now in his twilight years he was left to endure progressive physical decline and disillusionment.

Certainly, James's dislike of London, 'that filthy town', and Whitehall made it easy enough to abandon the seat of government at every opportunity – convenient or otherwise. Indeed, he would spend only about a third of the year in London, with predictable consequences. 'This is the cause of indescribable ill-humour among the king's subjects,' wrote the Venetian Molin, 'who in their needs and troubles find themselves cut off from their natural sovereign and forced to go before the council, which is full of rivalry and discord and frequently is guided more by personal interests than by justice and duty.' By 1 December 1604, indeed, Molin was reporting that posters had been fixed up in various locations around the capital, complaining that the king attended to nothing but his pleasures and left all to his ministers. Nor were such accusations by any means entirely groundless. Within two months of his accession James had issued a proclamation against illegal hunting in the royal forests and, with not a little sophistry, was soon attempting to justify his frequent absences. 'The king [...] finds such felicity in that hunting life,' wrote Chamberlain in January 1605, 'that he hath written to the council that it is the only means to maintain his health (which being the health and welfare of us all) he desires them to undertake the charge and burden of affairs.' In accordance with the king's wishes, the Privy Council issued orders prohibiting anyone other than members of the royal family from hunting with hounds within four miles of London, and ambassadors were rarely received at Royston in the way they had formerly been at Falkland during Elizabeth's time, since her successor did not like to be distracted while taking his leisure. 'He hath erected a new office and made Sir Richard Wigmore marshall of the field, who is to take order that he be not attended by any but his own followers, nor interrupted and hindered in his sports, by strangers and idle lookers on,' Chamberlain would report in 1609. And James's neglect of government was not the only cause for concern, since Lord Treasurer Buckhurst was soon uneasy, predictably, about excessive allowances for the royal buckhounds.

There was, moreover, much about James's actual hunting practice that made even contemporaries uneasy – his vindictive fury in pursuing and slaughtering the game, his dabbling in the dead animals' blood once the kill had been made, his rage when the quarry escaped, his low company and bad manners at the chase. 'Running hounds', as the king called them, would bring down and kill the stag, after which the king would dismount, cut the stag's throat and open its belly, thrusting his hands (and sometimes his feet) into the stag's entrails and daubing the faces of his courtiers. This, however, was the least offensive aspect of James's behaviour to local farmers and villagers who found themselves tormented by the indiscriminate rampaging of the king and his courtiers. 'As one that honoureth and loveth his most excellent Majesty with all my heart,' wrote the Archbishop of York to Cecil during December 1604, 'I wish less wastening of the treasure of this realm, and more moderation in the lawful exercise of hunting, both that poor men's corn be less spoiled, and other his Majesty's subjects be more spared.'

But James's delight in the 'sporting' destruction of animals was not an obsession he was willing to limit. Hawking, cockfighting, bull- and bear-baiting were all passions, and the bear-garden at Southwark, known as Paris Garden, which also housed lynxes and tigers, was owned by him personally. The Earl of Dorset, in fact, was 'brought into great grace and favour with the king', because of his love of cock-fighting. In addition, James kept cormorants to dive for fish and by 1605 had developed, according to Chamberlain, 'a great humour of catching larks'. A lion-baiting pit was also constructed at the Tower, and on one occasion, accompanied by young Prince Henry and several lords, he 'caused the lustiest lion to be separated from his mate, and put into the lion's den, one dog alone, who presently flew to the face of the lion, but the lion suddenly shook him off and grasped fast by the neck, drawing the dog upstairs and downstairs'. This, however, was not the end of proceedings, since the king 'now perceiving the lion greatly to exceed the dog in strength, but nothing in noble heart and courage, caused another dog to be put into the den, who proved as hot and lusty as his fellow and took the lion by his face'. And though this dog died, too, another one later recovered from its wounds, encouraging Prince Henry to order his servant Alleyne to look after the beast well, saying that since he had 'fought with the king of beasts', he 'should never after fight with any inferior animal'.

Not all such staged encounters were so predictable, however, since lions, it emerged, were generally unwilling to fight for entertainment purposes. Indeed, after a bear had killed a child and James arranged for the offending creature to be baited by a lion, the result was an embarrassing anti-climax in which the king of beasts chose to withdraw from the encounter without so much as a bared tooth or exposed claw. Nor, for that matter, were all the king's involvements with animals so violent. He was a keen patron of horse-racing, for instance, making it a royal sport after swift-footed

Spanish horses had been thrown ashore on the coasts of Galloway at the time of the wreck of the Armada, and establishing several race tracks, the most important of which was located at Newmarket. He also appears to have indulged a particular fascination for more exotic creatures, including Indian antelope and a flying squirrel from Virginia, as well as two young crocodiles, which were presented to him by Captain Christopher Newport after his journey to Hispaniola. The Prince of Orange sent him a tiger, while the Duke of Savoy presented him with another, as well as a lioness and a lynx – all of which died in transit. In 1623, meanwhile, King Philip of Spain would give James five camels and an elephant. At first, the camels were left to graze in St James's Park until stables were constructed for them at Theobalds, and though the elephant was for some reason kept from public view, its captivity was enlivened by a gallon of wine daily from September to April – a period during which its keepers considered it unable to drink water.

The chase, his 'sport', his curious pets, his love of country life were all, plainly, welcome distractions from the more tedious routines of everyday government. But even if the king's frequent excursions really were good for his health and thus the health of his kingdom too, the burden imposed on others – and not only Cecil – was now considerable. 'This tumultuary and uncertain attendance upon the king's sports affords me little time to write,' confided one official, while George Home, Earl of Dunbar, bemoaned the fact that he and his colleagues 'are all become wild men wandering in a forest from the morning till the evening'. But it was the Earl of Worcester, perhaps, who best encapsulated the problem. 'Since my departure from London,' he noted, 'I think I have not had two hours of twenty-four of rest but Sundays, for in the morning we are on horseback by eight and so continue in full career from the death of one hare to another until four at night; then for the most part we are five miles from home. By that time I find at my lodging sometimes one, most commonly two packets of letters, all of which must be answered before I sleep, for here is none of the council but myself, no, not a clerk of the council nor Privy Signet.'

Such packets of letters came down from the councillors in London daily and Salisbury sometimes sent messengers to talk with the king and report his wishes. But the dispatches from the capital might well be answered by any ranking official at hand, and it frequently fell to Sir Thomas Lake, who was in constant attendance upon James, to set down his master's thoughts and instructions after snatched consultations in the limited time available. Most often the king could be found at one of three houses – Royston, which he bought in 1604, Theobalds, which he acquired from Cecil in 1604 in exchange for Hatfield, and Newmarket – all of which were linked to London by private roads that he maintained for the purpose. But the council's business would often have to make its way to Thetford, Hinchingbrooke, Ware and Woking, too, rather than Windsor and Hampton Court, the two principal country palaces at the time of James's accession. And the expense as well as the inconvenience mounted further through the duplication of stables, deer parks, offices and living quarters, as even local farmers began to complain increasingly about the excessive burden imposed by the king's purveyors. When Jowler, one of James's favourite hounds, was lost near Royston in 1604, it was returned next day with a letter round its neck. 'Good Mr Jowler,' it read, 'we pray you speak to the king (for he hears you every day, and so doth he not us) that it will please his Majesty to go back to London, for else the country will be undone; all our provision is spent already and we are not able to entertain him longer.'

Meanwhile, the broader disruption caused by James's inattention to state affairs during the short periods he spent at Whitehall may well be imagined. Still inclined to rush in the way that George Buchanan had noticed so many years earlier, he could often exasperate those who had his best interests at heart. In 1603, for instance, Wilbraham noted that when Cecil presented James with patents for eight barons and two earls, 'the king signed them all at one time confusedly, not respecting who should have antiquity'. There were times, too, when haste could turn to outright flippancy – especially, it seems, when it came to the practice of dubbing knights, which, in his case, became such a common chore. On one occasion, Wilbraham tells us, he did not catch the long Celtic name of a Scot who was kneeling before him to receive his honour. 'Prithee,' James is said to have declared, 'rise up and call thyself Sir What Thou Wilt.'

Predictably, too, there was great irregularity in James's availability for more general business, with the result that suitors 'swarmed about his Majesty at every back gate and privy door, to his great offence'. Indeed, it was precisely because of the first Stuart's constant irritation at his loss of privacy, that Charles I soon decreed after his accession that suitors 'must never approach him by indirect means, by back stairs or private doors leading to his apartments, nor by means of retainers or grooms of the chambers, as was done in the lifetime of his father'. Rather than taking steps to deal with the problem himself, however, James had merely railed against his predicament, so that by the end of the reign he was still bedevilled by the same problem that his own informality had done so much to spawn. 'The king is not as her Majesty is,' Thomas Fowler had written upon a visit to James while Queen Elizabeth was still alive, 'any of his subjects being gentle or noble may speak his mind frankly to him'. And his inability to project a truly kingly aura now left all and sundry to pester him as they pleased. 'The king is much disgusted with it,' wrote a courtier in 1623, 'but knows not how to help it.' While he protested in private, moreover, he continued to think that no man was sincerely bound to him unless by a gift, which merely served to swell the throng – even confiding to Cecil on one occasion that he had been so prodigal in rewarding persons who had no claim upon him that he could not justly deny those who had.

And when Robert Cecil was not bailing his master out, he would continue to find himself bearing the tiresome burden of his master's deprecating familiarity for his trouble. In the king's own words, the tireless earl remained the little 'beagle' that 'lies at home by the fire when all the good hounds are daily running on the fields'. On another occasion, James wished that 'my little beagle had been stolen here in the likeness of a mouse, as he is not much bigger, to have been partaker of the sport which I had this day at hawking'. Even when weightier matters were involved, the tireless theme was maintained. In planning an embassy to France, for example, the king could not forego the opportunity to tell Cecil of his intention to dispatch a kennel of little beagles across the Channel and to ask him 'if ye mind to be of that number'. Rarely did the secretary bridle at either his workload or his nickname, though he certainly regretted the latter in particular. 'I see nothing that I can do,' he complained to Sir Thomas Lake, 'can procure me so much favour as to be sure one whole day what title I shall have. For from Essendon to Cranborne, from Salisbury to Beagle, from Beagle to Thom Derry, from Thom Derry to Parrot, which I hate most, I have been walked as I think by that I come to Theobalds I shall be called Tare or Sophie.'

Certainly, it is hard to escape the conclusion that there was a distinct element of unseemliness, if not outright spite, at the heart of much of James's playfulness, which extended to other confidants as well as Cecil. In 1605, on the eve of a visit to Greenwich, the king issued an apparently friendly warning to his most trusted councillors. 'If I find not at my coming to Greenwich,' he declared, 'that the big Chamberlain [Thomas Howard] have ordered well all my lodging, that the little saucy Constable [Cecil] have made the house sweet and built a cockpit, and that the fast-walking keeper of the park [Henry Howard] have the park in good order and the does all with fawn, although he has never been a good breeder himself, then I shall make the fat Chamberlain to puff, the little cankered beagle to whine, and the tall black and cat-faced keeper to glower.' Elsewhere, James pretended to be suspicious of their relations with his wife. 'I know Suffolk is married,' he writes, 'but for your part, master 10 [Cecil], who is wanton and wifeless, I cannot but be jealous of your greatness with my wife.' As for Northampton, meanwhile, 'who is so lately fallen into acquaintance with my wife … his part is foul in this, that never having taken a wife to himself in his youth, he cannot now be content with his grey hairs to forebear another man's wife'.

There were times, of course, when James acknowledged his servants' efforts. 'My little beagle,' he wrote on one occasion, 'although I be now in my paradise of pleasure, yet will I not be forgetful of you and your fellows that are frying in the pains of purgatory in my service.' 'Your zeal and diligence are so great,' he observed, 'as I will cheer myself in your faithfulness and assure myself that God hath ordained to make me happy in sending me so good servants, and the beagle in special.' But he was also quick to point out to the Archbishop of York that the absences so necessary for the maintenance of his health took up no more time than other kings spent upon feasting and visiting their whores. Besides which, such fair-weather praise as James sometimes deigned to offer his servants carried little substance when mixed with sneers and tantrums. 'Ye sit at your ease and direct all,' he would tell his ministers, while 'the king's own resolutions depend upon your posting dispatches, and when ye list, ye can (sitting on your bedsides) with one call or whistling in your fist make him to post night and day till he comes to your presence'. To such unthinking outbursts, the only sensible response was perhaps Cecil's, who indulged his master by flattery, begging James to send him copies of his writings or telling him how a kind word from the royal pen had cured a recent illness.

The court, meanwhile, was soon increasing in lavishness and cost while degenerating rapidly in orderliness and tone. According to the Venetian ambassador, it was soon fashionable for even the lesser nobles and councillors to appear in public with forty or fifty horsemen and sometimes with 200 to 300, so that 'the drain on private purses is enormous'. Vulgar ostentation became the order of the day, in fact, as great ladies sported ever more costly jewels at state functions and even the wives of ambitious civil servants would think it worthwhile to spend £50 a yard on the trimming of a dress. The value of presents became a subject for calculation and haggling too, it seems, for when de Beamont, the French ambassador, was recalled to his homeland in 1605, he complained bitterly that his parting present of plate weighed only 2,000 ounces. Quoting precedents and whingeing continually, he eventually received 500 ounces more. And when perquisites and commissions of this kind were the norm and even minor offices were for sale, the opportunities for corruption became legion – so much so that open peculation ran throughout the king's own household and, much worse still, every department of state.

The newly prevailing atmosphere of expenditure for expenditure's sake would, however, be epitomised most strikingly by the excesses of Sir James Hay who had arrived from Scotland with James at the time of the succession to be rapidly installed as a 'prime favourite' and gentleman of the bedchamber. A man of accommodating temper and good sense in most things, Hay was not without some diplomatic ability. But his extravagance and lavish expenditure, his costly entertainments and so-called 'ante-suppers' – at which banqueting tables were laden with splendid food that was then discarded without being consumed before the main, even more dazzling, dishes were brought in - became the theme of satirists and wonder of society, as his debts spiralled to more than £80,000. He left behind him, said the Earl of Clarendon, 'the reputation of a very fine gentleman and a most accomplished courtier, and after having spent, in a very jovial life, above £400,000, which upon a strict computation he received from the Crown, he left not a house or acre of land to be remembered by'. He had, wrote Clarendon, 'no bowels in the point of running in debt, or borrowing all he could' and 'was surely a man of the greatest expense in his person of any in the age he lived'.

Nevertheless, it was the personal extravagance of the king that fed as well as reflected this culture of brash display and conspicuous consumption. The number of gentlemen of the bedchamber, of gentlemen, ushers and grooms of the privy chamber and of the presence chamber, of carvers, cupbearers and sewers, of clerks of the closet and esquires of the body, of harbingers, yeomen, pages and messengers increased in line with the king's desire to advertise his exalted status. The gentlemen of the privy chamber, for instance, rose in number from eighteen to forty-eight, each with a fee of £50 a year, and before long there were 200 gentlemen extraordinary. Wardrobe costs, on the other hand, which averaged approximately £9,500 per annum in the last four years of Elizabeth's reign leapt to over £36,000 per annum in the first five years of James's. According to the records of the treasurer of the chamber, meanwhile, the cost of court ceremonial rose from £14,000 under Elizabeth to £20,000 under her successor. No man, it was soon being said, ate at court without costing the king £60 a year, while even the charge for one of the seamstresses or laundresses that teemed around the court was estimated at £86. And both the king and his queen spent extraordinary sums on jewels: James expending some £92,000 in the first four years of his reign alone, while Anne would run up a bill of £40,000 over ten years with the jeweller George Heriot. Only one year into the new reign, therefore, a royal commission had been appointed in the vain hope of curtailing household expenses.

In particular, the king's generosity to his favourites became a matter of widespread concern. 'I hear,' Chamberlain would write to Carleton in February 1607, at a time when all the City knew that money was low in the Exchequer, 'the king hath undertaken the debts of the Lord Hay, the Viscount Haddington and the Earl of Montgomery to the value of four and forty thousand pounds, saying that he will this once set them free, and then let them shift for themselves. In the meantime his own debts are stalled to be paid the one half in May come two years, the residue in May following.' Haddington was the same John Ramsay to whom James believed he owed his life on the day of Gowrie's conspiracy and therefore, arguably, had a special claim upon the king's generosity. But even in his case, let alone Hay's and Montgomery's, the merchants and tradesmen who had been left to wait over two years for money already long owing could well feel aggrieved – especially when James would go on, within twelve months, to mark Haddington's marriage with even more largesse. For along with the gold cup in which the king had drunk her health, the bride was also given not only 'a basin and ewer, two livery pots and three standing cups all very fair and massy, of silver and gilt' but a joint annuity of £600 out of the Exchequer to enjoy with her husband. Five years later, in the midst of an even worse financial crisis, James would give the Countess of Somerset £10,000 worth of jewels as a wedding present. And by 1610, a little over £220,000 had been given away in hard cash, with pensions granted by the king amounting to a further £30,000 a year. In other words, presents accounted for more than a quarter of James's total indebtedness, while annuities swallowed around 6 per cent of his yearly expenditure. The fact, moreover, that so many of his gifts had been lavished upon Scotsmen when the Scottish court itself had boasted only a 'small, unkingly household' merely added fuel to the fire.

For Englishmen, indeed, the transformation of the court into a hybrid entity, heavily and undesirably infested with barbaric foreigners, had become a cause of rancour and suspicion from the very moment of James's succession. The king, of course, had appointed his countrymen to bedchamber posts, paid their debts and proceeded to surround himself with them, and though he did so for good reason from some perspectives, there was little appreciation in the kingdom at large of either his motives or his predicament. It was only natural, of course, that he should want his friends around him, in court and elsewhere, and his friends in the first instance were inevitably Scots. Much more importantly still, however, there was now only one royal court, where before there had been two. In such circumstances, a hybrid court, accommodating both Scots and English was inescapable. But the former were swiftly branded greedy and uncivilised thugs, and the king would not help the situation by declaring to his first Westminster Parliament that Scotland, unlike England, had never been conquered by any outsider.

A key issue, predictably, was access to the king which appeared initially to be dominated by the newcomers. 'No Englishman, be his rank what it may, can enter the Presence Chamber without being summoned,' wrote the Venetian ambassador in May 1603, 'whereas the Scottish Lords have free entrée of the privy chamber.' But the greatest resentment continued to spring from the general perception that the Scots had a virtual monopoly on the royal bounty. In the parliamentary debate on purveyance in 1606, one speaker would describe the treasury as 'a royal cistern, wherein his Majesty's largesse to the Scots caused a continual and remediless leak'. And such hostility was ongoing. 'The Earl of Dunbar,' wrote Sir Dudley Carleton in 1608, 'is returned out of Scotland with a new legion of Scots worse than the former,' while in 1610 their influence upon James was once again being blamed for the kingdom's financial problems – this time in Parliament. 'The court is the cause of all,' declared Sir John Holles, 'for by the reception of the other nation that head is too heavy for this small body of England ... The Scottish monopolise his princely person, standing like mountains betwixt the beams of his grace and us.'

Even Cecil, for that matter, now found his operations complicated by the king's involvement with his countrymen. 'It fareth not with me now as it did in the queen's time,' he confided to Sir Henry Yelverton who had been forced to seek help from the Earl of Dunfermline and Dunbar after offending James, '... for then I could have done as great a matter as this without other help than myself; she heard but few, and of them I may say myself the chief; the king heareth many, yea, of all kinds.' Nor would James's attempts to amalgamate the Scottish and English upper classes by marriage always prove successful. When Sir John Kennedy did not get along with his wife, a daughter of Lord Chandos, she took up with Sir William Paddy, the king's physician, and triggered an explosive response. 'You have heard, I am sure,' wrote Dudley Carleton to John Chamberlain in 1609, 'of a great danger Sir William Paddy lately escaped at Barn Elms, where the house was assaulted by Sir John Kennedy by night with a band of furious Scots, who besides their warlike weapons came furnished ... with certain snippers and searing irons, purposing to have used him worse than a Jew, with much more ceremony than circumcision.'

As early as February 1604, James had made a desultory attempt to bar Scottish officials and noblemen from venturing south, on the grounds that the administration of Scotland would suffer by their flocking to London. In 1607, moreover, he went on to make an outright apology to Parliament for his excessive largesse to his countrymen. 'My first three years were to me as a Christmas. I could not then be miserable,' he pleaded. 'Should I have been over-sparing to them, they might have thought that Joseph had forgotten his brethren.' More significantly still, however, the hanging of Lord Sanquhar for the murder of an English fencing-master who had accidentally put out his eye finally convinced many that James was determined to be even-handed. And though hatred of the Scots did not diminish in the country as a whole, at court at least there was a gradual acceptance that, while Scots were a permanent fixture in office around the king, there was no unwritten Scottish monopoly as such. However obnoxious to English courtiers, the Scottish link had not, therefore, reached truly critical dimensions, and the arrival of the new king had at least brought one enticing consolation for any red-blooded man of his court: the emergence of a new gaiety and hedonism.

Now, indeed, that the restraining hands of Andrew Melville and his ilk had been removed, James could spread his wings in other ways, too, for whatever Whitgift, Bancroft and other leading members of the English episcopacy might think in private about their master's lifestyle, they would never take to the pulpit to compare him with Jeroboam. Court festivities therefore became not only more numerous and extravagant but sometimes more disorderly as well. Of the more innocuous entertainments on offer, however, James particularly enjoyed the spectacle of the elaborate masques, which, with their hugely ornate floats, costumes and scenery, so delighted his wife. In the process, no expense was spared on these elaborate entertainments, which came to occupy a place of special symbolic significance at the Jacobean court. *The Masque of Blackness*, which Ben Jonson and Inigo Jones staged for the queen in January 1605 to celebrate the creation of Prince Charles as Duke of York cost more than £3,000, for example, and typified the new atmosphere. The court, wrote Arthur Wilson, whose history of James's life and reign was published in 1653, became nothing less than 'a continued masquerade, where the queen and her ladies, like so many sea-nymphs or Nereids, appeared often in various dresses to the ravishment of the beholders, the king himself being not a little delighted with such fluent elegancies as made the night more glorious than the day'. At times, it is true, he appeared mildly distracted and uninterested in proceedings, but, on the whole, wherever there was revelry, especially the kind which involved any hint of immodesty in females – in whom he remained more interested than is often suggested – James could be guaranteed to respond with elevated spirits.

The king's attitude to women had never, in fact, been chivalrous or gentlemanly, and though the Scottish court had been an informal, masculine place where little attention was paid to etiquette in any case, there is much about his outlook that remains regrettable. Certainly, if his *A Satire against Woemen* is taken at face value, there seems little doubt that he regarded them as inherently flawed. 'Dames of worthie fame,' he believed, 'are to be congratulated for triumphing over their evil natures, since women of all kinds were inherently vain, ambitious, greedy, and untruthful.' And if his attitude did not actually worsen in England, it nevertheless seems to have provoked more of a reaction. 'He piques himself,' observed the French ambassador, de Beaumont, 'on great contempt for women. They are obliged to kneel before him when they are presented, he exhorts them openly to virtue and scoffs with great levity at men who pay them honour. You may easily conceive that the English ladies do not spare him but hold him in abhorrence and tear him to pieces with their tongues, each according to her humour.'

Yet for all his faults, it was at feasts and banquets that James nevertheless came into his own, rarely missing the opportunity to address the revellers, embrace the gentlemen, kiss the ladies and raise his glass in numerous toasts. In Scotland, the Venetian ambassador observed, the king had 'lived hardly like a private gentleman, let alone a sovereign, making many people sit down with him at table, waited on by rough servants who did not even remove their hats'. Now, however, he dined 'in great pomp' and entertained his guests with 'extraordinary bravery'. His first Christmas in England, for example, was celebrated at Hampton Court where there were feasts in honour of the ambassadors from France, Spain and Poland, involving many plays and 'dances with swords' and a number of masques, in one of which the queen and eleven of her ladies appeared as goddesses bringing gifts to the king. The costumes in each masque, thought Roger Wilbraham, must have cost from £2–3,000, and the jewels £20,000, while those worn by the queen he judged to be worth £100,000. A year later, Sir Dudley Carleton, the courtier and future diplomat described the celebrations at Whitehall when the king's favourite, the Earl of Montgomery, married Lady Susan Vere. 'There was no small loss that night of chains and jewels and many great ladies were made shorter by the skirts,' wrote Carleton. 'There was gaming for high stakes, too, and another grand masque for which the queen and three other ladies were seated in a shell upon a great float carrying figures of sea-horses and other large fish ridden by Moors.' The 'night's work' was then concluded, we hear, with a banquet in the great chamber 'which was so furiously assaulted that down went table and trusses before one bit was touched'.

But by far the most notorious example of overindulgence and debauchery came in the summer of 1606 with the visit of Christian IV of Denmark whose concept of happiness, like that of all his countrymen, began and ended, it seems, with the contents of a bottle. According to Sir John Harington, who provided a vivid description of what transpired for William Barlow, Dean of Chester, 'the sports began each day in such manner as persuaded me of Mahomet's paradise'. 'We had women and indeed wine too of such plenty,' he declared, 'as would have established each sober beholder,' adding that 'the ladies abandon their sobriety and roll about in intoxication'. It was at the performance of the *Masque of Solomon and Sheba*, however, that the boundaries were well and truly crossed. Following a magnificent banquet at which both James and Christian and all their retinues had far too much to drink, things rapidly spiralled out of control.

For the Queen of Sheba fell over the steps as she sought to offer the Danish king a present and deluged him in wine, cream jelly, cakes and spices. And when Christian, having been roughly cleaned up 'with cloths and napkins', then tried to dance with her, both fell down and had to be carried out and put to bed.

James, meanwhile, chose to sit through the rest of the spectacle as things went from bad to worse. Most of the other performers, Harington tells us, 'went backward or fell down, wine did so occupy their upper chambers', while those representing Hope and Faith fared worse still. Hope, it seems, 'did essay to speak, but wine rendered her endeavours so feeble that she withdrew', whereupon Faith, having been abandoned by her partner, 'left the court in a staggering condition'. Nor, it seems, did the more sober efforts of Charity entirely rescue matters, for she returned to the two other virtues, only to find them 'sick and spewing in the lower hall'. 'After much lamentable utterance', the young lady playing Victory was also 'led away like a silly captive and laid to sleep in the outer steps of the ante-chamber', while Peace, 'much contrary to her semblance', appears to have 'rudely made war' against those of her attendants who had attempted to prevent her getting 'foremost to the king'. Using her olive branch, it seems, she 'laid on the pates of those who did oppose her coming'.

Even allowing for Harington's flare for a witty tale and recognising the fact that he owed James a grudge for patronising him insufferably, this was hardly the sort of evening to cause anything but scandal among sober City merchants and taxpayers and in such remote manor houses – of which there was no small number – who might boast a London correspondent to keep them in touch with the gossip of St Paul's. The House of Commons, moreover, had only just voted the king the tidy sum of £400,000 and Harington is unlikely to have been the only man to point the obvious moral that Parliament 'in good soothe', had not stinted in providing the king 'so seasonably with money'. Not only would such stories undermine Robert Cecil's subsequent negotiations with MPs to rescue the king's finances, they would also have a more pernicious broader and longer-term affect. For they marked the beginning of the rift between the Jacobean court and the country at large, which was to be James's worst legacy to his descendants – a legacy which would arguably influence the entire character of politics throughout the century ahead.

And though the excesses witnessed in 1606 may not be considered typical, the junketing continued throughout the reign – with no expense spared. On 8 January 1608, for instance, John Chamberlain, described the kind of gambling session that became a commonplace of court life. 'On Twelfth eve there was great golden play at court, no gamester admitted that brought not £300 at least,' he wrote. 'Montgomerie played the king's money and won him £750, which he had for his labour.' Five years later, the marriage of James's daughter, Elizabeth, to Frederick V Count Palatine of the Rhine was a wonder of ceremonial magnificence even for that exceptionally extravagant age, costing £50,000 and almost bankrupting the Crown in the process. 'This extreme cost and riches,' remarked Chamberlain after describing what the princess wore at her wedding, 'makes us all poor'. And the cost to the government was mirrored in the expenditure of those around the court as clothes became richer – and more revealing – and gentlemen competed in vain against Sir James Hay's unimpeachable reputation as the great spender of his age. On Twelfth Night 1621, according to Chamberlain, the carefree Scotsman organised a feast for the entire court that employed 100 cooks over eight days in the creation of 1,600 dishes, costing over £3,300. It seemed likely, thought Chamberlain, that 'this excessive spoil will make a dearth of the choicest dainties, when this one supper consumed twelve score pheasants, baked, boiled and roasted'. Nor should it be forgotten, of course, that Hay's considerable fortune had come in the first place from the royal bounty – a fact, which made the largesse of a once-beggarly Scot all the more galling in some quarters.

This is not to say, of course, that the king's own 'generosity' was without a wider political purpose. Cecil himself would observe in Parliament in 1610 that 'for a king not to be bountiful were a fault', and such generosity not only tied important men to his cause but coincided with and confirmed notions of the godlike nature of the king. Nor did the Jacobean court lack an altogether more refined aspect, just as one might expect with such a learned occupant of the throne. It was, for instance, a place at which literary activity in particular was heartily encouraged. The king 'doth wondrously covet learned discourse', wrote Thomas Howard to Harington, and while Jacobean censorship may well have been tighter than that of Elizabeth – as Ben Jonson discovered when he found himself in hot water for mocking Scotsmen in *Eastward Ho!* – it was hardly oppressive. Lancelot Andrewes, bishop and scholar, became a particular celebrity at James's court, while Inigo Jones received rich patronage and John Donne was virtually coerced into the clerical career that eventually afforded him such distinction.

And though the king's interest in serious drama was so limited that even the genius of Shakespeare appears to have largely bypassed him; he was nevertheless prepared to provide official encouragement when able. As early as 19 May 1603, for instance, letters patent were issued altering the name of 'Lord Chamberlain's Men' – the theatre troupe of William Shakespeare and Richard Burbage – to the 'King's Men', and allowing them to perform 'as well for the recreation of our loving subjects, as for our solace and pleasure when we shall think it good to see them ... within their now usual house called The Globe, as well as all other boroughs and towns within the kingdom'. In the case of music, moreover, James was prepared not only to offer encouragement but to foot the bill. In 1596 the musicians of the chapel royal had petitioned the old queen for an increase in their stipend and were rebuffed. But what they lacked under Elizabeth, they gained under James and duly received an increase in their annual income from £30 to £40. Ultimately, the number of musicians on the payroll would more than double under James, as did the cost of the musical establishment in general, from £3,000 to £7,000 per annum.

Yet if the arts would flourish in James's time, the king's role was always primarily confined to largely passive recognition of established facts. And while James might easily have made his court so much more of an asset to his rule in England, it is no mere coincidence, perhaps, that the traditional picture of the Jacobean court took root so tenaciously: an endless stream of expenditure on a limitless parade of worthless people; a kaleidoscope of drunken maids of honour and effeminate young men produced by a complacent absentee king relentlessly pursuing deer. Greed, conspicuous consumption and sexual misbehaviour were all familiar features of the Whitehall scene before 1603, of course, and it was not for nothing that moral decay became a prominent theme of Elizabethan as well as Jacobean verse. However, the public face that Henry VIII and Elizabeth succeeded in projecting concealed a great deal of the baser behaviour which permeated their world, in much the same way that their personal gravitas would compensate for their own broader deficiencies. And while they, through careful stage-management and personal charisma, came to represent an ideal, James, for his part, remained an uncomfortable travesty of those very qualities he sought so desperately to embody. He had, of course, been brought up a king of Scots and ultimately his character and qualities had played out well enough upon the Scottish stage. But more than a border divided Scotland from England as James would soon discover. Curiously, therefore, his greatest fault as ruler of England would lie not in his laziness, his prodigality, his political theories or his unworthy favourites, but in something altogether more intangible and all the more irremediable: his plain, straightforward and frequently glaring lack of majesty.

# 12 ✔ Religion, Peace and Lucifer

'I could wish from my heart that it would please God to make me one of the members of such a general Christian union in religion, as laying wilfulness aside on both hands, we might meet in the middest, which is the centre and perfection of all things.'

*From James I's address to his first English Parliament,*
*19 March 1604*

B y the time that James I entered London for the first time on 7 May 1603, amid teeming crowds and lofty expectations, there were already two actively disappointed groups in the country: the long-suffering Catholic community and the motley crew of ambitious men on the fringe of official life who found themselves excluded from office by the triumph of the Cecil–Howard administration in which the king had placed his faith. For the former, in particular, James had seemed to offer fresh hope. Pursuing his dream of healing Europe's religious feuds, the new king had already resumed diplomatic relations with the papacy and drawn a careful distinction between those Catholics who accepted their duty of allegiance to the Crown and those that did not. To the loyal, at least, he was prepared to allow the free exercise of their religion in private, so long as they made no effort to increase their numbers. But even so eminently enlightened an approach to such an intractable problem was already inadequate, since in seeking Catholic support for his succession, James had encouraged still higher hopes that could now be neither satisfied nor extinguished. In 1599, the Scottish Jesuit, Robert Tempest, had boasted how he could produce evidence 'in the king's own hand' that he was a Catholic, and the pope's last informal communication to James before he left Scotland had been an inquiry whether Prince Henry might not be brought up in the faith of Rome. Penal laws and recusancy fines, or so it was widely believed, would soon be a thing of the past, while in Ireland, in particular, the old religion was making a new and strident show of confidence. 'Jesuits, seminaries and friars now come abroad in open show,' wrote one contemporary chronicler, 'bringing forth old rotten stocks and stones of images.'

It was hardly surprising, therefore, that frustrated Catholics would soon be indulging the penchant for conspiracy and intrigue that had already led to a number of spectacular incidents during the reign of Elizabeth – though the first attempt at subversion during James's rule would prove more farcical than threatening. Indeed, the so-called 'Bye Plot' revealed more about divisions within the Catholic priesthood between Jesuits on the one hand and 'secular' priests, who wished to remain loyal to the Crown, than about any mortal threat to the king. In fact, the plotters, led by the secular priest William Watson wished only for recusancy fines to be lifted and for Catholics to be accorded some posts in government – objectives which fell far short of the wishes of the leading English Jesuits, George Blackwell, John Gerard and Henry Garnet, who desired the complete restoration of the old religion and had only acquiesced in James's succession in the first place, since the King of Spain was not at that moment prepared to support opposition with armed force. When Watson laid plans for a Scottish-style kidnapping, therefore, the Jesuit leadership readily betrayed the plot to the authorities, with the intention of striking a telling blow against those 'loyal' Catholics who might interfere with their own more radical intentions. The result was a would-be attempt to 'take away the king and all his cubs' on St John's Day (24 June) 1603, which never materialised, and the subsequent execution of Watson and some of his confederates.

In the course of extinguishing this altogether pallid escapade, however, the government also exposed another, marginally more sinister conspiracy known as the Main Plot, which had apparently overlapped with Watson's intended venture at various points. The motives in this case, insofar as they can be gauged at all, were predominantly political, as men who found themselves frustrated by the continuance in office of Cecil and his friends decided to act. Lord Cobham, Warden of the Cinque Ports, had already attempted to intercept James on his journey south in a vain attempt to forestall the triumph of the Cecil–Howard grouping, and now – with the aid of a number of disaffected individuals including Sir Griffin Markham, a Catholic who already had links with the Bye Plot, and Lord Grey of Wilton, a leading Puritan – he attempted to persuade Count d'Aremberg, the ambassador of the Spanish Netherlands, to finance the landing of a Spanish force in Britain. The intention was to murder the king and his advisers, and to place Lady Arbella Stuart on the throne, after which she would be married in accordance with the wishes of the Catholic rulers of Spain and Austria.

The daughter of James's uncle, Lord Charles Stuart, Arbella had been left parentless early in life and was raised by her grandmother in the grandeur and seclusion of Hardwick Hall. She was a classical scholar whose learning was reputed to extend even to acquaintance with Hebrew and stood in high favour with the king. 'Nature enforces me,' James had written, 'to love her as the creature living nearest kin to me, next to my own children.' But, as a potential replacement for the king, she remained a painfully poor choice of candidate, for not only was she fundamentally loyal to his cause, she was also a Protestant. Under such circumstances, the Main Plot had no effect whatsoever other than to saddle its adherents with a clear-cut charge of treason. Almost incredibly, moreover, one of those implicated in this pro-Spanish daydream was none other than Sir Walter Raleigh whose career had been heroically anti-Spanish throughout the previous reign. Though despised by Cobham, who was the brother of Cecil's long-dead wife, Raleigh was nevertheless Cobham's cousin – and now would pay a heavy price for that connection.

Hitherto, only the personal favour of the former queen had kept Raleigh in office in the teeth of the enmity of the Cecils, both father and son. But the new king was known to favour peace with Spain, and two years earlier, Robert Cecil had informed James that Cobham, who was his own father-in-law, and Raleigh, 'in their prodigal dissensions would not stick to confess daily how contrary it is to their nature to resolve to be under your sovereignty'. Deprived of the captaincy of the royal guard and excluded from court, Raleigh is quite likely, moreover, to have played into his enemies' hands by bemoaning his mistreatment to Cobham, though the latter's evidence against him was later retracted and only delivered in any case at a moment when he had completely broken down. At the same time, Raleigh's passionate denials at his trial in the court room of Wolvesey Castle at Winchester in November 1603 continue to have a decided ring of truth about them.

The main charges were that he had conspired with Cobham 'to deprive the king of his government and advance the Lady Arbella Stuart to the throne', and set out to 'bring in the Roman superstition, and to procure foreign enemies to invade the kingdom' – all of which Raleigh vehemently and convincingly denied. But his protests were vain and his trial a carefully orchestrated sham, which proceeded with all the venom and virulence of a latter-day show trial. Prosecuted remorselessly by Sir Edward Coke, the great champion and exponent of the English Common Law, and denounced as 'a monster' and 'the greatest Lucifer that ever lived', Raleigh remained unwavering. 'Oh barbarous! ...' he declared, 'I was never any plotter with them against my country, I was never false to the Crown of England.' But when he denounced his accusers as 'hellish spiders', Coke's response encapsulated the obvious injustice of the proceedings. 'Thou hast an English face but a Spanish heart,' he retorted, 'and thyself art a spider of hell. For thou confesses the king to be a most sweet and gracious prince, and yet thou hast conspired against him.'

The guilty verdict delivered by Lord Chief Justice Popham that inevitably followed convinced nobody and guaranteed, ironically, a wave of sympathy for Raleigh that would hardly have been forthcoming otherwise. Indeed, it remains questionable whether most of the others implicated in the Main Plot, including Cobham and Grey, also needed to have been treated in quite the way that they were. Characteristically, James dissected the moral niceties involved with an intensity that ultimately cut common sense – and common decency – to pieces. It was to Englishmen one of the most surprising things about their new ruler that, in his overwhelming urge to demonstrate both his intellect and zeal for justice, he would go to such tangled lengths to unravel the truth – and think out loud in doing so. In this case, predictably, he was torn between his duty to make an example of traitors and his urge to be merciful. 'To execute Grey, who was a noble young spirited fellow,' he reasoned, 'and save Cobham, who was base and unworthy, were a matter of injustice.' However, 'to save Grey, who was of a proud insolent nature, and execute Cobham, who had showed great tokens of humility and repentance were as great a solecism.' Crucially, however, even when James eventually opted for clemency, he did so in a manner that not only robbed the act of much of its virtue but demonstrated a degree of spite that sat ill with the principles of justice he claimed to personify. For Grey and Cobham, along with Sir Griffin Markham were each brought to the scaffold twice before their reprieve was finally read to them – a decision, in Cecil's words, to which James had 'made no soul living privy, the messenger excepted'.

In Raleigh's case, meanwhile, the king's callousness was altogether more pronounced. Raleigh remained, of course, an intolerably proud, quick-tempered and violent man of action whom his Elizabethan colleagues had found no more tolerable than James now did, and the policies of war and piracy he stood for were in any case precisely those that the king had made it his mission in life to reverse. But to commute his execution, which had been fixed for 13 December only to consign him subsequently to the Tower and leave him there for thirteen years with a suspended death sentence hanging over him was to show a meanness of spirit which even James's own son found himself unable to condone. None but his father, Prince Henry was to say, would have found a cage for such a bird. And when peace with Spain quickly followed in August 1604, it fell as a further hammer blow to the prisoner who could now look forward, in his own words, only to the 'bribeless judgement hall' of Heaven, where Christ alone would be 'the king's attorney'.

If, however, the conclusion of hostilities with Spain represented the cruellest of cuts for Sir Walter Raleigh, it marked for James merely the first step in realising his most cherished vision of himself as a wise and benevolent mediator-king who, after uniting Scotland and England, would then bring peace to the warring nations of Europe and thereafter seal his illustrious place in history by reconciling that continent's conflicting faiths. Now, after all, he was ruler of Europe's most powerful Protestant state, strengthened by union with Scotland, and he enjoyed a number of advantages derived from his experience to date. His prestige was high, for instance, in Scandinavia and northern Germany, and he was also on good terms with the Catholic states opposed to Spain. To English friendship with France he could add the old tradition of Franco-Scottish alliance, and he had also cultivated Tuscany and Venice. But if he had maintained peace with foreign countries as ruler of Scotland, it was due in no small measure to the remoteness and insignificance of his kingdom. And if he had hitherto dabbled promiscuously across the religious divide with both Catholic and Protestant powers, he would not be able to do so with such impunity in England. Indeed, James's hope from this point onwards that he could be both a champion of Protestantism and friend of Spain would prove not only a cardinal error but a classic demonstration of his belief that he could square circles by force of intellect and goodwill alone.

As early as 1590, in fact, James had sent ambassadors to the German princes to organise a joint threat of economic boycott to force Spain to come to terms. But until the Spanish had finally lost hope of conquering Ireland, and their rebellious Dutch provinces were capable of resistance without foreign aid, all hope of an end to fighting was vain. And these conditions were only to be fulfilled at the very end of Elizabeth's life, leaving her successor, by sheer good luck, the chance to wield the olive branch successfully. One of James's first actions as King of England, therefore, was to recall the letters of marque that had enabled English privateers to prey so greedily upon Spanish commerce under his predecessor, and to order the cessation of all hostilities with Spain on the grounds that since he, personally, had never been at war with that country, he could not become so by inheriting the English Crown. It was specious reasoning of precisely the kind that James specialised in when determined to achieve his own ends. But the Spaniards, too, were ready for peace, and by the end of September a Spanish ambassador, Don Juan de Tassis, was presenting his credentials at Winchester with a view to opening preliminary negotiations.

Since the war had reached a state of utter deadlock, the best that either side could now hope for was merely a cessation of hostilities, and there was little need for prolonged negotiation. Yet there were endless delays, largely on account of what Sir Henry Wotton called 'Spanish gravity sake' and it was not until August 1604 that the Constable of Castile finally crossed the Channel with full powers to sign what would become known as the Treaty of London. In the meantime, however, James's previously friendly attitude to the Dutch now became lofty and condescending. Holding the so-called 'Cautionary Towns' of Flushing, Brille and Rammekens as security for the large sums of money advanced by Elizabeth, the king's unfamiliar status as creditor seems to have gone to his head. He believed, said the Venetian ambassador, 'that at a single nod of his the Dutch would yield him all the dominion that they had gained', and he also talked foolishly of their revolt from Spain as though it were a crime. When told that Ostend might fall if English aid were withheld, his answer was characteristically insensitive. 'What of it?' he declared, 'Was not Ostend originally the King of Spain's and therefore now the Archduke's?' And when Johan van Oldenbarnevelt,

Land's Advocate of Holland, was dispatched to plead the Dutch cause, he was unable to obtain audience with James until smuggled into a gallery where the king was about to walk.

James, it is true, stood his ground against the Spanish stalwartly, correctly judging that they were in greater need of peace than his own kingdom. But what is often construed as the one great triumph of James's statesmanship was not a definitive solution to the long-standing tension between the two countries. Above all, Spain remained at war with the Dutch, and while Spain, unlike her rebellious enemy, was denied facilities to raise money and volunteers in England, suspicion and mistrust was bound to remain. Nor was the Treaty of London, signed at Somerset House, a source of unalloyed relief for Englishmen as a whole. For while the king resolutely refused to denounce the Dutch rebels outright, maintaining that while they and the Spanish continued to fight he 'was resolved always to carry an even hand betwixt them both', and Cecil successfully defended English trading rights with both sides, there remained a bitter taste for many who felt the treaty a betrayal not only of their valiant co-religionists but of their own great past. Indeed, fanned by invective from Puritan pulpits and darkening storm clouds in Germany where the Counter-Reformation was already stoking the fires of what would become the Thirty Years' War, anti-Spanish and anti-Catholic feeling would continue, in spite of James's usual good intentions, to remain a prominent feature of his English kingdom for the next twenty years.

Yet the king would not lightly forsake the selfsame good intentions in religion either, which now became the next arena for his mediating efforts. To the end of his life, in fact, James could never rid himself of the illusion that it was possible to 'win all men's hearts' by reason, logic and purely intellectual persuasion. But when the truth was at issue he could only construe it as his to determine, and when resistance persisted, he could only perceive it as wilfulness. 'It should become you,' he would write to Archbishop Abbot some years later, when they had differed over a point of theology, 'to have a kind of faith implicit in my judgement, as well in respect of some skill I have in divinity, as also that I hope no honest man doubts of the uprightness of my conscience; and the best thankfulness that you, that are so far my creature, can use towards me, is to reverence and follow my judgement, except where you may demonstrate unto me that I am mistaken or wrong-informed.' Even archbishops, then, might expect to defer to the king in matters of religion and the same was true for Catholics, Calvinists and anyone else who found themselves at odds with his views. For it was one of many curious ironies that a ruler who saw himself as a mediator in all things was so rarely prepared to compromise on his own ideas.

Without doubt, the Church of England that greeted James upon his arrival in 1603 was an institution much to his taste. Moderate, placid, hierarchical and deferential, administered by upper clergy who, though sometimes worldly and arrogant, were learned scholars and able administrators, it was a far cry from the more notorious elements of the Kirk he had left behind in Scotland. Whitgift, the Archbishop of Canterbury at the time of Elizabeth's death, and Bancroft, the Bishop of London who replaced him in 1604, could not have contrasted more starkly with the likes of Andrew Melville and John Durie who had plagued him previously. And there were others, too, just as willing, it seemed, to underpin their 'Anglicanism' with 'High Church' principles and the so-called 'Erastian' conviction that the state should enjoy unqualified primacy in all ecclesiastical affairs. William Barlow, Dean of Chester, Thomas Bilson, the learned Bishop of Winchester, and John King, Bishop of London (whom James dubbed the king of preachers) were only some of the divines who now, often literally, surrounded the throne. In due course, too, he would find in George Abbot, the Master of University College, a man whose zest for the Early Fathers and mild Calvinism, tempered by careful study of St Augustine, exactly matched his own. But it was in the presence of Lancelot Andrewes especially that James's raucous mirth was most respectfully suppressed. Dean of Westmister at the opening of the reign before being elevated successively to the bishoprics of Chester, Ely and Winchester, Andrewes possessed all the qualities guaranteed to endear him to the king, combining learning with wit, piety with adroitness, and austerity with the ready tongue of the courtier. And, as always, when James was won over he was won over unreservedly, speaking of Andrewes' sermons as a voice from heaven and asking the bishop on one occasion whether his sermon notes might be laid at night beneath the royal pillow.

Yet if James revered his Anglican divines as men and scholars, it was the Church they represented that he valued above all. Episcopal in structure and Calvinist in doctrine, the Church of England represented, indeed, an ideal model from the king's perspective, and in Richard Hooker, whose *Of the Lawes of Ecclesiasticall Politie* was published in 1593, it had found, it seems, its ultimate apologist. For Hooker's book treated the Church of Rome as merely one part of the visible Church – like those of Jerusalem, Antioch and Alexandra – and skilfully contended that the unreformed Church had fallen into error, becoming unsound in doctrine and corrupt in its behaviour. By contrast, Hooker argued, the Church of England had returned to original truths and godly practices, while retaining its apostolic links by virtue of its emergence from its Roman predecessor. Most important of all, however, Hooker was convinced like his king that state and society were so intimately and integrally connected that there could be no question of ecclesiastical independence from secular authority. So it was hardly surprising, perhaps, that according to Hooker's biographer, Isaak Walton, James considered the *Lawes* such 'a grave, comprehensive, clear manifestation of reason, and that backed with the authority of the Scriptures, the fathers and schoolmen, and with all law both sacred and civil'.

If James loved the Church of England quite literally as his own, however, it is altogether more doubtful whether he ever fully understood it. Addressing his first parliament in 1604, he distinguished three religious elements within his new kingdom. On the one hand, there was the religion 'publicly allowed and by the law maintained', as opposed to what were 'falsely called Catholics, but truly papists', whom in spite of the Bye and Main Plots, he still proposed to conciliate by toleration. The third group 'lurking within the bowels of this nation' was, he maintained, 'a private sect', consisting of 'Puritans and Novelists, who do not so far differ from us in points of religion as in their confused form of polity and parity, being ever discontented with the present government and impatient to suffer any superiority, which maketh their sect unable to be suffered in any well governed commonwealth'. For James, these Puritans were a distinct and largely monolithic body, mirroring their radical counterparts north of the Border – men espousing a democratic theory of ecclesiastical government, and intent like their Scottish counterparts upon relegating him to the status of 'God's sillie vassal'.

But such a neat categorisation of what the king termed 'Purinisme' involved a considerable and costly oversimplification, which dogged his efforts to understand the religious status quo in England and swiftly dashed his attempts to foster unity. For in reality the very word 'Puritan' had gained currency as a catch-all term of derision, which obscured more than it enlightened and encompassed a wide spectrum of opinion ranging from those wishing to bring about minor modifications to the Church's everyday practice to those outright 'separatists' who sought the complete 'independency' of each congregation from the stranglehold of state authority. Those who broadly accepted the Church of England, as constituted by Elizabeth I's religious settlement, also varied considerably in matters of detail. Some desired a compromise with Presbyterianism which would combine a Council of Elders with the bishop in the administration of diocesan discipline. Others wanted the use of ceremonial and vestments to be left to the discretion of the incumbent in each parish, and there was debate, too, over the positioning of the altar. Some decried 'Romish' practices, such as bowing at the name of Jesus and there were objections in some cases to the use of the sign of the cross in baptism, to the ring in marriage, and the rite of confirmation.

For James, who found himself well satisfied with the Church of England's existing rituals and practices, such matters were of little significance in their own right. Writing, for instance, upon the subject of clerical dress, he made it clear that division over such an issue merely 'gives advantage and entry to the papists'. 'No,' he declared, 'I am so far from being contentious in these things (which for my own part I ever esteemed as indifferent), as I do equally love and honour the learned and graved men of either of these opinions.' But in downplaying the need for dispute, James was also determined to stress the need for obedience on the grounds that 'in things indifferent, they are seditious which obey not the magistrate'. 'There is no man half so dangerous,' he contended, 'as he that repugns against order.' In emphasising the necessity of obedience over non-essentials, moreover, he also sowed further confusion in more significant areas by straddling conflicting positions, particularly over matters of doctrine, and balancing contending groups against each other. Above all, while Queen Elizabeth had sensibly discouraged religious debate, James not only leapt in head first, but consciously chose the deep end for doing so. This, after all, was a king who was never happier than when exchanging essays with Lancelot Andrewes or Hugo Grotius over the issue of final damnation, notwithstanding his condemnation of 'vain, proud Puritans', who believed that they ruled the Deity 'upon their fingers'. And this, too, was a ruler who believed himself capable of determining at a stroke the precise worth of those ceremonies and rituals unsupported by Scriptural injunction – the key bone of contention between 'Puritans' and his own episcopacy.

By 1603, ironically, the clamour for a drastic revision of the Prayer Book and for the abolition of bishops in favour of some more democratic model of ecclesiastical government had largely died down. Furthermore, an overwhelming majority of Englishmen still acknowledged the mystical authority of the monarchy and accepted the inseparability of Church and

State. All agreed that pluralism should be abolished and that a 'preaching, Godly ministry' must somehow be established. All agreed, too, that stipends must be increased. In 1585, Archbishop Whitgift had estimated that more than half the beneficed clergy of England had meagre incomes between £8 and £10 a year, while less than half could be licensed to preach for lack of university degrees. But by the end of Elizabeth's reign little had changed, and it was no special surprise that when two of his ex-pupils went to see him at Drayton Beauchamp, even Lancelot Andrewes was found 'tending his small allotment of sheep in a common field' while reading the *Odes of Horace*. Had James confined his activities to remedying such ills, he might well have made genuine progress. Had he avoided the summits of theological debate and the pitfalls of dabbling so wilfully in ceremonial niceties, he might also have saved both himself and his kingdom a good deal of acrimony and frustration.

Yet keen intellect and high principles did not, in James's case, always sit well with common sense and sound man management. Even as he was travelling down from Scotland for the first time, a selection of prominent Puritan clergy had respectfully presented him with the famous Millenary Petition – a skilfully drafted and studiously moderate document, wholly acceptable in terms of content to the majority of English bishops. The petitioners were not, they emphasised, factious men like the Presbyterians, nor schismatics like the so-called 'Brownists', but loyal subjects of the king, whose Christian judgement they now sought. They desired the discontinuance of the use of the sign of the cross in baptism, of the ring in marriage, and of the terms 'priest' and 'absolution'. They requested, too, that the rite of confirmation be abolished, the wearing of the surplice made optional, and the employment of music in church moderated. The ministry, on the other hand, was to be recruited from more able and learned men, while non-residence and pluralism were to be ended, the ecclesiastical courts reformed, and the Sabbath more strictly observed. All in all, there was little, ostensibly, to offend or threaten, and the graciousness of James's initial response was enough to warm Puritan hearts. The petition's suggestion of a conference between Puritans and Anglicans accorded closely, moreover, with the king's much-advertised love of intellectual inquiry and rational discourse.

What the king did not grasp, however, was that such a conference, whatever its outcome, would give Puritans a recognition they had never been granted by his predecessor and raise hopes that he could never realistically fulfil – hopes that were soon leading to disturbances of the peace in Suffolk. Equally importantly, the proposed conference represented a direct challenge to the Anglican bishops in whom James placed so much faith and who were soon launching a counter-offensive which would mould his conceptions of Puritanism for the rest of his reign. Oxford had answered the Millenary Petition by branding its authors seditious and identifying them with the Presbyterian ministers of Scotland that the king so loathed – men whose aim, it was said, was 'the utter overthrow of the present church government and instead thereof the setting up of a presbytery in every parish'. In July, moreover, James held long and earnest conference with Bishop Bancroft, the self-avowed arch-enemy of Puritan innovation, at his palace in Fulham, with the result that by October a proclamation had been issued, prohibiting petitions concerning religion and asserting that the existing constitution and doctrine of the Church of England were agreeable to God's Word and in conformity with the condition of primitive Christianity. All men, he told Whitgift subsequently, must 'conform to that which we have by open declaration published', while any clergyman employing unauthorised forms of service was to be punished severely.

Yet when Bishop Bilson urged James to abandon plans for the conference, the response was predictable. 'Content yourself, my lord,' the king informed him, 'we know better than you what belongeth to these matters.' And so it was that, as soon as the festivities of James's first Christmas in England were over – while peace negotiations with Spain were still dragging on, and before he had even met his first parliament – a representative Puritan delegation from Oxford and Cambridge was commanded to meet a selection of bishops and clergy under his chairmanship at Hampton Court on 14 January 1604. Although now wrongly convinced of the nature of the Puritan 'threat', he remained equally convinced nevertheless that it could be banished once and for all by the cleansing effect of his superior intellect and solemn judgement. For it was clear to him, at least, that Puritan leaders like Dr John Rainolds, the President of Corpus Christi, Oxford, or John Knewstubs of St John's College, Cambridge, would never be able to demonstrate that he was 'mistaken or wrong informed'. 'I did ever hold persecution as one of the infallible notes of a false Church,' James had once told Cecil, adding that he would 'never agree that any should die for error in faith'. And now he would have the chance to exercise his healing influence as what he himself described as a 'good physician' on a suitably imposing stage.

With a more accommodating and less outspoken approach on the king's part the Hampton Court Conference might well have achieved much. Puritan demands were, after all, more moderate than they had been for the past twenty years and, most significantly of all, far more moderate than they would ever be again. Most of those protesting at this time were, it should be remembered, demonstrably the best educated, most zealous and conscientious of the parish clergy, the majority of whom had hitherto loyally accepted regulations of which they disapproved in the interests of Church unity as a whole. To have conciliated them, therefore, before they became irreparably embittered with the episcopacy was not only the obvious priority but a distinct possibility at this critical juncture. 'Religion is the soul of a kingdom,' James would assure his audience at the very outset of proceedings, 'and unity, the life of religion.' Yet it was no coincidence that the Puritan representatives were quite conspicuously excluded on the first day while the king addressed his bishops. And although his preliminary speech, which lasted some five hours, committed him to 'examine and try' complaints about the Church 'and fully to remove the occasions thereof, if scandalous', it also made clear that, unlike his predecessors, who 'were fain to alter all things they found established', he saw no reason 'so much to alter and change anything as to confirm what he found well settled already'. The religious status quo was therefore to remain intact in all fundamentals, while the conference was to serve, it seems, as a show case for the king's skill in divinity before a suitably submissive and amenable gathering of awestruck admirers who, when apprised of their errors, 'must yield to him'.

With these priorities fixed in advance, it was clear, of course, that the four Puritan leaders, led by Dr Rainolds, who faced the nine Anglican divines, eight deans of the Church and lords of the Privy Council ranged against them were unlikely to enjoy a truly impartial hearing when admitted to the conference on the second day. And if eye-witness William Barlow's account of proceedings, *The Summe and Substance of the Conference*, may well have exaggerated the abrasiveness of James's attitude towards the Puritans, it remains hard nevertheless to deny that the king was inclined to lapse on occasion into the kind of sharp-tongued pedantry and dismissiveness that has so often been associated with him. The four Puritan leaders had been hand-picked, in fact, by the Privy Council and were far from being the 'brainsick and heady preachers' that the king both feared and despised. In addition, to Rainolds and Knewstubs, there was Laurence Chaderton, whom even Bancroft considered a good friend, and Thomas Spark, a comparatively innocuous lecturer in divinity at Oxford. Yet when all four entered the king's presence on the second day, looking as if they wore 'cloaks and nightcaps', they would be facing an uphill struggle.

It was not without some justification, indeed, that Barlow referred to them as 'plaintiffs', and while Spark said little, Chaderton was 'mute as any fish'. Knewstubs, it is true, condemned the use of the cross in worship, only to be roundly quashed by Lancelot Andrewes, and Rainolds, too, was given short shrift over infant baptism, the use of the ring in marriage rites and, above all, the use of the phrase 'with my body I thee worship' in the Anglican wedding ceremony. Nor could the king, knowing full well that Rainolds was a confirmed bachelor, resist a jarringly patronising jibe in response to his pronouncements on the union of men and women. 'Many a man,' said James, smiling at his hapless target, 'speaks of Robin Hood who never shot his bow.' 'If you had a good wife yourself,' he continued, 'you would think all the honour and worship you could do for her well bestowed.' And if the famous aphorism of 'No bishop, no king', with which James concluded the later discussion on the ordination of bishops, was from his point of view little more than an emphatic statement of undeniable fact, it remained nonetheless another case of the type of cudgel phraseology, which ill accorded with what was supposedly the avowed intention of the whole conference. Where James did offer genial agreement, moreover – on the need for producing a revised catechism, raising stipends, recovering lost Church revenues, providing a better 'teaching ministry', discouraging pluralism and enforcing proper observance of the Sabbath – he repeatedly undermined any resulting goodwill by spasmodic outbursts of needless condescension. 'And surely,' he declared at one point, 'if these be the greatest matters you be grieved with, I need not have been troubled with such importunities and complaints as have been made unto me; some other more private course might have been taken for your satisfaction.' Whereupon, we are told, 'looking upon the lords, he shook his head smiling'.

It was not, however, until the subject of ecclesiastical discipline was raised that James's swagger and volubility turned to outright anger and provocation. Until this point, at least, the king had made some effort to leaven his jibes with attempts to play the honest broker. When, for instance, Bishop Bancroft interrupted Rainolds on the grounds that 'schismatics are not to be listened to against bishops', he had been put firmly in his place on the grounds that there could be no 'effectual issue of disputation, if each party be not suffered, without chopping, to speak at large'. Likewise, when Bancroft objected to the Puritan proposal for a new translation of the Bible, James appeared equally fair-minded. 'If every man's humour might be followed,' the bishop had protested, 'there would be no end of translating.' To which the king replied that he had 'never yet' seen a Bible 'well translated in English', before urging the creation of what would become the 'Authorised' or 'King James' version, which has been justly characterised as the great glory of his reign. Yet the harmony of these exchanges was undermined all at once when Rainolds suggested to his cost that episcopal synods, 'where the bishop with his presbytery should determine all such points as before could not be decided', would be less obnoxious than the existing system of archdeacons' courts,

The connotations of the very word 'presbytery', linked as it was so inextricably with the ecclesiastical polity of the Presbyterian Kirk of Scotland, made it intolerably offensive to James, and the use of it was, to say the least, ill-advised. But the innocence of Rainolds' intention could not have been clearer from the broader content of his proposal, and what followed demonstrated all too aptly how the king was unable to bridle his emotions when delicate negotiation was involved. No less importantly, it proved conclusively that a hectoring tone, particularly when resulting from an apparently inexplicable misapprehension, was unlikely to win converts even to the most valid cause. At this point, according to Barlow at least, 'his Majesty was somewhat stirred, yet, which is admirable in him, without passion or show thereof'. But Barlow had his eye on promotion and was shortly to be rewarded with the bishopric of Lincoln for his sympathetic account of the entire conference. The reality, therefore – as recorded in all the variously reported versions of James's words – could not have been more starkly different.

Plainly mistaking Rainolds' meaning, the king proceeded to tell him, in fact, how a 'Scottish' presbytery 'as well agreeth with a monarchy as God and the Devil'. 'Should such a body ever be permitted within the Church of England', he continued, 'then Jack and Tom and Will and Dick shall meet, and at their pleasure censure me and my council and all our proceedings ... When I mean to live under a presbytery I will go into Scotland again, but while I am in England I will have bishops to govern the Church'. And then, it seems, as a final flourish he turned to his Puritan audience and declared stridently that 'if this is all they have to say, I will make them conform themselves or I will harry them out of this land or else do worse'. So with all the insensitivity of a man utterly convinced of the transparency of his good intentions and correctness – as well as the unalloyed love of his subjects, which, he believed, entitled him to impunity even when making the most provocative statements – James demonstrated his failure to appreciate the fundamental principle that had underpinned his predecessor's whole system of government in Church and State. This principle rested on a series of delicate balances which could work in practice only so long as they were never subjected to any harsh dialectic and definition – or undermined by injudicious and needless threats. On this occasion as on others, James would eventually moderate his position and behave in practice altogether more pragmatically when his irrational blaze of anger had burnt itself out. But such unedifying outbursts were inevitably costly, not only because of the anger they provoked in turn, but far more importantly because of their impact upon the credibility of James's leadership and authority.

In this case, at least, the more damaging effects of the king's indiscretion were by no means immediately apparent. Indeed, the conference closed on the following day in an atmosphere of outward goodwill, with the Anglican divines heaping praise upon James, in spite of their misgivings that any concessions had been granted at all, and the king himself convinced that, since 'obedience and humility were the marks of honest and good men', the Puritans would thereafter toe the line. Archbishop Whitgift, it seems, was convinced that 'his Majesty spake by the special assistance of God's spirit', while Bancroft acknowledged 'unto Almighty God the singular mercy we have received at His hands in giving us such a king as since Christ his time the like he thought had not been'. In the meantime, Rainolds and his counterparts concealed their undoubted frustration with a humble plea that their brethren 'who were grave men and obedient unto the laws' might simply be given some time to determine whether they would conform to the new and more rigid enforcement of the Prayer Book ceremonies and the Thirty-Nine Articles.

But the fact remained that the underlying tone of Elizabethan government, however authoritarian in practice, had been altogether less abrasive. Lord Burghley, for instance, had considered it his duty to lessen the impact of Whitgift's ecclesiastical courts upon 'poor ministers' whom, he believed, were being made 'subject to condemnation before they be taught their error'. James, however, was the controversialist, thirsting for wordy and sardonic victory. 'The king,' wrote Sir John Harington, 'talked much Latin and disputed with Dr Rainolds at Hampton, but he rather used upbraidings than arguments', adding that if, as his bishops claimed, he had spoken 'by the power of inspiration', then 'I wist not what they meant, but the spirit was rather foul-mouthed'. And if Harington's or any other account should be doubted, there remain of course James's own comments to consider. 'We have kept such a revel with the Puritans here these two days as we never heard the like,' he informed the Earl of Northampton the day after the Hampton Court Conference closed. 'They fled me so from argument to argument without ever answering me directly, *ut est eorum mos*, as I was forced at last to say unto them that if any of them had been in a college disputing with their scholars, if any of their disciples had answered them in that sort, they would have fetched him up in place of a reply, and so should the rod have plied upon the poor boy's buttocks'. Thus, in the happy belief that he had 'peppered them soundly', James brought to an end the first part of his attempt to achieve a 'general Christian union in religion'.

James's indiscretions were capable, in fact, of wounding even his own avowed allies. The bishops, for instance, were less likely to have been assured of the divinity of the king's words when, not long after the Hampton Court Conference, he told Parliament that the Devil, sparing neither labour nor pains, was a busy bishop. But his sympathies remained unwavering. Only one month after the conference, moreover, Whitgift died and Bancroft was duly installed as Archbishop of Canterbury. And though the new archbishop's anti-Puritan sympathies have sometimes been exaggerated, he would nevertheless quietly choose to forget most of James's agreed concessions, as would the vast majority of his episcopal colleagues. Nor was the king's subsequent attention to his intended reforms ever sufficient to persuade Bancroft to do otherwise. Even if it was true, therefore, that only ninety of the Church of England's 9,000 incumbents eventually resigned in response to the Canons of 1604, this was mainly due to the fact that, in spite of their underlying hostility to reform, many bishops nevertheless proved reluctant to force so many of their best clergy into taking the ultimate step of defiance. 'The bishops themselves,' wrote Chamberlain, 'are loath to proceed too rigorously in casting out and depriving so many well reputed of for life and learning.' 'Only the king,' he added significantly, 'is constant to have all come to conformity,' while even an outspoken high churchman like James Montagu, Bishop of Bath and Wells and later Winchester, urged that recalcitrant ministers should be called to account gradually 'rather than all without difference be cut down at once' on the grounds that those who lost their places would gain more from pity than they would from their piety.

Yet, in spite of his detestation of Puritanism, James's good intentions for the Church as a whole were often evident. 'I have daily more and more cause to hate and abhor all that sect, enemies to all kings,' he informed Cecil in November after he had been presented with a Puritan petition while hunting near Royston. On the broader front, however he resented criticism of the Church in whatever form it came, supporting the ecclesiastical courts in their struggle with the courts of common law and defending his bishops when they came under fire from Parliament. Nor was he prepared to deepen ecclesiastical poverty by cynical alienations of property, and though simony was rife at court, he resisted it in clerical appointments. Indeed, he rejected with scorn a cynical scheme to reassess the evaluation of benefices in order to obtain large sums in first fruits, and on the other hand supported church leaders in their highly unpopular efforts to obtain enhanced revenues from tithes. In 1608, moreover, he was struck by the disgraceful condition of St Paul's Cathedral, and though he offered no money himself, nevertheless encouraged the Bishop of London to finance repairs.

James's appointments to the episcopacy were not, it is true, always judicious. Certainly, large numbers of royal chaplains who, like Robert Abbot, had won the king's favour for no especially compelling reason found themselves promoted. 'Abbot,' James informed this particular beneficiary of royal goodwill, 'I have had much to do to make thee a bishop; but I know no reason for it, unless it were because thou hast written a book against a popish prelate.' When, moreover, Lancelot Andrewes was passed over as Archbishop of Canterbury in 1611, amid general astonishment, the king explained his decision in terms that can only be described as eccentric. The successful candidate, George Abbot, a university man who had little experience of ecclesiastical administration, was told by the king that his appointment did not spring from his learning, wisdom or sincerity (though the king did not doubt that he possessed these qualities) but out of respect for the recently deceased Earl of Dunbar, who had recommended him. Yet, in other respects, James's ecclesiastical record was not without its merits. In this particular case, for instance, Abbot's appointment was undoubtedly less provocative to Puritan opinion than the elevation of Andrewes might have been and appears to have represented an incipient willingness on James's part to send signals of compromise. Much more important by far, however, was the ringing success of the publication of the 'Authorised Version' of the Bible, for which the king must be accorded all due praise.

If, as James himself acknowledged, the Hampton Court Conference had been intended 'to cast a sop into Cerberus's mouth that he bark no more', the upshot had clearly been less than wholly satisfactory. In 1610, indeed, the Commons' Petition on Religion, presented by the Puritan gentry in Parliament, plainly demonstrated the discontent of significant sections of the laity, and grumbling would only begin to abate at last with Bancroft's replacement in 1611. But John Rainold's proposal for a new English Bible on the conference's second day, which appeared at the time as an almost casual interjection, not only captured the king's imagination but brought out the best in him and, in doing so, prompted what was arguably one of his most significant achievements. Together with Bancroft and Cecil, he would engage the finest of Greek and Hebrew scholars to produce 'one uniform translation ... ratified by royal authority', which, if successful, was to represent both the lynchpin and crowning glory of James's quest for religious unity. Equally importantly, it would affirm in the process his own exalted conception of kingship, since the Geneva Bible, which was the version used by the majority of his common subjects, contained anti-monarchical margin notes which were, in his view, 'very partial, untrue, seditious and savouring too much of dangerous conceits'. If it could be superseded, along with the existing official Bible of the Church of England, the so-called 'Bishops' Bible', the slate could at last be wiped clean of contention and misapprehension. 'You will scarcely conceive how earnest his Majesty is to have this work begun,' Bancroft informed a colleague in June 1604 – and, as events would demonstrate, he did not exaggerate.

Bancroft himself had, of course, objected vehemently when a new translation was first mooted, only to change his tune completely upon realising the extent of the king's commitment. At Hampton Court, moreover, James's passion for sketching programmes and drafting directives had borne immediate fruit. The translation, he decided, should be made by the most learned linguists at Oxford and Cambridge, and thereafter reviewed by the bishops and other learned churchmen before being presented to the Privy Council and then 'authorised' ultimately by royal consent. Such, meanwhile, was Bancroft's sudden enthusiasm for the project that he was appointed general co-ordinator of the process of translation and by March, on the king's initiative, had asked Lancelot Andrewes to be one of the regional supervisors for three translation teams – the two others being Edward Lively and John Harding, both professors of Hebrew at the two universities. Fifty-four translators in all, each working separately within six groups would eventually confer with their group members before submitting their final translation to the scrutiny of the other groups. Ultimately, six men, two selected from each group, would review the work as a whole in London, after which Bishop Bilson and Miles Smith would give a last revision to the completed text.

Significantly, the translators were selected primarily for their linguistic expertise rather than their religious views, and while one Puritan, Hugh Broughton, was excluded for his radical opinions, Dr John Rainolds, who had been buffeted by the king at Hampton Court Conference, was nevertheless invited to contribute. James too, it seems, contributed directly by undertaking the translation of the Psalms in conjunction with Sir William Alexander, his friend and literary crony, who would nevertheless complain of the difficulty of working with the king, since 'he prefers his own to all else'. By the time of his death James had completed about thirty; at which point, as Bishop Williams remarked, he was called to sing psalms with the angels. Similarly, while James commanded that the Bishops' Bible be followed as far as possible, with words like 'church' retained in preference to 'congregation', he was responsible for making sure that the new version should be readily comprehensible even to the most humble of his subjects. The language of the Bishops' Bible had, after all, been not only inaccurate at times, but overly literal and in consequence lumpy, dense and difficult to navigate. Ecclesiastes 1:11, which was eventually rendered 'Cast thy bread upon the waters' in the King James Bible had, for instance, been presented in the earlier version as 'Lay thy bread upon wet faces'. Ultimately, indeed, although the Bishops' Bible was to be treated at the outset as what might be termed the 'default' version and left untouched if adequate, it would comprise only 8 per cent of the final Authorised Version.

To James's considerable credit, then, the bible which eventually saw the light of day in 1611 was, without any real question, the most significant and successful that had been produced to date. And though it was essentially a patchwork quilt, incorporating the finest elements of former translations, it would stand the test of time for good reason, since, as James seems to have appreciated, Jacobean culture was a culture of the word, and above all a listening culture, which gave the book both its clarity and poetic force. The bible was intended, after all, to be 'read in churches', and it was no coincidence therefore that the king had demanded that the words be 'set forth gorgeously'. Nor was it any coincidence that one of the last steps of the translation process was a 'hearing'. The result was a grand harmony, stateliness and splendour that in spite of its nature as a committee product and notwithstanding its high Anglican tone, largely superseded factional or sectarian divides. As such, the King James Bible was in many respects the embodiment of the highest and noblest of all his religious aims: the reconciliation of contending parties under the benevolent guidance of a wise, all-knowing and all-governing king.

It was not without some irony, therefore, that James remained in many respects more tolerant of English Roman Catholics than their Puritan counterparts. He harboured, it is true, a deep suspicion of Catholic priests and an outright abhorrence and terror of Jesuits, but he distinguished sharply between them and the Catholic laity. Indeed, before he left Scotland he had told Salisbury that he intended to seek a golden mean in dealing with English Catholics, on the one hand preventing them from rebellion and increasing their numbers until they were 'able to practise their old principles upon us', but asking at the same time for no more than outward conformity to the law by attendance at Church of England services. Even after the Main and Bye Plots, therefore, the Catholics were relieved of recusancy fines, though these had been collected in May 1603, primarily because the king was already stretched for money. The ideal of Christian unity was still, after all, far from dead at this time, and if Pope Clement VIII might somehow be persuaded to summon an ecumenical council of the kind that was still considered possible in some quarters, there remained hope that a middle ground could be established, so long as the Holy Father renounced temporal sovereignty and the political subversion of the Jesuits.

Certainly, James's own words speak eloquently enough of his intentions. 'We have always wished,' he wrote, 'that some good course might be taken by a general council, lawfully called, whereby it might once for all be made manifest which is the doctrine of antiquity nearest succeeding to the primitive Church.' He was, moreover, a declared 'Catholic' Christian, a member of the Ancient, Catholic and Apostolic Church, as constituted in the first five centuries of its existence. If, therefore, the papacy could be persuaded to purge itself of the unbiblical accretions that had emerged since that time, James was willing to accord Rome a high place in any newly united Church that might then arise. 'I would with all my heart,' he declared, 'give my consent that the Bishop of Rome should have the first seat. And for his temporal principality over the seignory of Rome, I do not quarrel it either. Let him in God's name be *Primus Episcopus inter omnes Episcopos*, and *Princeps Episcoporum*, so it be no otherwise but as Peter was *Princeps Apostolarum*.' James's only other requirement was that the pope should 'quit his godhead and usurping over kings'. 'I acknowledge,' he added for good measure, 'the Roman Church to be our mother church.'

It was typically dizzy oratory, fuelled in part by the kind of undiscriminating adulation which James was always unable to resist. 'We have a Constantine among us,' trilled George Marcelline, 'capable to preside as the other did in the Nicene assemblies, the presence of whom is able to dispose of differences, to soften the sharpest, to restore and place peace and concord among all good fathers, and to make them happily to finish such a design.' But Pope Clement was thinking along entirely different lines and by 1605 his hopes of James's conversion had reached their short-lived peak. Though he wished English Catholics to remain quiescent, he took few practical steps to lessen the threat from plots, and the queen's decision to urge a Catholic marriage for Prince Henry, coupled to her efforts to obtain office for her co-religionists, only served to undermine James's efforts further. Employing Sir Anthony Standen, the English ambassador in Italy, as her private agent in Rome, the queen also wrote to the Spanish infanta, imploring her to send two friars to Jerusalem to pray for herself and her husband. And by the time that James began to claw back the situation, the inevitable Protestant reaction had already outstripped him. Imprisoning Standen when he brought back sacred objects from the pope and commanding Anne's chamberlain, Lord Sidney, to exercise great care in the selection of her household would do little to quell the indignation of even moderate Anglicans that he himself had provoked by his attempts at reconciliation.

'It is hardly credible in what jollity they now live,' wrote one contemporary English Protestant of his Catholic counterparts. 'They make no question to obtain at least a toleration if not an alteration of religion, in hope whereof many who before did dutifully frequent the Church are of late become recusants.' Cecil, too, expressed concerns about the king's excessive clemency, which alienated the Anglican clergy and resulted in Catholic priests openly plying their trade in the country at large. And predictably there were the obvious comparisons drawn between the leniency extended to papists and the harsh treatment of Puritans. Indeed, even the Archbishop of York, Matthew Hutton, made this very point, while adding how Catholics 'have grown mightily in number, favour and influence'. Plainly, then, the king's irenic impulses found little sympathy with his councillors and bishops when it came to the Church of Rome. Nor did they impress his judges who continued to enforce the anti-Catholic laws wherever they could, for England as a whole, with its long tradition of hostility to the pope, was not ready for the toleration he proposed. On the contrary, the immediate results proved that the deep anti-Catholic prejudices of Englishmen, however irrational from some perspectives, were a sounder basis for policy than the theoretically high-minded sentiments of the king.

The revelation of the real numbers of Catholics in the country when they were allowed to disappear without penalty from the back benches of their Anglican parish churches, and the large numbers now attending Mass, startled even James. Previously, the returns which had been collected from every diocese of those who officially stayed away from church had led the government to estimate the total number of Catholics at about 8,500. When toleration allowed them into the open, however, it seemed that the papal claim to more than 100,000 was nearer to the mark – something which James believed, quite wrongly, could only be explained by widespread, rapid conversion resulting from his own policy of toleration. And it was increasingly clear, too, that while the majority of English Catholics were both loyal and peaceable, their leaders were inclined to think otherwise. The Jesuit Robert Persons, whom Sir Henry Wotton characterised with good reason as 'malicious and virulent', retained a place at the heart of papal policy in England and remained committed to the forcible restoration of Catholicism, even, if the need arose, at the price of assassinating a heretic ruler. Indeed, plans were already in hand for the imposition of censorship and the installation of an English Inquisition, which made nonsense of James's hopes that wounds could be healed and deals done.

Ironically, then, the admirable intentions of the 'British Solomon' were soon being replaced by growing irritation and impatience with those whom he sought to assist but could not help as a result of their own ignorance and fractiousness. Upon ascending the throne, he had suspended recusancy fines, allowed Catholics to worship in private as they pleased and turned a blind eye to the influx of Catholic priests. Now, however, only nine months after this first reversal of policy, a second was to occur. And at a meeting of the council in February 1605, he found himself venting his spleen against both Puritans and Catholics. As to the latter, he declared, he was so far from favouring their superstitious religion that if he thought his son would tolerate them after his death, he would wish him buried before his eyes. The only answer now, it seemed, was a restoration of the old Elizabethan measures and a rigorous execution of the laws against 'both the said extremes'. A proclamation duly ordered all Jesuits and priests to quit the country, and several were hanged in February, albeit without direct instructions from the government. Indeed, when the king learned of the hangings in Devon, he explicitly ordered that no executions should be carried out merely on grounds of religion.

From James's perspective, however, his first moves in favour of toleration had been foiled both wilfully and ungratefully by those who lacked his wisdom and vision, and the saboteurs would now have to reap the consequences. The fact that he had played with fire and, in doing so, exacerbated an already delicate situation seems to have escaped him – though, having excited Catholic expectations, the king too would now have a price to pay. For if Puritans wrote petitions when frustrated, there were those among the Catholic community who would express their discontent altogether more forcefully when no longer fed a diet of fair promises.

# 13 ⚹ Parliament, Union, Gunpowder

'We came out of Scotland with an unsullied reputation and without any grudge in the people's hearts but for want of us. Wherein we have misbehaved ourself here we know not, nor we can never yet learn ... To be short, this Lower House by their behaviour have periled and annoyed our health, wounded our reputation, emboldened all ill-natured people, encroached upon many of our privileges and plagued our purse with their delays.'

*James I's complaint to his Privy Council,*
*7 December 1610*

'Their Parliaments hold but three days, their statutes are but three lines.' So wrote Sir Anthony Weldon, that most caustic critic of the Scots, who on this occasion as on others both simplified and distorted broader realities north of the Border. The irony, however, was that Weldon's dismissive perception of the Scottish Parliament or 'Estates' as a submissive and ineffectual institution was propagated most effectively of all by the very King of Scotland who had now come to occupy the English throne. Some while before he headed for London, indeed, James VI, as he then was, had portrayed the Scottish Estates as little more than the chief court of the king and his vassals, and it was this institution - displaying, he suggested, no rash desires for liberty or innovation – that James I subsequently presented to his English subjects as the ideal parliamentary model. By 1603 his terminology had changed slightly, for he was now calling Parliament 'nothing else but the king's great council'. Yet his thinking was very largely unaltered. Parliament was assembled, or so he suggested, for the exclusive purpose of ratifying laws and punishing notorious offenders. And in holding up the Scottish Estates as an example to Westminster, he consciously chose to expound its operation in a way that suggested total subservience to the whim of the monarch. No member of that body, James told English MPs when the first parliament of the new reign assembled in March 1604, was allowed to speak without his chancellor's explicit permission, and any proposed new laws were automatically submitted to the monarch's scrutiny some twenty days beforehand, whereupon 'if there be anything that I dislike, they raise it out before'.

Yet the Scottish Estates had a long and often proud record of independent activity, which plainly contradicted the more obvious caricatures and suggested an altogether more subtle relationship with the monarchy than James saw fit to depict. Its origins dated back, in fact, to at least 1286 with the first use of the term 'the community of the realm', although William the Lion was recorded as having held a full parliament over a century earlier. Moreover, the notion that Scottish history was subsequently marred by a malign combination of supine parliaments, arbitrary rule and self-interested and remote noblemen is a simplification and misrepresentation firmly founded in Enlightenment historiography. By the early fourteenth century, the attendance of knights and freeholders had already become important, and from 1326 burgh commissioners also attended. Consisting of the three 'estates' of clerics, lay tenants-in-chief and burgh commissioners sitting in a single chamber, the Scottish Parliament thereafter acquired significant powers over particular issues. Most obviously it was needed to sanction taxation, and although taxes were raised only irregularly in Scotland during the medieval period, it also had a strong influence over justice, foreign policy and the conduct of war, as well as enacting a wide range of legislation on political, ecclesiastical, social and economic matters.

In reality, then, the underlying rationale of Scottish government had always been collaborative and inclusive, with the king-in-parliament at its apex, held in place by subtle checks and balances in a manner that seemed rare within a contemporary European context. Furthermore, this whole system actually reached its apogee in the 1590s under none other than James VI himself, which, in light of his modernising aspirations in other areas, was not as surprising as hindsight might appear to suggest. It was he, after all, who had encouraged the mapping of his country, the revamping of its weights and measures and the listing of its landowners and their estates by the Privy Council. And it was he, ironically, who appeared for so much of his reign to accept the theory and practice of the 'community of the realm' so readily. Indeed, whenever James convened the Estates, its physical layout could not have represented more tangibly the whole nature of the Scottish polity, with the king enthroned at the centre, the lesser barons and lairds in front of him, the nobility, barons and their guests, all appropriately ranked on his left, and, to his right, the burgh and shire commissioners, alongside the clergy.

Ultimately, moreover, the king could avoid calling Scotland's Parliament only if he had no need of money, was making no constitutional or religious changes, had no treaties to approve, no embassies to send or receive, no marriage to negotiate, no high-ranking miscreant to punish, and no need to finance or plan military expeditions. And though in the latter years of James IV, for instance, Parliament met rarely after the king had the comfort of a Tudor dowry, most monarchs in this relatively unwealthy country usually required both money and the parliaments to provide it. The Estates existed, in fact, precisely because Scottish monarchs were not absolute, and the very term 'community of the realm' implied not only that the realm was distinct from the ruler, but that Parliament was the guardian of the status of the kingdom and its people. By the end of the fifteenth century, it was accepted, indeed, that Parliament could directly restrain tyrannical monarchs, and by the time John Mair wrote his *Historia Majoris Britanniae* in 1521, he was able to assert that it could even frame laws of its own that were binding upon the monarch.

The question arises, therefore, just how far James's increasingly authoritarian views in the years directly before 1603 were actually encouraged by his impending succession to the English throne. Ironically, during the last decade or so before he became King of England, the Scottish Estates had enjoyed possibly its most effective period, undertaking in the 1590s its busiest and most wide-ranging legislative programme. Indeed, petitions to it had become so numerous that a vetting committee was proposed in 1594, though this did not prevent the overall explosion of law-making during the same period, as legislation poured forth on a broad spectrum of topics from consanguinity and divorce, through to property protection, legal guardianship, and the education of nobles abroad. Yet by 1600, James had nevertheless published *The Trew Lawe of Free Monarchies* and *Basilikon Doron* in which he clearly appeared to be advocating a move towards what can only be construed as some form of absolute monarchy.

There was no denying, of course, that the parliaments of the 1590s had been fractious, or that the king's imprisonment by a religious mob during a meeting of the Privy Council in Edinburgh's Tolbooth in 1596 prompted a desire on his part to redefine the theoretical authority of the monarch, as evidenced by his literary output thereafter. But it is equally likely that his transfer to London further stimulated his budding absolutism, for there was no developed notion of the monarch as *primus inter pares* south of the Border, where the nobles in particular were effectively on their knees before their ruler. More importantly still, there appeared no shortage of cash when compared to Scotland, permitting the king to indulge his absolutist fantasies in a way that could never have been possible otherwise. And when full account is taken of the recalcitrance of many MPs at Westminster under his predecessor, it becomes clearer than ever, perhaps, why James unwisely decided to assume the constitutional offensive in his new realm.

To all outward appearances, of course, England's Parliament had been growing steadily in status throughout the sixteenth century, and the reign of Elizabeth, above all, had certainly witnessed a number of spectacular incidents in which Crown and Parliament clashed over a variety of issues including free speech, her marriage, the succession and, of course, James's own mother, Mary Queen of Scots. In 1597, indeed, Elizabeth had been forced to accept that although she had formerly been 'exceeding unwilling and opposite' to all manner of innovations in ecclesiastical matters, there was now no choice but to give 'leave and liberty to the House of Commons to treat thereof'. More strikingly still, in 1601 there had been a vigorous and successful attack on her grants of trading monopolies, in the course of which one MP had dared to raise the whole issue of the balance between the authority of Parliament and the royal prerogative itself. 'To what purpose is it to do anything by Act of Parliament,' declared Francis Moore, 'when the queen will undo the same by her prerogative.' And by the time of the opening of her last Parliament, it was even being noticed that fewer voices than ever were being raised in the customary cry of 'God save your Majesty' as she passed among members.

James's task in managing his new English Parliament was therefore sure to be a delicate one. But in overestimating the scale of potential opposition and opting to assert his authority more vocally than his predecessor, he would actually instigate the kind of response that he had feared in the first place. In reality, the free speech debates of 1566, 1576 and 1571, the rancour over the queen's marriage and the succession, as well as the struggle over Mary Queen of Scots were actually far from typical, and insofar as real Crown–Parliament conflict occurred at all, it happened only once – over monopolies. On those occasions, moreover, when sections of the House of Commons had pressed a case against Elizabeth, the Crown was almost invariably victorious. In spite of religious controversy and attempts to establish a full-blown Presbyterian system in 1587, for example, the Elizabethan religious settlement stood fully intact in 1603, while the earlier free speech debates achieved nothing. Overall, indeed, harmony and co-operation had been the predominant feature of Parliament's relationship with Elizabeth, as no less than twelve out of a total of thirteen parliamentary sessions readily granted supply to the queen. Even more importantly, there had been little intrinsic interest in the general nature of the constitutional balance between monarch and subject until James himself raised it in arguably the most provocative manner possible. For although James's practice would frequently prove rather more subtle and accommodating than his rhetoric, his outspokenness on the question of his prerogative was almost invariably as jarring as it was unnecessary.

The catalogue of the king's indiscretions is, in fact, almost limitless. 'Hold no Parliament', he had informed his son in *Basilikon Doron*, 'but for necessity of new laws, which would be but seldom', and to force home the point he added later that any man desiring a new law should come to Parliament with a halter round his neck, so that if the law proved unacceptable he could be hanged forthwith. Fortified by Bancroft's flatteries in the wake of the Hampton Court Conference and secure in the knowledge of his own success in Scotland, he would also blandly inform his first Westminster Parliament of 'the blessings which God hath in my person bestowed upon you all', while in exalting his own status, he necessarily appeared to minimise Parliament's own. 'The state of monarchy,' James would tell the House of Commons, 'is the supremest thing upon earth. For kings are not only God's lieutenants upon earth and sit upon God's throne, but even by God Himself they are called gods.' Like God 'they make and unmake their subjects', he continued, having the power 'to exalt low things and abase high things and make of their subjects like men at chess, a pawn to take a bishop or a knight, for to emperors or kings their subjects' bodies and goods are due for their defence or maintenance'.

In truth, of course, such sentiments were neither new nor inherently offensive in their own right, and there is little doubt that many of James's more provocative utterances were delivered with all sincerity at times when he genuinely considered the royal prerogative to be under unjust threat from innovating MPs. But in spite of good intentions and a genuine wish to act in his subjects' best interests as a benevolent and paternal ruler, there remains little doubt that James's policy of employing attack as the best form of defence was often counter-productive. Above all, he simply talked too much. His predecessor's appearances in Parliament had, by contrast, been rare, judiciously timed and invariably regal in tone and execution. Confining herself to brief statements of policy at the start of a session, and the occasional rebuke or engaging appeal, which were all part of her armoury of 'love-tricks', Elizabeth I was able to obtain most of what she wanted without sacrificing those essential principles on which she knew she must stand firm. But she very rarely delivered substantial orations, such as that of 1601 in which she brilliantly covered her retreat over the issue of monopolies, while James, by contrast, was temperamentally unable, it seems, to curb his desire for publicity: to declaim at great length and in minute detail not only upon immediate policy but upon far-reaching philosophical issues encompassing Church and State. The result, in the opinion of one MP, was 'long oration that did inherit but wind'. Much more damagingly still, however, was discussion of matters best left alone and an extremity of expression in the heat of debate that led MPs, in turn, to deliver increasingly strident affirmations of their own rights and grievances.

Any successor to Elizabeth faced, then, an especially testing time, precisely because she had circumvented so much by saying so little. But if James inherited a challenging hand, there remains little doubt that he might have played it more skilfully at times, particularly with regard to matters of everyday procedure. During the last years of Elizabeth's reign, it was Robert Cecil who had managed the government's business in Parliament and he would continue to do so in the new reign, notwithstanding his promotion to the peerage. Yet the minister found himself increasingly frustrated not only by the king's meddling but by his tantrums, complaints and frequently misguided instructions. Royal messages, interruptions to debates and attempts by the monarch to dictate their course, which had been infrequent and weighty occurrences in the previous reign, now, in fact, became routine and an increasing disruption to royal business. Later, indeed, there would be outright complaints concerning the 'many intervenient messages' issuing from the throne that culminated in 1621 with a formal attempt to 'move the king that there be not so many interpositions'.

Likewise, James's passion for definitions often provoked dangerous counter-definitions from MPs, while his inability to resist poring over abstract technicalities at every opportunity could have only one outcome in an institution packed with common lawyers. And though these same lawyers continued almost without exception to reverence the lolling, ungainly figure in front of them, with his heavily padded clothes and thick Scottish accent, his general lack of majesty and charisma did nothing to enhance his cause or make his pedantry any more tolerable. Nor, for that matter, did his favouritism, his extravagance and his apparent inconsistency when it came to hard work. For any ruler so inclined to assert his own God-given authority as the Lord's Anointed could only hope for a truly amenable audience, if he seemed more obviously to personify it in his everyday habits and demeanour.

All in all, then, it was only inevitable, perhaps, that James found himself opposed in Parliament. Much more surprising, however, and much less forgivable was his bewilderment at this and his subsequent exasperation, which frequently manifested itself in either self-pity or pique. In 1604 he assured the Commons of his belief that their intentions were not seditious, before informing them nevertheless that they were rash, over-inquisitive and apparently distrustful. 'In my government bypast in Scotland, where I ruled among men not of the best temper,' he declared, 'I was heard not only as a king but, suppose I say it, as a counsellor.' Here, however, there was 'nothing but curiosity from morning to evening to find fault with my propositions'. All things, James concluded, were now 'suspected', and his bitterness and sense of hurt were even more intense six years later when he once again professed himself 'sorry of our ill fortune in this country', where his 'fame and actions' had been 'tossed like tennis balls' and 'all that spite and malice might do to disgrace and infame us hath been used'. Since he was a pious king, the father of his people and so manifestly willing to take such pains in redressing wrongs, the implication was clear. Only ill-intentioned men, bent upon perilous innovation could air 'grievances' against his government, established institutions such as the Court of High Commission, and above all the royal prerogative. And in taking such a firm stand against change of any kind, all appeals for alteration of existing practice became subsumed in James's mind under the same pernicious heading. 'All novelties are dangerous,' he asserted in 1610, 'and therefore I would be loath to be quarrelled in my ancient rights and possessions, for that were to judge me unworthy of that which my predecessors left me.' The fact that the Commons would ask for the same things over and over only convinced James further of their ill will.

But it was the new king's careless intervention in a matter he did not fully understand that prompted the first serious squabble in the reign, and one that would set the tone for a good deal of what was to follow. In a not altogether unprecedented attempt to influence the course of the forthcoming elections for the Parliament of March 1604, a proclamation had been issued to encourage sheriffs and electors 'to avoid the choice of any persons either noted for their superstitious blindness or for their turbulent humours other ways'. This attempt to categorise Catholics and Puritans as 'disorderly and unquiet spirits' and exclude them from election was not, however, the main bone of contention. Instead, it was the stipulation that election returns be made, not direct to the House of Commons, but to the Court of Chancery – a blatant reversal of the Elizabethan practice by which the House was the sole judge of cases involving disputed elections. In reality, the Crown's legal advisers were merely attempting to recover an item of Chancery's lost jurisdiction. Yet it was the sort of point on which MPs had already learned to be peculiarly sensitive, and the kind of issue, moreover, that required no direct intervention from the king, since, as the Commons respectfully indicated at the outset, it was 'an unusual controversy between courts about their pre-eminences and privileges' and therefore best treated as a matter 'between the Court of Chancery and our Court'.

When, however, the Commons seized upon a disputed election in Buckinghamshire involving a certain Sir Francis Goodwin, who had been chosen in preference to Sir John Fortescue, a councillor enjoying the backing of the government, the king swiftly plunged into the fray, treating the matter in effect as a test case for the status of royal proclamations as a whole. Chancery had, in fact, excluded Goodwin on grounds of outlawry, only to find after two days of debate that its decision had been nullified by the Commons, and this was enough, it seems, to provoke the full weight of royal displeasure. Informing MPs that their privileges depended on his goodwill, James also compared their complaints to the murmurings of the people of Israel and ordered them 'as an absolute king' to consult the judges on the legality of their proceedings. And though the response, according to Sir Henry Yelverton, the Attorney General, was merely 'amazement and silence', there remained little doubt that the king had overreacted. 'The prince's command is like a thunderbolt,' declared

Yelverton, 'his command upon our allegiance like the roaring of a lion. To his command there is no contradiction.'

There followed three weeks of deadlock, as the Commons reinforced their case with a selection of curious precedents, including one case of 1581 in which an election, voided on the grounds that the candidate had died, had eventually been declared valid by the House when the supposedly defunct MP finally appeared to claim his seat. The debate, it is true, was conducted respectfully and with its fair share of honeyed words. Sir Francis Bacon, representing the Commons at the conference called to resolve the matter, remarked, for instance, that he found James's voice the voice of God in man, and confirmed that his colleagues were ready to reconsider their position, something they had not done for any previous ruler. James, meanwhile, played his part in smoothing the waters by declaring that he would allow free rein to his kindly nature and decline to press his prerogative against his subjects' privileges. Yet it was only on 13 April that the Goodwin case was finally settled by a compromise of sorts which entailed the annulment of all previous proceedings and the calling of a fresh election. And when, thereafter, the House swiftly proceeded to decide two further cases without reference to either himself or Chancery, James was left with little choice other than to avoid unnecessary acrimony, as he should have done in the first place, since he had now not only lost but visibly lost.

There was little doubt, of course, that Parliament's intransigence reflected a sensitivity and assertiveness that, from the king's perspective, only served to justify his own dogmatism. Nor was it entirely his fault that a minor dispute with the warden of the Fleet Prison was subsequently dragged out by MPs for three weeks on the grounds that their 'privileges were so shaken before and so extremely vilified'. Certainly, if he had been better advised in January when the offending proclamation that had sparked the Goodwin dispute was first issued, the whole question of parliamentary privilege might never have exploded in the first place. But James had not only chosen the wrong battlefield, he had also failed in advance to prepare the chosen ground sufficiently carefully. For, while Elizabeth had achieved much of her success by cultivating the closest links between her Privy Council and Parliament, and by carefully managing parliamentary debates, the new king found himself personally defending his government's policy in a way that his predecessor would only deign to do in a position of crisis. And this situation, too, was the result of decisions whose consequences he had failed to foresee.

Above all, James had seriously weakened his influence in the House of Commons by the removal of key figures. His promotion of Robert Cecil to the earldom of Salisbury, for example, meant that Cecil could no longer exercise the same direct influence in the Commons that he had previously done, and the same applied to Sir Thomas Egerton, another wise and experienced supporter of the Crown, who became Lord Ellesmere at the start of the new reign. Indeed, in 1604 only two privy councillors, Sir John Herbert and Sir John Stanhope, were on hand in the Commons, though neither possessed much influence. Herbert, the Second Secretary of State was widely known, indeed, as 'Mr Secondary Herbert', and the existence of a significant body of support for the Crown in the shape of individuals like Sir Roger Aston, Sir Richard Levison and Sir Edward Hoby, not to mention the cadets of great courtly families such as the Howards and Sackvilles, would not make good the deficiency. For not only had the presence of talented councillors allowed Elizabeth to exert influence upon the Commons, it had also enabled her to gauge their mood. Herbert, however, had not even risen on the first day of the recent session to ask for the customary subsidy, and in the vacuum that this left, it was hardly surprising, perhaps, that MPs would ultimately find their own leaders and organise their business independently.

By the end of the session, in fact, an alarmed committee of the Commons had put together a document known as 'The Form of Apology and Satisfaction to be Presented to His Majesty', which amounted, in spite of its self-consciously respectful tone, to a bold lecture to a foreign king upon the constitution of his new country. Laden with cumbersome statements of love and loyalty, and never actually presented to the king because it was never accepted by the Commons as a whole, the 'Apology' nonetheless represents an interesting insight into at least one significant element of parliamentary opinion. Certainly, everything James had said in the previous session and every implication of the language he had used was seized upon, while each of the claims he had advanced was opposed by counter claims which, in the main, were every bit as sweeping, provocative and unhistorical as his own. Not only was the king told that he had been misinformed on several important points but he also heard how these 'misinformations' had been 'the chief and almost the sole cause of all the discontentful and troublous proceedings so much blamed in this Parliament'. Above all, it was suggested, James had threatened the Commons' privileges, as a result of which 'the liberties and stability of the whole kingdom' had been 'more seriously and dangerously impugned than ever (as we suppose) since the beginnings of Parliament'.

This, then, was strong stuff. Yet the 'Apology' did not stop here, for its authors went on to point out that MPs' privileges were not a matter of the king's grace, but a 'right and due inheritance' no less than lands or goods, and that the Speaker's formal request to the Crown at the beginning of each session was 'an act only of manners'. They reiterated, moreover, the same claim to complete freedom of speech which had stirred such coals in the previous reign, and famously complained how 'the prerogatives of princes may easily and do daily grow' while 'the privileges of the subject are for the most part at an everlasting stand'. Worse still, James's habit of making sweeping statements on fundamental aspects of constitutional theory broadened the authors' agenda into a summary of a further range of unremedied grievances – matters which, they said, they had previously refrained from pressing upon Elizabeth 'in regard of her sex and age'. They noted, for instance, the exasperation caused by the excesses of royal purveyors and complained bitterly of the burdens placed on landowners by the activities of the Court of Wards. More importantly still, they reminded James that Kings of England had no 'absolute power in themselves either to alter religion (which God defend should be in the power of any mortal man whatsoever), or to make any laws concerning the same otherwise than as in temporal causes, by consent of Parliament'. And while they agreed with the king and his bishops on the need for uniformity and obedience, they nevertheless requested that some of the points in dispute – the sign of the cross in baptism, the use of the surplice and of the ring in marriage – should be made optional, while any persecution of dissenters should be in the hands of Parliament rather than Convocation.

Nor did the fact that the 'Apology' was never ultimately presented to James prevent him from seeing a copy and reacting with predictable irritation. Ultimately, indeed, the flourish of loyal and affectionate assertions with which the document ended, would do nothing to temper the king's suspicions about its authors' overall intentions. 'The voice of the people,' they had asserted, 'in the things of their knowledge is said to be as the voice of God.' And when James returned to prorogue the session after he had gone off to hunt at Royston at the end of April, in order to gain some respite from his 'fashious and froward' opponents in the Commons, his final speech could not have been more forthright. He was sure, he said, that there were many dutiful subjects in Parliament, but the pertness and boldness of some idle heads had cried down honest men, and he would not give thanks where no thanks were due, since it was neither Christian nor kingly to do so. 'I cannot enough wonder,' he declared, 'that in three days after the beginning of Parliament, men should go contrary to their oaths of supremacy,' and in making this point he confirmed his preoccupation with Puritan opponents. 'I did not think the Puritans had been so great, so proud, or so dominant in your House,' he declared, though he ended, sensibly enough, with appeals and admonitions rather than threats. Acknowledging that he had seen no evidence of disloyalty, he told MPs nevertheless that they had done 'many things rashly'. He was, after all, 'a king as well born as any of my progenitors' who required respect and expected MPs to use their liberty 'with more modesty in time to come'.

However, the main business of James's first Parliament, apart from obtaining funds, had actually been the matter of the formal union of his two kingdoms, and here, too, he was roundly frustrated. The cold disdain of Englishmen for their Scottish counterparts had not, in fact, been extended to James personally, but fears that Scotland might, in effect, take over her southern neighbour as a result of his accession were common enough. The Scots, wrote one contemporary, 'were suffered like locusts to devour this kingdom, from whence they became so rich and insolent, as nothing with any moderation could either be given or denied them'. And Ben Jonson's *Eastward, Ho!* only confirmed the prejudice that James's ultimate objective, however nobly and sincerely held, would have to overcome. When one of the play's characters, a mariner described Captain Seagull, refers in Act III to the Scots residing in Virginia – 'a land so rich that even the chamber pots are made of gold' – he is quick to add a biting quip that would swiftly run the playwright into trouble with the king. 'I would a hundred thousand of them were there,' the captain continued, 'for we are one countrymen now, ye know, and we should find ten times more comfort by them there than we do here.' The result was a brief jail sentence for both Jonson and his co-authors George Chapman and John Marston.

Yet James, predictably enough, remained undeterred by all complaints concerning not only 'the effluxion of people from the northern parts' but the proposed union itself. From his perspective, after all, the case for amalgamation was eminently reasonable and therefore infinitely persuasive. 'Hath not God first united these two kingdoms both in language, religion and similitude of manners?' James suggested in his introductory speech to the Commons. 'Yea, hath He not made us all one island encompassed with one sea, and of itself by nature so indivisible as almost those that were borderers themselves on the late Borders, cannot distinguish, nor know, or discern their own limits?' Then came a reminder that previous trouble between the two realms had been 'the greatest hindrance and let that ever my predecessors of this nation gat in disturbing them from their many famous and glorious conquests abroad'. After which, James rounded off his appeal with a selection of Scriptural references, spiced with the kind of heavy humour and egocentricity that was unlikely to appeal to MPs far more preoccupied with the legal, constitutional and financial consequences of the king's proposal. 'I am the husband, and the whole island is my wife,' he told them after a brief historical survey of the development of the English monarchy. 'I am the head and it is my body; I am the shepherd and it is my flock,' he continued, and on this basis he therefore hoped that 'no man will be so unreasonable as to think that I, that am a Christian king under the Gospel, should be a polygamist and husband to two wives; that I, being the head, should have a divided and monstrous body; or that being the shepherd of so fair a flock (whose fold hath no wall to hedge it but the four seas) should have my flock parted in two.'

Any hope that MPs would countenance such a project proved wholly unrealistic, however. They gave way enough in the first session, it is true, to set up an Anglo-Scottish commission intended to consider how to 'make perfect that mutual love and uniformity of manners and customs' necessary to 'accomplish that real and effective union already inherent in his Majesty's royal blood and person'. But when in April 1604 the king announced his wish to assume the title of King of Great Britain and alter the name of England, a line in the sand was quickly drawn. It was a line, moreover, that the Commons, notwithstanding James's own talents in debate, had little difficulty in defending. Would not a change in the name of the kingdom, it was suggested, abrogate existing laws and necessitate their re-enactment? This, after all, was something that the judges, too, had suggested in spite of considerable pressure from the king. And should a change of name not, in any case, occur until union had actually been realised? A commission had been agreed – albeit it one that would never give the king what he wanted – and this commission should be allowed to submit its conclusions in due course.

Parliament, therefore, had defeated the king in his most favoured and revered preserve – the realm of abstract argument. But in spite of any apparent commitment to the rule of reason, James would still not give way.

He had presented the choice before Parliament with all the usual excess of vigour, telling its members how any rejection of his plans would be to 'spit and blaspheme' in God's face 'by preferring war to peace, trouble to quietness, hatred to love, weakness to strength, and division to union'. And when defeat followed, he would not retreat gracefully. 'I am not ashamed of my project,' he told MPs at the end of the session, 'neither have I deferred it out of a liking to the judges' reasons or yours.' Accordingly, in spite of a pledge that he would not for the time being alter his title, he duly began to style himself King of Great Britain by royal proclamation in October 1604 on the grounds that God had given this title to the island, and on the advice, it seems, of Sir Francis Bacon alone, since the council appear to have regarded the gesture as provocative. The king was determined, wrote the Venetian ambassador, 'to call himself King of Great Britain and like that famous and ancient King Arthur to embrace under one name the whole circuit' of the island. And, as Bacon suggested, by proceeding through proclamation rather than statute, he could now use the new style in letters, treaties, dedications, further proclamations and upon coins. The result, however, was a largely empty victory and one achieved, ironically, not only at the cost of offending Parliament but of jeopardising all further serious dialogue about the goal of union itself.

Nor would James's subsequent pressure prove any more fruitful. The English Parliament, for instance, continued to employ the old terminology and though the Scottish Estates were forced to adopt it, Scots themselves were, if anything, even more resentful. Councillors were also asked to consider the feasibility of reducing the laws of the two kingdoms to a single system and to consider the possible benefits that might accrue from free trade, while the king toyed, too, with the notion of making Archbishop Bancroft primate of Great Britain, though obstacles on all three counts proved insurmountable. The conference of English and Scottish commissioners, which assembled in London in October 1604, did not, for instance, even consider the creation of a single parliament, and English merchants, fearing competition, roundly rejected any prospect of free trade, as did English ship owners who protested that while a Scots sailor could live on nothing more than oysters, his English counterpart required beer and roast beef. Ultimately, the pacification of the Borders would continue steadily and there was eventually some progress towards a common currency. But while the College of Arms managed to devise a new flag by imposing the cross of St George upon that of St Andrew, James was still complaining in 1607 of the 'crossings, long disputations, strange questions, and nothing done' that had dogged proceedings.

Indeed, the only topic even to receive a full airing at the London conference was the intricate issue of whether the Scots might be naturalised as English subjects. The commissioners, following the English judges, had made a distinction between those Scotsmen born before James's accession to the throne of England (the so-called 'ante-nati') and those born afterwards (the 'post-nati'). The former, it was suggested, should be naturalised by statute, while the latter should be automatically naturalised by virtue of common law. Only the post-nati, however, were to be deemed capable of holding office – a distinction which James was prepared to accept, so long as it was recognised that his prerogative remained wholly intact. And if this additional proviso made bad reading for the House of Commons, MPs too gave no quarter on the claim that the post-nati were naturalised by common law. Common law precedents, it was suggested, were established in ancient times when nationalism was weak. In any case, argued Sir Edwin Sandys, 'unions of kingdoms are not made by law but by act express', and naturalisation should not be conferred so easily by the chance results of royal marriages.

As the debate unfolded, moreover, long-established prejudices against the Scots were swift to resurface. England was depicted as a rich pasture about to be overrun by herds of lean and hungry cattle, while Sir Christopher Piggott, despite royal rage, poured forth a torrent of abuse, deriding Scotsmen as proud beggarly, quarrelsome and untrustworthy. Even the post-nati, for that matter, were to be barred from holding office, and neither category of Scotsmen should be granted the full rights of English citizens. How, it was suggested, could the Scots be subservient to English law and held to the payment of English taxes without the establishment of a single parliament, which, in spite of the king's wishes, was not a practical option at this stage? And how could Scottish laws be prevented from diverging without one chancellor and one Great Seal. The only practical solution, the Commons suggested, was 'perfect union', which was itself impractical at this time, as even the king had come to recognise. Once again, it seems, James had been foiled by the very techniques upon which he himself set so much store, though this did not prevent him from telling MPs that those who now spoke up in favour of perfect union did so only with their lips rather than their hearts.

On almost all counts, then, the first Parliament of the new reign had been little more than an exercise in frustration – and a largely unnecessary one at that. But it was not only the royal prerogative and England's relationship with Scotland that raised hackles and conjured frowns of frustration. Nor was it only the authors of the 'Form of Apology and Satisfaction' who found themselves stirring. For in spite of James's early guarantees that he would 'never allow in my conscience that the blood of any man shall be shed for diversity of opinions in religion', the session also delivered ominous signs of a less tolerant approach to English Catholics. Hitherto, the king had suggested that he was disinclined to 'persecute any that will be quiet and give an outward obedience to the law', and where he had hinted at action at all, he had actually tended to suggest that exile might be a preferable solution to capital punishment. 'I would be glad,' he declared at one point, 'to have both their heads and their bodies separated from this whole island and transported beyond seas.' Likewise, while James made it clear in his opening speech to Parliament on 19 March that Catholics were not to 'increase their number and strength in this Kingdom', so that 'they might be in hope to erect their Religion again', he also spoke of a Christian union and reiterated his desire to avoid religious persecution, declaring his readiness 'to meet them in the midway, so that all novelties might be renounced on either side'.

Just one month earlier, however, on 19 February, shortly after he discovered that his wife had been sent a rosary from the pope via one of his own agents, Sir Anthony Standen, James had ordered all Jesuits and Catholic priests to leave the country and reimposed the collection of recusancy fines. Only a week after James's address to Parliament, moreover, Lord Sheffield informed him that over 900 recusants had been brought before the Assizes in Normanby in Yorkshire, and by 24 April a Bill was introduced in Parliament which threatened to outlaw all English followers of the Catholic Church. The very few Catholics of great wealth who refused to attend services at their parish church were fined £20 per month, while middle-class recusants were fined 1s a week and those of more modest means found themselves liable for a sum totalling two-thirds of their annual rental income. In the atmosphere of increased stringency that followed, the fact that James allowed his Scottish nobles to collect English recusancy fines proved doubly provocative, and 5,560 convictions for refusal to pay followed in 1605 alone. Nor did the otherwise haphazard and negligent collection of all these fines serve to lessen their predictable impact upon

Catholic opinion. To Father John Gerard, the king's speech was almost certainly responsible for the heightened levels of persecution the members of his faith now suffered, while for the priest Oswald Tesimond it was a clear rebuttal of the early claims that James had made – claims upon which the papists had built such confident hopes. More importantly still, however, James's decision to reimpose the collection of recusancy fines was the fuse that led directly to the gunpowder deposited 'under his Palace of Parliament House' in November 1605.

The monstrous simplicity of the Gunpowder Plot was, in fact, the secret of its potential success. For zealots like Robert Catesby and Sir Thomas Percy there was no hope for a triumphant *coup d'etat* unless some altogether extraordinary disaster should temporarily paralyse the governing class, and the destruction of king, queen, Prince Henry, bishops, lords and Commons in a single, devastating explosion was precisely the sort of event in a society so closely bound to ties of tradition and territorial loyalty at every level that might well ensure the necessary upheaval – at least long enough for the arrival of foreign aid. Certainly, the moral considerations that made the majority of Catholics shudder at such mass murder do not appear to have bothered the plotters themselves. When James eventually asked Guy Fawkes, for instance, if he did not regret his involvement, the answer was succinct and clinical. The only cause for sorrow was the plot's failure, he retorted. 'A dangerous disease requires a desperate remedy' was Fawkes's sole additional comment. And in expressing himself thus, he spoke no doubt for the whole gang who had employed him to plant and fire the thirty-six barrels of gunpowder that were intended to raise the English Catholic gentry to arms and install James's daughter, the Princess Elizabeth, upon the throne as the puppet of a Catholic government.

According to latest estimates, Fawkes had more than double the powder he needed and even accounting for deterioration of the explosive in storage, no one inside Westminster Palace – or outside to a distance of 100 metres – is likely to have survived. So long as the plot remained secret, moreover, the assassination plan, if not its intended after-effects, had a more than reasonable chance of success. The king's ministers, it should be remembered, had no complicated network of spies at their disposal of the kind that had allowed Walsingham to foil so comprehensively all the conspiracies centred round Mary Queen of Scots. And though Robert Cecil would eventually exploit the plot's occurrence with predictable opportunism, any notions that he somehow knew about the plot in advance remain wholly unsubstantiated. The habit of secrecy was, of course, deeply engrained in those Catholic households where a fugitive priest might lay hidden all day in a space behind a chimney, with his communion plate and vestments, and only dare to ride out to his humbler parishioners under cover of darkness. For their part, neither James nor Cecil had any reason to suspect an imminent plot and would in any case have had no real idea where to start looking for one, since Catesby and his fellow conspirators had set to work in May 1604 and slipped into the English countryside some six months before 5 November. Indeed, even at the time of the plot's accidental discovery, the other conspirators, with the exception of Percy, who was known to be Fawkes's employer, remained wholly unidentified.

Robert Catesby, moreover, was a particularly formidable figure in his own right. Born in or after 1572 at the family seat of Lapworth in Warwickshire, he was the son of a prominent recusant Catholic who had suffered years of imprisonment for his faith before being tried in the Star Chamber in 1581 alongside William Vaux, 3rd Baron Vaux of Harrowden, and his brother-in-law Sir Thomas Tresham, for harbouring the Jesuit Edmund Campion. Another relation, Sir Francis Throckmorton, had been executed in 1584 for his involvement in a plot to free Mary Queen of Scots. And if any confirmation of the younger Catesby's readiness to carry forward the family tradition of zealotry was required, it was furnished by his education. For in 1586, he entered Gloucester Hall in Oxford, a college noted for its Catholic intake, only to leave before taking his degree after refusing the Oath of Supremacy, an act which would have compromised Catesby's Catholic faith. Presumably to avoid this consequence, he may then have attended the seminary college of Douai in France.

But it was not until the death of his wife, Catherine, in 1598 and the death of his father earlier in the same year, that Catesby became fully radicalised, and reverted to a more fanatical Catholicism. In 1601, for example, he was involved in the Essex Rebellion. And although the Earl of Essex's purpose lay mainly in furthering his own interests rather than those of the Catholic Church, Catesby nevertheless hoped that if Essex succeeded, there might once more be a Catholic monarch. The rebellion was a failure, however, and the wounded Catesby was captured, imprisoned at the Wood Street Counter, and fined 4,000 marks by Elizabeth I. Thereafter, Sir Thomas Tresham helped pay a proportion of Catesby's fine, following which Catesby sold his estate at Chastleton. Yet his opinions did not moderate and as Elizabeth's health grew worse, he was probably among those 'principal papists' imprisoned by a government fearing open rebellion. Certainly, he funded the activities of some Jesuit priests, making occasional use of the alias Mr Roberts while visiting them, and in March 1603 he may also have sent Christopher Wright to Spain to see if Philip III would continue to support English Catholics after Elizabeth's death.

By the start of King James's reign, therefore, Catesby was already an experienced and devoted crusader for the Catholic cause who would not hesitate to use violence in support of his faith. More importantly still, however, he had not only the skills and tenacity but also the charisma to pose a real and substantial threat to the new monarch and his government. Writing after the events of 1604 to 1606, the Jesuit Father Tesimond's description of his friend was most favourable. 'His countenance,' wrote Tesimond, 'was exceedingly noble and expressive … his conversation and manners were peculiarly attractive and imposing, and that by the dignity of his character he exercised an irresistible influence over the minds of those who associated with him.' Fellow conspirator Ambrose Rookwood, shortly before his own death, also declared that he 'loved and respected him [Catesby] as his own life', while Catesby's friend, Father John Gerard, claimed he was 'respected in all companies of such as are counted there swordsmen or men of action', and that 'few were in the opinions of most men preferred before him'. His frustration, meanwhile, at the failure of Essex's adventure actually seems to have sharpened an already well-honed monomania, which could only have served to increase the potency of any attempt on the king's life considerably.

However, as a result of lack of funds, the grinding physical labour of tunnelling and the wider objectives of the whole enterprise, the number of conspirators in the Gunpowder Plot became dangerously enlarged and led to their well-known betrayal by Francis Tresham, a wealthy Catholic gentleman who had been enlisted to gather stores of arms and prepare the West Country gentry for the impending insurrection. Distressed, it seems, by the prospect that the Catholic peers attending Parliament would almost certainly perish, Tresham opted to inform his brother-in-law, Lord Monteagle, and expose the conspiracy in a note delivered on the evening of 26 October, some eleven days prior to the opening of Parliament. 'I would advise you,' the note warned, 'as you tender your life, to devise some excuse to shift your attendance at this parliament, for God and man have concurred to punish the wickedness of this time.' But by instructing that the message be read aloud by his servant when it arrived during dinner, Monteagle attempted at one and the same time – as Tresham seems to have intended – to both thwart the plot and ensure that the conspirators were informed of their betrayal before the news was conveyed to Cecil, who decided to postpone a search until the last possible moment, partly, it seems, out of genuine incredulity, but more importantly in the hope that any conspiracy might thus be exposed more fully.

The king, meanwhile, who was away hunting was not shown the letter until two days before Parliament opened, and agreed that the note was likely to refer to an imminent attack. He remembered, he said, that his father had died by gunpowder, but it was not until three in the afternoon of 4 November that a search was first conducted of the cellar, and not until 11 p.m. that Guy Fawkes was finally arrested, still waiting with his 'blinde lanterne' and the watch a friend had bought for him specially, so that he might time his explosion accurately next morning. Knowing already of the warning letter to Monteagle and having encountered Lord Suffolk during an earlier search that afternoon, he had nevertheless pressed forward upon the slenderest hope that the government might not gauge his purpose in time. Yet his eventual discovery by Sir Thomas Knyvet sealed both his own and the plot's fate, as a few wild-eyed Catholic conspirators continued to gallop westward to raise a stillborn insurrection that even their supporters could see was now quite hopeless. Before long, indeed, John Chamberlain was already recording the lighting of 'as great store of bonfires as ever I thinke was seen', as the king's loyal subjects heard news of his deliverance and the first 'Guys' were burnt in celebration.

The king, meanwhile, had already retired for the night, but was awoken after Fawkes's arrest and promptly instructed that the prisoner be placed under close guard to prevent the possibility of suicide. Transformed, quite literally overnight, into something of a national hero, it was not long either before James was claiming full credit for single-handedly uncovering the plotters' designs. By early 1606, indeed, James had arranged for the publication of a short tract entitled *A Discourse of the Maner of the Discovery of the Late Intended Treason*, in which he was directly presented not only as the saviour of his own sacred person but as the deliverer of his kingdom as a whole and Parliament in particular. It was his 'fortunate judgement', it seems, that had been responsible for 'clearing and solving' the 'obscure riddles and doubtful mysteries' associated with the so-called 'Powder Treason', as it was known to contemporaries, and it was he, too, who had remained indifferent to the 'many desperate dangers' confronting him throughout. Appearing eventually in the king's collected works, the tract nevertheless makes clear in its preface that it was written by a court official under instruction from James himself, and leaves no doubt of James's perception of the entire episode's broader significance.

Since his Protestant faith had been so conclusively confirmed by Providence and he himself had been rescued by his own God-given perspicacity, the conclusions were indeed indisputable from the king's perspective, and he made them abundantly clear to Parliament only four days after the plot had been foiled. Comparing his escape to that of Noah from the flood and elaborating the parallel between the redemption of mankind and his own miraculous preservation, he reminded MPs that kings bore hallmarks of divinity and, like the tallest trees of the forest, were exposed to the greatest dangers. He had survived the Gowrie Plot, when his destruction would have brought immediate ruin to Scotland and deprived England of its future king, and now he had triumphed over an equally dastardly enterprise. When Tresham's letter to Monteagle had been presented to him, he had, he claimed, detected 'upon the instant' certain 'dark phrases therein', though the plot itself was not, he emphasised, the responsibility of English Catholics as a whole but of a few fanatics. Had the plot succeeded, he concluded, he would at least have perished in the noble execution of his regal duties, 'for Almighty God did not furnish so great matter to His Glory by creation of the world, as He did by redemption of the same ...'

No speech, in fact, could have captured more aptly the curious mix of egotism, good intentions and naivety that lay at the heart of James's kingship. At a time when his subjects were baying for retribution and sensing Jesuits – those 'reverend cheaters', 'prowling fathers' and 'caterpillars of Christianity' – behind every panel, he was prepared nevertheless to absolve the loyal majority of his Catholic subjects. But he remained almost obsessively preoccupied, too, with the plot's deep significance for him personally – something that manifested itself ultimately in a curious fascination with the trials and executions that followed as he framed specific questions for the prisoners, commanded the use of torture to extract information and pestered the government's prosecutor, Sir Edward Coke. At one time, indeed, he had thought to interview the captured plotters himself, though the idea, it seems, was ultimately too intimidating for him. Catesby, for his part, had been lucky enough to be shot dead while resisting arrest at Holbeach House, and Percy had died of his wounds soon afterwards.

Tresham, on the other hand, who had been in poor health before he joined the conspiracy, would die in the Tower on 22 December. But Guy Fawkes and the other surviving captives would incur no such good fortune before they were finally hanged, drawn and quartered on 27 January 1606.

Even so, neither James nor his subjects would be easily rid of the nagging anxiety and revulsion generated by the conspiracy. 'The king,' observed the Venetian ambassador, 'is in terror,' refusing to 'take his meals in public as usual' and living instead 'in the innermost rooms with only Scotsmen about him'. 'His Majesty on Sunday last', wrote the same Italian in early 1606, 'while at chapel and afterwards at dinner, appeared very subdued and melancholy; he did not speak at all, though those in attendance gave him occasion', which was 'unlike his usual manner'. And though he later 'broke out with great violence' against the 'cursed doctrine' by which some were 'permitted to plot against the lives of princes', declaring that 'they shall not think they can frighten me, for they shall taste of the agony first', he would confine his action ultimately to palliatives and pamphlets. The Oath of Allegiance, for instance, which was proclaimed law on 22 June 1606, required those who took it to affirm that the pope had neither 'any power or authority to depose the king … or to discharge any of his subjects of their allegiance…' and to acknowledge that no prince 'excommunicated or deprived by the pope may be deposed or murdered by their subjects …'. But when Paul V denounced the oath and forbade Catholics to comply, James merely plunged with characteristic relish into a protracted battle of print that was soon to spread to every corner of Europe.

'Hardly a day passes,' wrote the scholar Isaac Casaubon, 'on which some new pamphlet is not brought to him, mostly written by Jesuits,' and as James surrounded himself with his favourite coterie of 'ripe and weighty' Anglican divines, 'ever in chase after some disputable doubts which he would wind and turn about with the most stabbing objections that ever I heard', he became increasingly convinced, in his own words, that 'the state of religion through all Christendom, almost wholly, under God, rests now upon my shoulders'. One result was *Triplici nodo, triplex cuneus*, or an *Apologie for the Oath of Allegiance*, produced (or so it was claimed by the Bishop of Bath) over only six days and published in 1607, in which James did his level but laboured best to refute his Catholic critics. Such was the dreariness of the book's 112 pages that Boderie, the French ambassador, was adamant that James's principal councillors would have preferred if he had never published it 'or at least not acknowledged it as his own'. Reiterating time and again the familiar theme that Scripture, church councils and patristic authorities alike all advocated the primacy of secular authority, the king's last resort was a futile blast dismissing his arch-adversary, Cardinal Bellarmine, as a liar and a madman – an outburst rivalling an even more drastic loss of composure in debate four years later which led to the burning of the Protestant radicals, Edward Wightman and Bartholomew Legate, who became the last men to be executed for heresy in England.

Overall, of course, English Catholics continued to fare better than might have been expected. In the *Apologie for the Oath of Allegiance*, indeed, James was at pains to emphasise 'the truth of my behaviour towards the papists'. 'How many did I honour with knighthood of known and open recusants?' he asked. 'How indifferently [impartially] did I give audience, bestowing equally all favours and honours on both professions [religions]? And above all how frankly and freely did I free recusants of their ordinary payments [fines]?' He was even ready to add, for that matter, how 'strait order' had been 'given out of my own mouth to judges to spare the execution of all priests', and in a later edition of the *Apologie*, issued in 1609 under the title *A Premonition to all most Mighty Monarchs, Kings, Free Princes and States of Christendom*, he felt no compunction in describing himself as a 'Catholic Christian' and declaring that though 'I may well be a schismatic from Rome ... I am sure I am no heretic'. For the time being, moreover, such professions would be indulged compliantly enough by the majority of James's Protestant subjects, for while the 'Powder Treason' had fuelled the anti-Catholic bile of most Englishmen to new heights, it had also placed them at one with their new ruler as never before or after. But if James VI of Scotland was now, by sheer good fortune, truly King of England in his subjects' eyes, he could only hope to prosper in the longer term by learning to govern his altogether less regal whims and inclinations.

# 14 ⚘ Finance, Favouritism and Foul Play

'... a Prince's court
Is like a common Fountaine, whence should flow
Pure silver-droppes in general; but if't chance
Some curst example poyson't neere the head
Death and diseases through the whole land spread.'

*John Webster*, The Duchess of Malfi, *Act I, Scene i*

Some two years or so after James I ascended the throne of England, a royal commission reporting on the sorry state of the royal finances informed him, with all due delicacy, of the profligacy and greed that were sapping his resources and poisoning his court. 'The empty places of that glorious garland of your crown,' the king was told, '... cannot be repaired when the garden of your Majesties Treasure shall be made a common pasture for all that are in need or have unreasonable desires.' What the commissioners may have thought of the £15,593 lavished upon Queen Anne's childbed for the birth of Princess Mary on 8 April 1605, we can only guess. But it was the steady flow of gifts and pensions to a seemingly endless list of servants, associates, hangers-on and outright blackguards that emptied the royal coffers most remorselessly. A certain Jon Gibb, one of the king's lesser Scottish servants, had, for example, been gifted all of £3,000 that same year, while one of Queen Anne's favourites, listed in the accounts as 'Mrs Jane Drummond', had benefited to the tune of £2,000. And these, of course, were but droplets in a growing torrent of waste and excess. In 1603 'divers causes and rewards' accounted for £11,741, the next year £18,510, and the year following £35,239, while over the same period Exchequer spending on 'fees and annuities' rose giddily from £27,270 to £47,783.

In the meantime, like many incorrigible spendthrifts of his kind, James continued to salve his conscience by occasional half-hearted gestures of reform. On 17 July 1604, for example, he had signed a book of *Ordinances for the Governing and Ordering of the Kings Household*, which laid down, among other things, that only twenty-four dishes of meat should henceforth be served at the royal table instead of the customary thirty. The sergeant of the cellar, on the other hand, was to limit his issue of sack to twelve gallons a day and to offer it only to those noblemen and ladies who desired it 'for their better health'. Yet any notion that the financial crisis inherited from the previous reign and fuelled by James's broader improvidence could be remedied by savings from the royal larder were blatantly misconceived, particularly when soaring inflation was already drastically eroding income from the Crown's estates and undermining the rapidly dwindling yield from direct taxation. By the end of her reign, Elizabeth had accumulated a debt of some £430,000, though the disparity in value between the Scottish currency and its English equivalent may actually have served in part to lessen James's appreciation of the scale of his predicament. Scotland's pound, after all, was only one twelfth the worth of England's and when James reflected from the alternative perspective that the income of the Scottish Crown in 1599 had been merely £58,000 (Scots) as opposed to the £110,000 (English) available at the time of his accession, this too may well have exaggerated his impression of the new funds available to him.

James could, of course, comfort himself with the sounder assumption that some of the financial outlays dogging his predecessor's government no longer applied to his own. There was, for example, the cost of maintaining the expensive Border garrison at Berwick which had largely evaporated upon his accession, while the huge sums that Elizabeth found herself obliged to expend upon the Irish rebellion had also been curtailed by Lord Mountjoy's successful campaign at the very time that James was securing his new throne. During the financial year ending at Michaelmas 1602, the war in Ireland cost £342,074, only to fall after four years of the new reign to little more than a tenth of that figure. And when it is remembered that peace with Spain had been achieved in 1604, there were added grounds for cautious optimism about the Crown's potential solvency, so long as suitable economies could be sustained in other areas.

But moderation was, it seems, no less inimical to James's nature than humility and by 1607 he was already complaining bitterly of 'the eating canker of want, which,' he maintained, 'being removed, I could think myself as happy in all other respects as any other king or monarch that ever was since the birth of Christ.' Moreover, neither the expense of supporting a comparatively large royal family, nor the necessity of sustaining regal splendour by lavish gifts and patronage can remotely excuse the full measure of the king's wastefulness. 'My first three years were to me as a Christmas, I could not then be miserable,' James told Parliament at this time. But the ramifications of such generosity cannot be underestimated. On the one hand, as the wages of the king's servants, high and low, fell into arrears, graft and peculation infected every corner of his court. While royal bakers cooked lightweight loaves and misappropriated what they saved, and members of the king's boiling house interpreted their perquisite of the 'strippings' so freely as to leave little meat on the fowl served to 'the kings poor officers', Lord Treasurer Dorset – who would himself 'have spared a life to gain a bribe' – looked on largely impassively, borrowing heavily at interest rates of up to 10 per cent. And as Sir Julius Caesar, Chancellor of the Exchequer, entreated Sir Thomas Chaloner, governor of the Prince of Wales' household, to pay the wages of the boy's embroiderer, who 'is redy to perish for want of money', so the moral tone at court continued to decline in parallel.

Only a year before the meteoric rise to prominence of Robert Carr, it seemed that the king might have at last outgrown the need to lavish inordinate affection upon some handsome young man or other. James Hay had, of course, arrived from Scotland with the reputation of a favourite, and in the early months of his reign, England's new ruler had also fawned over Philp Herbert, Earl of Montgomery. But Hay's relationship with his master, though resented, was always seemly, while Montgomery always smelt far too strongly of the stables to appeal at any deeper level to the king's fancy. For James required manners and refinement as well as good looks from his ideal companion, and if gratitude, docility and a dash of vulnerability could be added to the mix, the king's devotion was assured. When Montgomery's wenching and drinking finally cooled the king's ardour for him, therefore, it was only to make space for Carr – an altogether more eligible competitor whom Sir John Harington described as 'straight-limbed, well-favoured, strong-shouldered and smooth faced', with fair hair and a pointed beard, and who, in the words of Sir Anthony Weldon, had 'had his breeding in France and was newly returned from foreign travel'.

The youngest son of Sir Thomas Ker of Ferniherst, who had served as Warden of the Middle March and been a faithful friend of Esmé Stuart, Robert Carr, as the surname became spelt in England, had been born around the time of his father's death in 1586, making him around 21 at the time that he was first dangled under the king's nose by none other than James Hay in an attempt both to undermine the Cecil–Howard stranglehold on power and advance once more the Scottish interest at court. Having served as a page who ran beside the royal coach in Scotland – a post from which he had been dismissed for clumsiness, according to Queen Anne – Carr seemed the perfect instrument for Hay's purpose. Athletic, personable and apparently guileless, he could be shaped to need, or so it seemed, and guaranteed to enliven the king's paternal instincts, which might grow with suitable prompting into something more compelling still. To all intents and purposes, the only requirement was a suitable setting to lay the bait, and such an opportunity was duly forthcoming on 'King's Day', 24 March 1607, when the annual jousting event to celebrate James's accession was held in the Whitehall tiltyard.

It was Hay, in fact, who made the first flamboyant entrance that day, attended by a number of gentlemen and pages adorned 'in their richest ornaments', one of whom, on a high-bred horse, had been appointed to carry the courtier's shield, and present it to the king. But the handsome young stranger's mount was 'full of fire and heat', we are told, and, after encouraging it to prance and curvet, he was thrown to the ground with such violence that his leg was broken. With the stricken rider lying prone before the royal stand, there could be, in fact, only one outcome. For the king, 'whose nature and disposition was very flowing in affection toward persons so adorned', was overcome with compassion, and 'mustering up his thoughts, fixed them upon this object of pity, giving special order to have him lodged in the court, and to have his own physicians and chirurgeons to use their best endeavours for his recovery'. Thereafter, it seems, James visited Hay's young gentleman not once but several times, captivated by his modesty and ingenuousness when questioned about the progress of his recovery. 'And though', Arthur Wilson tells us, the king 'found no great depth of literature or experience' in the patient, 'yet such a calm outside him made him think there might be good anchorage and a fit harbour for his most retired thoughts'.

So it was, then, that Robert Carr became firmly lodged in his sovereign's affections. Possessing, no doubt, a native shrewdness that enabled him quickly to gage the character of the man from whom he might hope all things, the young Scot seems, nevertheless, to have exhibited genuine charm and grace of manners, since even those who eventually came to hate his influence acknowledged his 'gentle mind and affable disposition'. Before long, James was personally teaching Carr 'the Latin tongue' and laying a foundation 'by his daily discourses with him, to improve him into a capability of his most endeared affections'. With no less care and trouble, the king also attended to his new favourite's appearance and bearing, equipping him with the finery in which he liked to see his courtiers 'make a brave show'. 'The young man,' wrote Sir John Harington in *Nugae Antiquae*, 'doth much study art and device: he hath changed his tailors and tiremen many times and all to please the Prince. The King teacheth him Latin every morning and I think some one should teach him English too, for he is a Scotch lad, and hath much need of better language'.

Harington makes clear, moreover, that James not only cared for Carr but smothered him with the kind of cloying attentiveness that was soon exceeding the bounds of strict propriety and plain common sense. 'The Prince,' the author informs us, 'leaneth on his arm, pinches his cheeks, smoothes his ruffled garments, and when he looketh at Carr, directeth discourse to divers others.' But this was not the limit of James's indiscretion, for in indulging his infatuation ever more ardently, the king's control of the young man became increasingly unwholesome. Robert Carr was to be, as James liked to phrase it, his 'creature'. 'Remember,' James wrote later, 'that all your being except your breathing and soul is from me.' Nor, from some perspectives, was this an exaggeration as the king's new favourite enjoyed the broader generosity of his master's patronage. In a letter from John Chamberlain to Dudley Carleton, dated 30 December 1607, we read that Robert Carr, 'a young Scot and new favourite', was appointed Gentleman of the Bedchamber. Then, on 6 December a royal warrant was made out 'To Robert Carr, Groom of the Bedchamber, for a yearly rent-charge of £600, to be paid to him for fifteen years by John Warner and three others, in consideration of a grant to them of certain arrears of rent due to the Crown'. And on 22 March 1608, there is a further warrant to pay £300 to Henryck von Hulfen 'for a tablet of gold set with diamonds and the King's picture, given by the King to Robert Carr, Gentleman of the Bedchamber'.

It was not until January 1609, however, that Carr was finally afforded a gift that placed his position far above that of any private gentleman. Since any advancement to higher honours necessitated a substantial endowment, the king opted in the worst possible way to employ Sir Walter Raleigh's estate at Sherborne for Carr's benefit. The only property saved for his family's benefit from the wreck of his fortunes, Sherborne was now lost to Raleigh as a result of a criminally careless flaw in the deed conveying the land to trustees. When so much Crown land was being sold to meet his most pressing debts, the king had therefore acted without compunction and in the teeth of widespread public hostility, which was shared by both Queen Anne and Prince Henry. 'I mun hae it for Carr,' James insisted, and, in accordance with his growing habit of flagrantly disregarding opposition to his immediate wishes and brazenly ignoring unpalatable facts, Sherborne was indeed acquired. Early in 1610, moreover, Carr was duly created Viscount Rochester, a Knight of the Garter and a Privy Councillor, before becoming Keeper of the Signet and, in effect, the king's private secretary in May 1611. Two years later he was to complete his ascent by becoming Earl of Somerset and Lord Chamberlain.

How far this remarkable advancement was linked to the emotional needs arising from the disintegration of the king's family life will naturally remain a matter for speculation. James had never, of course, been able to share any worthwhile intellectual activity with the queen, and her propensity for intrigue, gnawing intolerance to opposition, and widely broadcast flirtations with Rome had made it necessary, ultimately, to exclude her altogether from politics. Nor had she had ever been able to reciprocate the pent-up romanticism and desire for love and sympathy that her husband required of her. In spite of any residual friendship and tolerance they still shared, therefore, the death of the Princess Sophia within twenty-four hours of her birth in June 1606, followed the year after by that of the other baby daughter, Mary, seems to have damaged the couple's relationship irreparably. Still only 33, Anne remained pretty, if only blandly so, and, apart from the twinges of gout that she shared with her husband, continued to enjoy good health. But having borne James several children and endured a number of miscarriages along the way, the queen seems to have decided once and for all to escape the roundabout of pregnancy and bereavement she had ridden for too long and give herself over to more gratifying pursuits, such as the masques in which she caused such indignation by acting herself.

Nor, more importantly still, was the king able to fulfil his emotional needs by the kind of intensely devoted relationship that he might have been expected to enjoy with his eldest son. Indeed, as Prince Henry began to exhibit a cool, clear mind of his own around the age of 12, the gulf in personality and tastes between the two became increasingly evident. 'He was a prince,' wrote Sir Simonds D'Ewes, 'rather addicted to martial studies and exercises than to golf, tennis, or other boys' play; a true lover of the English nation, and a sound Protestant, abhorring not only the idolatry, superstitions and bloody persecutions of the Romish synagogue, but being free also from the Lutheran leaven.' Much more typically English than his father, then, Henry also preferred the company of 'learned and godly men' to that of 'buffoons and parasites, vain swearers and atheists'. But he was both insular and immovable in his prejudices, and while James had wished him to be all that he was not – athletic, self-confident and attractive – he was nevertheless hurt when the boy proved incapable of sharing his bookishness and open-hearted demonstrativeness. For Henry preferred action to dialectic and the tales of Elizabethan heroism to any talk of peace. By the age of 14, indeed, the prince found greater inspiration in the company of Phineas Pett, the Master Shipwright at Woolwich, than that of his father, acquiring far more knowledge about naval administration and dockyard construction, we are told, than king and council combined. That Prince Henry's greatest hero, however, should have been none other than Sir Walter Raleigh must surely have been his father's most galling disappointment of all.

Lonely and starved of affection as he was, therefore, it was not perhaps altogether surprising that James should have found in Robert Carr an emotional prop of sorts and a delectable object for his sweeter nature. But the naivety which led him to believe that he could turn his favourite into a statesman, and his blindness to the political consequences of his infatuation, were again indicative of that self-same lack of majesty that would always negate the king's more admirable qualities. Emotionally vulnerable – sometimes truly pathetic so – and stricken by insecurities that he was never ready to confront sufficiently earnestly, this was nevertheless a ruler who was incapable of doubting his own wisdom or the status of his divinely ordained office and believed that this was enough in itself to make him a leader of men. Such, then, was the potent combination of conflicting characteristics that had long infected James's kingship and would now threaten to poison the fibre of his entire court.

For the time being, however, it remained the king's financial position that troubled his ministers most pressingly. When Lord Treasurer Dorset dropped dead at a meeting of the Privy Council in April, 1608, his last 'accompt' showed debts totalling more than £700,000 and revealed, according to his successor, the Earl of Salisbury, that James's expenditure exceeded his ordinary revenue by some £80,000 a year. Wishing to consolidate all areas of policy under his sole control for an attempt at root and branch reform, and partly because there was nobody other than himself obviously qualified to assume the post, the Secretary of State therefore added the killing burden of the treasurership to his already overwhelming workload. For unless the Crown could secure adequate revenue to govern in times of peace without parliamentary grants, the king's independence from the tax-voting House of Commons, as Salisbury well knew, must surely be compromised. All hinged initially, however, upon clearing the mountain of government debt and the crippling burden of annual interest payments resulting from it.

In this last respect, at least, the new treasurer seems to have been surprisingly successful. Selling Crown property to the value of £400,000 and retrieving old debts amounting to a total of £200,000, Salisbury also revived various lapsed dues and fees and uncollected fines, and raised in the process a further £100,000. Yet the annual deficit continued to snap at the treasurer's heels and even his more drastic efforts to drive home the need for stringent economies proved unavailing with the king. James had made clear in *Basilikon Doron* that it was the duty of any prince to 'use true Liberality in rewarding the good, and bestowing frankly for your honour and weal'. The use of patronage was, after all, a tried and trusted method of guaranteeing loyalty, and even Salisbury observed to Parliament in 1610 that 'for a king not to be bountiful were a fault'. But James remained a long-term addict to excess: impulsive and compulsive at one and the same time, and driven by a heady need to satisfy his sentimental urges by giving. And just as Thomas Fowler had reported in 1588, the king's largesse still exceeded all sensible bounds of generosity as 'vain youths' and 'proud fools' continued to be lavished with royal gifts and favours. Nor had confession of his faults saved him from their consequences. 'I have offended the whole country, I grant, for prodigal giving from me,' he told Maitland in 1591.

So when Salisbury resorted to pleadings and shock tactics the outcome was hardly surprising. In 1610 a *Declaration of his Majesty's Royal Pleasure in the Matter of Bounty* committed James to 'expressly forbid all persons whatsoever, to presume to press us, for anything that may … turn to the diminution of our revenues and settled receipts …' Five years later, however, commentators were still bemoaning the throng of self-seekers besieging the throne. 'The King hath borrowed £30,000 of the aldermen of this city,' wrote John Chamberlain. 'But what,' he added, 'is that among so many who gape and starve after it?' Even more childlike lessons, for that matter, appear to have had no lasting effect. According to an anecdote related by Francis Osborne, Salisbury resorted on one occasion to piling up in front of James the £20,000 he had ordered the Exchequer to pay out as a gift, whereupon, we are told, 'the king fell into a passion, protesting he was abused, never intending any such gift: and casting himself upon the heap, scrabbled out the quantity of two or three hundred pounds', swearing that the intended recipient 'should have no more'.

Tall tale or not, the implication was nevertheless entirely borne out by the hard facts, which left the exasperated treasurer to resort to an altogether more painful course for the king's hard-pressed subjects. For if fire sales and savings were not the solution, then the yawning gap between expenditure and income could only next be bridged by raising import duties – an opportunity for which had conveniently presented itself as a result of the collapse, early in the reign, of the Levant Company. To compensate itself for the handsome yearly sum that the company had been paying for its monopoly of trade in the Eastern Mediterranean, the treasury subsequently imposed an extra duty on imported currants, and when a merchant named Thomas Bate refused to pay, the outcome was a lawsuit in the Exchequer Court which raised the whole issue of the Crown's right to 'impose' extra duties of this kind. Backed by sound Elizabethan precedents and the firm support of the judges, however, Salisbury prevailed and, after careful discussion with leading City merchants as to what the trade could reasonably stand, imposed in 1608 a new Book of Rates calculated to yield a further £70,000. In consequence, as the total value of the kingdom's trade continued to increase, the Crown's revenue rose from £264,000 in 1603 to £366,000 in the very same year that the new rates were implemented.

Sadly, however, and all too predictably, expenditure continued to rise even more rapidly over the same period – from £290,700 in 1603 to £509,524 in 1610, the year in which Salisbury finally called upon Parliament in the hope of agreeing a 'Great Contract' that might render the Crown solvent by surrendering the more provocative methods of raising revenue in return for a guaranteed grant of £200,000 lasting for the duration of the king's life. Yet the most promising financial reform of the reign was never to materialise. By July the bargain had actually been struck, though Parliament hesitated, it seems, 'to engage themselves in any offers or promises of contribution to the King, afore they were sure of some certain and sound retribution from him', and the final details were left until Parliament was reconvened in November. In the intervening period, moreover, MPs came to like the arrangement less and less until some bristled with resentment, while the king, possibly in consequence of statistics submitted to him by Sir Julius Caesar, Chancellor of the Exchequer, had decided that the agreed annual grant from Parliament was incommensurate with the monetary concessions he had granted. Poisoned too, it seems, by the influence of Robert Carr who was justifiably alarmed by the palpable hostility to Scots in general and himself in particular, James duly decided to abort his treasurer's plans by irritably dissolving Parliament in January 1611.

The prior behaviour of the Commons had demonstrated, however, just how far their confidence in the king's ability to manage his affairs was already compromised. Under the leadership of men like Edwin Sandys certain members had plainly adopted the principle that redress of grievances should be linked to financial co-operation, and proposed accordingly that the king's predicament should be used as a lever to wring wider concessions from him. They urged, therefore, that laws against recusants should be properly enforced and that all grants to courtiers should be cancelled: the first clear sign that Carr's rise to prominence was now a matter of open disapproval. 'Where your Majesty's expense groweth by the Commonwealth we are bound to maintain it: otherwise not', warned Sir Henry Neville, before demanding to know 'to what purpose is it for us to draw a silver stream out of the country into the royal cistern, if it shall daily run dry from private cocks'. But it was the member for Oxford City's last comment that carried with it the most wounding barb of all when he added how he would never 'consent to take money from a poor frieze jerkin to trap a courtier's horse withal'.

And as debate expanded to encompass 'impositions' and, in particular, the fear that James might soon see fit to raise far more than the sum laid down by the current Book of Rates, his only response was to visit the

House on 21 March and deliver the most outspoken defence of his royal prerogative to date. 'The state of Monarchy,' he declared in what has rightly become one of his best known speeches:

> is the supremest thing upon earth; for kings are not only God's lieutenants upon earth and sit upon God's throne, but even by God himself they are called gods ... In the Scriptures kings are called gods, and so their power after a certain relation compared to the Divine Power. Kings are also compared to the fathers of families, for a king is truly *parens patriae*, the politic father of his people ... Now a father may dispose of his inheritance to his children at his pleasure, yea, even disinherit the eldest upon just occasions and prefer the youngest, according to his liking; make them beggars or rich at his pleasure; restrain or banish them out of his presence, as he finds them give cause of offence, or restore them in favour again with the penitent sinner. So may the King deal with his subjects.

There followed, it is true, a reassurance from James that God would punish all kings who nevertheless failed to govern according to the laws. But this did not deter a minority of MPs from producing a solemn Petition of Right in defence of free speech or prevent calls for the reinstatement of 300 ejected clergy and criticism of the activities of the ecclesiastical courts. Nor was Salisbury spared the sting of the king's tongue when James finally called a halt to proceedings after discussion had turned to the scandals of Scottish favourites and the iniquities of court extravagance. 'Your greatest error,' James told him in the aftermath of his decision to prorogue Parliament, 'hath been that ye ever expected to draw honey out of gall, being a little blinded with the self-love of your own counsel in holding together of this Parliament, whereof all men were despaired, as I have oft told you, but yourself alone.' And just how little the king had learned throughout the last sorry year of ill-tempered debate was amply demonstrated by his prompt elevation of Carr to the House of Lords as Viscount Rochester and the scattering of another £34,000 in indiscriminate gifts, mostly to Scotsmen.

Thereafter, relations between the king and Salisbury appeared, superficially at least, to resume their former course. James returned to his rural delights, while his principal minister received the usual flow of instructions and admonitions, delivered more and more often now in the hand of Viscount Rochester to whom most royal correspondence was dictated. Many of James's letters dealt, in fact, with trivialities. He was irritated, for example, by the felling of trees in the Forest of Dean which disturbed the hawks, and concerned about the treatment of an albino hind. He wrote, too, about foreign affairs and the queen's illness of 1611, as well as the Oath of Allegiance and fines from those who refused it. On another occasion, he told Salisbury how he had heard from his pastoral retreat that deprived clergy still preached in the vicinity of Peterborough, and required the hard-pressed minister to admonish the bishop. But this did not stop him either from objecting to Salisbury's draft of a commission which, he believed, might be exploited to limit the royal prerogative.

Nor does the banality of much of James's correspondence conceal the fact that he continued to blame his minister for the failure of his first Parliament, and was intent upon quietly dropping him as his chief adviser. There were no more 'Little Beagle' letters, the old jocularity disappeared, and royal messages were uncharacteristically formal and business-like. Increasingly, too, Salisbury found himself relegated to routine matters, while the king took counsel with Northampton, who hated him, and Rochester who conspired against him at every turn. 'I have seen this parliament at an end,' the waning minister reflected, 'whereof the many vexations have so overtaken one another as I know not what to resemble them so well as to the plagues of Job.' To add to his woes, a scheme was hatched to seal an alliance with Spain by marrying Prince Henry to a Spanish princess and granting toleration to English Catholics – a project which only the uncompromising hostility of the prince himself ultimately thwarted.

Not altogether surprisingly, therefore, by February 1612 Salisbury was seriously ill. The king, meanwhile, as a friend informed the stricken minister, was 'careful exceedingly of your lordship's health', the more so, it seems, because he had continued in spite of his pain to attend to instructions about a royal paddock. There was a visit from the king, too, which appears to have consoled the patient further. 'This royal voice of visitation (like *visitatio beatifica*),' wrote Salisbury, 'has given new life to those spirits which are ready to expire for your benefit.' Yet within a few days James delivered a complaining letter. He did not like the manner in which Salisbury had dealt with a problem in London where many Englishmen were attending Mass in the chapels of ambassadors from Roman Catholic states. If he himself had not been absent, James reflected, the matter would have been better managed.

That April, as a last desperate remedy for the dropsy which was gaining on him, Salisbury resorted vainly to the healing waters at Bath before finally dying in the parsonage at Marlborough on 24 May during his return to Hatfield. Northampton, unable to conceal his malicious satisfaction, spoke heartlessly of 'the death of the little man for which so many rejoice and so few do so much as seem sorry'. And John Chamberlain, too, left little doubt that similar sentiments were circulating widely. 'I never knew so great a man so soon and so generally censured,' he wrote, 'for men's tongues walk very liberally and very freely, but how truly I cannot judge.' Yet it was Sir Francis Bacon, no friend, it must be said, to the late secretary and treasurer, who probably encapsulated most effectively his achievements and limitations. 'Your Majesty hath lost a great subject and a great servant,' he told the king on 31 May. 'I should say,' he added, 'that he was a fit man to keep things from growing worse but no very fit man to reduce things to be much better.'

For James, meanwhile, the news of his minister's demise appears to have represented nothing less than a blessed relief from a long-standing and irksome tutelage. Upon hearing of it at Whitehall, he delayed his intended departure for the country, according to Bishop Goodman's memorial of the reign, only until after dinner. Plainly, the obligation of gratitude and deference had become tedious to the king, and the restraints upon his conduct galling. No new Secretary of State was therefore chosen or any Lord Treasurer appointed to restrain and criticise the lavishness of his impulses. Instead, the king would receive his financial counsel henceforth from a commission, of which Northampton was the most influential member, and which soon ascertained that the debt so assiduously reduced by Salisbury to £300,000 in 1610 had risen once again by two thirds. In November 1612, moreover, Prince Henry also sickened and died, calling for his friend, David Murray, and his beloved sister, Elizabeth, and subsequently leaving his father freer than ever to administer his kingdom entirely as he pleased. All correspondence now was to be conducted through Carr, as Keeper of the Signet, or 'bedchamber men' who were, in any case, Carr's nominees. And with this, the transfer of effective power to the new favourite whom Salisbury had unobtrusively, but on the whole effectively resisted, became complete.

Though not promoted to any higher office, Viscount Rochester nevertheless became the mainspring on which the king's entire style of government now largely depended. Too dull of wit to offer effective counsel on matters of state, he was nevertheless faithful, obedient and ever watchful, and this, above all, made him a formidable guardian of his master's interests. 'I must confess,' wrote James, 'you have deserved more trust and confidence of me than ever man did, in secrecy above all flesh, in feeling and impartial respect … And all this without respect either to kin or ally or your nearest and dearest friend whatsoever, nay, unmovable in one hair that might concern me against the whole world.' Nor was James's estimation of his favourite's better qualities by any means entirely unfounded. Bishop Goodman, for instance, would describe Robert Carr as 'a wise, discreet gentleman', and even Sir Anthony Weldon, crabbed and tainted witness that he was, acknowledged how the young Scotsman 'was observed to spend his time in serious studies, and did accompany himself with none but men of such eminences as by whom he might be bettered'. There was no denying, of course, that Carr took bribes, as did almost everyone else at court, but he was always ready to secure the king's approval in doing so. And though he was ready to benefit from others' misfortune, he did not in general deprive men of their posts and influence gratuitously.

Yet if Rochester was indeed discreet for the moment, shunning his Scottish compatriots and own kindred, he was inevitably under pressure to join one of the two factions into which the court was cleanly divided, and ultimately, like all favourites who have been pampered too long, he would become overweening, forgetful of his dependence upon the king and thereby invite disaster. Deprived of the Salisbury alliance on which their power had rested, the Howard grouping in particular, headed by the earls of Northampton and Suffolk, was bound to lose its primacy without the favourite's good offices. But while the Howards offered fawning blandishments, Rochester was also courted by his old friend, Sir Thomas Overbury, who sought to draw him to the anti-Spanish camp and the ranks of the 'parliamentary mutineers'. Holding aloof with good sense, the pig in the middle for some time made no move 'save where the king had his interest'. But love for the king was ultimately overwhelmed by ardour for another – the Earl of Suffolk's very own daughter. Thus, wrote Arthur Wilson, were Rochester's good and affable qualities finally swallowed up in a 'gulf of beauty'.

Already married to the young Earl of Essex, son of the former queen's own firebrand favourite, Lady Frances Howard was a bad lot – proud, headstrong and violent, and raised in an atmosphere of self-interest, self-indulgence and sexual and political intrigue that had left her capable of both flagrant immodesty and implacable hatred. Her marriage had occurred in January 1606, when she was still only 13 and her groom only a year older, and had resulted from a typically well-intentioned and misconceived attempt by the king to heal a long-standing feud dating back to the time that the Howards had helped deliver the young earl's father to the scaffold. James saw himself, after all, as *rex pacificus*, the bringer of peace and harmony to each and any situation, who had not only rescued his realm from war with Spain but had already engineered a marriage between Salisbury's son and a daughter of the Earl of Suffolk, and would now sow further concord by similar means. That the principals on this occasion were mere children was neither an obstacle nor a concern.

In the event, for two years after her nuptials the bride returned to her father's house while her husband left for the Continent to mature over two years of travel. But by 1609 the earl, a solid if humourless young man, was back in England and set in vain upon consummating his marriage in his country home at Chartley. Witness after witness, in fact, would later confirm that the two had repeatedly bedded together and the countess herself would testify how she had made every effort 'that she might be made a lawful mother'. Yet Essex, according to his own subsequent testimony, 'felt no motion or provocation, and therefore attempted nothing': a situation that persisted well beyond the compulsory period of 'triennial probation', after which a marriage could normally be nullified and the couple given their longed-for release. Whether, of course, the groom's impotence was natural or the consequence of drugs which, years later, it transpired his wife had secretly procured from quacks and ministered to him, will remain uncertain. But the countess's mounting aversion to her spouse was an open secret at court and by 1613 she had become the object of outright scandal, for it was widely rumoured that she was both angling for divorce and already Robert Carr's mistress. More salaciously still, it was also suggested that she had relieved Prince Henry of his virginity and that jealousy over her affection had been the real cause of his hostility towards the king's favourite.

The Howards, however, saw only opportunity in the countess's prospective marriage to Carr until, that is, they found themselves confronted by a formidable obstacle. For favourites have favourites of their own and Sir Thomas Overbury, poet, bosom friend and personal mentor to Robert Carr, would prove an implacable enemy to their designs. Clever, able and intolerably arrogant, Overbury had already made many enemies, but his ascendancy over the king's favourite made him a formidable entity at court. Enjoying the privilege of unsealing and reading reports from English ambassadors abroad before passing them on to Carr, complete with margin comments, it was said that Overbury knew more secrets of state than the Privy Council. And while a casual dalliance between his protégé, who would be created Earl of Somerset on 3 November 1613, and a Howard daughter might be borne, the prospect of their marriage was utterly unacceptable to him. 'Will you never leave that base woman?' Overbury is said to have asked his friend during a heated altercation at 1 a.m. upon Carr's return from a tryst with his loved one. After which, according to Henry Peyton who witnessed the exchange, 'they were never perfectly reconciled again'.

The king, meanwhile, who was always inquisitive in matters of sex and therefore particularly attracted by the more novel and tawdry aspects of this case, had immersed himself thoroughly in every detail of the wretched affair. In all likelihood, of course, he regretted his own part in encouraging the marriage initially, for at one point in the subsequent trial he inveighed against the risks of marrying too early. But he was surrounded, nevertheless, by men who favoured the countess's divorce and he was in no doubt either about the potential alliance between Carr and the Howards. The immediate result was the appointment in May 1613 of a commission to investigate the validity of the marriage, headed by the muddle-headed but scrupulously honest George Abbot who had succeeded Bancroft as Archbishop of Canterbury in 1611. Within a year, however, this same divorce case had assumed dimensions that the king could scarcely have imagined, as he waded thigh-deep into a stagnant pool of sexual scandal, intrigue, corruption, sorcery and, ultimately, poison.

'What a strange and fearful thing it was', wrote Abbot, '... that the judges should be dealt with beforehand, and, in a sort, directed what they should determine', and that the king should profess how he himself 'had set the matter in that course of judgement'. For James was utterly credulous from the outset to Frances Howard's lies and resolved that she should have her way. When, for instance, a jury of twelve matrons examined the countess and asserted her virginity, the king ignored that she had been allowed to wear a veil throughout the examination, and discarded claims that her cousin – a true virgin – or some other woman had impersonated her. As the case dragged on throughout the summer, moreover, James had tried to influence the commissioners' decision by inviting them to Windsor and browbeating them on theological issues relating to the case for more than three hours. Throughout, there had been much talk of witchcraft, though Abbot could find no mention in the Church Fathers of a link between 'maleficium' and impotence in marriage.

Such, indeed, was Abbot's perplexity at the king's accusations of prejudice against the countess, 'which prejudice is the most dangerous thing that can fall in a judge for misleading of his mind', that the archbishop dropped at one point to his knees and tearfully implored the king to relieve him from his role as chairman. Completely oblivious to the irony of his own advice, however, James merely urged the dumbfounded cleric 'to have a kind of faith implicit in my judgement, as well in respect of some skill I have in divinity, as also that I hope no honest man doubts of the uprightness of my conscience'. Whereupon, after discovering that the vote of the commissioners would be tied he duly added two more members, Bishops Bilson and Buckeridge, with the result that on 25 September the divorce was finally granted by a vote of seven to five. That Bilson's son was thereafter created a knight, and Lancelot Andrewes, another in favour of a nullity verdict, soon became Bishop of London did not, of course, escape the notice of the cynics.

Even now, however, the sorry episode had still not run its course, for, after her eventual re-marriage on 26 December 1613, Carr's new wife showed no trace of forgiveness towards Overbury for opposing her divorce in the first place. Worse still, she not only hated him but feared he knew too much about her murky dealings with the quack doctor, Simon Forman, 'that fiend in human shape' as he was described by Richard Nichol, a contemporary poet. By now, in fact, Overbury had already fallen foul of a trap laid by Northampton in April, which had left him a close prisoner in the Tower. Using his daughter's hold on Carr and Carr's hold upon the king, the Howards' leader had arranged for his enemy to be offered a mission abroad while encouraging him to believe that he could count upon Carr's protection, should he refuse. Thereafter, when Overbury did indeed reject the offer for fear of losing influence at court, his insolence was punished accordingly, leaving him mortally exposed to the further intrigues of the woman who was soon to become Countess of Somerset.

Before that title was even hers, however, Robert Carr's bride had indeed seen off Overbury once and for all. Sending poison through a certain Richard Weston whom she had arranged to serve as Overbury's keeper, her first attempt at murder was foiled when Weston's design was discovered and prevented by Sir Gervase Helwys, Lieutenant of the Tower. But though Helwys suspected the main culprit he dared not accuse her and chose instead to keep the matter quiet, leaving her free to send further poisons, including arsenic and mercury introduced into tarts and jellies and a brace of partridges, some of which were sent by Carr himself, though there is evidence, mainly in the letters of Northampton who was certainly aware of his daughter's skulduggery, that the king's favourite had no direct knowledge of the plot. Ultimately, in any case, the lethal dose appears to have been delivered by an apothecary's boy whose handiwork resulted in Overbury's death the next day. And though a posthumous poem by Overbury entitled 'The Wife', which had been written, it seems, to discourage the marriage, would sell out five editions in less than a year, the nature of his death would remain, for the time being at least, a secret, as feasting and revelry marked the wedding and the new Countess of Somerset became the recipient of jewels worth £10,000 gifted to her by the king himself.

For the next year, indeed, Robert Carr enjoyed the high watermark of his fortunes, as James's confidence in him continued unbounded. He was, wrote Sir Geoffrey Fenton, the king's secretary in Ireland, 'more absolute than ever any that I have either heard or did see myself', while John Chamberlain observed how all matters were conducted between the king and his favourite 'within the shrine of the breast'. 'The Viscount Rochester at the Council table,' reported Gondomar, the Spanish ambassador, 'showeth much temper and modesty, without seeming to press and sway anything. But afterwards the King resolveth all business with him alone, both those that pass in the Council and many others wherewith he never maketh them acquainted'. With Northampton's death in June 1614, moreover, the duties of Lord Privy Seal and Warden of the Cinque Ports were, for the time being, entrusted to Carr who was also installed as Lord Chamberlain. He was even lucky enough, in the process, to be spared the poison chalice of the treasurership, which was passed instead to the Earl of Suffolk at a time when debt stood at £680,000 and £67,000 of the anticipated revenue for 1614 was already spent.

But Somerset lacked, it seems, the intuitive skill to handle his now increasingly complicated relationship with the king. During that summer James continued to indulge his peculiar delight in the domestic intimacies of his favourites and fussed over the countess almost as much as he did over her husband. When she fell ill after a wedding banquet in May, Chamberlain wrote that there had been 'much care and tender respect had of her, both by her Lord and the King'. Yet James's love was essentially possessive, and for all his gushing sentiments and lavish presents, the independence of his royal will and ego was what he had fought most passionately to establish and maintain throughout his life. Though he might happily become a slave to his own infatuation, therefore, he would never subject himself to the mercy of another's whim, and when Somerset now became rude and exacting, taking for granted what he had so far earned by chance, his many enemies made ready to strike in the most effective – and ironic – way possible. For it was in August 1614, on a hunting visit to Sir Anthony Mildmay's estate at Apethorpe that the king first encountered another young newcomer to the court, and by September Sir Geoffrey Fenton was observing how this same bright light, a youth named Villiers, 'begins to be in favour with his Majesty'.

# 15 ✦ Favourite of Favourites

'I, James, am neither a god nor an angel, but a man like any other. Therefore I act like a man and confess to loving those dear to me more than other men. You may be sure that I love the Earl of Buckingham more than anyone else, and more than you who are here assembled. I wish to speak in my own behalf and not to have it thought to be a defect, for Jesus Christ did the same and therefore I cannot be blamed. Christ had John, and I have George.'

*Comment made by the King of England to his Privy Council in 1617*

On 5 April 1614, the king opened his second Parliament, expressing the hope that it might become a 'Parliament of Love'. By 7 June, however, the same assembly had been dissolved in general acrimony, to be dubbed by posterity the 'Addled Parliament'. And though James had displayed a studied moderation and respect for legality that belies his later reputation, the lack of trust and respect that dogged him was apparent throughout. 'Kings,' he had declared, 'that are not tyrants or perjured, will be glad to bind themselves within the limits of law ... For it is a great difference between a King's government in a settled state and what Kings in their original powers might do ...' Yet the Addled Parliament proved, it was said, 'more like a cockpit than a grave Council' as MPs refused to take Holy Communion in Westminster 'for fear of copes and wafer cakes' and a hot-headed Puritan minority railed against morris dances and games upon the Sabbath. 'The House of Commons,' James complained to the Spanish ambassador, 'is a body without a head' where 'nothing is heard but cries, shouts and confusion'. 'I am surprised,' he added, 'that that my ancestors should ever have permitted such an institution to come into existence.' And such, indeed, was the disorder leading to its dissolution without the desired grant of taxation that one of the House's members, Sir Thomas Roe, thought he had witnessed the end 'not of this, but of all parliaments'.

Henceforward, impositions would continue to be raised without parliamentary consent, and when James called for a 'benevolence' or free gift from his subjects in 1614 against the advice of his Lord Chief Justice, Sir Edward Coke, ripples of resistance were predictable. Though Archbishop Abbot had donated a selection of plate to the treasury and Coke himself had come forward with £200, humbler folk like Oliver St John of Marlborough, who was prosecuted in Star Chamber for his protests, were not so willing to comply. In the case of Edmund Peacham, meanwhile, the king became personally involved and ordered that the elderly Somerset rector be consigned to the Tower for daring to warn of the possibility of rebellion and his sovereign's death within eight days like Ananias and Nabal. When, moreover, James suggested to Sir Francis Bacon, his Attorney General, that the realm's leading judges might be consulted singly on the issue of whether Peacham had committed high treason, Chief Justice Coke again intervened to warn that 'such particular and auricular taking of opinion' was in breach of English common law. Even so – and in spite of Coke's overall conclusion that Peacham's outbursts, though scandalous, were not treasonable, since he had not impugned the king's title – a treason verdict was indeed delivered in Taunton by King's Serjeant Montagu and Chief Baron Tanfield of the Exchequer Court. In the event, only the foul air of the local jail, which quickly killed him, prevented the execution of Peacham's sentence.

Rober Carr, meanwhile, had become increasingly prone to what the king himself described in a letter as 'streams of unquietness, passion, fury and insolent pride, and a settled kind of induced obstinacy'. The same letter, addressed to the favourite personally in 1615, also details how he had raised complaints with his royal master at unseasonable hours, as if on purpose to vex him, and how the court had become increasingly conscious of their angry exchanges and the king's sadness thereafter. Why, James complained, was Carr now refusing to sleep in the royal bedchamber and continuing to trouble him with so many idle and unfounded concerns? 'Do not all courtesies and places come through your office as Chamberlain, and rewards through your father-in-law as Treasurer? Do not you two as it were hedge in all the court with a manner of necessity to depend upon you?' the letter continued. And the same tone of slighted affection and sincere sorrow was to climax in a profession that the king was writing 'from the infinite grief of a deeply wounded heart' that he can bear no longer.

Amid such utterances, however, there also lurked more ominous sentences of the deepest significance for Carr's future prospects. 'For the easing of my inward and consuming grief,' James appealed, 'all I crave is, that in all the words and actions of your life you make it appear that you never think to hold me but out of love, and not one hair by force.' As the letter unfolds, furthermore, there are threats as well as entreaties. 'I told you twice or thrice you might lead me by the heart and not by the nose', the favourite is reminded, and 'if ever I find you think to retain me by one sparkle of fear, all the violence of my love will in that moment be changed into as violent a hatred'. 'God is my judge,' the message concludes, 'my love hath been infinite towards you, and only the strength of my affection towards you hath made me to bear these things and bridle my passion ... Let me never apprehend that you disdain my person and undervalue my qualities; and let it never appear that your former affection is cold towards me. Hold me thus by the heart, and you may build upon my favour as upon a rock.'

Arguably, no words of James capture more aptly so many essential features of his personality: the potency of his passions, the underlying insecurity that made them such a political liability, and his residing need for both unreserved affection and utter control, which inclined him so often to treat criticism as disloyalty, and equate opposition with enmity. Such traits could both endear and enrage, and, as Robert Carr and George Villiers would now find, both sweep to prominence and wash away in the same flood tide. For the latter's emergence was 'so quick', as Clarendon later observed, 'that it seemed rather a flight than a growth'. By April 1615, George Abbot and other enemies of Carr, knowing the king's curious rule of seeking the queen's approval for his favourites, were soliciting her for Villiers' appointment as gentleman of the bedchamber. And though she temporised, noting with commendable foresight that the young cupbearer would soon prove a plague to 'you that labour for him', she nevertheless proved willing on St George's Day 1615 to visit the king's bedchamber with the most fateful of consequences. Telling James that she had a new candidate for the honour of knighthood worthy of St George himself, she then asked Prince Charles to hand her his father's sword unsheathed, and proceeded to compensate for the king's well attested fear of naked steel by guiding his hand as the blade was duly applied to Villiers' shoulders.

For some time already, the new gentleman of the bedchamber had been in constant attendance upon James as royal cup-bearer and had shown, on one occasion at least, a feistiness that equalled his grace and good lucks. For when one of Somerset's followers had previously spilt a bowl of soup onto his magnificent white suit, he had been stung to anger and struck the man in the king's presence: an offence which could have led in theory to the loss of his right hand. But James refused to take the matter further and was soon idolising his new favourite in the all too familiar fashion. In the same year of his knighthood, indeed, the young man who had started life as the younger son of a Leicestershire squire also became Viscount Villiers, and in 1617, Earl of Buckingham, while 1618 witnessed his promotion to marquis and appointment as Lord High Admiral. Five years later the king bestowed his highest accolade of all by creating him the only duke of non-royal blood in the kingdom.

'The Duke,' wrote Clarendon many years later, 'was indeed a very extraordinary person; and never any man, in any age, nor, I believe in any country or nation rose, in so short a time, to so much greatness of honour, fame and fortune, upon no other advantage or recommendation than of the beauty and gracefulness of his person.' The Puritan memoirist, Lucy Hutchinson, however, expressed an altogether earthier verdict on the same theme when she reflected how 'a knight's fourth son' had been raised 'to that pitch of glory … upon no merit but that of his beauty and his prostitution'. What Hutchinson meant by this may well be imagined, and the nature of the king's sexuality has, of course, been a residing source of speculation across the centuries. Writing in 1617, the politician John Oglander observed how he 'never yet saw any fond husband make so much or so great dalliance over his beautiful spouse as I have seen King James over his favourites, especially the Duke of Buckingham'. The MP Edward Peyton, moreover, was another who noted how 'the king sold his affections to Sir George Villiers, whom he would tumble and kiss as a mistress'. Nicknaming Villiers 'Steenie' after St Stephen who was said to possess the 'face of an angel', James would also end a now famous letter of 1623 by affirming their relationship in the most striking manner. 'God bless you, my sweet child and wife,' the king declared, 'and grant that ye may ever be a comfort to your dear father and husband.'

As the king's 'great dalliance' proceeded, moreover, Villiers reciprocated in kind. In reply to James, he confessed how 'I naturally so love your person, and adore all your other parts, which are more than ever one man had'. 'I desire only to live in the world for your sake,' he continued, and 'I will live and die a lover of you'. Writing many years later, Villiers also pondered if the king loved him now 'better than at the time which I shall never forget at Farnham, where the bed's head could not be found between the master and his dog'. Whether the incident at Farnham was, of course, an isolated incident or merely part of a short-lived phase in the relationship, as some commentators have suggested, can never be known for sure. James was, after all, in vigorous middle life at the age of 48 when the young Villiers was first presented to him, though his health would deteriorate steadily beyond the age of 50. In similar fashion, it has sometimes been argued that the ardent friendship between Villiers and Prince Charles might have been rendered impossible by any sexual relationship between the favourite and the prince's father. But Charles would always show a rare capacity for blinding himself to a situation he did not wish to face, and the discovery during restoration work at Apethorpe Hall in 2004–08 of a secret passage linking Villiers' bedchamber with the king's state apartment appears particularly compelling.

Nevertheless, during the earliest days of Villiers' ascent James still seemed anxious to reassure his other favourite. Indeed, he was at pains to guarantee that Robert Carr's position was in no way threatened by the new arrival, even hoping to engineer the rise of the newcomer under the earl's own mantle and protection, so that they could all three be happy and harmonious together. Sir Humphrey May, renowned for his tact, was therefore dispatched to Carr to convey that Villiers would be calling to offer his services, and according to Sir Anthony Weldon, Villiers presented himself precisely as required. 'My Lord,' Carr was told, 'I desire to be your servant, and your creature, and shall desire to take my Court preferment under your favour, and your Lordship shall find me as faithful a servant unto you as ever did serve you.' But the olive branch was brusquely rejected, it seems, as Carr gave vent to the wrath that was already undermining his status in the king's affection. 'I will have none of your service, and you shall have none of my favour,' he is reported to have raged. 'I will, if I can, break your neck, and of that be confident.'

These, however, were the rantings of a thoroughly beleaguered man. In November 1614, he had initially thwarted Villiers' appointment as a gentleman of the bedchamber by installing a bastard kinsman of his own. Yet by the end of that month, it was known that the king was once again ignoring the parlous state of his treasury by donating £1,500 towards the expenses of a Christmas masque, 'the principal motive whereof is thought to be gracing of young Villiers and to bring him on the stage'. As his enemies circled, moreover, Carr saw fit in July 1615 to inquire from Sir Robert Cotton whether a pardon might be issued under the Great Seal of England exonerating him from any and all offences he had committed in the past. In the meantime, while Sir Henry Yelverton, the Solicitor-General, and Lord Chancellor Ellesmere refused to sign the necessary documents, the pardon was actually supported during a heated debate in the Privy Council by none other than the king himself. 'And so, my Lord Chancellor,' James commanded, 'seal the pardon immediately, for that is my will.'

But in spite of a royal tantrum accompanied by a stormy exit from the council chamber and a subsequent flight to the country in search of peace of mind, the pardon was never sealed, as Ellesmere continued to demur. And as the king, in Gondomar's view, listened to further tales against him and courtiers openly cut him, Carr grew increasingly ripe for the final blow, which was duly delivered in September 1615 when further revelations about the death of Sir Thomas Overbury finally issued from a dying Englishman in Brussels, named William Reeve. Smitten by conscience, Reeve confessed to servants of William Trumbull, James's ambassador, that as a former apprentice to the London apothecary William de Lowbell, he had been charged to administer to Overbury an enema contaminated by a mercury sublimate, and that he had been paid a sum of £20 for the murder by Carr's own wife, the Countess of Somerset.

When Sir Ralph Winwood, no friend of the pro-Spanish Howards, was subsequently informed and the countess was further implicated by the suspicions of Sir Gervase Helwys, the ensuing scandal threatened to expose in one fell swoop the full scale of the moral depravity and corruption for which the royal court had long become a byword. Under such circumstances, the king made before his council, we are told, 'a great protestation before God of his desire to see justice done, and that neither his favourite, nor his son himself, nor anything else in the world should hinder him'. Appointing commissioners to determine 'whether my Lord of Somerset and my Lady were procurers of Overbury's death, or that this imputation hath been by some practised to cast an aspersion upon them', James issued instructions 'to use all lawful courses that the foulness of this fault be sounded to the depth, that for the discharge of our duty both to God and man, the innocent may be cleared, and the nocent may severely be punished'. But in doing so he plainly appreciated that any attempt at concealment carried far more danger still than thorough inquiry, as he would indeed tell Somerset later. If, he informed his desperate favourite, 'I should have stopped the course of justice against you in this case of Overbury, who was committed to the Tower and kept there a close prisoner by my commandment, and could not have been so murdered if he had not been kept close, I might have been thought to be the author of that murder and so be made odious to all posterity'.

Even so, James's conduct as the trial proceeded, notwithstanding his intense interest and constant interference, was generally creditable. The countess confessed her crime beforehand and some time after October 1615 she and her husband were arrested. And though Carr's direct responsibility for the murder remains uncertain, his destruction of Northampton's many letters to him and attempts to falsify the dates of Overbury's correspondence did little for his credibility. Seizing other letters in the possession of a certain Mrs Turner, the depraved woman from whom the countess had obtained charms and poisons, he even seems to have made a vain resort ultimately to blackmail the king himself. 'It is clear,' James wrote, 'that he would threaten me with laying an aspersion upon me of being in some sort accessory to his crime.' To his credit, however, James stood firm against Carr's 'scribbling and railing', informing him how he would never 'suffer a murder (if it be so) to be suppressed and plastered over' nor spare, 'I vow to God, one grain of vigour against the conspirators'.

At 9 a.m. on 24 May 1616, therefore, the Countess of Somerset was duly conveyed to Westminster Hall to face trial for murder, the headsman's axe – its blade turned away – preceding her as she entered. Though she had recently given birth to a daughter Anne and now stood – 'in black Tammel, a Cypress Chaperon, a cobweb lawn ruff and cuffs' – with downcast, gently weeping eyes, Sir Edward Coke would soon be defaming her as a whore, a bawd, a sorcerer, a witch, a felon, a devil and a murderer: the very incarnation, no less, than the seven deadly sins. Spectators, moreover, had paid dearly to behold the Lord Chief Justice at his trade. 'I know a lawyer', wrote John Chamberlain, 'who had agreed to give £10 for himself and his wife for two days', while another buyer 'gave £50 for a corner that could hardly contain a dozen'. For Coke's reputation preceded him: so powerfully indeed that when the elderly judge requested Somerset's presence in London, the accused vowed he would not go and, in doing so, called upon the king's support. But while James, in spite of all that had passed between them, still gushed sentiment for his fallen favourite and feared that he might 'never see his face more', the response was nevertheless unfaltering: 'Nay, man, if Coke sends for me, I must go.'

Meanwhile, in spite of the Lord Chief Justice's imprecations and a plea of guilty by Somerset's wife, which the countess acknowledged could not 'extenuate my fault', there was never any doubt that Lord Ellesmere's subsequent sentence of death against her would be overturned by the king's prerogative of mercy. One Italian observer, Eduardo Pallavicino, had been wholly overcome by the countess's nobility, grace and modesty, leaving him in no doubt that she had been led into crime by her husband. In Chamberlain's view, however, the accused had won pity both by her shows of tears and otherwise sober demeanour, 'which in my opinion was more curious and confident than was fit for a lady in such distress'. For her sins, moreover, she was spared the anguish of lengthy trial, since the entire process was over within no more than two hours. She had come, seen, sobbed and conquered, it seems – the beneficiary of the best possible legal solution from the king's own point of view: both clear-cut and bloodless, since confession and contrition were considered powerful mitigators of guilt, and proferring, under these most awkward circumstances, the best available hope of minimising public outrage.

Carr, for his part, had also been urged by the king to 'honour God and me' by confessing his guilt, after which the royal prerogative might likewise be employed to guarantee a pardon. In contrast to his wife, however, the embattled earl would face a trial at Westminster Hall on 25 May lasting from nine in the morning until ten at night, and one that did not run nearly so smoothly from his royal master's perspective. Wearing 'a plain black satin suit laid with satin laces in a seam' and 'a gown of uncut velvet', Carr obstinately refused, in fact, to confess his guilt. 'I am confident in mine own cause,' he declared, 'and am come hither to defend it.' On the other hand, his prosecutor, the Attorney General, Sir Francis Bacon, was unable and even a little unwilling, it seemed, to press home any truly convincing case that the accused had been a knowing accomplice to his wife. 'For the poisonment', he declared, 'I am sorry it should be heard of in our kingdom', since such a crime was not *nostri generis* but 'an Italian comfit for the Court of Rome'.

Defending himself, as Bacon observed, both 'modestly and wittily', Carr gave James, in fact, one of the most miserable days of his life as he watched the landing stage at Whitehall eagerly for any boat that might bring him news of the outcome. No courtier, it was said, had ever seen him 'so extreme sad and discontented' and his efforts to distract himself by discussing with Gondomar possible terms for a marriage treaty had proved wholly ineffective. That night, indeed, he would neither dine nor sup until he had learnt the result he most desired: a unanimous vote of guilty by the Lords that vindicated the king's justice and left him conveniently unentangled in the whole sorry episode. Hereafter, he could not only intervene once more to spare the fallen favourite's life but duly ensure that that life, as well as the life of the convicted countess, would be wholly worth living. For, even as the Somersets were promptly conveyed to the Tower under sentence of death, their long-term comfort and well-being was already assured.

In all, the couple would remain in the Tower for some six years, during which time, it seems, a 'great falling out' occurred between them, which continued for the rest of their days, leaving them, we are told, 'though in one house as strangers one to another'. The revulsion that Carr now felt towards the convicted murderess who had robbed him of his friend eventually became overwhelming, it seems, as the countess exploited her comparative liberty within the Tower's walls to conduct an affair with the so-called 'Wizard Earl' of Northumberland. But she would live a further sixteen years before finally meeting what appears to have been, if Arthur Wilson is to be believed, a particularly unwholesome end. 'Her death,' wrote Wilson, 'was infamous ... for that part of her body which had been the receptacle of her sin, grown rotten (though she never had but one child), the ligaments failing, it fell down and was cut away in flakes, with a most nauseous and putrid savour, which to augment, she would role herself in her own ordure in her bed [and] took delight in it.'

Before that time, however, both she and her husband had nevertheless been allowed to retire to Lord Knollys's house at Rotherfield Grays in Oxfordshire, a secluded country residence, on condition that they would confine their movements to within 3 miles of it. Still unable to cast off his sentimental attachment to his former favourite, the king had also ignored Carr's conviction and allowed him to remain a member of the Order of the Garter. The fallen favourite had saved his life in the first place with nothing more than a letter to his royal master requesting that he be hanged rather than beheaded and that his daughter might be maintained out of the income from his forfeited lands. And the king's goodwill did not end here, for on 7 October 1624, a little over five months before his death, a formal pardon was produced under the Great Seal, which allowed Carr to retain an income of £4,000 per annum until his own death in 1645.

In return, James would incur untold damage, both to his own reputation and that of his court, notwithstanding Archbishop Abbot's dutiful refrain that the king's life was 'so immaculate and unspotted from the world … that even malice itself could never find true blemish in it'. James's own relief that the emotional turmoil of Carr's downfall had been negotiated without major political disaster was palpable. But rumours of undisclosed facts persisted and moralists continued to complain that the king had appeared to condone such wickedness among those close to him. Indeed, the impression which Carr's rise had created among the country gentry and the nobility who did not frequent the court was as nothing compared with that created by the manner of his fall. Nor, it seems, were such misgivings confined to the more sober elements of Jacobean society, for John Chamberlain relates how Queen Anne and the Countess of Derby were mistaken, while driving in a coach, for the Countess of Somerset and her mother, and subjected to fierce abuse by a mob of Londoners.

As contempt and disillusionment bore in upon James, moreover, the effects upon his health and personality became increasingly marked. Arthritis, combined with gout, became chronic during the winter months from 1616 onwards and he found himself plagued, too, by abdominal colic, sleeplessness, frequent diarrhoea and, after a serious attack of jaundice in 1619, acute kidney pain resulting from nephritis – all of which depressed his spirits and rendered him either fractious or morose. 'He is of exquisite sensitiveness,' wrote Sir Theodore Turquet de Mayerne, his physician, 'and most impatient of pain; and while it tortures him with violent movements, his mind is tossed as well, thus augmenting the evil.' Passing urine 'red like Alicante wine' and refusing access to individuals like Sir Ralph Winwood, in order to avoid government business, the king turned increasingly to every available expedient to ease his anguish and relieve his symptoms. 'He demands relief from pain,' wrote de Mayerne, 'without considering the causes of his illness.'

In such circumstances, therefore, it was hardly surprising that James's dependence upon George Villiers, Earl, Marquis and later Duke of Buckingham, should now have increased to unprecedented levels. 'The king,' it was said, 'is not well without him, his company is his solace.' And Buckingham would indeed provide a powerful antidote for the misery and humiliation of Robert Carr's long-drawn-out ruin: a private haven of beauty, grace, sympathy and gaiety, in which the king could forget both his own ailments and the censure of the outside world. Tall, comely and handsome, with a fine forehead, clear blue eyes, dark chestnut hair and a pointed beard of golden brown, as well as long, slender legs which made him renowned for his elegance as a dancer, Buckingham exhibited an ideal combination of masculine strength and female delicacy to make him irresistible to the king. 'I saw everything in him full of delicacy and handsome features,' wrote Sir Simonds D'Ewes, the contemporary antiquarian, 'yea, his hands and face seemed to me especially effeminate and curious,' while another observer noted how 'from the nails of his fingers, nay from the sole of his foot to the crown of his head there was no blemish in him'. Yet he was manly in his tastes, excelling in sports and nurturing a roving eye for the opposite sex, for, as Arthur Wilson put it, 'if his eye cull'd a wanton beauty, he had his setter that could appoint a meeting'.

Nor, it seems, was Buckingham prone to the naivety or presumption that had finally put paid to his predecessor. On the contrary, the new favourite was altogether more formidable, because more able, than the old. 'No one dances better,' wrote Arthur Wilson, 'no man runs or jumps higher.' 'Indeed,' continued Wilson, 'he jumped higher than ever Englishman did in so short a time, from a private gentleman to a dukedom.' And the secret of this giddy ascent was his expert reading of the king's own needs and expectations. Pawed, petted, pampered and puppied, Buckingham pleased his royal master by diligent attendance at divine service and when James decided to produce a meditation on the Gospel of St Matthew, the favourite was quick to ask that he might act as amanuensis. 'How can I but write merrily when he is so I love best and beyond the world,' he told James in response to a request for merry letters, and after a gift from the king, his response was even more effusive. 'I am now,' he wrote, 'going to give my Redeemer thanks for my maker.' No hyperbole, indeed, was too much for Buckingham's pride or beyond his master's satisfaction: so much so that he was also capable of cultivating a witty, playful impudence that both enhanced and emphasised his hold upon the king. 'And so I kiss your dirty hands,' he wrote later.

But Buckingham was also prepared to be instructed. When, for example, James proffered a New Year's gift in 1619, dedicating *The Meditations upon the Lord's Prayer* to him, it was accompanied by a message declaring how 'I dayly take care to better your understanding to enable you the more for my service'. And it had not been long either before Queen Anne, always so hostile to Carr, was duly charmed by the handsome courtier's winning ways. Referring to him in several letters as 'my kind dog', she thanked him on one occasion for 'lugging the sow's ear' and urged him to remain 'always true' to her husband, in response to which Buckingham confirmed that in obedience to her desire, he had pulled the king's ear 'until it was as long as any sow's'.

And with the good offices of the royal family assured, Buckingham duly sought to employ his own relatives to bolster his position, fully aware, it seems, that Robert Carr's failure to do so had left him cruelly isolated when the moment of crisis arrived. Buckingham's brother John, for example, notwithstanding temporary bouts of insanity, was duly joined in marriage to the daughter of Sir Edward Coke and created Viscount Purbeck, while 'Kit', his other brother, who was widely regarded as little more than an amiable fool, would find himself created Earl of Anglesey in 1623. The favourite's sister Susan, meanwhile, was married to Sir William Fielding, who subsequently became Earl of Denbigh and the father of a daughter who was in her turn betrothed at the age of seven to the Marquis of Hamilton. Finally, as if to seal the happy nexus, Anne Brett, a cousin of the Villiers family, was married to Lionel Cranfield, the future Lord Treasurer and Earl of Middlesex.

But if Buckingham's grasping tribe of kinfolk caused ill-concealed murmurs in the country at large, it was the influence of his mother – an overweening and predatory old termagant - that evoked the most unremitting outrage. Descended from an impoverished branch of a great medieval family, she had been born Mary Beaumont and served as a waiting-gentlewoman to her cousin, Lady Beaumont of Coleorton, before attracting an offer of marriage from Sir George Villiers, a widowed Leicestershire knight of no exceptional means. Even after her husband's death, moreover, subsequent marriages, first to Sir Thomas Rayner and then Sir William Compton, an alcoholic nonentity, brought her little advantage. Yet if her choice in husbands belied her true ambition, the rise of her son was squeezed for every opportunity. Parading her 'numerous and beautiful kindred' before a string of wealthy husbands, she exploited what amounted to a vicious system of blackmail, in which the king frequently connived. It was she, it seems, who had set her mind upon Frances Coke as a bride for John Villiers in the spring of 1616, and it was her pressure for a generous marriage portion that ultimately sealed the Lord Chief Justice's dismissal for initially resisting the match. Such, indeed, was her reputation that even Buckingham himself discouraged her presence at court upon her elevation to the rank of countess in 1618.

Buckingham's own marriage in 1620, meanwhile, was arguably the final step in securing his status in English society. A Florentine observer living in England at the time considered it 'very dangerous for such a powerful courtier to marry at all', but the choice of Lady Katherine Manners, daughter of the Earl of Rutland, was an impeccable one, irrespective of the Catholicism of her family, and the union proved fruitful on all counts. Reputedly the richest heiress in England, her great wealth would provide Buckingham with precisely the independence he required, while her simplicity and devotion guaranteed that she would never become the kind of political liability that had finally undone Robert Carr. Indeed, her gentleness and womanly tenderness, devotion and purity of life, became conspicuous amid the almost universal corruption and immorality of the Court. 'There was never woman loved man as I do you,' she wrote her husband during one of his absences, and she doted equally tenderly upon the son and daughter that the marriage eventually produced.

But it was not only his choice of bride that distinguished the Duke of Buckingham so markedly from his predecessor. Above all, unlike Robert Carr, he wanted to enjoy the exercise of real political power and to employ it, wherever possible, in his royal master's interests. The former he certainly achieved on an extraordinary scale until he became, in the words of the Earl of Clarendon, 'the man by whom all things do and must pass' and '… entirely disposed of the wealth of the three kingdoms'. However, Clarendon's further claim that, in dispensing of patronage, Buckingham was guided 'more by the rules of appetite than of judgement …' remains harsh. For, while no promotion was ever granted without financial sweeteners, the king's favourite was neither oblivious to the broader interest nor devoid of judgement in his choice of men. 'I never saw a young courtier,' James told Parliament in 1624, 'that was so careful for my profit without any respect as Buckingham was.'

Nowhere was this plainer, moreover, than in the appointment in 1614 of Lionel Cranfield as head of a commission to examine government expenditure in an effort to free the king from his perpetual bondage to debt. A thrusting man of business who had begun as a mere city apprentice before winning rapid success as a cloth merchant and member of the Mercers' Company, Cranfield had become a farmer of various royal revenues and eventually Surveyor General of the Customs in 1613. 'The first acquaintance I had with him,' James later recalled, 'was by the Lord of Northampton, who often brought him unto me as a private man before he was so much as my servant.' But it was Buckingham who 'fell in liking with him and brought him into my service' and Buckingham who 'backed him against great personages' and 'laid the ground and bare the envy'. One result was stringent savings: in the royal household, in the exchequer, in the wardrobe, in the navy, and in Ireland. The other was a mortal struggle between Buckingham and his vengeful Howard enemies who could neither forget his upstart origins nor forget how his rise had coincided with Robert Carr's disastrous ruin.

The corrupt practices of the treasury under the Earl of Suffolk's administration were, of course, a byword, and all, with the possible exception of the king himself, acknowledged that in order to gain payment for a bill or settle overdue expenses, a hard bargain must first be driven with Sir John Bingham, the sub-treasurer, and indeed Lady Suffolk. Only when these two expressed their satisfaction would the earl himself approve, and only when the whole unholy trio had been duly ripped from influence, therefore, was the king's financial predicament likely to amend. Possessing no administrative expertise, Suffolk's personal extravagance had led him to spend some £200,000 on his Audley End estate, and he had displayed the same insouciant disregard for economy in government. So now, as Buckingham's attitude switched from careless tolerance to hostility and stories of Lady Suffolk's transactions reached the ears of the king in June, the whole rotten tree of Howard influence, with all its branches, became ripe for cutting.

Within the month, in fact, Lady Katherine had been ordered from the capital, and by the end of July her husband's resignation was also demanded by the king. And though it would require a further eighteen months to complete the Star Chamber inquiry which culminated in imprisonment for both and a crippling fine of £30,000, the rout was comprehensive. Not without good reason, Sir Francis Bacon compared the countess to a woman who kept shop while her creature Sir John Bingley cried 'What d'ye lack?' And accordingly the Attorney General left no worm a hiding place. The Suffolks' two sons lost their court appointments and Lord Wallingford, as a son-in-law, his office as Master of the Wards. Sir Thomas Lake, meanwhile, who as Secretary of State had so far been able to fend off the more dangerous attacks upon his Howard cronies, now found himself tainted and doomed to resignation by a vicious and indefensible accusation of incest launched against the Countess of Exeter by his wife. The king himself, indeed, had exposed the perjury of a maid involved in the case by taking her to the room at Wimbledon and demonstrating that the arras, behind which she falsely claimed to have heard a compromising conversation, fell far short of the floor.

So it was, then, that by November 1619 a popular jest came to be in general circulation throughout the capital. For now, it was said, the entire Howard faction were at liberty to set up a Privy Council of their own within the Tower, with Suffolk as treasurer, Carr as chamberlain, Lake as secretary, Lord Wallingford as Master of the Wards, and the hapless Lord Howard Walden, who had also been sucked into the vortex, as Captain of Pensioners. Ultimately, indeed, only the senile Earl of Nottingham was left his freedom, though he too had been forced into resignation by the threat of an inquiry into Admiralty and Dockyard accounts. Perhaps because of his reputation as commander of the ships that defeated the Spanish Armada, he had been allowed to retire unmolested, though only to be replaced in February 1619 by Buckingham himself.

Now, however, there would be no new faction at the heart of government, since Buckingham, unsurprisingly, desired no near rivals. For replacements, he sought only talented and industrious servants who would neither seek to fashion policy nor aspire to control patronage. As secretaries, Sir Robert Naunton and George Calvert were ideal prototypes, since the king himself preferred 'conformable men with but ordinary parts', while Cranfield would be left to go about his business unimpeded, shaving Admiralty costs from £57,700 to £30,000, and reducing costs for the king's velvets, silks, saddle costs and other items from £28,000 to £20,000. For the time being, then, the wings of the great nobility, notwithstanding the wealth and territorial independence that had given them a certain independence from the Crown, were safely clipped – leaving only Buckingham and the ailing king himself to steer the ship of state.

# 16 ⚘ Faraway Realms

'Even such is Time, which takes in trust
Our youth, our joys, and all we have,
And pays us but with age and dust ...'

*Sir Walter Raleigh's 'Epitaph', written in the Gatehouse at
Westminster Palace the night before his death*

The Castilian nobleman Don Diego Sarmiento de Acuña, better known
as the Count of Gondomar, first cast eyes upon England as Spain's
new ambassador in the spring of 1615, though he was to bring with him
neither fresh new shoots of friendship nor even buds of compromise. On
the contrary, conceiving his embassy as a sortie into enemy territory and
taking for his motto the maxim *aventurar la vida y osar morir* – 'risk your life
and dare to die' – he had brazenly refused to strike the colours of Spain
upon his warships' entry into Portsmouth harbour, whereupon only an
appeal to the king himself averted an exchange of cannon fire that was
certain to have sunk the ambassador in his vessel. At the time, a marriage
between Prince Charles and Christine of France, sister to Louis XIII, was
already under negotiation and Gondomar had fostered little hope that an
alternative bride in the shape of Philip III's daughter, the Spanish infanta
Maria, could be plausibly presented to his English hosts. But the man whom
Thomas Middleton characterised as the Black Night in his play *A Game
at Chess* knew his craft and, more important still, the way to ply and prime
the increasingly penniless king who now confronted him.

James chose for the intricate diplomacy in store a man of intelligence
and exceptional honesty, Sir John Digby, who relayed from Madrid in

May 1615 what appeared to be wholly unrealistic Spanish demands. The children of any marriage were, on the one hand, to be baptised and educated as Catholics, and guaranteed the right to succession. In the meantime, moreover, the infanta was to be granted Catholic servants and a chapel which was to be a place of public worship for English Catholics, against whom all penal laws were to be rescinded. And while such terms were shocking by any standards, the bait of a Spanish dowry amounting to some £600,000 and James's fear that the King of Spain 'had many kingdoms and more subjects beyond comparison' proved sufficient to fix him in a dangerous scheme of deception. Though clearly unacceptable, as his margin comments upon the Spanish proposals make clear, James nevertheless refused to reject them out of hand, preferring instead to play the kind of double game at which he considered himself so skilled – all of which perplexed his subjects and undermined their confidence in his commitment to the Protestant cause. His hankerings for peace, dislike of rebels and republicans, and prejudice against the Dutch, not to mention his vanity, fears of assassination and growing indolence, were all, it seemed, eminently exploitable by one such as Gondomar, and when the wily Spaniard returned to his homeland in 1618, accompanied by 100 Catholic priests whom James had seen fit to release as a gesture of goodwill, the worst fears of many appeared confirmed.

Nor was James's dalliance with Spain the last eccentricity imposed upon him by financial necessity. Indeed, his decision to release Sir Water Raleigh from the Tower of London in March 1616 was not only equally astonishing in its way, but driven even more directly by a vain hope that his problems might somehow be solved at a stroke through bold action and a timely gust of good fortune. For some time, in fact, James had been under pressure from the anti-Spanish faction led by George Abbot, Sir Ralph Winwood and the earls of Southampton and Pembroke, to liberate Raleigh and permit him to make a return voyage to the Orinoco River that he had first visited in 1595. Such a strategy could be guaranteed to outrage the King of Spain and also proffered a vast store of gold, which, according to Raleigh himself, lay only a few inches below the ground 'in a broad slate, and not in small veins'. 'There was,' he claimed, 'never a mine of gold in the world promising so great abundance.' And though he was now over 60, grey, lame and malaria-ridden, he remained ready to enact his dreams of creating an English empire in Guiana, centred on the Orinoco delta, that might eventually destroy Spanish power in the Indies.

In *The Discoverie of the Large, Rich and Beautiful Empire of Guiana*, he told how he was still haunted by 'the strange thunder of the waters' and how he seemed to hear these same same huge waters – 'each as high over the other as a church tower' – while lying in his prison cell. And now he had a king to share his fantasies.

On 27 March 1616, therefore, John Chamberlain informed Dudley Carleton how 'Sir Water Raleigh was freed out of the Tower the last week and goes up and down seeing sights and places built or bettered since his imprisonment'. In the meantime, however, James had extracted the most solemn pledges from Raleigh that no Spanish subject should be molested without forfeit of his own life, and had also attempted to smother the Count of Gondomar's protestations with a series of further pledges of his own. The notorious sea-captain would be sent to Madrid bound hand and foot, he assured the ambassador, if a single Spaniard were harmed, and there were further assurances that Raleigh's release had involved no free pardon or any revocation of the death sentence imposed upon him in 1603. In effect, the great Elizabethan, who now trudged the streets of a capital made unfamiliar to him by years of incarceration, had therefore been granted a one-way ticket to disaster, though it was one he had both purchased and stamped with glowing confidence.

The king, in all fairness, may well have had some rightful grounds for sharing that confidence or at least wagering at reasonable odds upon a potential windfall of considerable proportions. Already, in 1604, Charles Leigh had established a settlement on the banks of the river Wiapoco, which lasted two years, while another party, encouraged by Prince Henry, had sailed soon afterwards under Robert Harcourt and survived until 1613. Likewise, a third expedition to Guiana – partly financed by Raleigh to the tune of £600 – had been led by Sir Thomas Roe in 1610. And there were grounds, however slender, for believing, too, that Raleigh's expedition of 1595 had established some kind of English claim in Guiana, notwithstanding the fact that the Spanish were firmly established upon the coast of modern-day Venezuela and at the very site of the mines of San Thomé that had captured James's imagination in the first place.

But if James was indeed making a genuine bid for treasure and for territory that might rightfully be his, and in the process granting Raleigh a chance to gain his freedom and achieve his dreams, the likelihood of success actually remained minimal. In the event, Raleigh's ship, *The Destiny*, would not leave Plymouth until June 1617 since the storms at sea were the worst since the sinking of the Spanish Armada almost thirty years earlier, and from the outset the expedition was dogged by misfortune. Initially compelled to land at Kinsale harbour in southern Ireland, Raleigh's seven warships and three pinnaces were subsequently forced to weigh anchor at Lanzarote in the Canary Islands, where Gondomar had little difficulty in persuading his government that the heavily armed vessels were planning to attack the Spanish fleet. Deaths among his leading officers from a strange sickness were then followed by the desertion of Cyrus Bailey, who ultimately returned to England to spread rumours that Raleigh was turning pirate. And though the coasts of Guiana were finally sighted in mid-November, the expedition's leader was by that time stricken by fever, cared for by his son Walter and nephew George.

The Indians who eventually greeted Raleigh's men were, however, friendly. 'To tell you that I might be here King of the Indians were a vanity,' he informed his wife in a letter, 'but my name hath still lived among them. Here they feed me with fresh meat and all that the country yields; all offer to obey me.' Yet Raleigh was unable to lead the subsequent search for gold and appointed Laurence Keymis, who, in spite of specific instructions not to provoke hostilities with the Spanish, nevertheless misjudged his landing spot and arrived too near the fortified village of San Thomé. While Raleigh himself was therefore waiting in Trinidad with other sickly members of his crew, hoping to trade with the Spaniards, Keymis found himself under a surprise attack, in which Raleigh's son was killed while leading a gallant stand at the head of a group of pikemen.

'God knows, I never knew what sorrow meant till now,' Raleigh wrote soon afterwards. But though he was heartbroken and infuriated with Keymis, who subsequently committed suicide, he nevertheless resolved upon one last attempt to reach the mines, which his captains refused to support. The result was a wretched return to England, where arrest by his kinsman Sir Lewis Stukeley, Vice-Admiral of Devon, awaited him soon after his arrival at Plymouth on 21 June 1618. He had been tempted to sail *The Destiny* to Brest and subsequently contemplated escape to France after being placed under house arrest. But he was betrayed and subsequently placed in the Tower, and when Roger North, one of his captains, denied the existence of any South American mine, Raleigh became the object of the king's cold, vindictive fury. Shortly after Bailey's return with tales of piracy, Sir Thomas Lake had noted how 'his Majesty is very disposed and determined against Raleigh and will join the King of Spain in ruining him'.

But now, considering himself the victim of a despicable hoax, and prey increasingly to Gondomar's taunts and goading, James determined to strike.

On 18 August, therefore, Raleigh was duly summoned before a commission to answer for his misdeeds. The king, much to his discredit, had already submitted to Gondomar's insolence in demanding that any sentence might be carried out in Spain, and in doing so had ridden roughshod over the opposition of his council, declaring that he could take what course he pleased 'without following the advice of fools and badly disposed persons'. He had also taken pains to guarantee that Raleigh should not be afforded the opportunity of a public hearing, since 'it would make him too popular, as was found by experience at the arraignment at Winchester [in 1603], when by his wit he turned the hatred of men into compassion for him'. Yet the commission itself, which included Archbishop Abbot and Sir Edward Coke, had the appearance at least of equity, even if no such gathering could have been unmindful of the king's express will. For Raleigh stood accused not only of disloyalty and deceit but of compromising the king's avowed policy of international peace. And though Prince Henry had admired him, and the queen would intercede on his behalf by means of her 'kind dog' Buckingham, the accused was by no means popular at court. Ultimately, therefore, Raleigh's death sentence, inevitable as it was, represented not so much the sacrifice of a national hero by a weak, embittered Scottish king as a calculated act of realpolitik. If, in the final analysis, an expendable liability might be offered up in the broader interests of peace and personal credibility, then James, as ever, was equal to his kingly obligations.

And the sentence, which at Philip III's behest was eventually carried out in England rather than Spain, duly resulted in a fitting addition to national folklore. 'He was the most fearless of death that ever was known,' wrote one observer, 'and the most resolute and confident, yet with reverence and conscience.' For on 28 October 1618, as he was about to be taken from the Tower to the Gatehouse at Westminster, where he was to pass the night before his execution, Raleigh encountered an old servant who noticed his untidy hair and tearfully offered him a comb. 'Let them kem it that are to have it', replied the condemned man in his broad West Country accent. 'Dost thou know, Peter,' he continued, 'of any plaster that will get a man's head on again when it is off?' And when delivered to the block itself the following day, there was similar bravado. Feeling the edge of the axe, Raleigh could not resist a remark to the presiding sheriff. 'This is a sharp medicine, but it is a physician for all diseases,' he quipped before laying his head on the block and uttering a final exhortation to the headsman: 'What dost thou fear? Strike, man, strike.'

The Americas, meanwhile, would remain for James a faraway and meagerly exploited realm. The first of the East India Company's voyages had set out in 1601 and returned under Sir James Lancaster a few months after James's accession, with cargoes yielding a 100 per cent profit. And though contemporary commentators condemned the trade for emptying the kingdom of gold bullion in exchange for goods, it was much too profitable to be suppressed entirely. There had therefore been another expedition in 1604, and from 1607 onwards similar excursions occurred annually. The richer peers, courtiers and politicians joined Sir Thomas Smith and his City colleagues as regular patrons of these ventures, and when, in 1609, the charter was renewed and the company reorganised, the earls of Salisbury and Nottingham, as well as the Earl of Worcester, all joined the board. In the meantime, as English interest in the Indian trade and Persian Gulf expanded, so Portuguese control of these same areas slackened, largely through lack of support from a Spanish government painfully overcommitted elsewhere.

But progress in establishing American settlements was altogether slower and less spectacular, and the Virginia Company, which received its first charter in 1606, proved a regrettably neglected enterprise which failed to realise its true potential before the King of England finally abandoned control of it in 1612. Raleigh and his half-brother Humphrey Gilbert, had dreamt of an overseas empire in strikingly modern and purely nationalist terms, and had used the hope of finding gold primarily as a lever to secure official support and financial backing. But this early imperialism had resulted in a series of costly failures at Roanoke, and King James, like most of his contemporaries, showed little imagination from 1603 onwards beyond treating American settlements as convenient repositories for the kingdom's surplus population. Moreover, the inexperience and jealousies of the pioneers and the hostility of the native Indians all but wrecked a second series of plantations in 1606, where only the adventuring genius of Captain John Smith and the friendship of Princess Pocahontas, daughter of the most important of the local chiefs, averted total disaster. Even in 1609, indeed, when James and Salisbury took a hand in the London Company's wholesale reorganisation, the results were limited. Though the Jamestown settlement on the Delaware River became secure, the colony succeeded ultimately for what all the backers considered at the time to be the wrong reason: tobacco. And while James, in particular, would rail against the spread of a filthy habit, which led to a rise in consumption from £20,000 in 1617 to £50,000 by the end of the reign, the windfall both to the government and the venture's backers remained comparatively paltry, and represented, like the rest of the king's colonial policies an uninspiring case of squandered possibilities.

Of more abiding interest to James, however, was another far-off realm – albeit one that was altogether more familiar to him personally and one that entailed no ocean-going perils to reach. By the end of 1616, as Raleigh pressed on with preparations for his impending voyage, the king had resolved, in fact, to make a long-deferred return to the homeland he had previously promised to visit every three years. Obeying a strong impulse, and taking advantage of an opportunity that he claimed to be the first, but sensed might be the last, he therefore informed the Scottish Privy Council on 15 December 1616, of his intended visit. 'We are not ashamed to confess,' he wrote, 'that we have had these many years a great and natural longing to see our native soil and place of our birth and breeding, and this salmon-like instinct of ours has restlessly, both when we were awake, and many times in our sleep, so stirred up our thoughts and bended our desires to make a journey thither that we can never rest satisfied till it shall please God we accomplish it.'

Yet the king's homing instincts were unpopular with his courtiers, none of whom relished a progress of unprecedented to a length to a land where only cold, discomfort and barbarism appeared to await them, and there were more pressing concerns, too, about the likely cost. Indeed, the whole council, including Buckingham in the first instance, had implored James not to go, though the favourite's opposition had melted soon enough for him to be granted his earldom once the journey was underway. Even the queen, for that matter, had balked at the prospect of accompanying her husband and was granted leave to stay at home. But no such indulgence was granted to the noblemen and clerics who sallied forth from Theobalds on 14 March 1617. Three English bishops – Andrewes of Ely, Neile of Durham and Montagu of Winchester, who had edited the king's *Collected Works* the year before – were all in James's entourage, along with a bevy of his Scottish kinsmen, such as the Duke of Lennox and the Marquis of Hamilton, and the English earls of Pembroke and Montgomery.

On the one hand, the leisurely journey along the Great North Road, which took him all of two months to complete, represented a nostalgic attempt to recapture his lost youth and experience anew the kind of euphoria that had greeted him on his journey to London in 1603. Attended by hundreds of gentlemen ushers, grooms and other officers, James enjoyed the hunting so much around Lincoln that he neglected to meet the county's sheriffs, who had gone to welcome him. But he nevertheless lapped up in full the state reception offered to him once more at York, where the kneeling Lord Mayor presented him with a cup of silver double gilt and a purse containing 100 double sovereigns. Neither the illness of the queen, whose physicians, according to John Chamberlain 'feared an ill habit of body', or the fever of Secretary Winwood, who also found himself 'much vexed with the perpetual visits of great folks' after being left in charge of English affairs, could dim the king's enthusiasm when he finally reached Berwick on 13 May and crossed the Border into the land of his birth.

Three days later, the royal entourage entered Edinburgh itself and it was here, on 19 June, that James celebrated his fifty-first birthday. The English, Chamberlain had informed Dudley Carleton just over a fortnight earlier, were 'much caressed' in Scotland, while the king himself continued to exhibit the kind of generosity that had been such a feature of his journey south over a decade earlier. 'So many knights are made,' wrote Chamberlain, 'that there is scarce a Yorkshire esquire left to uphold the race, and the order has even descended to the Earl of Montgomery's barber and the husband of the Queen's launderess.' Even so, many courtiers found their quarters along the Royal Mile inadequate and uncomfortable, and while James enjoyed his hunting enormously at Falkland and Kinaird, he was nevertheless more critical of his fellow Scots than previously, wishing that they might imitate their English counterparts more in their worthy habits than in drinking healths, 'tobacco takin' and 'glorie of apparel'.

There were darker memories, too, that James had plainly failed to displace. Left alone with a guide in a mineshaft at Culrose in Fifeshire, he suddenly suspected an assassination attempt and dissolved into cries of 'Treason', which were only calmed with considerable difficulty. It seems likely, too, that his visit to Stirling on 30 June was conducted with mixed emotions as he once again surveyed the site where his grandfather had met a violent end, and that his stay at Perth was also tinged by recollections of his narrow escape from death at the hands of the Gowries. Not dissimilarly, there were also echoes of a sullen Scottish Kirk, soaked in cynicism and suspicion regarding their king's religious plans. For James had already gone as far as he could in bringing Scotland candles and choristers, and when an organ from the Chapel Royal of Whitehall arrived for use at Holyroodhouse, the reaction was predictable. 'The organs are come before,' grumbled one Calvinist divine, 'and after comes the Mass.'

It was the restless Scottish Kirk, moreover, that had partly prompted James's visit in the first place. Still beguiled by dreams of unification, he hoped to enforce upon Scotland the notorious 'Five Articles of Perth' which required, amongst other things, kneeling at Holy Communion and the administration of Confirmation by bishops. Thus, or so he believed, might the Kirk's practices be brought into line with the Church of England. But though the Articles were indeed formally adopted by the General Assembly which met at Perth in 1618, James was nevertheless sufficiently attuned to the resulting opposition to realise that they could not be imposed as vigorously as he might have wished. And heady rhetoric of the sort produced by William Hay upon the king's visit to Glasgow belied a range of deeper realities. Eulogising James as 'that great peacemaker' and 'only Phoenix of the World' – the man who had achieved what others 'neither by wit, nor force, nor blood' had been able to accomplish – Hay not only proceeded to describe him as the king who had 'united two [of] the most warlike nations of the world' and 'made a yoke of lions' but entirely ignored the very insensitivity on James's part that exacerbated the gaping rifts that still remained. For when even the Scottish bishops objected to the gilded figurines of the apostles and patriarchs that had accompanied the king's organ to Holyrood, he could not resist a half-sneering letter in which he deplored the ignorance of the native clergy and suggested that his English doctors should give them instruction. Had he ordered figures of dragons and devils instead, he quipped, the Scots would have raised no objection.

And if James's own prejudices against his countrymen were not proof enough of the gulf between his kingdoms, the now notorious account attributed to Sir Anthony Weldon, who accompanied him on his northern odyssey, rammed the point home with all the cudgel-bluntness so characteristic of English commentators in general. 'For the country,' the author begins, 'I must confess it is too good for those that possess it, and too bad for others … There is great store of fowl – as foul houses and shirts, foul linen, foul dishes and pots, foul trenchers and napkins, with which sort we have been forced to fare …' But the tirade mounts in intensity as it unfolds. 'The country,' we hear, 'affords no monsters but women … To be chained in marriage with one of them were as to be tied to a dead carcase and cast into a stinking ditch; formosity or a dainty face are things they dream not of …' 'And therefore to conclude,' the account ends, 'the men of old did no more wonder that the great Messias should be born in so poor a town as Bethlem in Judea, as I do wonder that so brave a prince as King James should be born in so stinking a town as Edinburgh in lousy Scotland ...'

So much, then, for any lofty expectations that a union of Crowns might really lead to unity of vision and purpose, and as the king's entourage rode southward again, leaving Carlisle in August and traversing once more 'that wild northern country, which no other English sovereign had passed for centuries', the relief was palpable – not least, it must be said, for many Scots themselves. For the official welcomes and pageants in every city, not to mention the cost of refurbishing seven royal palaces, and the mere expense of housing the English court for four months had taxed Scottish resources to the limit, though this, it seems, was of less than pressing concern to James himself as he continued on his way towards Preston and Hoghton Tower, home of Sir Richard de Hoghton, where on 17 August, at a feast given by the host, he would famously knight 'Sir Loin of Beef' for services to his palate. 'Swan roste, Quailes 6, Redd Deare Pye, Duckes boyld and Shoulder of Mutton roste' were also made available for the king's delectation, for, according to the Venetian diplomat Antonio Foscarini, he generally preferred meat to more exotic food, though fruit, and especially melons and cherries, had been specially transported from England during his Scottish stay. Such was Sir Richard's hospitality, moreover, that he eventually found himself consigned to the Fleet Prison for debt.

By the time that Hoghton Tower was behind him, however, the nagging realisation that his holiday idyll was almost over was doubtless weighing upon James increasingly heavily. On his journey south, he had revelled in the Lancashire countryside, hunting the stag and making a special visit to Hoghton's alum mines. Knowing of Sir Richard's involvement in the famous Pendle witch trial, he had also, it seems, encountered a group of witches, though he had gone on to insist that they should be kept beyond the outer walls. But his homeward journey through Coventry, Warwick and Compton Wyngates, and arrival at Windsor on 12 September offered little consolation for the ennui, sickness and vexation in store. According to the Venetian ambassador, Giovanni Batista Lionello, James was eventually met in London by 'five hundred of the leading burgesses on horseback and a countless number of people, who shouted for joy at his return'. But he was soon poorly with arthritis and gout and succeeded, it seems, in spraining his leg in bed. The queen, moreover, was now constantly ill and Christmas at Whitehall appears to have been particularly dull and dreary for the king, though he took some pleasure in a gift from the Czar of Russia, since it was richer than any given to Queen Elizabeth. 'I am sorry to hear,' wrote Chamberlain, 'that he grows every day more froward.'

Even the news of the birth of his second grandson in the Palatinate did not cheer James, and by May of 1618 altogether more ominous news from the same far-off land would plunge all Europe into crisis. The marriage of his daughter Elizabeth to Frederick V of the Palatinate on St Valentine's Day 1613, had in Robert Allyn's words seen England 'lend her richest gem, to enrich the Rhine', and the subsequent celebrations matched the significance of an event that would not only result in the creation of the Hanoverian dynasty a hundred years later but carry England into the very heart of European politics. Though the outstandingly extravagant Lords' Masque was considered 'long and tedious' by one observer, all else went so well, it seems, that public officials and ambassadors subsequently applying to the treasury for their 'Extraordinarys' were met with the not unfamiliar response that 'the King is now disfurnished of money'. Spectacular shows and fireworks, costing £9,000 had been staged along the Thames, while the Inns of Court excelled themselves in elaborate entertainments, and Lord Montague lavished £1,500 on frocks for his two daughters. And though the groom had been dismissed by some as 'a slight edifice on a small foundation' and was considered 'too young and small timbred' by Chamberlain, he was nevertheless a figure of considerable importance: the leader of the Evangelical Union of German Calvinist rulers and one of the so-called electoral princes of the Holy Roman Empire, whose privilege it was, along with the Princes of Saxony and Brandenburg, the King of Bohemia and the Archbishops of Mainz, Treves and Cologne, to elect the Holy Roman Emperor himself when that throne fell vacant.

By 1618, however, a marriage which had initially seemed to confirm James's credentials as what one contemporary tract termed the 'King of peace', was presenting him with the kind of challenge that few could have envisaged when his charming, vivacious, auburn-haired daughter – 'Th'eclipse and glory of her kind' – had taken her wedding vows. 'I have ever, I praise God, kept peace and amity with all,' James had told his first English parliament before informing MPs how of all 'the blessings which God hath in my person bestowed upon you', the first is peace. In flowing from James's dominions, moreover, peace was to become universal, it seems, as he boasted in 1617 that he had established harmony in all neighbouring lands. 'Come they not hither,' asks a tract entitled *The Peace-Maker, or Great*

*Brittaines Blessing* and written mainly by Lancelot Andrewes with small additions by the king himself, 'as to the fountain from whence peace springs? Here sits Solomon and hither come the tribes for judgement. O happy moderator, blessed Father, not father of thy country alone, but Father of all thy neighbour countries about thee.'

But the King of England's chosen self-image of *'rex pacificus'*, like so many of his high-flown aspirations, conformed poorly with earthier realities. Friendly with all nations, allied with Protestant states, on peaceful terms with Spanish territories, the King of England intended to survey from on high an imposing vista of peace and concord on a European scale fashioned by his own hand. But no champion of Protestantism could realistically hope to flirt with Spain, and while James had succeeded with such diplomatic promiscuity in Scotland, he could not hope to do so in his southern realm, where both people and Parliament were hostile to a strategy that was never adequately explained, and where the impending convulsions on the Continent were beyond all hope of mediation. Rightly or otherwise, James was described by the country gentleman, Sir John Oglander, as 'the most cowardly man I knew', but the more general claim about his aversion to war as an instrument of policy remains indisputable. 'He could not,' wrote Oglander, 'endure a soldier or to see men drilled' and 'to hear of war was death to him'. But now, as Counter-Reformation Germany fractured into armed religious camps, hard-headed Dutchmen pursued their implacable enmity with Spain, and Spain herself assumed the offensive after the assassination of the French king, Henry IV, in 1610, war was not merely the best but the only policy available in the longer term. Indeed, having chosen *'Beati Pacifici'* – 'Blessed are the peacemakers' – as his personal motto, James would swiftly discover that peacemakers like himself were much more likely to feel themselves accursed rather than 'blessed', since the marriage alliances he had arranged with concord in mind had actually linked the English Crown to the very ruler who would now become one of the main protagonists in the outbreak of Thirty Years' War.

When the childless Holy Roman Emperor, Matthias, who also held the electoral throne of Bohemia, instructed in 1617 that his Catholic Habsburg cousin, Archduke Ferdinand of Styria, should be nominated as his successor to the Bohemian throne, the native Protestant lords were committed by May of the following year to rebellion. And when the King of England's son-in-law, Frederick of the Palatinate, pitched into the struggle with wild-eyed promises of support and hopes of wrecking Habsburg power throughout Germany, the blue touch paper was finally lit for the renewal of a general war of religion that had previously been abandoned in exhaustion in 1555. Since the Spanish Habsburgs were bound to come to the rescue of their cousins in Vienna, the danger was already critical, but when Frederick himself subsequently accepted the Crown of Bohemia from the rebels in October 1619, he not only unleashed a catastrophic conflict that had been looming for at least a decade, but at once plunged his father-in-law into the thick of a European maelstrom that he scarcely apprehended.

James's ignorance had not, however, prevented him from already leaping at a cynical suggestion made by Gondomar as a means of keeping England inactive for a few vital months, that he should mediate between the Bohemians and Ferdinand. 'The vanity of the present King of England is so great,' wrote Gondomar, 'that he will always think it of great importance that peace should be made by this means, so that his authority will be increased.' And with his mind thus clouded, James had duly dispatched a grandiose mission to Prague, headed by James Hay, who swiftly discovered that his king's mediation was considered wholly inappropriate not only by Ferdinand, whose military position had improved, but by the Bohemians too, who declared their preference for armed assistance rather than olive branches. And while Hay's 150 strong entourage vainly crossed and re-crossed Europe at a cost of £30,000, the election of Ferdinand as emperor and his deposition as King of Bohemia unfolded regardless.

Even so, James would be afforded an opportunity for more decisive action in August 1619 when Frederick sent Baron von Dohna to London to seek his advice on the offer of the Bohemian Crown – a prospect which rightly filled the English king with mortal dread. For not only would his personal honour now oblige him to intervene on behalf of his son-in-law, he would be hotly encouraged to do so by his own subjects. His daughter Elizabeth, after all, had already endeared herself to English men and women alike, and her marriage to a sound Calvinist was now a potential rallying call to all seeking firm action in defence of the Protestant cause in Europe. Only if swift action was forthcoming and war was somehow averted by Frederick's rejection of the Bohemian Crown might James therefore avoid the descent of his own kingdom into the abyss. But he remained, as he himself admitted, 'in a great strait, being drawn to one side by my children and grandchildren, my own flesh and blood, and to the other side by the truth and by my friendship to Philip [of Spain] and to the House of Austria'. And while James studied the niceties of the Bohemian constitution in quest of an answer, others drew their own conclusions. 'It seems to me,' wrote Tillières, the French ambassador, 'that the intelligence of this king has diminished. Not that he cannot act firmly and well at times and particularly when the peace of the kingdom is involved. But such efforts are not so continual as they once were. His mind uses its power for a short time, but in the long run he is cowardly. His timidity increases him day by day until old age carries him into apprehensions and vices diminish his intelligence.'

And by the time the King of England had finally made up his mind, it was, indeed, already too late. For in October 1619, while his father-in-law hesitated, Frederick duly arrived in Prague to claim his new throne, and in doing so committed what James had already admitted to Baron von Dohna would be an act of insupportable aggression. Since his subjects, James told Frederick's emissary, were as dear to him as children, he would not 'embark them in an unjust and needless quarrel'. But while James protested that the Bohemians had committed an outrageous act of rebellion against Ferdinand, their rightful king, and that any assistance to his son-in-law would both wreck his reputation as peacemaker and necessitate a summons of Parliament, he appeared to resort once more to the double game that came so naturally to him. Confiding to the Venetian ambassador his ongoing fear of Catholic plots, James declared how he could not even remain alive except by peace with Spain, and once more blamed Frederick as the usurper of a kingdom not his own. Yet on other occasions, it seems, he was inclined to express himself altogether differently. 'The king is taking great pains at present,' the ambassador informed the Venetian Senate on 22 November, 'to make everybody think so, showing displeasure at the election and at the Palatine's acceptance without his consent, but those who converse familiarly with him, tête à tête, easily perceive his delight at this new royal title for his son-in-law and daughter.'

Ultimately, James would neither directly discourage Frederick's acceptance of the Bohemian Crown nor actively support it – merely preferring to imply that he would acquiesce in the accomplished fact. And this was something that even his daughter's most earnest entreaties could do nothing to alter. Proclaimed upon her arrival by a wildly enthusiastic populace, Bohemia's 'Winter Queen' had given birth on 17 December 1619, to her third son, who was to be known to history as 'Rupert of the Rhine'. But by September of the following year, she was imploring her brother Charles to be 'most earnest' with their father about his 'slackness to assist us'. And after a year of playing cheerfully and incompetently at their roles as

King and Queen of Bohemia, both Elizabeth and her husband were indeed plunged into headlong flight amidst the remnants of their army which had been ripped to shreds by the troops of the Catholic League on 8 November 1620 at the Battle of the White Mountain. Bereft and beleaguered, and facing the imminent conquest of the Palatinate itself by Don Ambrosio Spinola Doria's Spanish troops, they were now no more than homeless pretenders to a Crown that would never be theirs again.

# 17 ⚕ 'Baby Charles' and 'Steenie'

'What sudden change hath darked of late
The glory of the Arcadian state?
The fleecy flocks refuse to feed,
The lambs to play, the ewes to breed;
The altars smoke, the offerings burn,
Till Jack and Tom do safe return.'

*From a pastoral poem written by James I in response to the*
*Prince of Wales' journey to Madrid in 1623*

'The King,' wrote Sir Anthony Weldon, 'was ever best when furthest from his Queen.' And her death on 2 March 1619, appears to have been greeted with precisely the kind of broad equanimity that might have been anticipated in such circumstances. While Anne lay ill at Hampton Court with dropsy, James visited her dutifully twice a week, but it was Prince Charles rather than he who took up residence in an adjoining bedroom in an effort to provide some modicum of comfort and support. In the meantime, the king remained preoccupied with the issue of his wife's will, for fear that she might leave the majority of her jewels to her Danish maid Anna – a worry that would remain unfounded – though the gems, valued at £30,000, nevertheless found their way into the maid's possession illicitly with the help, it seems, of a Frenchman named Pierrot. Equally regrettably, before James left for a hunting trip at Newmarket in February, there had been angry words between the royal couple over the Catholic priests that hovered continually around the queen's sickbed. Ultimately, indeed, only Prince Charles and his sister Elizabeth, who wrote from Heidelberg on 31 May to express her inexpressible sorrow at 'so great a misfortune', appear to have been genuinely stricken by the queen's demise, and the verses that James later penned in her memory were largely coloured by more general reflections on the mortality of princes, 'who, though they run the race of men and die' were further ennobled by their passing, since 'death serves but to refine their majesty'. In the words of one observer, the king took his wife's death 'seemly', neither weeping publicly, nor managing, for that matter, to brave the illness that kept him from her funeral. Only later, arguably, as his own health collapsed and he reflected more broadly upon the transience of human affairs, would a deeper despondency descend upon him.

The malady that kept the king from Anne's obsequies at Westminster on 13 May was, however, real enough. It had begun with 'a shrewd fit of the stone' coupled with arthritis during his stay at Newmarket, and by the time of his arrival at Royston in mid-March his condition had deteriorated significantly. He was weak and faint, could neither eat nor sleep and was debilitated further, we are told, by a 'scouring vomit'. 'After the Queen's death,' wrote James's French physician, he suffered 'pain in the joints and nephritis with thick sand, continued fever, bilious diarrhoea, hiccoughs for several days, bitter humours boiling from his mouth so as to cause ulcers on his lips and chin, fainting, sighing, dread, incredible sadness, intermittent pulse'. At one point, indeed, there had been fears for his life, and though he had recovered sufficiently to be removed from Royston to Ware on 24 April, he was nevertheless carried part of the way in a Neapolitan chair provided for him by Lady Elizabeth Hatton, and the rest of the way in a litter. Even by the time of his midsummer hunting trip to Oatlands, Woking and Windsor, moreover, James was still seeking to strengthen his legs and feet by bathing them in the bellies of slain deer.

Upon his arrival in London on 1 June, however, it was already clear that any residual shadow cast by Queen Anne's death would not be long-lived. Riding through the city in a suit of pale blue satin, and sporting a hat of blue and white feathers, the king was received with such enthusiasm that the whole scene surprised and perplexed an embassy of condolence, sent by the Duke of Lorraine, which had arrived at the same time. The kingdom had thrown off mourning, it seems, to celebrate James's recovery to good health, though the twenty-four Frenchmen, clad in unrelieved black, remained less than impressed by the apparent lack of sensitivity. Convention, if nothing else, required more of the king than what had amounted, in effect, to a fleeting and partially self-centred fit of dolour. Nor could his own undoubted fears of death and dissolution justify his current wish to return to normality so rapidly, especially at a time when any outward play at celebration was marred so obviously by growing pressure both abroad and at home. For, as events in Europe continued to darken, the need for action on all fronts increased daily.

Above all, the marriage of the king's son was becoming a matter of particular urgency. Prince Charles, the sickly child who had never been expected to survive infancy and whose weakness of limb at the age of almost 5 made it necessary for the Earl of Nottingham to carry him at his investiture as Duke of York, had grown up shy, sensitive, obstinate, priggish and dull – mollycoddled and overawed by his father, and completely overshadowed his elder brother, who would often taunt him till he wept, 'telling him that he should be a bishop, a gown being fittest to hide his legs'. Yet with Henry's death in 1612, his own creation as Prince of Wales four years later, and the king's declining health, Charles had gradually emerged, for all his deficiencies, as a figure of increasing importance both at court and in the kingdom at large. And though in the opinion of the Venetian ambassador writing in 1617, he remained 'very grave', exhibiting 'no other aim than to second his father, to follow him and do his pleasure', he had nevertheless developed into a young man 'of good constitution so far as can be judged from his appearance', who enjoyed theatricals, rode excellently and delighted in hunting.

The prince's devotion to his father was, moreover, fully reciprocated by the king himself, though the glow of paternal pride and affection for his delicate, studious child was too often tainted by spiteful displays of annoyance. In consequence, the heir to the throne had come to harbour a particularly warm resentment towards the dazzling favourite whom all adored and who exercised such a hold over his father's affections. That Charles should have manifested such antagonism towards George Villiers was hardly surprising in view of the rivalry he posed, but it was magnified significantly by James's apparent partiality whenever the two fell out. In 1616, for instance, after Charles had tried on and mislaid one of Villiers' rings, the king is said to have called for his son and 'used such bitter language as caused his Highness to shed tears', before banishing the prince from his presence until the elusive ring was eventually found in his breeches by a valet. The heir to the throne would feel the sting of his father's anger, too, in a later incident that resulted from a childish prank where, by turning the pin of a fountain, he had caused water to spurt onto the favourite's splendid clothes. Witnessing the mischief, James this time not only spoke angry words but gave the prince two boxes on the ear, though this would not, it seems, prevent a more serious quarrel between the prince and Villiers over a game of tennis in 1618, when the latter, who was the prince's senior by eight years, is alleged to have raised his racket in anger. 'What, my lord,' said Charles with all the self-satisfied superiority of a young man enjoying an unfamiliar taste of victory, 'I think you intend to strike me!'

James, however, wished only for peace between his 'sweet babies', and the result was a characteristically lavish and honeyed attempt at reconciliation staged by the favourite after the king had called the two together and urged them, upon their allegiance to him, to love one another. In consequence, at a sumptuous banquet held in June 1618 to signal to the entire court the dawning of a new entente between the freshly created Earl of Buckingham and the Prince of Wales, the older of the two duly courted his young rival, and not only mended the rift but established himself at one and the same time as the prince's lasting idol. Held out of doors at Buckingham's new estate at Wanstead, the so-called 'Prince's Feast' was a triumph, in fact, for the host's grasp of political necessities and timing, and a testament to his intuitive grasp of Charles's underlying needs and inclinations. For their widely opposed temperaments, as Buckingham fully appreciated, could easily be turned to rich and permanent profit, if only the less prepossessing of the two could be offered the opportunity to share the glow of adulation in which he himself basked. Both, after all, were devoted to hunting, riding, poetry, music and painting, and for an introvert like the prince, who shared his father's emotional volatility, the gulf between antipathy and hero-worship might be easily bridged – as indeed it was.

By the time, therefore, that James had risen from his table at the end of the feast at Wanstead and made his way over to the place where Buckingham and his kin were seated, the man who had previously been firmly entrenched as rival, enemy and embodiment of all the virtues lacking in the prince himself had already been transformed into mentor, role model and oracle – a radiant being from whom a formerly delicate, awkward and taciturn individual could draw unfamiliar sources of confidence and allure. Drinking a toast to each of the Villiers family in turn, James then swore that both he and his descendants would 'advance that House above all others'. 'I live,' he declared, 'to that end.' And now, he added gratefully, he harboured no doubt that his heir would do the same. For Charles, too, would henceforth be turning to 'Steenie' for guidance and succour – writing to him as 'your true constant friend' and praying him 'to commend my most humble service to his Majestie'.

But while the best behaviour of the prince, whom Buckingham would soon dub 'Baby Charles', was now assured, the same was hardly true of the MPs that James was eventually compelled to summon for only the third time in his reign. In the aftermath of the Addled Parliament of 1614, the king had intended to cope for as long as possible by means of forced loans and benevolences, by the sale of honours and monopolies, and by the judicious levying of impositions. With the assistance of Cranfield as treasurer, moreover, such expedients had for some time proved adequate for his ordinary financial needs. But the Spanish invasion of the Palatinate in August 1620, which had driven his English subjects to fury, also jarred James's own sense of justice to such a degree that he was even prepared to vent his anger to Gondomar. He would never trust a Spanish minister again, the ambassador was told, amid angry tears, during an audience with James at Hampton Court in September. Nor, declared the king, would he permit either his children or his religion to perish. Instead, he would go in person to defend the Palatinate.

The outburst, it is true, was followed by misgivings and hesitation, as Gondomar presented the Spanish invasion as a pathway to long-term peace. Should Frederick renounce his claims in Bohemia, the ambassador contended, then the Palatinate could be duly restored to him. And such was Gondomar's persuasiveness that James was even prepared to dismiss Sir Robert Naunton, his secretary, from office. He remained uneasy, too, not only at the popular hatred of Spain but at the freedom with which it was expressed, as Puritan preachers found time, amid their personal attacks upon Buckingham, to fume at Spanish perfidy and the plight of Frederick and his English wife. Complaining that his subjects were becoming too republican, James therefore issued a proclamation forbidding contentious discussion of state affairs, while Buckingham did all he could to throw in his weight with Gondomar. 'The Puritans have rendered Buckingham Spanish,' wrote Tillières, the French ambassador, 'for seeing that they mean to attack him, he knows no way of securing protection against them except by the Spanish match.'

But however James handled the problem which faced him at Christmas 1620 he could not hope to do so unarmed and on an income barely sufficient even for his peacetime needs. For even if his role as mediator was to continue, as he always hoped, his efforts would carry little weight unless backed by at least the potential for military action. In the summer, therefore, when Spinola first threatened the Palatinate, James had permitted a force of 2,000 English volunteers under Sir Horace Vere to go to the defence of his son-in-law's hereditary lands. And their presence at once provided a convenient toe-hold, which could be used to provide leverage, should war become desirable or unavoidable. At the same time, since Vere's force was composed of volunteers, it could be disavowed or reinforced as advantage served. Under such circumstances, a summons of Parliament might prove a popular, comparatively low-risk strategy of precisely the kind that the king was so often inclined to favour.

The House of Commons that confronted James in January 1621, however, was soon threatening to prove even more truculent than its predecessor. Initially, there had been grounds for optimism. Both Crown and Parliament shared, after all, the same broad objective: the restoration of Frederick and his wife to at least their hereditary dominions in the Palatinate. Even Buckingham, for that matter, had temporarily joined the majority on the king's council that clamoured for war, and James's opening speech showed less of the customary cudgel bluntness and rather more humility than usual. He had been carried to the House in a portative chair and there were whispers that he might not walk again, but, in acknowledging the threat of war and his need for money, he appeared amenable and ready to curry sympathy. In speaking of his efforts to keep the peace, he also spoke graciously of his predecessor: 'I will not say that I have governed as well as she did, but I may say we have had as much peace in our time as in hers.' And his request for money was also couched in anything but strident tones. 'I have laboured as a woman in travail,' he declared, '... and I dare say I have been as sparing to trouble you not with monopolies or in subsidies as ever King before me, considering the greatness of my occasions and charges.'

It was the very issue of patents and monopolies, however, that soon dominated debate and led directly to criticism of those close to Buckingham. For, although the favourite had garnered little personal profit from the sale of exclusive marketing rights, he had nevertheless supported them for the benefit of his relations. Sir Giles Mompesson, one of the worst offenders, was connected to the Villiers nexus by his sister-in-law, while Buckingham's brother Kit and half-brother Edward had also benefited handsomely. Such, indeed, was the favourite's alarm that he initially sought the dissolution of

Parliament, only to be rebuffed by the king, before opting to turn tail and disown his corrupt kinsmen. In the process, he spared himself by posing as nothing less than the champion of reform, though Francis Bacon proved less fortunate. Pleading guilty to a charge of corruption on 3 May and describing himself as 'a broken reed', the Lord Chancellor was debarred from office, subjected to a fine of £40,000 and consigned to the Tower at the king's pleasure. 'Those who will strike at your Chancellor,' Bacon had warned James, 'it is much to be feared will strike at your crown.' But the king had insisted all the same upon an impartial hearing, before ultimately mitigating the sentence. Bacon had claimed, after all, that he had done no more than 'partake of the abuses of the time', and the king seems to have accepted as much. 'In giving penalties,' he declared, 'I do always suppose myself in the offender, and then judge how far the like occasion might have tempted me.'

Such admirable sentiments did not, however, prevent James from making a foolish and costly lapse into dishonesty of his own. Though the Parliament of 1621 had been summoned in response to the crisis on the Continent, the issue of foreign policy remained, in accordance with tradition, beyond its remit, and the king himself had touched on the European situation in only the vaguest terms. Assuring MPs that peace remained his objective, James nevertheless made clear that he must negotiate 'with a sword in his hand', and that, if necessary, he would spare no personal cost to recover the Palatinate. But in asking for funds, he concealed the real sum required – a sum that had been made clear to him by a report he had specifically requested from a council of war some months earlier. Advised at that time that an army of at least 30,000 men was required for intervention on the Continent and that such a force would necessitate a down payment of £250,000, followed by further payments of around £900,000 per year thereafter, James nevertheless chose to ask for £500,000. The result was an interim grant of only £145,000 from MPs made suspicious of their king's motives by his insistence on further negotiations with Spain rather than an immediate declaration of war.

When Parliament adjourned in June, therefore, the scope for optimism was limited. But when MPs reassembled in November, dissent was soon broadening ominously – not least because James had taken the remarkable decision to leave London for a Newmarket hunting trip before the new session began, leaving his ministers hopelessly exposed in their efforts to contain the situation. 'His Majesty seems to hope,' wrote the Venetian ambassador, 'that the Parliament will readily afford him every means of making war with little trouble on his part.' Yet before long, MPs were employing what they now declared to be their 'ancient and undoubted right to free speech' to encroach upon areas of policy that had always been closed to them. Some called for sea war with Spain rather than 'pottering and pelting in the Palatinate', as John Chamberlain put it, 'only to consume both our men and means'. Others called for broader action against Catholics at home, as enemies of the commonwealth. There were demands, too, that Prince Charles should be 'timely and happily married to one of our own religion'. And in the meantime James merely saw fit to dispatch from Newmarket an angry letter to the Speaker, which would pour oil from afar upon already troubled waters.

In suggesting that his absence – through what he termed an 'indisposition of health' – had emboldened certain 'fiery spirits' to debate matters 'far beyond their reach and capacity', James also complained how this had tended towards 'our high dishonour and breach of prerogative royal', and forbade that MPs should hereafter meddle in 'deep matters of state, nor deal with our dearest son's match with the daughter of Spain, nor touch the honour of that king'. But it was the threat accompanying the letter that forced the issue, for James also added 'that we think ourselves very free and able to punish any man's misdemeanours in Parliament as well during their sitting as after; which we mean not to spare hereafter upon any occasion of any man's insolent behaviour there that shall be ministered unto us'. Instead of sticking to the point at issue, on which he was certainly within his constitutional rights, James had therefore plunged knee deep into questions that had best been left unmentioned. Worse still, he had chosen his battleground poorly, for in lacking the natural gravitas and finesse of his predecessor, he had nevertheless resorted to intimidation and threats of force that he was ill-equipped to deliver. An assault on individual members was likely to leave him not only penniless and discredited but at the mercy of Spain, and the tone of his warning represented in any case a challenge that the more recalcitrant elements of the House of Commons were hardly able to ignore.

The result was a meeting at Newmarket between the king and a parliamentary delegation that demonstrated all too palpably the scale of the rift between the two sides. Though the MPs had arrived ostensibly with the intention of expressing their requests more moderately than before, James was not only unbending but brashly dismissive. Greeting the delegates with a patronising request that his servants 'bring stools for the ambassadors', he balked once more at the suggestion that freedom of speech was 'our undoubted right and an inheritance received from our ancestors'. 'We are an old and experienced king,' he objected, 'needing no such lessons.' He resented too, it seems, the petitioners' 'great complaints of the danger of religion within this kingdom, tacitly implying our ill-government on this point', and rejected what he deemed to be their desire to 'bring all kinds of causes within their compass and jurisdiction' like 'the Puritan ministers in Scotland' had supposedly done before them. Almost inevitably, moreover, he concluded with a further declaration that Parliament's privileges were solely 'derived from the grace and permission of our ancestors and us'.

By equating the present with his Scottish past and treating an opportunity for conciliation as a reckoning, James had therefore inflamed and exacerbated, and, in doing so, called forth a formal protestation from MPs on 18 December, since there was now no point in further debate. Instead, the privileges of the House of Commons would be set forth in writing, with neither hint of retreat nor trace of apology, leaving James to broil in the kind of frustration and fury that had marred his reign in Scotland. 'The plain truth is,' he wrote, 'we cannot with patience endure our subjects to use such anti-monarchical words concerning their liberties, except they had subjoined that they were granted unto them by the grace and favour of our predecessors.' And though councillors urged that he should not dissolve Parliament, Gondomar added counter-pressure of his own by suggesting that Spain could not negotiate while such a body remained in existence. Nor was the Spaniard alone, it seems, in his desire for dissolution, for he recorded gladly how 'the king was being valiantly urged on by the Marquis of Buckingham and other good friends'.

Finding that his bid for popularity by throwing his monopolist cronies to the wolves was unsuccessful, Buckingham had indeed returned to his former pro-Spanish course and predictably carried the heir to the throne with him. For now Charles too had begun to fancy himself as anxious for a Spanish match as his father. Cajoled, therefore, by those whose sympathy he most valued, and faced, apparently, with the intolerable prospect of surrendering to the House of Commons the very authority he had fought so strenuously to defend against the Scottish Kirk, James found diplomatic dependency upon Spain the only acceptable option. Nor, in doing so, could he temper his bile or resist the urge for a vacuously imperious gesture. Sending for the *Journal of the House of Commons*, he tore out, in the presence of his council, the very page on which the offending protestation had been recorded – an act which for Gondomar was 'the best thing that has happened in the interest of Spain and the Catholic religion since Luther began to preach heresy a hundred years ago'. 'It is certain,' he added, 'that the king will never summon another Parliament as long as he lives.' With no prospect of English intervention now at hand, moreover, the Emperor Ferdinand would duly confirm by the end of 1622 that the Palatinate was to be presented to his cousin Maximilian of Bavaria, leaving James to dangle on the end of a Spanish hook, baited with lukewarm prospects of a marriage that was, in any case, bitterest wormwood to all red-blooded Protestant Englishmen.

The interminable negotiations for Charles's union with a Spanish bride had already assumed a curious air of unreality since their initiation a decade earlier under the aegis of Carr and the Howards, when Gondomar had first arrived at the English court. Now, indeed, it was not only a different infanta, but a different prince: less passive, more influential, and, most significantly of all, altogether more passionate about the enterprise. For at the same time that Gondomar nursed dreams of the English royal family's conversion to Catholicism, and James held fast to the notion that a Spanish alliance might yet confirm his status as the 'peacemaker king', Charles harboured boyish visions of emulating the romantic feat of his father by bringing back his bride from over the sea. While professional diplomats like John Digby, Earl of Bristol, tediously trod water in Madrid, and his father offered unconvincing half-promises of relaxing the penal laws and allowing a Catholic upbringing for the infanta's children, the prince himself now dreamt of breaking the deadlock by direct action of the most effective kind. He would cross to France, incognito and without a formal pass, to win his bride in person. He would do so, moreover, in the sole company of his guide and mentor, Buckingham, who, seeing the opportunity for a spectacular personal victory, now encouraged the venture at every turn.

When Gondomar returned home in 1622, therefore, and while James and his council were seriously discussing a military expedition to recover Heidelberg, the prince and his hero were secretly writing to the absent ambassador of their intention. The Spaniard, indeed, had hatched the plan initially and, as negotiations flagged once more in February 1623, Charles and Buckingham mooted it with the king himself. As James well knew, the escapade would expose his son to the very real and largely unnecessary dangers of a journey halfway across Europe, unguarded and unattended. Equally recklessly, it would destroy at a stroke the delicate bargaining position built up by the Earl of Bristol and allow the Spaniards the opportunity to raise their terms as high as they pleased, while the king's English subjects seethed with indignation. But ill health, despondency and a doting fondness for his two 'sweet babies' seem to have sapped James's resolve until his consent was finally wrung from him during a surprise visit late one night. And though, according to Clarendon, he fell next morning 'into a great passion of tears, and told them he was undone, and that it would break his heart if they pursued their resolution', Charles and Buckingham were nevertheless allowed to prevail.

In consequence, on 18 February 1623 two heavily disguised young men using the names of Jack and Tom Smith crossed the Thames by ferry at Gravesend, leaving the King of England to brood and fret upon the dangers ahead of them. Wearing false beards, which in one case fell off inopportunely, they grossly overpaid the ferryman and further excited his suspicion by requesting that they be set ashore just outside Gravesend instead of at the usual landing place within the town. When, moreover, the local magistrates were subsequently informed of the likelihood that two suspicious travellers were slipping out of the country for the purpose of fighting a duel, an attempt was made to intercept them at Rochester, though they had left before they were apprehended. Escaping arrest, too, at the hands of Sir Henry Mainwaring, Lieutenant of Dover Castle, as well as the Mayor of Canterbury, the two would nevertheless reach Dover unscathed and arrive in Paris on 21 February after a wretchedly seasick crossing to Boulogne.

Just over a fortnight later, at 8 p.m. on 7 March, the Earl of Bristol was attending to important business at his embassy in Madrid when a mysterious Mr Smith demanded immediate access. The visitor, in fact, was Buckingham, and though Bristol disapproved of the venture, he had little option but to inform a jubilant Gondomar of the heir to the English throne's arrival. Hastening to his superior, the Count of Olivares, who conveyed the news in turn to King Philip, Gondomar had achieved, it seems, the ultimate diplomatic coup. For, while Philip chose to remain cool until the attitude of the Pope had been clarified, the bargaining position of his government had been enhanced immeasurably, since any return to England without the infanta in tow would involve such an intolerable loss of face for the two visitors. From Gondomar's perspective, meanwhile, the prince's arrival could only be explained by his imminent conversion to Catholicism – a belief that both Charles and Buckingham foolishly saw fit to encourage initially.

So far, however, at least James's worst fears for his son and Buckingham had not been realised. Though he had agonised over their safety, he had hailed them, nevertheless, as 'dear adventurous knights worthy to be put in a new romanso', and in spite of the Earl of Bristol's misgivings, the first signs appeared encouraging. 'I must confess ingenuously,' Bristol informed James, 'that if Your Majesty had been pleased to ask my advice concerning the prince his coming in this fashion, I should rather have dissuaded than given any such counsel, especially before the coming of the dispensation.' But though the papal dispensation sanctioning the marriage, to which the earl referred, was never likely to prove acceptable, the Spanish king, at least, had hidden his reservations admirably and made himself suitably agreeable, riding beside Charles as he escorted him to a suite in the royal palace. Nor was the prince disappointed upon meeting his 16-year-old inamorata for the first time. Fair haired with languorous eyes and full lips, the infanta Maria made such an impression, in fact, that Buckingham wrote home to James to tell how his son was 'so touched at the heart that he confesses all he ever yet saw is nothing to her'.

The initial meeting had been long delayed, however, as every conceivable device of the rigidly formal Spanish court was employed to keep the couple apart. And when at last an audience was granted, Charles was left in no doubt as to what he must wear and what he might say. Never allowed to speak to the infanta alone, he resorted ultimately to leaping the wall of her orchard – an act which not only breached the bounds of Spanish decorum but also succeeded in frightening his quarry uncontrollably. When permitted to see her at court theatricals, moreover, the prince found her impenetrably aloof as he watched from afar, 'half an hour together in thoughtful posture' and with an intense concentration that reminded Olivares of a cat watching a mouse. Convinced by her confessor of the eternal peril to her soul which would be entailed by each night sharing her bed with a heretic, the young girl was predictably repelled by the prospect of her sacrifice on the altar of marriage. And when Buckingham informed his wife of Charles's predicament, she responded, in all good faith it seems, with a characteristically ingenuous attempt at helpfulness that captured the underlying oddity of the situation. 'I have sent you some perspective glasses,' she told her husband, 'the best I could get. I am sorry the prince is kept at such a distance that he needs them to see her.'

In England, meanwhile, James was faced with increasing pressure as his councillors headed by the Earl of Arundel, the only Howard who now still held power, expressed consternation at the proposed marriage and dissatisfaction that the king had ever permitted the prince's journey in the first place. When pressed, the king was merely inclined to heap all blame upon both Buckingham and his son's high passion. 'The king,' wrote John Williams, Dean of Westminster, to the favourite, 'would seem sometimes, as I hear, to take it upon himself (as we have advised him to do by proclamation); yet he sticks at it and many times casts it upon you both.' And in sparing himself further opprobrium, James had clearly exposed Buckingham to the kind of hostility that he was less able than ever to ignore. 'Detestation of the Marquis', wrote the Venetian envoy Valaresso, 'has increased beyond all measure', while Dean Williams pulled no punches in informing the favourite how 'all the court and rabble of the people lay the voyage upon Your Lordship'. Such, indeed, was the outcry in London churches that James had no choice but to forbid all prayers for the prince's soul 'now that he was going into the House of Rimmon'.

In spite of James's misgivings about his predicament, however, the same flow of cloying correspondence to his two 'sweethearts' in Madrid continued unabated. 'God bless you, my sweet baby,' one letter ended, 'and send him good fortune in his wooing, to the comfort of his old father, who cannot be happy but in him. My ship is ready to make sail, and only stays a fair wind.' On another occasion, there were assurances to Buckingham that the king was wearing his picture 'in a blue ribbon under my wash-coat, next my heart'. And while he expressed concern on 25 March about an implication from Buckingham that he might lean further towards Rome, the king was nevertheless prepared to interpret the suggestion in the most positive light possible. 'I know not what ye mean,' he wrote, 'by my acknowledging the pope's spiritual supremacy … but all that I can guess at your meaning is that it may be ye have an allusion to a passage in my book against Bellarmine, where I offer, if the pope would quit his godhead, and usurping over kings, to acknowledge him for the chief bishop, to which all appeals of churchmen ought to lie en dernier resort … For I am not a monsieur that can shift his religion as easily as he can shift his shirt when he comes from tennis.'

There were paternal concerns, too, about the prince's welfare, as James begged his son not to exert himself in hot weather, 'for I fear my baby may take fever by it', and suggested that Charles and Buckingham should keep themselves fit by private dancing, 'though ye should whistle and sing to one another, like Jack and Tom, for fault of better music'. The prince's spending, it is true, continued to perplex the king, and after jewels valuing some

£80,000 were dispatched in what proved to be a fruitless effort to dazzle the Spaniards, Charles was urged to be 'as sparing as ye can', for 'God knows how my coffers are already drained'. But in other respects James's fussing and indulgence remained unstinting. Garter robes and insignia, for example, were specially dispatched, so that they could be worn on St George's Day, 'for it will be a goodly sight for the Spaniards to see my boys in them'. Nor, it seems, were Buckingham's relations neglected in James's thoughts at this trying time. He had written letters to the favourite's mother, consoling her during her son's absence, and kept a particularly close watch, it seems, upon his wife, believing that she might be pregnant. 'And, my sweet Steenie gossip,' the king wrote, 'I must tell thee that Kate was a little sick these four or five days of a headache, and the next morning, after a little casting, was well again. I hope it is a good sign that I shall shortly be a gossip over again, for I must be thy perpetual gossip.'

In spite of their futility, there were even arrangements for the prince's heroic homecoming with his bride. Eight great ships and two pinnaces were to sail, and planning was in hand for the infanta's magnificent cabin. A wing of St James's Palace was also enlarged and refurbished and specially equipped with an oratory for her use. And all the while, the prince remained as wide-eyed and artlessly optimistic as ever. 'I, your baby, have since this conclusion been with my mistress, and she sits publicly with me at the plays, and within this two or three days shall take place of the queen as Princess of England.' But it was now two years since the pope had first been asked to sanction the marriage and the condition that James should grant full liberty of worship to English Catholics seemed as implausible as ever. 'I have written a letter,' James had already told his two 'sweet boys', 'to the Conde de Olivares as both of you designed me, as full of thanks and kindness as can be devised as indeed he well deserves, but in the end of your letter ye put in in a cooling card anent the nuncio's averseness to this business.' 'The pope', James added ominously, 'will always be averse' unless the infanta be given 'free exercise of her religion here'.

But when the precise terms for the dispensation were finally brought from Madrid by Sir Francis Cottington in the late summer, they proved far more shocking still. Not only was the infanta to control the education of her children, she was also to be permitted to open her chapel in London for public worship, and retain a fully Catholic household, the members of which were to be personally selected by none other than her brother, Philip IV. The religious life of this household, moreover, was to be administered by a bishop at the head of twenty-four priests, all of whom, though resident in England, were not to be subject to English law. Worse still, it was stipulated that there must be complete freedom of worship for Catholics, and that the Oath of Allegiance should be altered to accommodate them. To cap all, James and his entire council were to agree on oath to all clauses of the dispensation, while the king was to swear additionally that he would obtain parliamentary agreement to these terms within a year. Even the date of the infanta's arrival remained doubtful, for that matter, since Spain remained reluctant to dispatch its princess until the year in question was finally over.

To suggest that Cottington's tidings fell as a hammer blow upon the King of England's pipe dreams would be an understatement. Vain hope gave way at once to despair, which dissolved soon after into hysteria, as James became convinced that his babes were now prisoners, and determined to recover them by signing anything that the Spanish government might now ask of him. The news that Cottington had brought him, he wrote to Buckingham and Charles on 14 June, 'hath stricken me dead. I feel it shall very much shorten my days, and I am the more perplexed that I know not how to satisfy the people's expectation here, neither know I what to say to the council ...' And as the true scale of his essentially self-inflicted predicament dawned upon him, James's main concern remained personal: on the one hand, his own credibility, but above all the safety of the two individuals around whom his world centred. 'But as for my advice and directions that ye crave,' he wrote, 'in case they will not alter their decree, it is in a word, to come speedily away, and if ye can get leave, give over all treaty ... alas, I now repent me sore that ever I suffered you to go away. I care not for match nor nothing, so I may once have you in my arms again. God grant it, God grant it, God grant it; amen, amen, amen!'

All, then, depended from James's perspective upon the swiftest possible conclusion of the marriage treaty, in the hope that something might be salvaged from the wreckage as a prelude to the return of his beloved boys. And accordingly, on 20 July, James formally ratified it in the Chapel Royal at Whitehall. For some time, in fact, the marriage had lost what little appeal in Spain it had ever had, but even the most outrageous Spanish demands, which had been raised largely with the intention of killing all negotiations once and for all, were now accepted without demur. The secret clauses, which had earlier reduced the King of England to a state of horrified incomprehension, were duly sworn by him in private before the Marquis of Inojosa, who had succeeded Gondomar as ambassador, and Don Carlos de Coloma, Philip IV's special envoy. But there was no reference to the £600,000 dowry, to which the Spaniards had already whittled the intended windfall that had been dazzling James's imagination for the past ten years, and without which, as he assured Buckingham, he must surely go bankrupt. Guarantees of Spanish help to bring about the restoration of Frederick and Elizabeth in the Palatinate also went unmentioned – leaving James to confess that he was 'marrying his son with a portion of his daughter's tears'. That same evening, however, in the new Banqueting House which Inigo Jones had built specifically for the wedding of the Prince of Wales, James and his Spanish ambassadors still dined in particular splendour from plates 'of pure and perfect gold'.

# 18 ⩔ Dotage, Docility and Demise

'All good sentiments are clearly dead in the king. He is too blinded in disordered self-love and in his wish for quiet and pleasure, too agitated by constant mistrust of everyone, tyrannised over by perpetual fear of his life, tenacious of his authority as against the Parliament and jealous of his son's obedience, all accidents and causes of his almost desperate infirmity of mind.'

*From a report made by Alvise Vallaresso, Venetian ambassador to England, 1622–24*

The last official portrait of Scotland's King of England dates to 1621. Painted by Daniel Mytens the Elder, it depicts him bare-headed and seated in a posture of limp repose, with a plumed hat on the table beside him, but no other symbol of regality beyond the Garter robes he wears and a Tudor rose woven into the tapestry behind his chair. His hands droop wearily over the arm of his chair. His tired eyes lie far back in their sockets beneath heavy lids. His lips are pursed in indifference and resignation. It is the face, in fact, of a deeply disillusioned man who has lost his vigour and abandoned all real desire to prevail against adversity. *Beati pacifici* – 'blessed are the peacemakers' – is there above the king's head, but hardly noticeable and only then, it seems, as a dutiful afterthought from an artist whose main intent was to capture an eloquent image of worn-out authority. Fit only to be cosseted by the ladies of the Villiers family and in particular Buckingham's mother, who according to Archbishop Mathew now 'fulfilled a double function as a middle-aged gossip and as a nurse', James, it seems, was already slipping into careworn lethargy, born of ill-health, the relentless passage of time and a world that had consistently refused to conform to his more noble aspirations.

There were, it is true, bright spots ahead for him, not the least of which was the much anticipated return of his beloved son. Towards the end of September, the prince and Buckingham embarked for England, sailing from the north-west coast of Spain and landing at Portsmouth on 5 October to a rapturous reception as bonfires blazed all along the Portsmouth road and hundreds of hogsheads of wine were emptied to fuel the ardour. At Cambridge, church bells rang out for two solid days, while at Blackheath the people were said to have been 'so mad with excess of joy that if they met with any cart laden with wood they would take out the horses and set cart and all on fire'. Not far from Tyburn, meanwhile, Prince Charles took time to reprieve a group of felons, and upon the heroes' arrival at Royston, the king clambered downstairs, in spite of crippling gout, to fall upon their necks in heartfelt thanks and affection. For some days, moreover, the high spirits continued. 'The prince and my lord of Buckingham,' wrote Sir Edward Conway, 'spend most of their hours with his Majesty, with the same freedom, liberty and kindness as they were wont.'

But the bonfires and bells in the country at large were at least partially deceptive, since they were not so much expressions of joy at the prince's safe return as of relief that he had returned both as a Protestant and without a Catholic bride in tow. And even as the Villiers ladies, whom the king had hastily summoned, were being regaled with rousing tales of romance and derring-do, an anthem from the 114th Psalm was being solemnly sung at St Paul's, which could not have reflected more eloquently the feelings of most Englishmen now that the 'house of Jacob', just like Israel before it, had been delivered from 'a people of strange language'. Even more perplexingly, perhaps, both the prince and his companion had now returned from Spain thoroughly disenchanted with the marriage and alliance they had hitherto striven so strenuously to forge. In particular, Buckingham's vanity seems to have played a crucial part in this curious transformation, for he had been created a duke by his adoring master to place him on an equal footing with the Spanish grandees confronting him, only to find that his hosts would neither defer to his affectations nor tolerate his sensitivity. In bridling at Spanish 'trickery and deceit', he was told directly by Olivares that negotiations would have been better left to a professional like the Earl of Bristol. And when Buckingham incurred displeasure by remaining hatted in Prince Charles's presence, and Charles's Protestant attendants were ejected from Madrid after Sir Edmund Verney had struck a priest, any grains of enthusiasm for a lasting treaty were swiftly dissipated.

Henceforth, indeed, nothing would satisfy Buckingham's wounded ego other than war itself. And war, of course, offered the added attraction of maintaining the gust of popularity he had so surprisingly experienced upon his return. To that end, therefore, the king would have to be either deceived or coerced: Parliament must be called at once and all doves on the council duly silenced or ruined. With the aid of the prince and the reliable offices of the Villiers ladies, moreover, James was to be effectively debarred from contact with any Spanish representative, and as Charles assumed more and more influence with the declining energy of his father, the last condition for the plan's success became a formality. For, while James had by no means entirely lost his native shrewdness and remained as averse to war as ever, he was unable to thwart the ongoing pressure for a postponement of the planned proxy marriage until an understanding had been reached with Spain concerning restitution of the Palatinate. Buckingham was now no longer the 'humble slave and dog' of former times, and travel had plainly broadened not only the princes's mind but his shoulders, too. 'The Prince,' wrote one courtier indeed, 'is now entering into command of affairs by reason of the King's absence and sickness, and all men address themselves unto him'. Accordingly, in January 1624 he felt sufficiently confident to inform his father categorically that he would not hear of either friendship or alliance with Spain.

And while the momentum of the newly resurgent war party gathered pace, a beleaguered and befuddled king duly wrecked his long-held dreams of peace by wearily insisting upon Spanish action in the Palatinate that could never be forthcoming, and by summoning a Parliament that was bent on conflict on all fronts. Telling his son that 'he would live to have his bellyful of Parliaments', James was now a largely broken reed, though when Buckingham eventually determined to organise the impeachment of Lionel Cranfield, Earl of Middlesex, for supporting the king's peace policy, the failing ruler was still capable of exercising a testy prescience. 'By God, Steenie,' came his retort, 'you are a fool and will shortly repent this folly and will find that in this fit of popularity you are making a rod with which you will be scourged yourself.' But, as so often, the king's resolve did not match his wisdom, and Cranfield would nevertheless fall to charges of financial corruption and find himself imprisoned in the Tower by the very Parliament which duly assembled on 19 February 1624.

Stricken, it seems, by a brooding conviction that those closest to him had, in effect, deserted him, James approached the session with minimal energy, mainly occupying himself beforehand with inconsequential matters like the importation of some Spanish asses from the Netherlands, 'making great estimation of those asses, since he finds himself so well served with the mules to his litter'. And as he engaged in any available displacement activity to avoid the looming realities surrounding him, he descended further into self-pity and listlessness. When he should have acted, he chose instead to delay or demur, and where he might have led he merely lamented. 'The king,' wrote the Venetian ambassador, 'seems practically lost … He now protests, now weeps but finally gives in.' And Tillières' observations were even more striking, as he described how James was descending 'deeper and deeper into folly every day, sometimes swearing and calling upon God, heaven and the angels, at other times weeping, then laughing, and finally pretending illness in order to play upon the pity of those who urge him to generous actions and to show them that sickness renders him incapable of deciding anything, demanding only repose and, indeed, the tomb'.

Under such circumstances, therefore, James's opening speech to the last Parliament of his reign proved uncharacteristically faint-hearted and defensive. Casting himself upon the compassion of his audience, he pleaded that he had worked consistently to preserve his people's love and was now to seek their advice on the very matter that he had hitherto barred from all discussion so tenaciously: the issue, namely, of war or peace. 'Never soldiers marching the deserts and dry sands of Arabia,' he claimed, '… could thirst more in hot weather for drink than I do now for a happy ending of this our meeting.' And so, it seems, the doors were suddenly to be opened wide to previously forbidden territory. 'The proper use of a parliament,' James continued, 'is … to confer with the king, as governor of the kingdom, and to give their advice in matters of greatest importance concerning the state and defence of the king, with the church and kingdom …' 'Consider of these,' came the king's remarkable concluding appeal, 'and upon all give me your advice … you that are the representative of this my kingdom' and 'my glasses to show me the hearts of my people.'

Still wishing in the first instance to stage a final rearguard defence of his peace policy by using Parliament's bellicosity to lever Spain into action over the Palatinate, the king's tired rhetoric was nevertheless outshone by Buckingham's, who now personified the nation's war fever and experienced little difficulty in swinging the House of Commons to his cause. When, moreover, Parliament did indeed call for an end to the marriage treaties with Spain and promised assistance in the event of war, the king reacted petulantly, retreating to Theobalds and refusing to see either Buckingham or Charles who, he claimed, had misrepresented his intentions. Thereafter, he sent secret instructions to his councillors in the Commons, intercepted a dispatch on its way to Spain to announce the ending of the marriage treaty and bemoaned the impeachment of Cranfield, which he proved powerless to prevent. And it was no small irony either that while three subsidies amounting to more than £400,000 were willingly delivered – 'the greatest aid which was ever granted in Parliament to be levied in so short a time' – the money was only proffered for a cause that ran directly counter to the king's wishes. The marriage treaties were dissolved and peace teetered in the balance. Nor, indeed, was any money to be disbursed on any authority other than that of a council of war nominated by the House of Commons.

In the meantime, Buckingham and Charles consolidated their niche as popular heroes, with all that had gone wrong in Madrid duly laid upon the Earl of Bristol's buckling shoulders. Throughout the spring and summer of 1624 they beavered at the task of forging a grand alliance against Spain and made extravagant commitments without the funds to enact them. £360,000 was promised to the King of Denmark and a further £240,000 to pay for the conscription of an English army of 12,000 men to fight under the notorious German mercenary Ernst von Mansfeld. Above all, however, Charles and Buckingham pursued agreement with France and in particular marriage to the French princess Henrietta Maria – a policy which caused additional embarrassment for James, since in spite of his promises to Parliament that no future marriage treaty would entail concessions to English Catholics, the French now proceeded to demand precisely that. For three days during September, in fact, the king resisted the pressure of his favourite who had made a confidant of the French ambassador, the Marquis d'Effiat. But his capitulation followed, and while the resulting treaty was presented innocuously enough to his subjects, James privately accepted the French formula and, in a letter, promised to fulfil it. The royal signature which was appended to the treaty with a stamp, since the king's arthritic fingers could no longer move a pen, effectively purchased nothing, for while France had entered the Thirty Years' War in resistance to Habsburg power, their interest did not extend to the restitution of the Palatinate to its rightful ruler. In consequence, England found itself at last at war, but for a cause that the king had never espoused and without prospect of achieving the only specific foreign policy objective that had encouraged him to engage in war talk initially.

As the reign drew to its end, however, at least no military action was underway, since the remnants of the English volunteer force in the Palatinate had by now ceased to exist and its leader, Sir Gerald Herbert, had been killed. Yet the absence of fighting was of little consolation to James as he fell ill once more during December 1624. Wholly pre-occupied with his dwindling pleasures, his ailments and his tantrums, he was further disheartened by the death in February of the old Marquis of Hamilton. Only the year before, James had lost his cousin Lennox – a fine figure of a man who had 'shared his pleasure with many ladies' before becoming the constant 'servant' of that imperious beauty, the Countess of Hereford. When her husband, a curious old eccentric, had died in 1621, Lennox had stepped into the breach and lost no time in marrying her. But after retiring to bed in perfect health one night, he had been found dead in the morning – the consequence, according to one court wag, of an overdose of aphrodisiac. Seeing his friends and contemporaries thus disappearing into the grave, James now told his courtiers how he would be next.

Nevertheless, he had recovered enough by July to gorge himself with melons and even to set out on a modified royal progress not long afterwards. But he failed to emerge from his chamber at Christmas, 'not coming once to the chapel, nor to any of the plays', and experienced a more serious decline in the New Year, as a result of his own stubbornness and indiscipline. The malaria, or so-called 'tertian ague' that attacked him in March 1625 was by general agreement 'without any manner of danger if he would suffer himself to be governed and ordered by physical rules'. He refused the advice of his doctors, however, and not only drank vast quantities of cold beer but opted resolutely for the remedies of the old Countess of Buckingham who fussed continually at his bedside.

Stricken by a series of painful convulsions and fainting fits, the king then succumbed, it seems, to a minor stroke which left him unable to control the muscles of his face and choking upon vast quantities of his own phlegm. Even at the time there was largely groundless talk of skulduggery, and rumours that poison may have been involved were soon fuelled by one of James's Scottish physicians, George Eglisham. In a pamphlet published in Latin at Frankfurt in 1626, Eglisham suggested that the Duke of Buckingham had administered a white powder to the king which made him very ill, after which his mother had applied a plaster, also unbeknown to the royal doctors. In a Jacobean court already tainted by the Overbury scandal, it could hardly have been otherwise, of course, and such scandalous gossip found ready propagators in the small handful of Robert Carr's former servants who still tended the king.

Certainly, on 14 March James is said to have drunk a posset prepared by a country doctor named Remington, who had been warmly recommended by Buckingham and his mother, whereupon the king's physicians reacted angrily, refusing to proceed until Remington's medicine and the countess's plasters were discontinued. According to Eglisham's account, indeed, there was an unsavoury scene at the king's bedside when some of his doctors declared outright that poison was involved and a furious Buckingham drove them at once from the bedchamber. When, moreover, the countess begged James to clear both her son's and her own name from such slanders, the king, it seems, fainted from shock at the very mention of the word 'poison'. Yet even disregarding the fact that James's condition had improved sufficiently within the week for him to request further remedies from the good countess, the evidence for foul play remains slender, to say the least. The potential profit for Buckingham from foul play was in any case minimal, for the king had effectively ceased to rule long before he ceased to live. And it was actually a violent attack of dysentery that delivered the killing blow, bringing the king's misery and degradation to a merciful close.

By that time James's malarial fits were lasting for up to ten hours, and shortly before the end finally arrived, he had called for Lancelot Andrewes, though his favourite bishop was now himself a sick man, and John Williams, Dean of Westminster and Bishop of Lincoln, was therefore summoned to deliver the last rites instead. Whenever conscious, the king's talk was of repentance, remission of sins and eternal life – a faint and fading echo of his lifelong interest in all things theological. But when Prince Charles arrived, his father was already beyond speech and unable to deliver the last message he had intended. Upon hearing of his imminent death, James had shown no sign of disquiet. In life, of course, he had been prey to every conceivable apprehension, but the king, whose fear of shadows had played such a part in shaping both the man and his rule, was stalwart, it seems, when the shadow of death itself finally descended. It did so shortly before noon on Sunday 27 March at Theobalds – far from his Scottish homeland. He was 58 years old, and for only one of those years, as a cradled infant, he had not borne the heavy burden of a royal crown.

# 19 🌱 Ruler of Three Kingdoms

'This I must say for Scotland. Here I sit and governe it with my Pen, I write and it is done, and by a Clearke of the Councell I governe Scotland now, which others could not do by the sword.'

*James I to the English Parliament, 1607*

Some twenty-six years before his death, when the Crowns of England and Ireland were still by no means guaranteed him, James had earnestly urged his heir 'once in three yeares to visit all your kingdomes', and in the early months after his succession to the English throne, this precept appears to have remained fixed in his thinking. In 1603, indeed, he had left his homeland amid farewells rather than goodbyes, and in August of the same year negotiated the purchase of the manor of Southwell from the Archbishop of York, making it clear that he required a half-way hunting and resting spot on his regular journeys north. Yet within a year of his accession, the postal service to Edinburgh had been greatly improved by proclamation, and thereafter some sixty royal letters a year, laden with directives, inquiries, exhortations and admonitions, were soon being diligently dispatched along the rugged Great North Road. Ruling a far-off land that was now inclined to peace did not, it seems, entail the king's personal presence in the way that he himself had envisaged, though the Scottish council register leaves no doubt of James's ongoing absorption in Scottish affairs. Even with the passing of the years, moreover, as his councillors became more and more adept at tempering or blunting his less judicious instructions, they continued to do so under close supervision and in their sovereign's best interest. For, as the Earl of Mar told James's successor in 1626, 'a hundred times your worthy father has sent down directions to us which we have stayed, and he has given us thanks for it when we have informed him of the truth'.

Certainly, upon his arrival in London in 1603, James had no intention of appointing a Lord Deputy for his native realm, though his need for worthy assistants remained paramount. At first, his right-hand man was Alexander Seton, Earl of Dunfermline, who had served as one of his cost-cutting 'Octavians' in 1596, though when faced with the revival of Presbyterian opposition in 1606, James duly opted for the firmer hand of George Home, Lord Treasurer of Scotland from 1601 and later Earl of Dunbar. By travelling between the English capital and Edinburgh at least once a year, and attending the king on his summer hunting expeditions, Home kept his master closely apprised of events north of the Border, regardless of the personal inconvenience and tedium that the royal passion for 'sport' imposed upon him. Ultimately, such distractions would be remedied by tireless effort and clarity of purpose, as the Scottish privy council became under his guidance a loyal, cohesive and potent instrument of centralised control. Indeed, until his death in 1611, Home remained the lynchpin of the king's avowed policy of maintaining an integral political connection between his two kingdoms. And the reinstatement of the Earl of Dunfermline thereafter did nothing to undermine the ongoing process of consolidation. On the contrary, James could not have been more fortunate in enjoying the services of two such talented and selfless assistants, as he demonstrated the art of absentee kingship with a degree of efficiency and finesse he would rarely achieve in England itself.

To its very great credit, the Scottish Privy Council, dominated as it was by a dedicated core of office holders, not only maintained but extended the ambit of royal control. On the one hand, laws imposing heavy penalties for 'the ungodly and barbarous and brutal custom of deadly feuds' were strikingly affirmed in 1613 by the execution of Lord John Maxwell after the murder of the Laird of Johnstone who had been slain by a gunshot to the back some five years earlier. Border raiding, too, declined as James strove 'utterlie to extinguishe as well the name as substance of the bordouris' in an effort to create a peaceful region of 'middle shires'. Dunbar's influence as a Borderer himself proved particularly invaluable in this respect, and in 1605 a joint Anglo-Scottish commission was established to stabilise the six Border counties, employing a small cavalry force which was ultimately dissolved in 1621 as a result of Cranfield's economies. Meanwhile, in an attempt to add teeth to his general policy of pacification, James also determined to introduce Justices of Peace on the English model, though by the time of his death they were still present in less than a quarter of Scottish territory.

As James governed from London, there were further efforts, too, to extend the sway of central authority in the Western Isles. In 1608 an expedition under the command of Lord Ochiltree, which had been dispatched to collect royal rents, resulted in the peaceful apprehension of a number of Highland chiefs, after which the Scottish council pursued a policy of co-operation encapsulated in the Statutes of Iona, whereby the chiefs' authority over their followers was recognised in return for an agreement that they would act as agents of royal jurisdiction within their domains. Restrictions on alcohol and the size of lords' households further undermined the time-honoured Gaelic pastime of fighting and feasting, and, most importantly of all for the longer term, the same lords were not only encouraged to abandon their residual Catholicism for the Protestant Kirk, but to educate their eldest sons in Lowland schools. The influx of Gaelic-speaking Protestant clergy, who made a reality of the parish system and turned it into a powerful agent of social order, was merely one more factor assisting the crucial process of Scottish state building, which had still been very much a work in progress when James first ventured south in 1603.

Only on Orkney, in fact, where sheer distance from Edinburgh allowed Earl Patrick, a distant royal cousin, to rule as a princeling, was there determined resistance to the Scottish privy council. After disregarding the Statutes of Iona, the earl was soon at odds, too, with Bishop James Law, a commissioner for the Northern Isles since 1610 who became a royal revenue collector in 1612. Ultimately, however, even 'Black Patie' would find himself bridled by the hangman's noose in 1615, after a brief attempt at rebellion, which was systematically crushed by the Earl of Caithness and followed by the earldom of Orkney's retention by the Crown. When, moreover, Caithness fell into debt and contemptuously disregarded the Edinburgh legal proceedings brought by his creditors, he too was driven into exile by the Scottish council in 1623.

Such victories were not, it is true, achieved without cost. Certainly, the last years of James's reign witnessed the onset of autocratic tendencies that boded ill for the future. In particular, the Five Articles of Perth, introduced in 1617 and forced through Parliament in 1621, were a direct affront to religious feeling in the king's homeland. 'I am ever for the medium in every thing', James had professed characteristically in 1607. 'Between foolish rashness and extreme length there is a middle way.' But the understanding of Scottish problems which had done so much to compensate for the potential difficulties of absentee rule was gradually deserting him in his declining years. As early as 1607, in fact, James was remarking of his countrymen north of the Border how 'I doe not already know the one halfe of them by face, most of the youth now being risen up to be men, who were but children when I was there'. And by 1621 a new political divide had emerged in the king's northern Parliament after the Earl of Rothes and Lord Balmerino, among others, found their complaints against the Five Articles and the effects of heavy taxation blocked by an unsympathetic phalanx of royal appointees.

In the third and most troubled of his kingdoms, meanwhile, James had neither personal links nor any trace of direct experience to guide him. At the time of his succession, Ireland lay stricken and seething, with links to Rome as strong as ever and any prospect of economic recovery rendered all the more unlikely by a grievous debasement of the currency that had fractured commerce and impoverished the populace. So when news of Elizabeth I's death reached Waterford, Cork and Clonmel, Ireland's principal towns, the resulting euphoria had been palpable. The books of Protestant clergymen were, one Irish Jesuit reported, summarily burned 'and the ministers themselves hunted away', whereupon 'masses and processions were celebrated as frequently and upon as grand a scale as in Rome itself'. In the wake of the queen's death, moreover, Irish men and women had continued to nurture exaggerated hopes of her prospective successor who was, of course, himself the son of a Catholic martyr and king of a land which during the 1560s and 1590s had supplied some 25,000 fearsome 'gallowglass' mercenaries to serve across the Irish Sea in the conflict against English expansionism.

By June 1605, in fact, fervent calls for freedom of worship had been roundly thwarted by a royal proclamation in which James made clear that he would never 'confirm the hopes of any creatures that they should ever have from him any toleration to exercise any religion than that which is agreeable to God's word and is established by the laws of the realm'. Yet James's other early actions remained laudable, since he knew full well that his orders for Catholic priests to leave his realms were nowhere more unenforceable than in Ireland where 'every town, hamlet and house was to them a sanctuary', and in practice gave scant encouragement to those elements in the Dublin government favouring wholesale repression. 'He would much rejoice,' he professed, 'if the Irish Catholics would conform themselves to his religion, yet he would not force them to forsake their own.' And in the meantime there was a broader attempt at Anglo-Irish reconciliation, delivered by Elizabeth's victorious general Lord Mountjoy who restored the rebel Earl of Tyrone to his lands and bestowed upon his ally, Rory O'Donnell, the earldom of Tyrconnell in September 1603. More importantly still, perhaps, James had promptly decided to override Cecil's worries about a silver shortage, to upgrade the Irish coinage in the same month. For the first time, therefore, a fixed rate of exchange was established, with the result that English coinage was soon circulating freely in former enemy territory and facilitating a marked improvement in both internal and external trade.

Once more, predictably, the king's success was by no means unalloyed. Sir Arthur Chichester, who became Lord Deputy in 1604, disagreed profoundly, for example, with Mountjoy's earlier moderation and opted instead for colonisation. Despising the Irish as 'beasts in the shape of men', he sought, it seems, to 'civilise' the land by demolishing the local power bases of Tyrone and Tyrconnell, and encouraging the plantation of Protestant settlers. The result was the so-called 'flight of the earls' in 1607, and the clumsy application of a policy of settlement that James had previously applied with limited success in his own Scottish Isles. Despite a substantial flow of English and, above all, Scottish tenant farmers into Ulster from 1609 onwards, in which James took a close personal interest, numbers were never sufficient to corral the native Irish into restricted areas which could be easily controlled, and the king's first practical project to push forward his ideal of a 'greater Britain' succeeded only in creating a hostile class of Catholic under-tenants who remained cheaper for the settlers to employ than further migrants. Thus were the seeds of a disastrous future conflict sown.

In the meantime, the central problem of Ireland, as far as the English government was concerned, remained finance. Between 1604 and 1619, the annual subsidy sent from England stood at over £47,000 and Cranfield's boast to Buckingham that he would make the land self-sufficient was never made good. On the contrary, the English Parliament of 1622 attacked corruption and royal prodigality in James's third kingdom, and when Viscount Falkland became Lord Deputy in 1622, and opted to enforce the recusancy laws as a means of raising money, the threat to internal stability, fuelled by declining relations with Spain, escalated ominously. 'Ireland is such,' wrote the Venetian ambassador in the year of Falkland's appointment, 'that it would be better for the king if it did not exist and the sea alone rolled there.' And a month before James's death, in spite of his personal intervention to curtail the Dublin government's escalation of religious persecution, John Chamberlain reflected London opinion all too aptly by describing Ireland as 'tickle and ready to revolt'.

Yet a commission appointed to assess the state of the country in 1622 still made clear the changes of James's reign. English law was steadily replacing Irish, English counties had been introduced as units of local government, and the merchant companies of the City of London were building new ports at Derry and Coleraine. And although the commission's report was never published, largely because of its criticisms of Buckingham's ravenously self-interested clients, there was other evidence, too, of a genuine Jacobean achievement in Ireland. 'The love of money,' observed Oliver St John, Chichester's replacement as Lord Deputy in 1615, 'will sooner effect civility than any other persuasion whatsoever.' And where famine had stalked the country during the 1590s, Irish towns now enjoyed greater prosperity than for centuries, as significant communities of artisans, merchants and moneylenders took root, and itinerant pedlars forged networks of internal trade between urban centres and the surrounding countryside. Through wardships and intermarriage, meanwhile, many leading families such as the Fitzgerald earls of Kildare acquired English educations and English wives, as well as links with Scottish noble families.

Overall, the administration of an imperial monarchy encompassing three kingdoms inevitably entailed considerable structural tensions at the heart of government. Resentment at the king's absence, problems over the disposal of offices and the sharing of war costs, conflicts over trade and colonies, foreign intervention and above all religion were all, in fact, ongoing problems for James to grapple with. Some, indeed, played no small part in triggering the civil war that ultimately consumed his heir and may yet, four centuries later, put paid to his unifying aspirations once and for all. For if conflicts of faith have thankfully receded, modern-day resentment at centralised control from long distance and antagonism over fiscal propriety now rankle with new vigour. Under such circumstances, James's absentee kingship of his Gaelic realms may well seem increasingly impressive across the centuries, though there is still, perhaps, no small irony in this, since it was precisely because he ruled his outlying kingdoms from afar that those personal indiscretions and inadequacies, frequently so damaging to his English dealings, were unable to compromise his nobler, wiser aspirations. If England, in truth, never consistently warmed to its resident Scottish king, his fellow countrymen and their Irish counterparts experienced no few benefits in his absence. And it remains one of the more curious features of British history that the descendants of 'the king's barbarians' – those savage Irish and brutish Highlanders whom he equally despised – would become, in the fullness of time, the most loyal supporters of his Catholic grandson and that grandson's ill-starred heirs.

# Source Notes and Bibliographical Information

## Contemporary, near-contemporary and later printed material

Though they must be treated with caution, the accounts of those contemporaries who boasted a personal knowledge of James VI and I remain the starting-point for any study of the man and ruler. *The Court and Character of King James: written and taken by Sir Anthony Weldon, being an eye and eare witnesse* which was published in 1650 and reprinted in 1651 under the title *Truth brought to Light*, is still the classic account and essential reading for any general student of the reign wishing to return to primary sources. It was answered in a work usually attributed to Thomas Sanderson, entitled *Aulicus Coquinariae*, and both the original book and Sanderson's response were reprinted in *The Secret History of the Court of King James* (Edinburgh, 1811, two vols), which was edited by Sir Walter Scott and went on to create what has become the traditional perspective on England's first Stuart monarch. Other famous accounts include M. de Fontenay's, which is to be found in *Calendar of State Papers relating to Scotland and Mary Queen of Scots, 1547–1603* (Edinburgh, 1913, vol. 7), ed. W. K. Boyd, and Arthur Wilson's *The History of Great Britain, being the Life and Reign of King James the First* (London, 1653). Nicolo Molin's description, meanwhile, which was presented to the Venetian government in 1607, is located in ed. H.F. Brown, *Calendar of State Papers and Manuscripts Relating to English Affairs, existing in the archives and collections of Venice, and in other libraries of Northern Italy*, vol. 10, (London, 1900), and there is also the well-known summary provided by Sir John Oglander in *A Royalist's Notebook: The Commonplace Book of Sir John Oglander of Nunwell, 1622–1652*, ed. F. M. Bamford (London, 1936).

The general reader wishing to broaden his or her understanding may also wish to consult the contemporary material available in: ed. G.P.V. Akrigg *The Letters of King James VI and I* (University of California Press, 1984);

ed. R. Ashton, *James I by his Contemporaries* (London, 1969); and ed.
J.R. Tanner, *Constitutional Documents of the Reign of James I AD 1603–1625*
with an historical commentary (Cambridge University Press, 1930).

## Other relevant sources for the reign

Baker, L. M. ed., *The Letters of Elizabeth Queen of Bohemia* (London, 1953).

Bannatyne, R., *Journal of the Transactions in Scotland, during the contest
between the Adherents of Queen Mary, and those of her son* (Edinburgh,
1806).

Bannatyne, R., *Memorials of Transactions in Scotland, AD MDLXIX – AD
MLXXIII* (Bannatyne Club, Edinburgh, 1836).

Barlow, William, *The summe and substance of the conference, which it pleased his
majestie to have with the lords, bishops and other clergie, at Hampton Court*
(London, 1604).

Bell, R. ed., *Extract from the Despatches of M. Courcelles, French Ambassador at
the Court of Scotland MDLXXXVI – MDLXXXVII* (Bannatyne Club,
Edinburgh, 1828).

Bowes, Robert, *The Correspondence of Sir Robert Bowes, Esquire* (Surtees
Society, 1842).

Bruce, J. ed., *Correspondence of King James VI of Scotland with Sir Robert Cecil
and others in England* (Camden Society, 1856).

Bruce, J. ed., *Letters of Queen Elizabeth and King James VI of Scotland*
(Camden Society, 1849) .

Buchanan, George, *Opera Omnia* ed. T. Ruddimann, 2 vols (Edinburgh,
1715).

Buchanan, George, *The Powers of the Crown in Scotland*, trans. C. F.
Arrowood (University of Texas Press, 1949).

Calderwood, David, *A History of the Kirk of Scotland*, ed. T. Thomson (8
vols), (Edinburgh, 1842 (first published 1678)) .

Chamberlain, John, *The Letters of John Chamberlain*, ed. N. E. McClure,
2 vols (Memoirs of the American Philosphical Society, xii,
Philadelphia, 1939).

D'Ewes, Simonds, *The Autobiography and Correspondence of Sir Simonds
D'Ewes, Bart., during the reigns of James I and Charles I* (London, 1845).

Eglisham, George, *The Fore-runner of Revenge, Being Two Petitions: The one
To the Kings most Excellent Majesty, the other to the most Honourable*

Houses of Parliament. Wherein is expressed divers actions of the late Earle of Buckingham; especially concerning the death of King James, and the Marquesse Hamelton, supposed by Poyson (London, 1642).

Ellis, H. ed., *Original Letters Illustrative of English History*, first series, 3 vols (London, 1825).

Ellis, H. ed., *Original Letters Illustrative of English History*, third series., 4 vols (London, 1846).

Foster, E. R. ed., *Proceedings in Parliament 1610* (Yale University Press, 1966).

Goodman, Godfrey, *The Court of King James*, ed. J. S. Brewer, 2 vols (London 1839).

Green, M. A. E. ed., *Calendar of State Papers, Domestic Series of the Reign of James I:* 1603–1610, 1611–1618, 1619–1623, 1623–1625 with Addenda 1603–1625 (London, 1857–1859).

Harington, John, *Nugae Antiquae: being a miscellaneous collection of original papers*, 2 vols, ed. T Park (London, 1804).

Historical Manuscripts Commission, *Calendar of the Manuscripts of the Most Honourable Marquess of Salisbury preserved at Hatfield House, Hertfordshire* (London, 1883–1965).

Hyde, Edward, Earl of Clarendon, *The History of the Rebellion and Civil Wars in England begun in the year 1641*, ed. W Dunn Macray, 6 vols (Oxford, 1888).

Laing, D. ed., *Original Letters relating to the Ecclesiastical Affairs of Scotland*, 2 vols (Ballatyne Club, Edinburgh, 1851).

Maidment, J. ed., *Letters and state papers during the reign of King James the Sixth, chiefly taken from manuscript collection of Sir James Balfour of Denmyln* (Edinburgh, 1838).

Mayerne, Theodore Turquet de, *Opera Medica* ... ed. J. Brown (London, 1703).

Melvill, James, *The Autobiography and Diary of Mr James Melvill ... with a Continuation of the Diary*, ed. Robert Pitcairn, 2 vols (Wodrow Society, Edinburgh, 1842).

Melville of Halhill, Sir James, *Memoirs of his Own Life*, ed. T. Thomson (Bannatyne Club, 1827).

Moysie, David, Memoirs of the Affairs of Scotland, 1577–1603, ed. J. Dennistoun, (Edinburgh, 1830).

Nau, Claude, *The History of Mary Stewart from the Murder of Riccio until her Flight into England*, ed. J. Stevenson (Edinburgh, 1883).

Nicholls, E., *Proceedings and Debates in the House of Commons in 1620 and 1621*, 2 vols (Oxford, 1776).

Nichols, J., *The Progresses, Processions and Magnificent Festivities of King James I*, 4 vols (London, 1828).

Normand, L. and Roberts, G. eds, *Witchcraft in Early Modern Scotland: James VI's Demonology and the North Berwick Witches* (University of Exeter, 2000).

Sawyer, E. ed., *Memorials of Affairs of State in the Reigns of Q. Elizabeth and K. James I. Collected (chiefly) from the original papers of the right honourable Sir Ralph Winwood*, 3 vols (London, 1725).

Spottiswoode, John, *The History of the Church and State of Scotland*, 4th ed. (London, 1677).

Stevenson, J, ed., *Correspondence of Robert Bowes, the ambassador of Queen Elizabeth in the court of Scotland*, Surtees Society (1842).

Strickland, A. ed., *Letters of Mary Queen of Scots* (London, 1844).

Thomson, T. ed., *A Diurnal of Remarkable Occurents that have passed within the Country of Scotland since the death of King James the Fourth till the year MDLXXV* (Bannatyne Club, Edinburgh, 1833).

Wilbraham, Roger, *The Journal of Sir Roger Wilbraham*, ed. H. S. Scott in *The Camden Society*, Volume the Tenth (London, 1902).

Wotton, Henry, *Letters of Sir Henry Wotton to Sir Edmund Bacon* (London, 1661).

Wotton, Henry, *The Life and Letters of Sir Henry Wotton*, ed. L. P. Smith, 2 vols (Oxford, 1907).

## Modern biographies

The most recent full-length biographies are A. Stewart, *The Cradle King: A Life of James I* (London, 2003) and the more concise P. Croft, *King James* (Basingstoke, 2003). For a more thematic approach, there is also M. Lee, Jr, *Great Britain's Solomon: James VI and I in his Three Kingdoms* (University of Illinois Press, 1990) and R. Lockyer, *James VI and I* (Harlow, 1998). Three slightly older works by Caroline Bingham remain valuable: *James VI of Scotland* (London, 1979); *James I of England* (London, 1981); *The Making of a King: The Early Years of James VI and I* (London, 1968). Nor are two other biographies, which are now considered largely out-of-date, without interest: D.H. Willson, *King James VI and I* (London, 1956) and W. McElwee, *The*

*Wisest Fool in Christendom: The Reign of James VI and I* (New York, 1958). Mention should be made too of the following important article: J. Wormald, 'James VI and I: Two Kings or One', *History*, 68 (1983).

## Modern works relating to the rule of James VI in Scotland, 1567–1603

Brown, K.M and MacDonald, A.R. eds, *The History of the Scottish Parliament, Volume 3: Parliament in Context: 1235–1707* (Edinburgh, 2010).

Brown, K.M. and Mann, A.J., eds, *The History of the Scottish Parliament, Volume 2: Parliament and Politics, 1567–1707.*

Brown, K.M. and Tanner, R.J., eds, *The History of the Scottish Parliament, Volume 1: Parliament and Politics, 1235–1560* (Edinburgh, 2004).

Brown, K.M., *Kingdom or Province? Scotland and the Regal Union 1603–1715* (Basingstoke, 1992) .

Brown, K.M., *Bloodfeud in Scotland, 1573–1625* (Edinburgh, 1986).

Burns, J.H., *The True Law of Kingship: Concepts of Monarchy in Early Modern Scotland* (Oxford, 1996).

Donaldson, G., *Scotland, James V to James VII* (Edinburgh, 1965).

Duncan, A.A.M 'The early parliaments of Scotland', *Scottish Historical Review*, 15 (1966).

Goodare, J., *The Government of Scotland, 1560–1625* (Oxford, 2004).

Hewitt, G.R., *Scotland Under Morton 1572–80* (Edinburgh, 1982).

Jones, C., ed., *The Scots and Parliament: Parliamentary History*, (Edinburgh, 1996).

Lang, A., *James VI and the Gowrie Mystery* (London, 1902).

Law, T.G., 'The Spanish Blanks and the Catholic Earls, 1592–1594', *Scottish Review* 22 (1893).

Lee, M., Jr, *Government by Pen: Scotland under James VI and I* (University of Illinois Press, 1980).

Lee, M., Jr, *John Maitland of Thirlestane and the Foundation of the Stewart Despotism in Scotland*(Princeton University Press, 1959).

Lynch, M., *Scotland: A New History* (Edinburgh, 1991).

MacDonald, A.R. 'Deliberative processes in the Scottish Parliament before 1639: multi-cameralism and the Lords of the Articles', *Scottish Historical Review*, 81 (2002).

MacDonald, A.R., *The Jacobean Kirk, 1567–1625: Sovereignty, Polity and Liturgy* (Aldershot, 1998).

Macinnes, A.I., *Union and Empire; The Making of the United Kingdom in 1707* (Cambridge University Press, 2007).

Mason, R.A., 'Rex Stoicus: George Buchanan, James VI and the Scottish Polity' in Dwyer, J., Mason, R.A. and Murdoch, A. eds, *New Perspectives on the Politics and Culture of Early Modern Scotland* (Edinburgh, 1982).

Mason, R.A., ed., *Scots and Britons: Scottish Political Thought and the Union of 1603* (Cambridge University Press, 1994).

Rait, R.S., *The Parliament of Scotland* (Glasgow, 1924), passim.

Riley, P.W.J., *The Union of Scotland and England* (Manchester, 1978).

Robertson, W., *The History of Scotland, during the Reigns of Queen Mary and James VI till his Accession to the Crown of England*, 2 vols, (London, 1759).

Stevenson, D., *Scotland's Last Royal Wedding: The Marriage of James VI and Anne of Denmark* (Edinburgh, 1997).

Tanner, R.J., *The Late Medieval Scottish Parliament* (East Linton, 2001).

Whatley, C.A., *The Scots and the Union* (Edinburgh, 2006).

Wormald, J., *Court, King and Community: Scotland, 1470–1625* (London, 1981).

## Modern works relating to the rule of James I in England, 1603–1625

Akrigg, G.P.V., *Jacobean Pageant* (Harvard University Press, 1962).

Barrol, L., *Anna of Denmark, Queen of England: A Cultural Biography* (University of Philadelphia Press, 2001).

Bellany, A., *The Politics of Court Scandal in Early Modern England: News, Culture and the Overbury Affair, 1603–1660* (Cambridge University Press, 2001).

Bergeron, D., *King James and Letters of Homoerotic Desire* (University of Iowa Press, 1991).

Bingham, C., *Darnley: A Life of Henry Stuart, Lord Darnley Consort of Mary Queen of Scots* (London, 1995).

Birch T., *The Life of Henry Prince of Wales, Eldest Son of King James I* (London, 1760).

Borman, T., *Witches: James I and the English Witch Hunts* (London, 2014).

Bradshaw, B. and Morrill, J., eds, *The British Problem c. 1534–1707: State Formation in the Atlantic Archipelago* (London, 1996).

Bradshaw, B. and Roberts, P., eds, *British Consciousness and Identity: The Making of Britain, 1533–1707* (Cambridge University Press, 1998).

Carter, C. H., 'Gondomar: Ambassador to James I', *Historical Journal* vol. 7 (1964), p.189–208.

Collinson, P., 'The Jacobean Religious Settlement: The Hampton Court Conference' in H. Tomlinson ed., *Before the English Civil War.*

Croft, P., 'Fresh Light on Bate's Case', *Historical Journal* vol. 34 (1991).

Croft, P., 'The Reputation of Robert Cecil: libels, political opinion and popular awareness in the early seventeenth century', *Transactions of the Royal Historical Society*, 6th ser., I (1991), p. 43–69.

Curtis, M., 'The Hampton Court Conference and its Aftermath', *History* vol. 46 (1961), p.1–16.

Cust, C. and Hughes, A., eds, *Conflict in Early Stuart England: Studies in Religion and Politics 1603–1642* (London, 1989).

Daiches, D., *The King James Version of the English Bible* (University of Chicago Press, 1941).

Durston, C., *James I* (London, 1993).

Ellis, S.G. and Barber, S., *Conquest and Union: Fashioning a British State, 1485–1725* (Harlow, 1995).

Fincham, K, *Prelate as Pastor: The Episcopate of James I* (Oxford, 1990).

Fincham, K. and Lake, P., 'The Ecclesiastical Policy of James I', *Journal of British Studies*, vol. 24 (1985).

Fischlin, D., Fortier, M. and Sharpe, K., eds, *Royal Subjects: The Writings of James VI and I* (Detroit, 2002).

Fraser, A, *The Gunpowder Plot: Terror and Faith in 1605* (London, 1996).

Galloway, B.R., and Levack, B.P., *The Jacobean Union: Six Tracts of 1604* (Edinburgh, 1985).

Galloway, B.R., *The Union of England and Scotland, 1603–1608* (Edinburgh, 1986).

Gibbs, P., *King's Favourite: The Love Story of Robert Carr and Lady Essex* (London, 1909).

Goodacre, G., and Lynch, M., eds, *The Reign of James VI* (East Linton, 2000).

Hammer, P.E.J., *The Polarisation of Elizabeth Politics: The Political Career of Robert Devereux, 2nd Earl of Essex, 1585–1597* (Cambridge University Press, 1999).

Harris, T., *Rebellion: Britain's First Stuart Kings, 1567–1642* (Oxford, 2013).

Haynes, A., *The Gunpowder Plot: Faith in Rebellion* (Stroud, 1994).

Jansson, M., *Proceedings in Parliament, 1614* (University of Philadelphia, 1988).

Larkin, J.F. and Hughes, P.L., *Stuart Royal Proclamations: Royal Proclamations of King James I, 1603–1625* (Oxford, 1973).

Larner, C., *Enemies of God: The Witch-Hunt in Scotland* (London, 1981).

Lindley, D., *The Trials of Frances Howard: Fact and Fiction at the Court of King James* (London, 1993).

Lindquist, E., 'The Failure of the Great Contract', *Journal of Modern History* vol. 57 (1985).

Loades, D., *The Cecils: Privilege and power behind the throne* (The National Archives, 2009)

Lockyer, R., Buckingham, *The Life and Political Career of George Villiers, First Duke of Buckingham 1592–1628* (London and New York, 1981).

Loomie, A.J., *Spain and the Early Stuarts, 1585–1655* (Aldershot, 1996).

McFarlane, I.D., *Buchanan* (London, 1981).

Moodie, T.W., Martin, F.-X. and Byrne, F.J., eds, *A New History of Ireland vol. 3: Early Modern Ireland 1534–1691* (Oxford, 1991).

Newton, D., *The Making of the Jacobean Regime: James VI and I and the Government of England, 1603–1605* (Woodbridge, 2005).

Nicholls, M. and Williams, P., *Sir Walter Raleigh: in Life and Legend* (London, 2011).

Nicholls, M., *Investigating Gunpowder Plot* (Manchester University Press, 1991).

Nicholson, N., *When God Spoke English: The Making of the King James Bible* (London, 2011).

Opfell, O., *The King James Bible Translators* (Jefferson, 1982).

Parry, G., *The Golden Age Restored: The Culture of the Jacobean Court* (Manchester, 1981).

Patterson, W.B., *King James and the Reunion of Christendom* (Cambridge University Press, 1997).

Pawlisch, H.S., *Sir John Davies and the Conquest of Ireland: A Study in Legal Imperialism* (Cambridge University Press, 1985).

Peck, L.L., ed., *The Mental World of the Jacobean Court* (Cambridge University Press, 1991).

Peck, L.L., *Court Patronage and Corruption in Early Stuart England* (Routledge, 1993).

Peck, L.L., *Northampton: Patronage and Policy at the Court of James I* (London, 1982).

Perceval-Maxwell, M., 'Ireland and the Monarchy in the Early Stuart Multiple Kingdom', *Historical Journal* vol. 34, 1991.

Perceval-Maxwell, M., *The Scottish Migration to Ulster in the Reign of James I* (London, 1993).

Prestwich, M., *Cranfield: Politics and Profits under the Early Stuarts* (Oxford, 1966).

Rait, R. S. and Cameron, A. I., *King James's Secret: Negotiations Between Elizabeth and James I. Relating to the Execution of Mary Queen of Scots, From the Warrender Papers* (London, 1927).

Rees, G. and Wakely, M., *Publishing, Politics and Culture: The King's Printers in the Reign of James I and VI* (Oxford, 2009).

Rhodes, N., Richards, J. and Marshall, J., *King James I/VI: Selected Writings* (Farnham, 2003).

Rickard, *Authorship and Authority: The Writings of James VI and I* (Manchester University Press, 2012).

Ruigh, R.E., *The Parliament of 1624: Politics and Foreign Policy* (Harvard University Press, 1971).

Russell, C., *King James VI/I and his English Parliaments*, eds Cust, R. and Thrush, E. (Oxford, 2011).

Russell, C., *Parliaments and English Politics, 1621–1629* (Oxford, 1979).

Seddon, P.R., 'Robert Carr, Earl of Somerset', *Renaissance and Modern Studies*, vol. 14 (1970), p. 48–68.

Sharpe, K., *Politics and Ideas in Early Stuart England: Essays and Studies* (New York, 1989).

Smith, A.G.R., *The Reign of James VI and I* (Basingstoke, 1973).

Somerset, A., *Unnatural Murder: Poison at the Court of James I* (London, 1998).

Sommerville, J.P., *King James VI and I: Political Writings* (Cambridge University Press, 1994).

Stewart, A., 'Boys' Buttocks Revisited: James VI and I and the Myth of the Sovereign Schoolmaster', in *Sodomy in Early Modern Europe*, ed. T. Betteridge (Manchester University Press, 2002).

Strong, R., *Henry, Prince of Wales and England's Lost Renaissance* (London, 1986).

Treadwell, V., *Buckingham and Ireland, 1616–1628* (Dublin, 1998).

Welsby, P., *George Abbot* (London, 1962).

Williams, E.C., *Anne of Denmark: Wife of James VI of Scotland: James I of England* (London, 1971).

Young, M.B., *James VI and the History of Homosexuality* (Basingstoke, 2000).

Zaller, R., *The Parliament of 1621: A Study in Constitutional Conflict* (University of California Press, 1971)

# Author's Note

Biographers old and new, academic and otherwise, have been instrumental in shaping this book. Its earliest influences were David Harris Willson, George Philip Vernon Akrigg, William McElwee and Caroline Bingham. Later, as perspectives on its central character evolved, the book drew added inspiration from the work of a long list of others, but most notably Maurice Lee, Jr.

No writer is an island, and least of all this one. My thanks, therefore, are due to all those who have paved the way in their writings, as well as the smaller group of people who have supported me more personally in my efforts. In this latter respect, the help and encouragement of Mark Beynon, Juanita Hall and the team at The History Press has been unstinting, while Barbara, my wife, has continued throughout to hearten, uplift and cheer. To all concerned, I raise my glass.

# Index

# Index

# Index

# Index

# Index